CDMA
CAPACITY AND QUALITY
OPTIMIZATION

CDMA CAPACITY AND QUALITY OPTIMIZATION

Adam Rosenberg

Sid Kemp

McGRAW-HILL

New York   Chicago   San Francisco
Lisbon   London   Madrid   Mexico City
Milan   New Delhi   San Juan   Seoul
Singapore   Sydney   Toronto

The McGraw-Hill Companies

Library of Congress Cataloging-in-Publication Data

Rosenberg, Adam (Adam N.)
  CDMA capacity and quality optimization / Adam Rosenberg, Sid Kemp.
      p.   cm.
  Includes bibliographical references and index.
  ISBN 0-07-139919-4 (alk. paper)
  1. Code division multiple access.   I. Kemp, Sid.   II. Title.

  TK5103.452.R67 2003
  621.3845—dc21   2002045212

1 2 3 4 5 6 7 8 9 0   DOC/DOC   0 9 8 7 6 5 4 3

ISBN 0-07-139919-4

*The sponsoring editor for this book was Judy Bass, the editing supervisor was Daina Penikas,
and the production supervisor was Sherri Souffrance. It was set in New Century Schoolbook by
ATLIS Graphics.*

Printed and bound by RR Donnelley.

This book is printed on recycled, acid-free paper containing a minimum of 50
percent recycled de-inked fiber.

McGraw-Hill books are available at special quantity discounts to use as premiums and
sales promotions, or for use in corporate training programs. For more information,
please write to the Director of Special Sales, Professional Publishing, McGraw-Hill,
Two Penn Plaza, New York, NY 10121-2298. Or contact your local bookstore.

Dedications

I would like to dedicate this book to my family and to the amazing folks I worked with in the Cellular Systems Engineering Department at Bell Telephone Laboratories in West Long Branch, New Jersey, from 1982 to 1986.

—Adam Rosenberg

To Guilliamo Marconi for starting us on the way of wireless, to Arthur C. Clark for writing about communications with lucidity, and to all of the engineers and technicians who make wireless a reality every day.

—Sid Kemp

Acknowledgments

Adam Rosenberg
I starting working with Sam and John in 1982 when I joined Bell Telephone Laboratories in the Cellular Systems Engineering Department, and their contribution to this book has been major.

Samuel W. Halpern, Telecommunications Research Associates (TRA)
John J. Schubel, Acompas

These people made significant contributions to the technical content of this book, making it complete and correct.

Donald J. Barnickel, AT&T Labs

Henry L. Bertoni, Polytechnic University

Ric Biederwolf, Nextel

Ken Christensen, University of South Florida

Jim Davis, TRA

Steve Dick, InterDigital Communications Corporation

Matt Dooley, PrimeCo

Reiner A. Hoppe, AWE Communications GmbH

C. C. Lee, Northwestern University

Michael J. Luddy, Agere Systems

Jeff Polan, Advanced Systems Concepts

P. Subrahmanya, Qualcomm

Andreas Wachter

Robert W. Willis, Bell Telephone Laboratories

Chuck Wheatley, Qualcomm

And Judy, Steve, and Patty were essential in turning my flow of words and drawings into my first book.

Judy Bass, McGraw-Hill

Steve Kemp, University of North Carolina, Chapel Hill

Patty Wallenburg, TypeWriting

My thanks to all of them for all their help.

Sid Kemp

First of all, I would like to thank my coauthor, Adam N. Rosenberg, Ph.D. In addition to being a precise thinker, he has a gift for clarifying ideas for others. His decades of experience in wireless and his understanding of the underlying mathematical and engineering principles are the heart of this text. Second, I would like to thank Steven M. Kemp, my brother and Adam's friend, for suggesting this collaboration to both of us and also for careful and insightful editing of the text. My wife, Kristen H. Lindbeck, provided wonderful support throughout the months in which this book was written, but her Ph.D. was not relevant to those efforts. Many engineers and technicians added their experience and knowledge to the text. Most particularly, I would like to thank John Schubel for his review of the text and substantial contributions to the part on telephony, Paul Bedell and Ric Beiderwolf for their insight into the practical day-to-day engineering and management of wireless systems, and Mark Siedler for his information about the TCP/IP specification. Lastly, I would like to thank Judy Bass, senior editor at McGraw-Hill, for her wizardly midwifery in guiding this text into production.

Contents

Acknowledgments   v
A Note on Terminology   xvii
How to Use This Book   xix

1   Key Radio Concepts   1

1. Radio Engineering Concepts   3

1.1. Radio / 3
1.2. Frequency / 4
1.3. Multiple Access / 7
1.4. Bandwidth as Real Estate / 9
1.5. Amplitude and Power / 9
1.6. Decibel Notation / 10
1.7. The Radio Path / 13
1.8. Antenna Gain / 14
1.9. The Link Budget / 17
1.10. Modulation / 18
1.11. Adding Radio Signals / 23
1.12. Conclusion / 24

2. Radio Signal Quality   25

2.1. Radio Impairments / 25
2.2. Radio Measurement Errors / 29
2.3. Signal Quality Measurement / 30
2.4. Conclusion / 31

3. The User Terminal   33

3.1. Mobile and Portable Telephones / 33
3.2. Other User Terminals / 40
3.3. Conclusion / 47

4. The Base Station   49

4.1. The Antennas / 49
4.2. The Tower and Cable / 52
4.3. The Power Amplifier / 53
4.4. The Radio Receiver / 56
4.5. The Power Supply and Environmental Controls / 56
4.6. Microcells, Picocells, and Repeaters / 56
4.7. The Base-Station Controller (BSC) / 57
4.8. Component Reliability / 58
4.9. Conclusion / 61

5. Basic Wireless Telephony   63

5.1. The Wireless Signal Path / *63*
5.2. Radio Architecture / *66*
5.3. Conclusion / *69*

6. Analog Wireless Telephony (AMPS)   71

6.1. The Early Days of Cellular / *71*
6.2. The Cellular Principle / *72*
6.3. Managed Interference / *73*
6.4. The AMPS FM Channel / *74*
6.5. AMPS Call Setup / *76*
6.6. AMPS Call Maintenance / *77*
6.7. Conclusion / *79*

7. TDMA Wireless Telephony (GSM)   81

7.1. GSM Architecture / *81*
7.2. The TDMA Channel Concept / *81*
7.3. The GSM TDMA Channel / *82*
7.4. GSM Signaling / *83*
7.5. Conclusion / *86*

8. The CDMA Principle   87

8.1.   Spread Spectrum / *87*
8.2.   Processing Gain / *88*
8.3.   Pseudonoise (PN) Codes / *91*
8.4.   Multipath and the Rake Filter / *96*
8.5.   CDMA Power Control / *100*
8.6.   Varying Data Rates / *102*
8.7.   Soft and Softer Handoff / *103*
8.8.   The Forward (Downlink) Direction / *105*
8.9.   The Reverse (Uplink) Direction / *105*
8.10. Conclusion / *106*

2.   Standards for Cellular Systems   109

9. General Cellular Standards   111

9.1. The Role of Standards / *111*
9.2. Generations of Cellular Standards / *118*
9.3. Conclusion / *123*

10. Worldwide CDMA Standards   125

10.1. cdmaOne (IS-95) / *125*
10.2. cdma2000 / *130*

10.3. W-CDMA (UMTS) / *135*
10.4. TD-SCDMA / *140*
10.5. Conclusion / *142*

3   Key Telephony Concepts   143

11. The PSTN and Telephone Switching   145

11.1. The Public Switched Telephone Network (PSTN) / *145*
11.2. Telephone Switch Technology / *151*
11.3. Telephone Switch Functions / *152*
11.4. Mobile Switch Functions / *154*
11.5. Conclusion / *161*

12. Telephony Engineering Concepts   163

12.1. Telephone Call Sequence / *163*
12.2. Quality of Service / *164*
12.3. Reliability and Redundancy / *167*
12.4. Wireless Telephone System Architecture / *167*
12.5. Wireless Telephony Engineering Issues / *168*
12.6. Conclusion / *173*

13. Telephone Transport   175

13.1. Telephone Transport Protocols / *176*
13.2. Wireless System Transport / *178*
13.3. Conclusion / *182*

14. Signaling with SS7   183

14.1. Introduction / *183*
14.2. SS7 Network Architecture / *183*
14.3. SS7 Protocol / *186*
14.4. Logical and Physical Models / *188*
14.5. Conclusion / *190*

15. ANSI-41   191

15.1. Inter-MSC Handoffs / *195*
15.2. Automatic Roaming / *198*
15.3. Short Message Service (SMS) / *201*
15.4. Operations, Administration, and Maintenance (OA&M) / *203*
15.5. Over-the-Air Service Provisioning (OTASP) / *203*
15.6. Interaction with Other Networks / *204*
15.7. Conclusion / *205*

16. Call States   207

16.1. Defining Call States / 207
16.2. PSTN Call States / 209
16.3. Wireless Call States / 213
16.4. Wireless Radio States / 215
16.5. Conclusion / 218

4 Key Data Concepts   219

17. Quality of Service (QoS)   221

17.1. The Customer Experience of Quality / 221
17.2. Capacity / 222
17.3. Latency, Jitter, and Loss / 226
17.4. Error Detection and Correction / 228
17.5. Conclusion / 233

18. Speech Coding   235

18.1. Pulse Code Modulation (PCM) / 236
18.2. Linear Predictive Coding / 243
18.3. Speech Coding Performance / 244
18.4. Conclusion / 245

19. Hybrid Voice-Data Networks   247

19.1. Quality of Service (QoS) / 247
19.2. Voice and Data Sharing Physical Transport / 248
19.3. Voice and Data Sharing ATM / 250
19.4. Voice over IP (VoIP) / 251
19.5. Conclusion / 255

20. Short Message Service (SMS)   257

20.1. History / 257
20.2. Benefits for Subscribers and Vendors / 258
20.3. Technical Specifications / 259
20.4. Advanced Uses of SMS / 262
20.5. EMS and MMS: Newer Short Message Standards / 263
20.6. Conclusion / 264

21. Wireless Data Services   267

21.1. QoS over the Air Interface / 267
21.2. Changes at the Base Station and MSC / 269
21.3. Conclusion / 270

5 Capacity and Quality Principles   271

22. Capacity and Quality Tradeoffs   273

22.1. Business and Engineering Tools / 274
22.2. Quantitative Optimization / 276
22.3. Tradeoff / 278
22.4. Pipes, Bottlenecks, and Alternate Routes / 279
22.5. Constrained Optimization / 280
22.6. Optimization Variables / 282
22.7. Conclusion / 284

23. Traffic Engineering for Voice and Data   285

23.1. Voice Calls Assuming Fixed-Size Channels / 285
23.2. Effects of Adding ISDN Service / 293
23.3. Traffic Engineering for Packet Data / 296
23.4. Network Routing for Increased Capacity / 299
23.5. IP Services (DSL) / 300
23.6. Reliability and Redundancy / 301
23.7. Conclusion / 302

24. Switching Capacity   303

24.1. Main Processor Capacity / 303
24.2. The Flow of Bits Through the Switch / 305
24.3. Circuit Switching / 306
24.4. Packet Switching / 306
24.5. Speech Coding / 307
24.6. Conclusion / 307

25. ANSI-41 Signaling Capacity   309

25.1. Inter-MSC Handoffs / 309
25.2. Roaming / 310
25.3. Call Processing / 311
25.4. Short Message Service (SMS) / 312
25.5. Other Signaling Activity / 313
25.6. Adding It All Up / 314
25.7. Conclusion / 315

26. Capacity Calculations for Cellular Networks   317

26.1. Backhaul Design Principles / 317
26.2. Cellular System Components / 324
26.3. Cellular Network Issues / 326
26.4. Conclusion / 329

27. Conventional Reuse Principles

331

27.1. The AMPS Channel (FDMA) / *331*
27.2. The GSM Channel (TDMA) / *333*
27.3. Signal-to-Interference Performance / *335*
27.4. Regular Channel Reuse / *336*
27.5. Sectors / *339*
27.6. Power Control / *340*
27.7. Growing for Increasing Demand / *342*
27.8. Power Balancing / *348*
27.9. Traffic Engineering / *349*
27.10. Cookie-Cutter Hexagons / *350*
27.11. Capacity Rules of Thumb / *351*
27.12. Conclusion / *353*

28. CDMA Principles for Multicellular Systems

355

28.1. The CDMA Equations / *355*
28.2. Applying the Equations to Both Directions / *363*
28.3. Applying the Equations to the Forward Direction / *366*
28.4. Applying the Equations to the Reverse Direction / *368*
28.5. Applying the Equations Statistically / *369*
28.6. Conclusion / *371*

29. CDMA Data Capacity Principles

373

29.1. Bit Error Rate (BER) and E_b/N_0 / *373*
29.2. E_b/N_0, Rate, and Capacity / *376*
29.3. cdmaOne Data Capacity (2G) / *378*
29.4. 3G Data Capacity / *378*
29.5. Conclusion / *380*

30. Capacity Issues Specific to CDMA

381

30.1. Sectorized Cells / *381*
30.2. Soft and Softer Handoff / *382*
30.3. Multipath / *383*
30.4. Power Control and Pilots / *383*
30.5. Speech Coding / *384*
30.6. Soft Blocking / *385*
30.7. CDMA Radio Capacity Estimation / *385*
30.8. Conclusion / *393*

6   Planning for CDMA Capacity

395

31. Estimating Wireless Telephone Demand

397

31.1. Estimating Voice Call Demand / *397*
31.2. Estimating Data Service Demand / *400*
31.3. Estimating Demand for Other Services / *401*
31.4. Estimating Cell Count / *402*
31.5. Conclusion / *404*

32. Planning Locations for Base Stations   405

32.1. Pilot Pollution / *405*
32.2. Keeping the Grid / *408*
32.3. Cell Splitting for High Traffic Density / *409*
32.4. Microcells for Highly Concentrated Demand / *414*
32.5. Good Cell Locations / *415*
32.6. Conclusion / *418*

33. Base Station Planning   419

33.1. Coverage / *419*
33.2. Downtilting and Lowering Antennas / *421*
33.3. Sector Plans / *423*
33.4. Carrier Assignment / *424*
33.5. Handling Peak Traffic Demand / *426*
33.6. Migration from FDMA or TDMA to CDMA / *428*
33.7. Migration from cdmaOne to cdma2000 / *432*
33.8. Quality and Cost Issues in Air-Interface Planning / *433*
33.9. Reliability in Base Station Design and Layout / *435*
33.10. Conclusion / *436*

34. Mobile Switching Center (MSC) Planning   437

34.1. Estimating MSC Count / *437*
34.2. Location Planning for MSCs / *440*
34.3. Equipment Provisioning for MSCs / *440*
34.4. Assigning Base Stations to MSCs / *441*
34.5. Inter-MSC Borders / *442*
34.6. Auxiliary Component Planning / *443*
34.7. Anticipating System Growth / *445*
34.8. Conclusion / *446*

35. Backhaul Planning   447

35.1. Cost Issues in Backhaul Planning / *447*
35.2. Fan-out Backhaul Networks / *448*
35.3. Backbone Backhaul Networks / *449*
35.4. Data Direct from the Base Station / *451*
35.5. Reliability Engineering / *451*
35.6. Anticipating Growth / *452*
35.7. Other Backhaul Technologies / *453*
35.8. Conclusion / *453*

36. Signaling Capacity Planning   455

36.1. Handoffs / *455*
36.2. Call Processing / *457*
36.3. Roamers / *458*
36.4. Short Message Service (SMS) / *458*
36.5. Other Signaling Functions / *459*
36.6. Conclusion / *460*

37. MSC Transport Planning   461

37.1. MSC-to-MSC Transport Planning / *461*
37.2. MSC-to-Network Transport Planning / *463*
37.3. Conclusion / *467*

38. Special Situations   469

38.1. Asymmetric Demand / *469*
38.2. Market-Entry Planning Issues / *471*
38.3. Conclusion / *473*

7   Increasing Capacity   475

39. Measuring System Performance for Growth   477

39.1. Radio Signal Path Measurements / *477*
39.2. Forward Power Levels / *478*
39.3. Reverse Power Levels / *479*
39.4. Handoff Activity / *480*
39.5. Backhaul Congestion / *483*
39.6. MSC Performance / *483*
39.7. Signaling Capacity / *484*
39.8. Transport Utilization and Congestion / *484*
39.9. Conclusion / *486*

40. Turning User Complaints into Useful Data   487

40.1. Symptoms and Causes / *487*
40.2. Ineffective-Attempt Complaints / *488*
40.3. Lost-Call Complaints / *489*
40.4. Lousy-Call Complaints / *490*
40.5. Slow-Data-Link Complaints / *491*
40.6. Data-Error Complaints / *492*
40.7. Reasons for Change and Growth / *492*
40.8. Conclusion / *494*

41. Increasing Capacity of a Base Station   495

41.1. Determining When to Add Capacity / *495*
41.2. Adding Power / *496*
41.3. Adding Carriers / *497*
41.4. Adding Sectors / *498*
41.5. Installing Repeaters / *499*
41.6. Conclusion / *500*

42. Adding Cells to a CDMA System   501

42.1. Determining When to Add Cells / *501*
42.2. Cell Splitting / *502*
42.3. Keeping the Grid / *503*

42.4. Downtilting Antennas / *504*
42.5. Microcells for High Traffic Density / *505*
42.6. Picocells for Specific Local Areas / *506*
42.7. Handoff Management / *506*
42.8. Reliability Issues for Growth Cells / *507*
42.9. Conclusion / *508*

43. Mobile Switching Center (MSC) Growth   509

43.1. Determining When to Upgrade MSC Equipment / *509*
43.2. Assigning Base Stations to a New MSC / *511*
43.3. Dealing with New MSC Borders / *512*
43.4. New MSC Placement / *513*
43.5. Equipment Provisioning for a New MSC / *514*
43.6. Auxiliary Equipment Growth / *515*
43.7. Conclusion / *516*

44. Adding Transport   517

44.1. Adding Backhaul / *517*
44.2. Adding MSC-to-MSC Capacity / *519*
44.3. Adding MSC-to-PSTN Capacity / *520*
44.4. Adding MSC-to-PPDN Capacity / *520*
44.5. Conclusion / *521*

45. Regional Growth in a Specific Area   523

45.1. Determining Where to Encourage Demand / *524*
45.2. Targeting Growth / *527*
45.3. Worldwide Marketing Plans / *530*
45.4. Conclusion / *530*

8 Modeling for CDMA   531

46. Business-Case Models   533

46.1. Fixed and Variable Costs / *533*
46.2. Cost-per-Subscriber Models / *537*
46.3. Return-on-Investment (ROI) Models / *541*
46.4. Conclusion / *542*

47. Propagation Models   543

47.1. CDMA Issues and Challenges / *543*
47.2. Variation in Radio Propagation / *544*
47.3. Distance Models / *546*
47.4. Terrain-Based Models / *551*
47.5. In-Building Models / *553*
47.6. Ray-Tracing Models / *554*
47.7. The Statistical View / *555*
47.8. The Signal Matrix / *557*

47.9.   Radio Propagation Maps / *558*
47.10. Conclusion / *559*

48. Subscriber Traffic Modeling   561

48.1.   Traffic Tally Calculations / *561*
48.2.   CDMA-Specific Factors / *562*
48.3.   Simulation Models / *563*
48.4.   Analytical Models / *565*
48.5.   Pilot Delta Maps / *566*
48.6.   Measurement-Based Decisions / *566*
48.7.   Conclusion / *567*

9   Conclusion   569

49. CDMA Now and in the Future   571

49.1.   CDMA Now / *571*
49.2.   The Future of CDMA / *578*

Appendix A: Acronyms   587
Appendix B: Physical Units   589
Glossary   593
References   601
Index   603
About the Authors   629

A Note on Terminology

To the best of our ability, we have tried to use terminology consistently and to define each term clearly in the text and glossary. However, outside narrow subfields, the English language is imprecise, and we also have sought to recognize the general uses of terms in the wireless field. This has created three challenges for us in our effort to provide precise and unambiguous use of language.

The first challenge results from wireless being a technology that arises from the convergence of four existing technologies: radio, telephony, data communications, and television. For example, the adjective *virtual* has a general meaning: having the functional qualities of a thing without being that thing. For example, a virtual circuit functions as a circuit but is not a circuit. However, the term *virtual circuit* has a wider meaning in telephony than it does in data transmission. Circuits built on time division multiplexing are considered virtual circuits in the field of telephony but not in the field of data communications. As much as possible, we have identified these discrepancies in the text.

The second challenge has to do with our effort to avoid cumbersome language. For example, engineers often speak of the signal-to-noise ratio or the signal-to-interference ratio when what is being discussed, to be precise, is the signal-to-(noise-plus-interference) ratio. In these cases, we have explained common variations of terminology and defined the one that we will use most regularly.

The third challenge arises from the process by which technology develops over time. It is often easiest to understand a technology in comparison with what came before. However, the first technology that was developed was named without reference to the term used for the later technology. For example, it would be logical to think that the opposite of *forward error correction* would be *backward error correction,* but this is not the case. When error correction was first developed, it relied on detection and retransmission, but it was simply called *error correction* because it was the only type in existence. After this type of error correction was in use, engineers developed another method that did not rely on retransmission and distinguished it by calling it *forward error correction.* At that moment, there was no precise term for the original type of error correction. Engineers and authors have options when this occurs, such as coining a logical term such as *backward error correction* or a precise term such as *error detection and retransmission.* Different people come up with different solutions, and multiple terms for the same thing come into being. These cases arrive frequently as new technologies develop. [For example, did anyone ever refer to landline telephony as landline before the advent of wireless? No, it was simply telephony. And plain old telephone service (POTS) was just telephone service until other telephone services were developed.] We explain these events and define the terms as they arise.

If our terminology is awkward for you at times, please know that it is comfortable for someone else, and when you find it comfortable, some other reader will find it awkward.

How to Use This Book

To grow wireless systems today, we need a deep understanding of wireless technology. This draws on the disciplines of physics, mathematics, radio, telephony, data technology, television, economics, quality management, and business management. To move this technology into the future, we need to understand its roots in the past. Each of our readers undoubtedly has a great deal of expertise in some of these disciplines, some in others, and little or none in the remaining ones. We want all our readers to be expert in some of these areas and conversant in all of them. As a result, we have structured the text to provide each reader with all that is needed to gain basic knowledge where it is needed and develop advanced knowledge at the cutting edge of wireless technology, the third-generation (3G) systems that are being deployed today.

The first four parts of the text provide the core knowledge necessary to understand wireless technology, including "Key Radio Concepts," "Standards for Cellular Systems," "Key Telephony Concepts," and "Key Data Concepts." We recommend to each reader that you evaluate your depth of understanding in each of these areas. If you are expert in an area, feel free to skim or skip that part of the text. If you came to wireless through another area and are less familiar with a particular part, we recommend that you read it thoroughly. It should be both engaging and informative.

For example, one of the authors (Kemp) came to wireless technology from the field of data systems management. In developing the parts of the text on radio and telephony, I learned a great deal and also found that people in these fields face fundamentally different problems and find different approaches than do those in data systems. Developing an understanding of all the approaches to engineering problems in the root technologies that make up wireless systems is a key to becoming a genuine expert and leader in the field. If, in reading later parts of the book, you find some concepts or terms unclear, we suggest that you return to the relevant "Key Concepts" part to clarify the issue.

Part 5, "Capacity and Quality Principles," introduces fundamental concepts such as tradeoff and optimization in precise language. We are always seeking to develop the best system for a particular purpose using limited resources while facing underlying physical constraints. The theory we need to define these problems and develop solutions is laid out in Part 5.

Part 6, "Planning for CDMA Capacity," applies the principles in Part 5 to the situation where engineers are deploying entirely new systems or performing upgrades of existing systems with a large budget for new technology. It is designed primarily for planners and engineers on these large-scale projects and is also valuable for less technical administrators who need to understand the technology for their work.

Part 7, "Increasing Capacity," applies the principles from Part 5 to the situation of maintaining and improving existing CDMA networks. This part is written for engineers and technicians who maintain and grow systems at times when the budget is smaller and the changes being made are not as extensive as they would be in the rollout of a complete network upgrade or new network.

Part 8, "Modeling for CDMA," introduces the various modeling theories and methods that help us predict CDMA capacity and quality of service. It also explores methods of using models as simulation tools for troubleshooting network problems. It will be of value to planners, engineers, and technicians developing and improving CDMA networks.

Part 9, "CDMA Now and in the Future," gives a picture of how wireless technology is operating across the globe and changing our society. It presents the challenges and opportunities we face in improving communication for a mobile, global society.

CDMA
CAPACITY AND QUALITY
OPTIMIZATION

Part

1

Key Radio Concepts

Part 1 of this text, "Key Radio Concepts," is provided for readers who are not already familiar with the engineering principles of radio and how they apply to cellular systems. It also will benefit radio experts in two ways. First of all, it will help our readers explain these concepts to people in other fields and to businesspeople. Second, today's code division multiple access (CDMA) wireless technology is built on a series of developments going back over 30 years. It is easy to be expert in a system and not to know where it came from or to have an in-depth knowledge of how it works. However, in understanding the history of our field and the challenges faced by our predecessors, we gain a deeper expertise, improving our ability to handle the problems we face today.

Chapter 1, "Radio Engineering Concepts," defines the fundamentals of radio, including frequency, amplitude and power, and modulation. It also includes explanations of multiple access and modulation, a description of how a radio signal is altered by an antenna and by the space between the transmitter and receiver, and how we calculate signal power through those changes.

In Chapter 2, "Radio Signal Quality," we discuss impairments to the radio signal, such as noise, interference, distortion, and multipath. Chapter 2 also covers the measurement of radio signals, errors in those measurements, and the measurement of both analog and digital radio signals.

Chapters 3 and 4 describe the components at the two sides of the radio-air interface, the user terminal and the base station. Chapter 3, "The User Terminal," describes the components of a user terminal, commonly known as a cell phone. Chapter 4, "The Base Station," describes the cellular base station: the antennas that receive the signal through the tower and cable, the power amplifier, the receiver, the components that transmit cellular signals, those which send telephone calls through the link to the mobile switching center, and the base-station controller that manages the operations of the base station. There is also a discussion of component reliability modeling.

Chapter 5, "Basic Wireless Telephony," provides a picture of how all the parts of a cellular network work together to create the wireless signal path and how the whole cellular system is laid out, i.e., its architecture.

In Chapters 6 and 7 we describe the early analog and digital cellular radio technologies. In Chapter 6, "Analog Wireless Telephony (AMPS)," we describe the original Advanced Mobile Phone Service (AMPS) analog cellular technology that pioneered the cellular architecture as the first radio system relying on managed interference. In Chapter 7, "TDMA Wireless Telephony (GSM)," we introduce the world's first and largest digital cellular system, Europe's Global System for Mobility (GSM) time division multiple access (TDMA) technology, which is serving about 700 million users worldwide in 2002.

Having built a solid background in the fundamentals of radio and the evolution of cellular telephony, we turn to code division multiple access (CDMA) in Chapter 8. In Chapter 8, "The CDMA Principle," we discuss the underlying concept of CDMA, called spread spectrum, *the mathematical derivation of the CDMA method of managed same-cell interference, and the principles of key CDMA components such as the rake filter and power control. We also describe how CDMA operates in both the forward and reverse directions and how it performs handoffs as subscribers move from cell to cell.*

The CDMA cellular networks our readers support embody both concepts developed for the first analog cellular systems and also the latest digital chipsets and technologies. With the background provided in Part I, "Key Radio Concepts," cellular engineers will be well prepared to understand the latest CDMA technology so that we can design and optimize today's CDMA networks.

Radio Engineering Concepts

We all know what radio is, at least enough to get by. This chapter is for our readers who came to cellular from landline telephony or information technology and for those who want a refresher in the basics of radio engineering.

1.1 Radio

Radio is electromagnetic radiation, a changing electric field accompanied by a similarly changing magnetic field that propagates at high speed, as illustrated in Fig. 1.1. A radio wave is *transmitted* by creating an electrical voltage in a conducting *antenna,* by putting a metal object in the air and sending pulses of electricity that become radio waves. Similarly, a radio signal is *received* by measuring electrical voltage changes in an antenna, by putting another metal object in the air and detecting the very tiny pulses of electricity generated by the varying electrical field of the radio waves.

The technology of radio transmission is developing the ability to transmit a radio signal containing some desired information and developing a receiver to pick up just that particular signal and to extract that desired information. One of the latest technologies to do this is code division multiple access (CDMA), a long way from the dit-dah Morse code transmission of the earliest wireless equipment. Both Morse code and CDMA, however, are digital radio technologies.

We use radio to get some kind of information, a *signal,* from one place to another using a radio wave. We put that signal onto the radio medium, the *carrier* we call it, with some kind of scheme that we call *modulation.* The Morse code sender uses the simple modulation scheme of a short transmission burst as a dit and a longer transmission burst as a dah. The demodulation scheme does the reverse: The telegraph receiver makes audible noise during a radio burst, and the listener hears short and long bursts of noise as dits and dahs. Morse code is simple and elegant, and it used the technology of its day efficiently.

All the components of the process of radio communications were already present in the telegraph. There is a meaningful message to be sent that was coded into a specific format, the letters of the alphabet. The formatted message, the *signal,* is then *modulated* by the telegraph operator into a radio message that is then *demodulated* into something that looks like the original signal. The receiver restores the message's meaning, reading

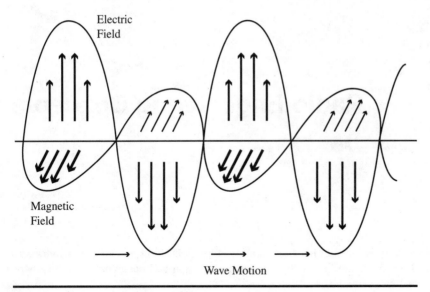

Figure 1.1 An electromagnetic radio wave.

the letters to form words. In this case, the formatted message is a sequence of letters, a digital signal.

The meaningful message in telephony is primarily spoken voice. The formatting stage is done with a microphone and amplifiers to form an electrical voltage over time that represents the speech, an analog signal in this case. If this signal is fed into an amplifier and a speaker, then we hear meaningful voice output.

1.2 Frequency

In addition to their magnitude, analog signals such as audio, electricity, and radio also have the attribute of *frequency*. We all know frequency as the pitch of a sound or the station numbers on a car radio, but frequency is a deep, basic, fundamental, primal mathematical concept that deserves some attention.

The simplest view of frequency is that it is the number of waves that pass a given point at a given time. In this simplified view, wavelength is in inverse ratio to frequency, with the speed of transmission as the constant.

We measure frequency in cycles per second, or *hertz* (Hz).[1] Frequencies we use in real life vary considerably. We have the very low 50 or 60 Hz of electrical power from the wall outlet. Sound we hear is air pressure waves varying from 20 Hz to 20 kHz.[2] Our AM radio stations operate from 500 kHz to 1.6 MHz, and FM and older broadcast

---

[1]More mathematical texts often measure frequencies in *radians* per second and usually use the Greek letter omega ω for radian frequency values, where ω = 2π*f*.

[2]While almost everybody reading this knows that kHz stands for *kilohertz,* 1000 hertz, some of the other prefixes may be more obscure. The entire list is in the "Physical Units" section at the end of this book.

television stations are in the very high frequency (VHF) band around 100 MHz. Aircraft radios operate in the VHF band as well.

Electromagnetic radiation propagates at the speed of light c, which is about 3×10^{10} cm/s. Therefore, Advanced Mobile Phone Service (AMPS), the U.S. analog cellular system, at 900 MHz frequency, has a wavelength of about 33 cm, and the primary cdmaOne frequency, 1.9 GHz, has a wavelength of about 16 cm. Wavelength is a major element in determining the types of attentuation that we will need to manage. For example, in the upper microwave bands used for satellite transmission, wavelength is a fraction of a centimeter, and raindrops can cause attenuation. However, rain is not a problem for cellular systems. Attenuation tends to occur when the intervening objects are of a size about equal to one-half the transmission wavelength.[3] To put radio frequencies into perspective, the visible red laser light used in fiber optic cable is around 500 THz, 500,000,000,000,000 cycles per second, with a wavelength of about 0.00006 cm.

In a sound wave, there is some atmospheric pressure at every instant of time, so we can say that the atmospheric pressure is a *function* of time, and we can describe that function as the *time response* of the sound wave. In our human experience, sound usually comes in periodic waves, and the number of waves per unit time determines the pitch, the frequency of the sound. Normal sound is a mixture of many frequencies, and its *frequency response* is often more informative than its time response.[4]

A radio wave has a voltage at every instant of time, so its time response is voltage rather than air pressure. Radio also is usually transmitted in periodic waves with an associated frequency. As in the case of sound, radio waves usually contain many frequencies, and their frequency response is important.

A function can be represented as $f(x)$. In the case of electrical voltage over time, we can represent the voltage v at each time t as $v(t)$. The mathematical concept of a *function* tells us that there is one $f(x)$ for each x or, in our electrical case, one specific voltage $v(t)$ for each time t.

Fourier analysis tells us that we can think of the same $v(t)$ in another form as $V(s)$, where s is one particular *frequency* rather than an instant of time. The function $V(s)$ is a little more complicated than $v(t)$ because it contains not only the amplitude of frequency s but also its *phase*. The relationship between time response $v(t)$ and frequency response $V(s)$ is a pair of integrals from college calculus.

$$V(s) = \int_{t=-\infty}^{\infty} v(t)e^{ist}\, dt \qquad\qquad (1.1)$$

$$v(t) = \int_{s=-\infty}^{\infty} V(s)e^{-its}\, ds \qquad\qquad (1.2)$$

---

[3]The coauthor who lives in Texas notes that this could mean that 3-in hail interferes in the CDMA band. Frankly, we're a lot more concerned about equipment damage than about radio interference when the hail is the size of tennis balls.

[4]The audible difference between an oboe and a violin playing the same steady note B♭ is in their response at higher frequencies. Those higher frequencies are called *harmonics*. The audible difference between an oboe and a piano, on the other hand, is not only their frequency response but the percussive time response of the piano.

While our time and frequency intervals in real life do not go from minus infinity $(-\infty)$ to plus infinity $(+\infty)$, an important message from these two integrals is that frequency response $V(s)$ depends on the time response $v(t)$ over an extended period of time and, conversely, that the time response $v(t)$ is determined by knowing enough about the frequency response $V(s)$. Equations (1.1) and (1.2) for time and frequency are nearly symmetric, and they tell us that there is a duality in the time-frequency relationship. A signal $v(t)$ consisting of a single continuous unchanging wave $v(t) = \sin(2\pi ft)$ has only one frequency f, as shown in Fig. 1.2.

An important asymmetry in the time-frequency duality is the notion of phase shift. Consider the waveforms shown in Fig. 1.3. In each case we have two frequencies, one twice the other. However, the phase relationship between the two waves is different in the two cases, and their pictures are quite different.

Frequency is often a more natural representation of our radio world than voltage amplitude. To put this another way, it is often easy and natural to work with frequency in radio system design. For example, we can build frequency filters that restrict a receiver to a certain range of frequency so that the received signal is not affected by activity at other frequencies. This all seems very natural today, hardly worth going over, but the core technology CDMA, the subject of this book, pushes radio technology very hard and tests these basic ideas. Thus, understanding the fundamental concept of radio frequency is a prerequisite to having a thorough understanding of CDMA.

Technology has changed, and VHF has become an anachronistic acronym. The ultrahigh frequency (UHF) band is the upper hundreds of megahertz, and it was allocated in the United States to television stations. As cable television has reduced the need for 70 UHF TV stations, the UHF band has been reallocated to other services, including the first North American cellular telephone service.

Since then, the competition for the UHF band has become severe, and wireless telephony has moved into the microwave band, above 1 GHz. It has been a constant

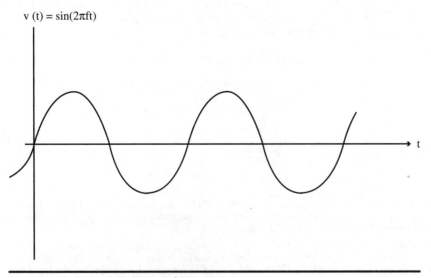

$v(t) = \sin(2\pi ft)$

Figure 1.2 A sine wave is just one frequency.

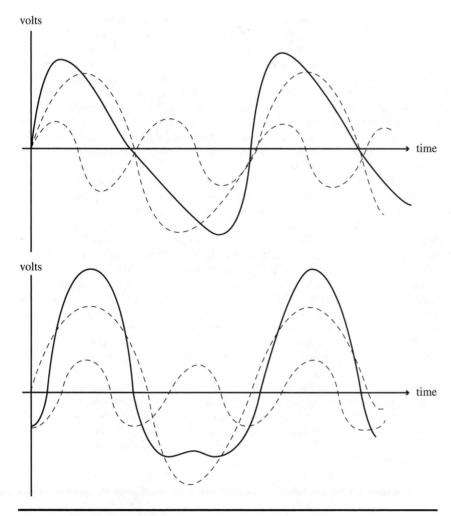

Figure 1.3 Phase relationships.

challenge to design cost-effective radio transmitters and receivers in the microwave band, and it becomes more difficult as frequencies get higher.

1.3 Multiple Access

In the very first days of radio, it sufficed to get a signal from here to there over the radio airwaves. We can imagine the listeners' excitement the first time they heard a live voice carried across the ocean on a radio wave and received for their ears. We also can imagine the desire to carry more than one radio signal. While radio link users can wait their turns in an ordered sequence, radio is only really useful when many users can use it simultaneously. The ability to send more than one signal at the same time is called *multiple access*.

Radio has been used for both broadcast and two-way communication. In broadcast, a single signal is meant for a large community of receivers, whereas we typically picture two-way radio as having two individual stations communicating with each other.

The usual picture of broadcast is a commercial radio station, but there are private broadcast channels of distribution. Pagers are a form of broadcast radio; a single source sends data over the airwaves for a large community of receivers.

The two-way walkie-talkie has evolved into sophisticated communications systems used in aviation, trucking, railroads, police, and the wireless telephone systems of today. Some systems often have one broadcast direction, a dispatcher talking to all the taxicabs or an air traffic controller talking to all the airplanes, with individuals replying on a common frequency with no privacy.

Wireless telephone systems require another level of sophistication because they manage separate two-way communication links in the same system. Unlike airline pilots, wireless callers do not want to be bothered by other telephone conversations on the same system. Wireless telephone users take their privacy seriously, and maintaining separate and confidential calls is an important component of system design.

In the earliest days of radio, we used frequency to discriminate among radio signals, and we called the system *frequency division multiple access* (FDMA). Each radio user gets a frequency range, although there may be other users on other frequencies. Commercial radio stations (both AM and FM) are assigned frequencies in their large geographic areas, and our receiver sets easily discriminate among the stations and allow us to exclude all but the one to which we are listening. Like broadcast radio, the first mobile telephone systems and, later, the first cellular systems were FDMA-based.

When the leap was made from analog to digital modulation described in Chap. 7, it was more efficient to use a larger frequency band and to divide it up among several signals using time slots. The system is synchronized so that each receiver knows which transmitted time slots belong to that receiver's signal. This *time division multiple access* (TDMA) is more complicated than FDMA, but it uses the radio frequency more efficiently. The Global System for Mobility (GSM), which started in Europe and became a worldwide standard, is a TDMA system with eight time slots aggregated into a single larger frequency band.

FDMA and TDMA have some kind of absolute protection from other broadcasts on the signal channels in the radio medium. A single frequency band or time slot gets minimal interference from other frequency bands or time slots because at the exact moment of reception, no other transmitter is broadcasting on the specified frequency. However, in the world of *spread spectrum,* a single stream of radio is shared simultaneously. Code division multiple access (CDMA), the subject of this book, is a spread-spectrum system that transmits many signals in the same radio band at the same time.

Allow us to use our favorite analogy for explaining CDMA. Consider a few dozen people in a small room all talking in pairs. Each listener knows his or her speaker's voice and can tune out the other voices that interfere with his or her own conversation. This tuning-out ability has limitations and the ability of these listeners to understand their speakers runs out if we have too many people talking at the same time. In CDMA, we assign each digital stream its own distinct voice in the form of a digital code, and all of these streams coexist on the same radio channel all at the same time.

Spread spectrum came from military research where resistance to enemy jamming was the major design issue. There are several spread-spectrum technologies, and CDMA happens to be resistant not only to deliberate outside interference but also to other users on the same channel.

1.4   Bandwidth as Real Estate

We define radio territory by land area and frequency range. A transmission license authorizes its owner to transmit only within a specified geographic region and between lower and upper frequency bounds set in the license. These licenses are regulated by governments in just about every country in the world today. Here in the United States, the Federal Communications Commission (FCC) gives out broadcast licenses based on the service being offered and the technology being used. In the United States, it would be fair to say that regulation of the radio spectrum is tighter and more restrictive than mineral rights but looser and freer than airspace, which is controlled minute by minute by the Federal Aviation Administration (FAA).

The allocation of radio frequency carries with it the obligation not to transmit on any *other* frequency bands. More frequency bandwidth means more capacity, in the form of more channels for FDMA and TDMA systems and more bits when using CDMA technology.

We refer to radio frequency ranges as *bandwidth* or *spectrum*. The usual discussion is about the frequency range, with the geographic coverage region assumed. There are hot debates over where one region ends and a neighboring region begins, but we will concentrate on the bandwidth issues in our CDMA discussions. Readers should keep in mind, however, that negotiating with geographic neighbors for compatible service is every bit as important as making sure radio transmission is contained within allocated spectrum. We will discuss the technical issue of calls being served by different wireless systems in Sec. 12.5.5.

As in any other acquisition, some frequencies are more desirable than others. Just as the downtown real estate commands a premium in most big cities, so also, in radio, lower frequencies are less demanding to operate. Lower-frequency amplifiers and antennas are simpler, and radio coverage is broader at a given power level. The lower frequencies are already firmly claimed by radio and television stations, police communications, and other long-established users.

In the early days of cellular, we were lucky enough to get radio spectrum allocations in the 900-MHz band, the upper end of "the UHF wasteland." We called it that because cable television was clearly alleviating the need for 70 UHF TV stations. Even before cable, most of us remember there being only a few U.S. TV stations numbered 14 through 83.

More recently, wireless has been pushed up into the microwave band. Much of this push is a consequence of our own success. As the demand for wireless telephone service has increased, we have become hungry for more bandwidth to satisfy that demand, and that bandwidth is out there in the microwave band.

1.5   Amplitude and Power

The *amplitude* of a radio wave is the electromagnetic voltage level as it propagates through space. Similarly, the amplitude of a sound wave is the pressure variation as it propagates through the air or other medium. Alas, amplitude and power are different, and this creates more than a little confusion.

The *power* of an electromagnetic (or audio) event is the energy per unit time. We are all familiar with power in units of watts or horsepower. In a car engine, for example, it is the driving force or thrust being applied multiplied by the speed at which it is being applied. In an electromagnetic event, power is the voltage multiplied by the magnetism or current. In an audio event, power is the air pressure multiplied by the air velocity. In both radio and sound, this means that the power level is proportional to the *square* of the amplitude.

To expand this idea a little more fully: In an electromagnetic wave, the amplitude is the voltage, and the current is proportional to that voltage. Power is voltage times current, so the power is proportional to the square of amplitude. In an audio event, amplitude affects both the air pressure and air velocity, so power is proportional to the square of amplitude there as well.

Amplitude is measured in volts, which we almost never use in discussions of radio systems, and power is measured in watts, which we use constantly.[5]

A typical base-station radio transmitter in our mobile telephone world has about 100 W of *effective radiated power* (ERP). A typical broadcast radio or television (TV) station might have 1 million W ERP, so our mobile radio stations are in the lightweight division. The telephone itself is usually limited to 1 W, the bantamweight division. However, this comparatively puny signal reliably gets hundreds of millions of calls through every day.

An effective medium for sending information, radio is not an efficient medium for sending energy. The mobile telephone signal transmitted at 1 W is typically 0.0000000000001 W, or 100 fW, at the base-station receiver. Losing 99.99999999999 percent of the energy sounds wasteful, but our receivers are able to demodulate this tiny signal to recreate the signal modulated at the transmitting end. We can demodulate and understand such a weak signal by engineering our wireless systems so that the other signals competing with it are even weaker.

The crucial issue in getting a signal through is not the amount of power; it is the ratio between the power of the signal and the power of the noise or interference, the *signal-to-noise* (S/N) *ratio* or the *signal-to-interference* (S/I) *ratio*. As long as this ratio is high enough at the receiver, the amount of power received is irrelevant.

Since every electrical amplifier has its own noise, the receiver has some internal noise level, and we want our signal to be stronger than the noise. Maximizing the S/N ratio is the key to good radio design. Electrical engineers are designing superb low-noise receivers, and it is the job of the wireless telephone system planners to get as much signal and as little that is *not* our own signal to the receiver as possible.

1.6   Decibel Notation

The power ratios in radio are often huge. We broadcast 100 W of power only to have 10 fW get into the receiver. Throwing numbers like 10,000,000,000,000, 10^{13}, around gets tiresome and confusing. The Americans and British do not even agree on what to call such a large number; it is called 10 *trillion* in the United States, but it is called 10 *billion* in England.

There is a more fundamental point, however. Ratios are often the *essential* matter, and we need some notation for describing ratios as ratios and articulating the differences between ratios. Fortunately, the decibel scale does the job nicely.

Named after Alexander Graham Bell, the *bel* is defined as a factor of 10, so 2 bels is a factor of 100 and 3 bels is a factor of 1000. As shown in Table 1.1, bels add where the

---

[5]A summary of units and notation is in the "Physical Units" section at the end of this book.

TABLE 1.1   Decibel Conversion to Ratios

Bels	Decibels	Factor (exact ratio)	Typical rounding	Percentage increase
0.01	0.1	1.02329		2.33
0.1	1	1.25892		25.89
0.2	2	1.58489		58.49
0.3	3	1.99526		99.53
0.30103	3.0103	2	3 dB	100.00
0.4	4	2.51188		151.19
0.47712	4.7712	3	5 dB	200.00
0.5	5	3.16227		216.23
0.6	6	3.98107		298.11
0.60206	6.0206	4	6 dB	300.00
0.7	7	5.01187		401.19
0.77815	7.7815	6		500.00
0.8	8	6.30957		530.96
0.9	9	7.94328		694.33
1	10	10		900.00
2	20	100		9,900.00
3	30	1,000		
4	40	10,000		
5	50	100,000		
10	100	10,000,000,000		
13	130	10,000,000,000,000		

factors multiply, so a 2-bel change followed by a 3-bel change is a 5-bel change because 1000 times 100 is 100,000. Negative bels are reduction factors, so −1 bel is a factor of 0.1 and −4 bels is a factor of 0.0001.

Fractional bels require a little high-school mathematics. One-half of a bell would be the *square root* of 10, approximately 3.16228 on a calculator. Using a logarithm table or the LOG function on a calculator, we can see that a factor of 2 would be 0.30103 bel and a factor of 3 would be 0.47712 bel. Using the addition rule, we see that a factor of 6 is 0.30103 + 0.47712 = 0.77815 bel.

The bel scale becomes more intuitive when we multiply the values by 10 and use *decibels* (dB) instead of bels. A factor of 10 is 10 dB, and a factor of 100,000 is 50 dB. And a path loss of 10,000,000,000,000 comes out as 130 dB.

The fractions are easier to deal with in decibels, too. A factor of 2 is 3 dB. Sure, the *exact* value for a factor of 2 is 3.0103 dB and the *exact* value for 3 dB is 1.9952, but it is a very rare occasion where the 0.2 percent difference is going to be important enough to matter. After living in the engineering world for a few weeks, most of us use 3 dB for a factor of 2 and 6 dB for a factor of 4 as freely as 20 dB for a factor of 100. After 6 months, most of us do not even remember which of these is the approximation and which is the mathematically derived exact value.

Using 5 dB instead of 4.77 dB for a factor of 3 is a little bit looser, but much of the time the difference between a factor of 3 and a factor of 3.16228 is not that important. We are not endorsing sloppy arithmetic, only using notation that articulates what needs to be communicated and is typical in the field of radio engineering.

Table 1.1 shows some decibel values converted to their ratios. After a while in radio, or just about any other part of electrical engineering, the decibel scale seems easy and natural.

So far we have a ratio scale of decibels, but we have no units. The most familiar application of decibels in everyday life (for most of us anyway) is sound pressure level (SPL). A committee decreed that a pressure wave of 0.0002 dynes per square centimeter is the unit of sound pressure level, the 0-dB point. We denote this very quiet level as 0 dB SPL. A subway train that is 1,000,000,000 times louder, 90 dB louder, is therefore 90 dB SPL. And if I buy a pair of earplugs that say "15 dB attenuation" right on the box, I can expect the resulting subway sound level in my ears to be 75 dB SPL, which is 90 dB SPL minus 15-dB earplug attenuation.

In the radio world, our 0-dB point, 0 dBm, is 1 mW of ERP. A 1-W radio signal is 30 dBm, a 100-W radio signal is 50 dBm, and a received power level of 10 fW is −110 dBm. Losing 99.99999999999 percent between the transmitter and the receiver is expressed as 130 dB of *path loss* in our decibel notation.

Most of the time in radio engineering we stay in the decibel-ratio world of units. The magnitude range is one reason, but there is an even more compelling reason to think in terms of decibels rather than absolute amounts. The performance of a radio link is almost always defined in terms of the S/N or S/I *ratio*.

These concepts are important in any radio engineering, but CDMA makes it particularly important because small changes in signal make comparable changes in capacity. The connection is often direct: A 0.5-dB change in signal quality is a 0.5-dB change in system capacity, which should be a 12 percent difference in revenue. (Before you scoff at 12 percent, ask yourself what a 12 percent change in your own salary is worth to you.)

And there is a more subtle reason to think in terms of ratios. There is a statistical notion, the law of large numbers, that tells us when we add a large number of random variables the sum tends toward a particular statistical shape called a *normal distribution,* the famous bell-shaped curve we read about.[6] Another term for a normal distribution is a *gaussian distribution,* named after the mathematician Karl Friedrich Gauss (1777–1855). In a radio path, the random variables are path losses from line-of-sight distance, buildings, trees, hills, and even rainy weather, and these path losses multiply in absolute numbers and *add* in decibel-ratio units. If distance path loss averages 80 dB, for example, buildings add another 15 dB of path loss, trees add 10 dB, and hilly terrain another 25 dB, then the total path loss will average 130 dB with a normal, bell-shaped, statistical variation in the decibel scale. We call this a *log-normal distribution* because it is normal in the logarithmic decibel scale.

Because of the fundamental difference between voltage and power, we have two equations for decibels that look different but are really telling us the same thing:

$$d = 10 \log \left(\frac{w_2}{w_1} \right) \tag{1.3}$$

$$d = 20 \log \left(\frac{v_2}{v_1} \right) \tag{1.4}$$

Equation (1.3) tells us the decibel expression for the ratio of two *power* levels w_1 and w_2 in watts, whereas Eq. (1.4) tells us the decibel expression for the ratio of two

---

[6]We are glossing over a great deal of deep and interesting statistical theory here, but the point is still a valid one.

voltage levels v_1 and v_2. The assumption being made in Eq. (1.4) is that the electrical load, which we call *impedance,* is not changing as the voltage changes from v_1 to v_2.

In the world of radio, we sometimes do not care about an extra 30 dB. So long as S/N ratios or bit error rates are low enough, an extra factor of 1000 in a signal strength does not matter. Other times we are concerned about 0.1 dB because a 2 percent difference in a particular factor is crucial to system capacity.

1.7 The Radio Path

Our physics textbooks make it clear that electromagnetic fields and electromagnetic radiation radiate through open space in a vacuum with the same energy in the same spherical angle. Since the surface area of a spherical angle increases as the square of the distance from the source at the center, this tells us that the intensity (power) p of a radio wave decreases as the square of the distance r:

$$p = \frac{\alpha}{r^2} \qquad (1.5)$$

(The constant term α takes into account all the factors besides distance.) When we all live in an unobstructed vacuum, we can start using this kind of propagation model.[7]

As usual, real life is far more complicated than Physics 101 would have us believe. Let us start with the two basic assumptions in Eq. (1.5): no air and no obstructions.

At frequencies up through 1 GHz and distances up to 10 km, it is pretty safe to ignore air losses. As we move up in frequency, the absorption of air becomes more important, but cellular systems are operating at short enough distances that we believe that it is a minor factor in these systems, even on rainy days.[8]

Terrain obstruction is more important to our wireless system design than air and weather. Our planet's surface is curved and hilly and covered with varying kinds of plant life and buildings. When Bell Telephone Laboratories engineers made their measurements, they found that radio power came down far faster than Eq. (1.5) suggests. After making many measurements and doing the best statistical analysis they could, they got an exponent of 3.84, as shown in Eq. (1.6):

$$p = \frac{\alpha}{r^{3.84}} \qquad (1.6)$$

In conversation, we refer to this relationship as "38.4 dB per decade," a loss of 38.4 dB (a factor of 7000) for each factor of 10 in distance. Considering how variable different places are and how much other factors affect a radio signal path, we have no problem

---

[7]But how are we going to speak and hear our telephones in the vacuum of space?

[8]As we recall from conversations back in the early days of cellular vision, *circa* 1970, using 60 GHz for cellular radio was talked about because 60 GHz was an absorption frequency for atmospheric oxygen and would provide excellent cell-to-cell isolation for that reason. Even Bell Telephone Laboratories did not have 60-GHz amplifiers that were powerful enough or cheap enough for a telephone market such as cellular that might someday approach 1 million subscribers, so they retreated to the more practical frequency band around 900 MHz.

calling the typical terrain loss an inverse fourth power, 40 dB per decade, as shown in Eq. (1.7):

$$p = \frac{\alpha}{r^4} \qquad (1.7)$$

The higher-than-free-space loss rate is a good thing for wireless capacity rather than a bad thing. Our interferers tend to be further from our receivers than our desired signal transmitters, so a tendency of the signal to deteriorate quickly with distance helps us discriminate between desired and undesired radio waves.

While most engineers think and speak about *path loss,* we find it easier to think of the *radio path* or even *path gain* with the understanding that the number is going to be far less than 1 (negative decibels). We will refer to a radio path gain of −130 dB or a higher radio path gain of −110 dB. In this way, we do not have to remember which number to subtract, a mistake far easier to make in a spreadsheet than on a piece of paper. The notion of path gain rather than path loss also avoids the constant hassle of remembering that greater path loss means less radio signal.

Life gets more complicated, however. The 40 dB per decade is an *average* over varying terrain and surface clutter. Rolling hills make the radio path gain higher on the side near the antenna and lower on the far side. Wooded areas have lower path gain than clear areas, and this difference is greater in the summer when the trees have leaves. Buildings not only reduce the path gain on average, but they also greatly increase the signal variability as one moves from one place to another nearby place. *Signal variability* is the variation in received signal strength from a fixed transmitter as a mobile unit is moved around a relatively small area.

While the earth's surface is almost always curved, it is not necessarily curved the same way in all places. Los Angeles and Salt Lake City are two U.S. cities built in valleys, so their ground curves up instead of down. This means that cellular interference in these cities, and others with similar topography, is going to be different and very likely more difficult to manage than elsewhere.

Bodies of water also affect propagation. Radio waves that would be absorbed or diffused by shrubbery can bounce off a water surface. This can form local radio "hot spots" near lakes and rivers.

Because of the multiplicative nature of path gain reductions from terrain, buildings, and shrubbery, the statistical effect is a log-normal signal distribution, as described in Sec. 1.6. Typical standard deviations are about 10 dB.[9] Understanding radio propagation is essential to CDMA capacity and quality planning. Chapter 47 goes into the subject in depth.

1.8 Antenna Gain

We see the term *antenna gain,* and we have to wonder: How can an antenna have gain? How can a piece of metal with no external source of power make a radio signal stronger? Well, an antenna can concentrate the radio energy in a specific direction and make that direction more powerful than it would have been from omnidirectional radiation. Antenna gain makes radio stronger the same way a cone-shaped megaphone makes a speaker louder in an intended direction.

---

[9]This gives a statistical variance of 100 dB^2. In all our various engineering travels we have encountered few dimensions as odd as decibels *squared.*

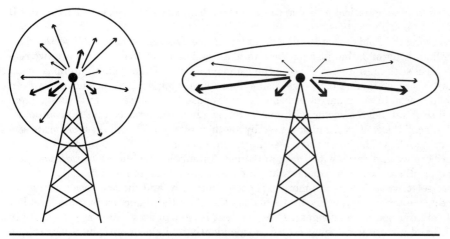

Figure 1.4 Antenna with no gain and with 10 dB of gain.

Consider a perfectly omnidirectional radio transmitter. The surface area of a sphere is $4\pi r^2$, so we sometimes call this spherical symmetry a *4π radiator*. If we focus the energy in a circular pattern out of the sides of the antenna system, then we have a horizontally omnidirectional transmitter with much less energy coming out its top and bottom. This is illustrated in Fig. 1.4. By squeezing nearly all the energy into one-tenth

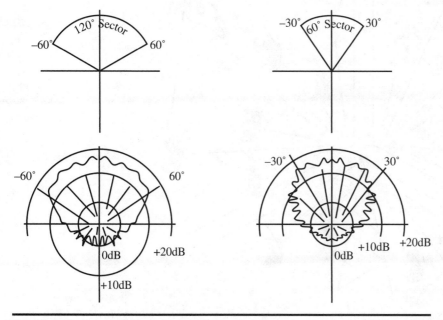

Figure 1.5 Antenna radiation pattern.

the surrounding spherical area, we can achieve a factor of 10 increase in ERP, a 10-dB antenna gain. This is shown in the figure.

Narrowing the field of this 10-dB gain antenna to a 120-degree horizontal sector gives us another factor of 3, that is, 5 dB, for a total antenna gain of 15 dB. There are two reasons to use such an antenna. First is the radio gain issue: Higher-gain antennas mean that we can reach further (coverage) using lower-power amplifiers (efficiency). However, a second reason to use a narrower beam is that we can be *selective* in our coverage area. If we use a 120-degree sector with one antenna, then the area behind that antenna, which we call the *backlobe,* can be served by another antenna. We call these separate service areas *sectors.* The lower the cross-interference between the two antennas, the more subscribers we can serve with those antennas. Antennas have different *radiation patterns* typically drawn on a circular scale in decibels, as shown in Fig. 1.5.

More selective antennas are more expensive, naturally, and the *backlobe attenuation* of a sector antenna is typically 15 dB. This is a substantial reduction in radio level but hardly elimination of radio signal. Thus, while it is easy to draw three or six sectors in Fig. 1.6 and picture each being served independently by its own antenna, there is substantial backlobe interference. And the sector boundary has significant cross-interference as well because antennas do not have perfect directional filtering, as shown in Fig. 1.7.

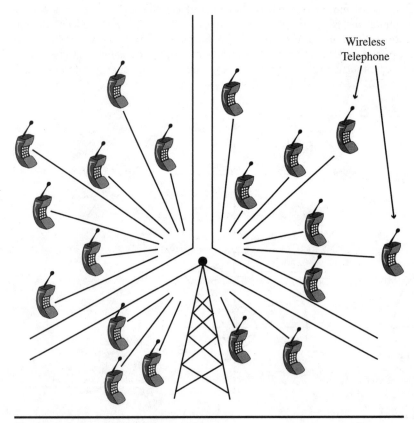

Wireless
Telephone

Figure 1.6 Antennas for 120-degree sectors.

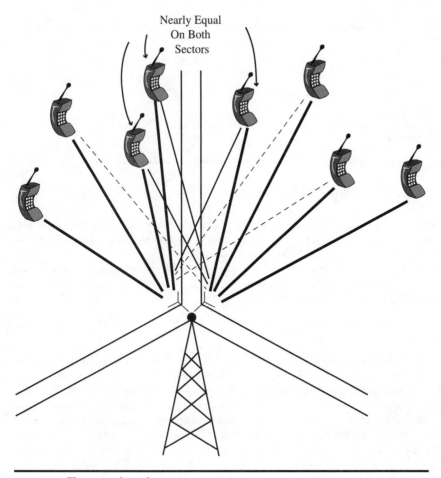

Nearly Equal
On Both
Sectors

Figure 1.7 The sector boundary.

1.9 The Link Budget

We sometimes express the combination of path gains and losses in a *link budget,* as shown in Table 1.2. In this simple link budget, the power level at the receiver is −79 dBm. The manufacturer's specifications or actual measurements might tell us that the receiver noise level is −105 dBm. The resulting S/N ratio (26 dB) is obtained by subtracting the noise (−105 dBm) from the signal strength (−79 dBm). In this example, a required S/N ratio of 20 dB was used. Surplus S/N is the extra signal strength above the minimum requirement that the communications link has to spare.

As the radio environment gets more complicated, we can add components to the link budget to reflect this. Of course, a link budget can be either an estimate of what we expect or a display of actual, measured results. Sometimes the link-budget concept is an oversimplification, but other times it gives a clear picture and brings a radio environment into perspective.

TABLE 1.2   A Link Budget

10-W amplifier	40 dBm
Transmit antenna gain	+10 dB
Radio path	−130 dB
Receive antenna gain	1 dB
Signal level (total)	−79 dBm
Receiver noise level	−105 dBm
Signal-to-noise (S/N) ratio	26 dB
Required S/N ratio	20 dB
Surplus S/N	6 dB

1.10   Modulation

Analog and digital signals are added to radio transmissions through very different types of modulation. For analog radio, the electrical signal modifies the carrier wave. For digital transmission, encoded bits alter the carrier.

The heart of a wireless telephone technology is its radio modulation scheme. We often refer to the full specification of the wireless modulation scheme, complete with all the signaling rules, as the *air interface.*

1.10.1   Analog modulation

We can send analog electrical signals (audio or video) on a radio link in two basic ways, amplitude modulation (AM) or frequency modulation (FM). AM means that the electrical voltage is matched by the *amplitude* of the radio wave. The receiver simply converts varying radio levels back to voltage that goes into an audio amplifier, a television or videocassette recorder (VCR), or some other terminal equipment. As shown on the left side of Fig. 1.8, AM is a continuous sine wave with its amplitude varying as the input signal changes. Should some other, interfering radio wave come along with its frequency close to the AM radio wave we are currently receiving (and demodulating), that wave will have its content added as a noise component to our received signal.

As its name suggests, FM means the electrical voltage is matched by the *frequency* of the radio wave, and the receiver converts varying frequency back to signal voltage. As shown on the right side of Fig. 1.8, FM is a continuous sine wave with its frequency varying as the input signal changes. An interfering radio wave is less likely to affect the frequency component at our receiver, so the FM receiver tends to produce a less noisy reproduction of its signal from a noisy radio environment. If we have two FM signals modulated around the same frequency, then there is some ratio between them that ensures that the stronger signal is the one we hear. We call this the *capture ratio* of an FM receiver, and a typical capture ratio is 1.0 dB. That means if the desired signal is 60 percent of the power at the receiver and somebody else's signal is 40 percent, then a typical FM receiver should produce the desired signal.

1.10.2   Digital modulation

AM and FM are *analog* modulation schemes in that some physical property of the radio wave is modulated in some proportion to a physical property of the meaningful

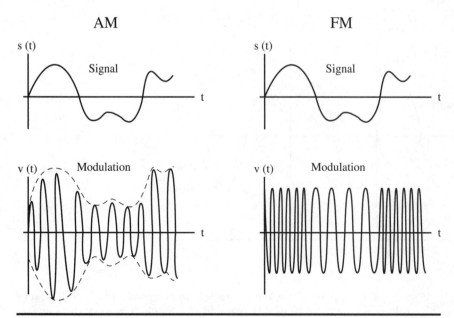

Figure 1.8 AM and FM modulation schemes.

signal. The contrast, of course, is with *digital* modulation, where the signal is turned into a sequence of binary digits, *bits,* and those bits are transmitted by modulation of the radio wave. Since CDMA, the subject of this book, is digital modulation, let us concentrate on digital modulation.

In digital modulation, we are adding bits to the radio signal. A *bit* is the smallest unit of information, represented by one of two states at a particular location. In writing, we usually represent bits as having a value of 0 or 1. Actually, the meaning associated with a bit could be a number (built from 0s and 1s) or a choice such as true/false or a representation of anything, including the human voice, through some particular code.

There are two general approaches to digital modulation, coherent, and noncoherent. In *coherent modulation,* a pilot or reference signal is used. A *pilot signal* is a broadcast signal that is modulated with a known pattern sent from the same transmitter. The receiver already knows what to expect in the pilot signal, so it can adjust its own internal parameters using the pilot as a reference.

In systems that use pilots, the signal is demodulated at the receiving end by a comparison of the two waveforms, the pilot signal and the modulated signal, at a single time *t*. In *noncoherent modulation,* a single signal is used, and the demodulator at the receiving end notes the changes in the signal over time. In noncoherent modulation, low amplitude might be a 0 and high amplitude might be a 1. Or the absence of a phase change might represent a 0 and the presence of a phase change might represent a 1.

One of the simplest digital modulation schemes is a coherent scheme called *binary phase-shift keying* (BPSK). The rule is simple enough: We broadcast a single frequency, and at predetermined time intervals, we broadcast the wave in the same phase as the pilot for a 0 bit and in opposite phase for a 1 bit, as shown in Fig. 1.9. Each of these predetermined time periods is called a *symbol.*

v (t)

symbol

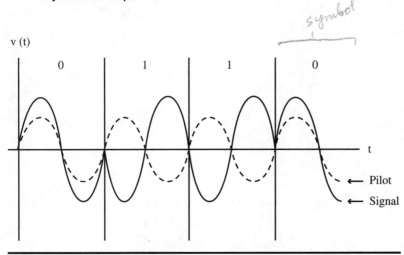

Figure 1.9   Binary phase-shift keying (BPSK).

The next level of digital sophistication is called *quadraphase phase-shift keying* (QPSK). The rule is not quite so simple: We broadcast a single frequency the same as BPSK, we compare to the pilot phase at predetermined time intervals the same as BPSK, but QPSK allows four different phase relations instead of two. The four phase shifts are 0, 90, 180, or 270 degrees for 0, 1, 2, or 3, as shown in Fig. 1.10. Using QPSK, we send 2 bits per symbol instead of just one. (Our four options, 0, 1, 2, and 3 in decimal, are represented as 2 bits in binary notation: 00, 01, 10, and 11.) QPSK makes twice as much use of the radio bandwidth as BPSK, and we can send twice as much information. We get a free lunch by using the full phase space that BPSK does not.

We use another type of graph called a *constellation* to depict digital modulation schemes. These radial graphs depict amplitude as the distance from the center of the graph and depict degrees of phase radially, from 0 degrees along the X axis to the right,

v (t)

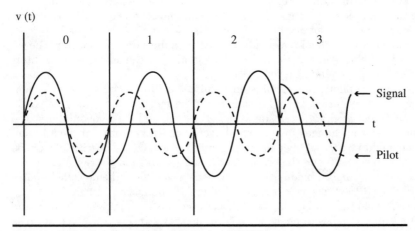

Figure 1.10   Quadraphase phase-shift keying (QPSK).

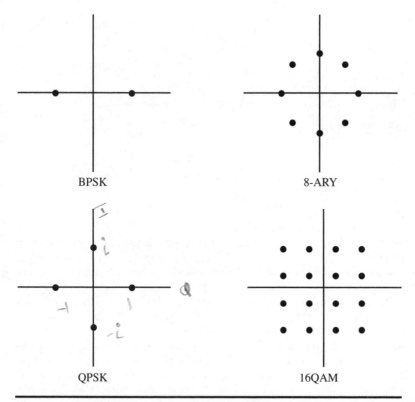

Figure 1.11   BPSK, QPSK, 8-ary, and 16QAM constellations.

90 degrees vertically up on the Y axis, 180 degrees horizontally to the left, and 270 degrees vertically down. Each point on the graph represents an amplitude and phase relation where digital information can be transmitted. In radio we do not call the two dimensions x and y or *real* and *imaginary* as we do in mathematics class. Rather, we call the same two axes *in-phase* and *quadrature*. Occasionally there will be some reference to I and Q for in-phase and quadrature.

BPSK has two points in its constellation, $+1$ and -1.[10] In mathematical terms, the frequency-phase space is the two-dimensional complex-number space, and the sophistication of QPSK takes advantage of the full two-dimensional modulation opportunity. QPSK uses the four complex numbers 1, i, -1, and $-i$.[11] There are more complex schemes that send more bits per time period. For example, there are m-ary schemes and quadrature amplitude modulation (QAM). Figure 1.11 shows BPSK, QPSK, 8-ary, and 16QAM constellations.

---

[10]We call the digital modulation states *points*. The astronomy analogy stops short of calling them *stars*.

[11]Mathematicians (like one of us, Rosenberg) use the letter i for the square root of -1. Electrical engineers use the letter j for the same concept.

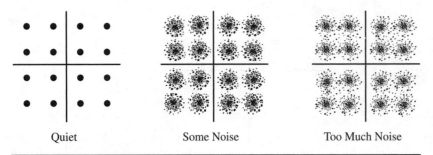

Quiet   Some Noise   Too Much Noise

Figure 1.12   Quiet and noisy 16QAM received constellations.

The factor of 2 gain from 1-bit BPSK to 2-bit QPSK is a one-time efficiency gain because BPSK simply ignores half the information opportunity. The gain in bit rate going up to 3-bit 8-ary or 4-bit 16QAM is not free. The effect of radio noise is to wiggle the points on the constellation picture, and more noise wiggles them further, as shown in Fig. 1.12. As the constellation points get closer together in more complicated modulation schemes, the noise component must be lower for the same bit-accuracy rate. A denser constellation gets its higher bit rate at the expense of a tougher S/N requirement.

There are other versions of QPSK, variations on the same four-point concept. There is *offset* QPSK (OQPSK), shown in Fig. 1.13, that displaces the four points by 45 degrees to $1 + i$, $1 - i$, $-1 + i$, and $-1 - i$. There is *differential* QPSK (DQPSK), a noncoherent modulation scheme in which the four phase angles are taken relative to the previous symbol. The GSM system uses gaussian minimum phase-shift keying (GMSK), a version of QPSK in which the carrier amplitude rises and falls in a gaussian-shaped impulse response to minimize radio transmission out of the desired frequency band.

In general, coherent modulation schemes are somewhat more resistant to noise than noncoherent schemes for the same bit rate. In a world with perfect phase alignment, perfect separation of in-phase and quadrature components, we would expect 3 dB better performance. Actual coherent systems tend to be about 2 dB better than similar noncoherent modulation schemes.

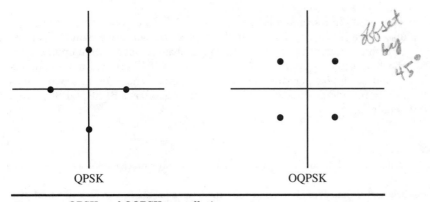

QPSK   OQPSK

Figure 1.13   QPSK amd OQPSK constellations.

The coherent-modulation performance improvement comes at the cost of having the extra pilot signal for coherent phase alignment. Thus we usually use coherent schemes when one transmitter is sending many signals so that they can share the pilot and non-coherent schemes otherwise. This is an important difference between the forward and reverse directions in a wireless telephone system, as discussed in Secs. 8.8 and 8.9.

Older digital systems used amplitude shift keying (ASK) and frequency shift keying (FSK), where amplitude and frequency changes were used instead of the phase changes we have discussed here.

1.11 Adding Radio Signals

Two radio waves combine by adding amplitudes, their voltages levels. (In the same way, audio sounds combine by adding their amplitudes, atmospheric pressure.)

This sounds like an easy concept, but it is not always so simple. Sometimes one plus one is two, sometimes four, sometimes zero, and sometimes one. To borrow a line from George Orwell's *1984,* sometimes it's all of them at once.

The statistical average of adding uncorrelated voltages is a *power sum.* To put this in plainer English, if radio signals are not closely synchronized in some way, then their combined signal is the sum of their power levels. If a receiver receives 10 pW (pico-watts) from source A, 13 pW from source B, and 32 pW from source C, then the combined signal from these three uncorrelated sources will be the total, 55 pW (10 + 13 + 32 = 55.[12] One plus one is two for uncorrelated addition.

In CDMA, we find ourselves paying a lot of attention to correlation and how signals are correlated. If two waves are of the same frequency in perfect phase, then adding their voltages increases power as the square of the voltage. We call this a *coherent sum.* Thus a 10-V, 2-W signal coherently added to another 10-V, 2-W signal yields a 20-V, 8-W radio signal. For coherent addition, one plus one is four.

Should the frequencies be the same but the phases be opposite, a 180-degree phase shift, then we get subtraction rather than addition of voltage. In the case of equal level and opposite phase, we get complete cancellation, no signal at all. An important example of such a cancellation is a signal that takes two paths to the same place (described in Sec. 2.1.4), where one path is half a wavelength longer than the other. If the two paths have similar signal path gains, then received signal is dramatically reduced. When we are out of phase, one plus one is zero.

Two signals can be arranged to be *orthogonal*[13] so that they are in phase as often as they are out of phase. The result is that a receiver looking for one of these signals sees no net contribution from the interfering signal. The most common U.S. CDMA system, cdmaOne (previously Qualcomm's IS-95 standard), uses orthogonal codes to create 64 signals that, at least in theory, have no interference with each other. For orthogonal signals, one plus one is one.

---

[12]This is one occasion where decibel notation obscures rather than illuminates what is happening: −80 dBm plus −79 dBm plus −75 dBm comes out to −72.5 dBm, but that is not terribly intuitive.

[13]The term *orthogonal* is a sophisticated mathematical term, and we are using it casually here. When we design orthogonal codes in CDMA, the orthogonality of those designs is sophisticated and quite mathematical, as we shall see in Sec. 8.3.2.

When the phase shift is 90 degrees, then the voltage sum is a power sum, and one plus one is two. In Sec. 1.10 we mentioned the in-phase I and quadrature Q components of a modulated signal, and these are 90 degrees out of phase with each other. Thus, while the total power is their sum, the I and Q components are orthogonal to each other, and a receiver can discriminate between them—and one plus one is one.

A few paragraphs cannot give an intuitive sense of how radio signals combine. The best we can do is help you understand that adding radio signals is a tricky business. Engineers doing this for decades still forget factors and components from time to time.

1.12   Conclusion

In this chapter we have reviewed the basics of radio engineering and their application to wireless technology. In the next chapter we will look at the challenges radio engineers face in making wireless systems work.

Radio Signal Quality

This chapter briefly defines the theoretical issues behind major challenges for cellular engineers. We look at noise, interference, and related issues, as well as issues related to radio signal measurement. The practical consequences of these issues will be addressed in depth throughout the remainder of this book.

2.1 Radio Impairments

Between a modulated and transmitted signal at one end and the received and demodulated version at the other end, a lot of bad things can happen. Since code division multiple access (CDMA) demands more of our radio links than most earlier technology, these impairments have to be understood and managed more carefully than before.

We can divide radio impairments into four basic categories: noise, interference, distortion, and multipath.

2.1.1 Noise

Every radio receiver has some internal electrical activity that has nothing to do with the electromagnetic radiation it is designed to receive. There is a quantum-level atomic component called *thermal noise* that increases with temperature and frequency bandwidth.[1] Other than operating our radio telephones in Dewer flasks of liquid helium, there isn't much we can do about thermal noise. Noise is also added by radio emissions from the sun and outer space.

Receiver amplifiers have their own sources of thermal noise from their own components that usually overwhelms the natural radio thermal noise. These can be reduced by better amplifier design or by putting the amplifiers closer to the antenna terminals so that the cable losses come *after* a stage of amplifier gain. This alone can reduce the receiver noise by several decibels.

Most radio noise is *additive white gaussian noise* (AWGN), so let us examine each of these four words in turn.

---

[1]Temperature is measured from absolute zero.

Additive noise is simply added to the signal. As elementary as this sounds, there are other things that happen to radio signals. We use amplifiers to multiply radio signals, processing equipment to modulate a radio carrier, and other equipment to filter frequencies. A medium such as air, or air with raindrops in it, reduces the radio signal, often reducing some frequencies more than others. The characterization of radio noise as *additive* is an important aspect of its behavior.

White noise is signal with equal energy at all frequencies, statistically speaking. In the sound medium, white noise has a hissing *shhhh* quality. One audio example of a nonwhite noise is hum, clearly concentrated at 60 Hz (in the United States). Audio engineers use pink noise that has equal energy per musical octave. It has a more rushing sound than white noise because it has comparatively more low-frequency content.[2]

Gaussian noise follows the statistical normal or gaussian distribution. The mathematical properties of gaussian distributions are an important part of radio engineering theory. Hum is an example of nongaussian noise, but we would be hard-pressed to find a good example of white noise that is not gaussian.

Noise is something that is not signal. In our CDMA models, radio noise is something introduced after the modulation phase and before demodulation.

2.1.2   Interference

Noise is something that is not signal. Not only is noise not *our own* signal, but it is also not *somebody else's* signal either. When somebody else's signal becomes an impairment to our own radio link, we call it *interference*. It may be additive but is not white or gaussian. In wireless telephone networks, calls have interference that consists of other calls using the same frequency and modulation.

Before cellular, interference was simply avoided for a long time. Systems were designed so that no two broadcasters in the same area were transmitting signals at the same time at the same frequency. The cellular principle, in a nutshell, is the idea that the tradeoff between interference and capacity is something we can *manage* through engineering and planning. Managing interference, rather than simply avoiding it, gives us much more flexibility, allowing us to offer new kinds of services. It is a deep and powerful truth in CDMA that interfering CDMA signals can be treated as noise, statistically speaking. This is the subject of Chap. 8.

2.1.3   Distortion

Both noise and interference are additive, superimposed on an unchanged underlying signal. Change in the shape of the signal itself is called *distortion*. A signal is not distorted if it is merely amplified or attentuated; the change has to be more substantial.

A radio amplifier that amplifies loud and soft signals differently is called *nonlinear*. A nonlinear amplifier produces a waveform other than the input signal and, therefore, creates distortion. An amplifier without enough power distorts the radio wave by *clipping* it (see Fig. 4.4 in Sec. 4.3).

Radio frequency (RF) filters are used to keep transmissions within allotted frequency bandwidth, but they also add distortion in narrowing the frequency distribution.

---

[2]The discussion of hum and pink noise is here as a contrast to white noise. To understand what something is, it is often useful to know something it is not.

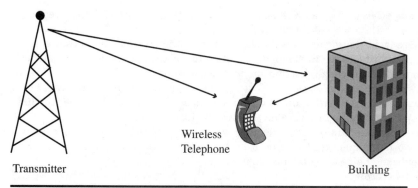

Figure 2.1 Fading due to multipath transmission.

The general term *distortion* is frequently used in the audio world, but one hears it less frequently in radio.

Analog telephone listeners know noise as a rushing or hissing sound on the line, interference as hearing fragments of another conversation, and distortion as garbled speech. These same three impairments affect radio signals as well.

2.1.4 Multipath

Radio waves can go in straight lines, and this is how we usually visualize them traveling. However, they often travel on the bounce, and we use the term *multipath* for the phenomenon of receiving several copies of the transmitted signal.

There are two multipath issues to deal with. First, the combination of the same radio wave at different phases causes dramatic shifts in the cumulative amplitude, and second, the signal may be received repeatedly at different times.

We use the term *multipath fading* for the varying amplitude due to multipath transmission. Fading is easy to understand if we picture a single-frequency broadcast from a single source being received on two paths, as shown in Fig. 2.1. Where one path is half a wavelength longer, the two signals will tend to cancel each other out because they are out of phase. The same is true if the difference in path length is 1.5 wavelengths or any other difference that adds the multiple paths out of phase with each other. A collection of many paths of similar radio path gains at a single frequency produces a *Rayleigh distribution*, and we call the single-frequency multipath-fading phenomenon *Rayleigh fading*.

For the mathematically curious, here are a few facts about Rayleigh fading. The distribution is a Rayleigh distribution in amplitude and an exponential distribution in power. Relative to mean power level, signal will be below -10 dB one-tenth of the time, below -20 dB 1 percent of the time, below -30 dB 0.1 percent of the time, and so on. The autocorrelation distance is about half a wavelength, so it is a pretty safe bet that moving a wavelength away from a low point will give you a better signal.[3] Finally, the

---

[3]Multipath fading has some cyclic behavior with high and low points about a wavelength apart, but we will not be too far astray if we think of this fading as an autoregressive process in three dimensions. And no, you do not need to know much about three-dimensional autoregressive processes to be a good radio engineer.

fading depends on frequency as well as on location, with an autocorrelation frequency change related to the range of the multiple paths.[4]

Where there is a significant line-of-sight path from the transmitter to the receiver, the distribution shifts from a Rayleigh distribution to a *Ricean distribution.* The Ricean distribution is just a Rayleigh distribution with a constant term added to it. A Ricean-distributed signal can get close to zero, just as a Rayleigh-distributed signal can, but it is less likely. Ricean multipath fading is less likely to be severe, and the larger the line-of-sight component, the more favorable the Ricean statistics get.

There are several techniques for dealing with fading built into cellular systems. There are two techniques that rely on *diversity,* where two receivers are placed far enough apart that their fading is uncorrelated.

When the two receivers are only a few wavelengths apart on the same tower, we call it *microscopic diversity.* There are two basic forms of microscopic diversity, *switched diversity* and *selection diversity.* In switched diversity, we use one receiver until its performance goes below some threshold and then switch to the other one. In selection diversity, we monitor both receivers and select the better one instant by instant. Switched diversity can be designed with two antennas and just one receiver, whereas selection diversity requires two complete receiver systems so that it is possible to compare the two signals and select one.

When the two receivers are in geographically distinct locations, we call it *macroscopic diversity.* The CDMA soft handoff described in Sec. 8.7 has two cells in two different places receiving the mobile telephone's signal; this is a form of macroscopic diversity. Macroscopic diversity combats multipath fading as well as larger sources of signal fading such as buildings, trees, and small hills.

While the fading phenomenon depends on the multipath environment, broader bandwidth always offers more protection from multipath fading. Resistance from multipath fading is a tremendous advantage for spread spectrum. The 25- or 30-kHz channel of analog cellular is very susceptible, the 200-kHz channel of the Global System for Mobility (GSM) is more resistant, and the 1.25-MHz of cdmaOne is better still. The 5-MHz or greater wideband CDMA is almost immune to multipath fading.

We will now turn to the second issue created by multipath, which is the spreading out of the signal across time at the receiver. We call the range of time delay containing most of the received power the *delay spread.*[5] The frequency autocorrelation is the inverse of the delay spread, so a delay spread of 1 μs requires about 1 MHz of bandwidth to allow us to be reasonably comfortable that most of the bandwidth will not be affected.

Spreading of the signal across time is not an important issue for analog transmission and audio frequencies.[6] AM and FM signals with sufficient amplitude should see no ill effects from delay spreads less than 20 μs, a very long path length difference of 6 km.

---

[4]The same caveat applies here. Fading is not a true autoregressive process in frequency or space, but it is close enough for us to treat it like one.

[5]This has nothing to do with how long it takes the signal to travel from transmitter to receiver, only to the *variation* in that travel time.

[6]Video is a different story, and many readers may remember an urban childhood where broadcast channels had ghosts in the picture from multipath reception. The left-to-right time in American NTSC television is 64 μs, quite short enough for a typical 1-μs delay to show up as a second image, usually fainter, on our television sets.

The high speed that makes CDMA attractive makes it vulnerable to multipath delay spread. If we consider a wideband CDMA system with an internal bit rate of 5 megabits per second (Mbps), then each digital bit is only 200 ns. Smearing the reception over 1 μs means that each digital bit is received over five time periods.

When the delay spread is greater than the length of time we use to transmit a single symbol, then delayed transmission of one symbol can interfere with reception of the following symbols. We have traded signal path fading for intersymbol interference when the delay spread is long enough and our symbol rate is high enough that alternative paths of our own symbols interfere with each other.

In CDMA, we have specialized *rake filters* (described in Sec. 8.4) to deal with this problem. These *adaptive matched filters* create an inverse image of the signal delay characteristics. The rake filter allows a CDMA receiver to resolve symbols shorter than the delay spread.

2.1.5 Doppler shift

Any source of radiated energy that is moving toward or away from the receiver creates a frequency change called the *Doppler effect*. This is easiest to experience in the pitch of sound from a moving source. As a race car moves past the spectators, the pitch of the engine whine drops noticeably. The frequency shift is the ratio of the speed of the source to the speed of sound.

The same effect occurs in light and radio. The astronomical red shift is caused by distant galaxies moving away from us at speeds close to the speed of light. Similar to the audio case, the size of the Doppler shift in radio is the ratio of the speed of the source to the speed of light. The Doppler shift also occurs if the receiver is moving toward or away from the transmitter.

While it is hard to visualize a moving automobile being close to the speed of light (even with some of the drivers we know), the effect can be significant at closely spaced high frequencies. A car moving at 100 km/h is moving about 30 m/s, one ten-millionth of the speed of light. At 2 GHz this works out to a 200-Hz shift, enough to affect a narrow channel.[7]

There are other causes of frequency shift, including calibration differences between radio transmitters and receivers. These have nothing to do with relative motion, but radio engineers often refer to these differences as Doppler shift anyway.

2.2 Radio Measurement Errors

A dynamic radio system has to measure its performance instant by instant and figure out what to do next. The base station measures the received power level from the mobile unit and instructs the mobile unit to adjust its broadcast power in a process called *power control*. Most calls require power control most of the time, and many calls require handoffs. The cellular system begins to consider handoff when the power received

---

[7]When one of us (Rosenberg) was doing narrowband digital channel work at Bell Laboratories, we were considering a 5-kHz channel at either 900 MHz or 2 GHz. The relative Doppler shift of two subscribers moving in opposite directions on the Garden State Parkway could easily be 500 Hz. The *guard bands* needed to deal with those shifts were an important design issue for this channel.

from a particular phone is greater at an antenna other than the antenna currently carrying the call.

In analog cellular development, power was measured in received signal strength indicator (RSSI) units, 0.78125 dB. Our understanding is that somebody told somebody that we needed to cover 100 dB of range, a reasonable interval. Having 7 bits available, somebody decided to divide the 100-dB range into 128 RSSI units with each unit being, therefore, 0.78125 dB.[8]

Our mathematical radio models usually assume nearly perfect response to perfect measurement. In the analog cellular world, we found that we were happy when our measurements were within a few decibels of the true value. In CDMA, we are trying to manage our power levels to within a few *tenths* of a decibel. CDMA systems access S/I ratios many times per second and adjust power up or down according to the S/I ratio measurement. This measurement can be wrong, too, and the effect on CDMA power control of these measurement errors is significant.[9] The effect of these errors on CDMA capacity is discussed later in Secs. 8.5 and 27.6.

2.3   Signal Quality Measurement

The two basic measures of radio signal quality are the signal-to-noise (S/N) ratio for analog and bit error rate (BER) for digital. While neither is a perfect measure of radio signal quality, both are excellent performance specifications.

2.3.1   Analog: The signal-to-noise ratio

Whatever the analog signal being sent, it will almost certainly sound better with less competition for the receiver's attention. This competition can be internal noise (N), or it can be interference (I) in the radio environment. The signal-to-whatever ratio is expressed as S/N or S/I or even S/(I + N). In more casual conversation, we have heard it called the "carrier-to-crud ratio."

Different modulation schemes have different S/N requirements. AM modulation sounds pretty terrible with a 10-dB S/N ratio, whereas a 3-dB S/N ratio exceeds an FM receiver's capture ratio and sounds fine to a listener. The use of the signal affects the S/N requirement as well: A telephone conversation tolerates much higher noise levels than the analog satellite feeds they used to use to transmit National Public Radio symphony orchestra broadcasts from the concert hall to the radio station.

The natures of the signal and the interference also change the S/N requirement. Analog cellular FM signals have a 1-dB capture ratio and sound fine with a 3-dB S/N ratio so long as nobody is moving. However, since cellular is specifically a *mobile* telephone technology, it behooves us to have a performance specification robust enough to ensure

---

[8]We see no reason why they could not have added 14 dB on either side of the 100-dB range they were trying to cover and left the units at 1 dB each. Now, three decades later, cellular engineers learn about 0.78125 dB RSSI units. But those decisions were made a long time ago.

[9]This is another study one of us (Rosenberg) did. There is a very nice and interesting mathematical model of power control that takes into account communication errors in power control commands. The additional effect on power control when the measurement errors were taken into effect and found to be quite significant.

high signal quality for a moving radio source, as discussed in Sec. 2.1.4. It takes another 20 dB of S/I protection, more or less, to overcome the handicap of motion.

2.3.2   Digital: Bit error rate

Rather than boast about a bit accuracy rate of 99.9999 percent, we describe digital link performance in terms of a 10^{-6} *bit error rate* (BER). BER is the defining parameter of digital link performance; two digital links with comparable BERs are equally good.

BER is simply the number of wrong bits divided by the total bit count over time. BER is expressed as a number between 0 and 1, so a channel with one bit wrong in a thousand has a BER of 0.001, or 10^{-3} in scientific notation. In casual usage, one hears of a BER of "3 percent" or "one in a million," but these refer to the fractional value of errors divided by the total.[10]

BER can vary widely from several percent on a noisy radio channel to some very small values in computer networks. A cellular telephone call voice channel has different digital accuracy requirements than a bank's computer network sending accounting data. Most digital systems send data grouped into *frames,* and one may hear the term *frame error rate* (FER), which is the fraction of frames with errors.

Different architectures have different BER requirements. Robust error-correction algorithms may use more radio bits for the same signal data but tolerate higher error rates. And different modulation and error-correction schemes may be less tolerant of bit errors that are close together. Since mobile telephones tend to produce their bit errors close together in time, in clumps, we can use a technique known as *interleaving* to spread the critical error-correcting bits over a longer period of time. One of the pleasant side effects of error-correction systems is that they give a good estimate of the BER because they know how many corrections they made. Error correction is discussed in more detail in Sec. 17.2.

Another issue with highly processed digital signals is the time delay created by processing. Callers may not notice a half-second delay from speaker to listener, but they almost certainly will notice 1.5 seconds of delay. It may take them a little while of "What? What?" to figure out that a round-trip radio link to the moon is faster than their cellular telephone call. Wireless telephone users will notice the delay and will not like it.

There is *always* some significant delay in digital mobile systems due to signal processing. This is why telephone companies always demonstrate long-distance calls when they show new equipment at conventions. If the call were made within the same room, the delay would be noticeable and irritating. If you wish to test this, simply stand next to a friend and have a mobile-to-mobile conversation while you are watching one another. You will notice a delay of about half a second. This is irritating when you can see each other but acceptable when you cannot see the other person.

2.4   Conclusion

This chapter ends our theoretical discussion of radio engineering. We will now turn to the equipment and systems that turn theory into reality for hundreds of millions of cellular users worldwide.

---

[10]This is *not* the same as the errors divided by the correct bits. On a very noisy channel, the wrong-divided-by-right value will be a larger number than the BER.

3

The User Terminal

A *user terminal* may be defined as any device in a cellular system that receives forward (downlink) transmissions from a base station and sends reverse (uplink) transmissions to a base station. Of course, the vast majority of these devices are mobile telephones, but with the possible growth in wireless local loop (WLL) and the advent of digital wireless, this will be changing soon. In this chapter we discuss mobile phones as components of the cellular system and then go on to look at some other types of user terminals.

3.1 Mobile and Portable Telephones

In the early days of Advanced Mobile Phone Service (AMPS), mobile telephones in vehicles were distinct from portable telephones, which were light and convenient enough to be carried by hand. Now the two technologies have merged, and a mobile telephone in a car is simply a physical holder for the subscriber's hand-held portable telephone with an external power supply drawing on the car's electrical system. These days, cellular engineers refer to a user terminal that functions as a telephone as the *mobile* or the *mobile telephone,* and we will do so here. Subscribers commonly refer to them as *cell phones.* Subscribers, however, have adopted the term *portable telephone* for household cordless telephones that do not use cellular technology at all.

The mobile telephone is a significant component of the cellular system, and its technology is changing constantly. Mobile telephones are becoming smaller, cheaper, and more common every year. Even though each phone is a low-cost item, the sheer number of phones means that the cost of user terminals is a major component of the cost of cellular systems. The engineering design of the radio link is a crucial element that governs the capacity of any cellular system. Yet the size and power of the battery are the key issues of the engineering design of the portable telephone itself.

3.1.1 Components and engineering

Figure 3.1 illustrates the components of a portable telephone. The external power source charges the battery and usually receives its power from a household or car electrical system. Customers are likely to own more than one battery charger for each cell

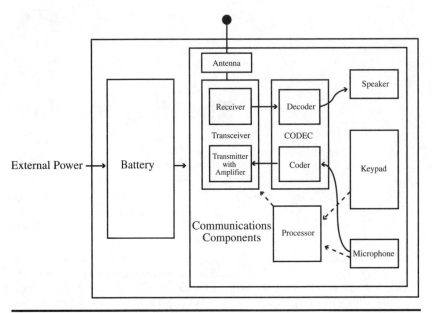

Figure 3.1   Portable phone components.

phone. In addition, innovative technologies include hand-powered generators that recharge cell phones.

The battery is a crucial component of mobile telephone design. Increased power in batteries of smaller size and lower weight has been a critical component of the technological revolution in mobile electronics, including battery-powered laptops, personal digital assistants (PDAs), and mobile telephones. The market for more compact, lighter sources of power has driven a remarkable engineering effort for more than a decade. For mobile telephones, the result has been a decrease in size and weight of about 25 percent per year. If this trend continued for another century, mobile telephones would become subatomic in size. More realistically, we are reaching the point where the battery is no longer the primary limit in the size of the mobile telephone. For phones to shrink any further, the keypad will have to be replaced with voice-activated dialing and controls, and indeed, this is beginning to happen. The major drive in battery technology probably will be used to deliver larger battery capacity at roughly the current weight and size. This larger capacity will be used primarily to increase the recharge interval of the cell phone rather than to increase radio power for greater system coverage.

The battery powers all components of the mobile unit; however, only the transmitter amplifier uses a great deal of power. This is why a typical mobile telephone can be active and ready to receive calls for 72 hours on one battery charge but only lasts for 3 hours of actual call time. Improved battery life, therefore, depends primarily on developing better batteries and increasing the efficiency of transmission through the transmitter, amplifier, and antenna.

The subscriber uses the keypad to direct the mobile telephone to perform its various functions. Sophisticated microprocessors in the mobile telephones provide a choice of features. Mobile telephones use preorigination dialing; that is, the number is entered first, and then the call is initiated when the SEND key is pressed.[1] Preorigination dialing was conceived when the market was vehicular. The driver could enter two or three digits at a time, with breaks to look up and check traffic. Several other advantages have been possible due to this design. Battery use is lower because the transmitter is not activated during dialing. In addition, since keystroke management is local to the telephone, the keypad can have a backspace or erase function and also can be used for a number of other alphanumeric data entry activities, such as maintaining a speed-dial list.

The microphone picks up the user's voice and sends it through the modulator to the radio transmitter during a call. The sound of the voice is also sent to the speaker so that the subscriber can hear his or her own voice. This deliberate recreation of the speaker's voice on a telephone call is called *sidetone*.[2] The microphone also can receive voice commands, such as voice-activated speed dialing.

The modulator turns the electrical audio signal from the microphone into a radio signal, analog FM modulation for AMPS and digital modulation for Global System for Mobility (GSM) or code division multiple access (CDMA), as we discussed in Sec. 1.10.

The transmitter amplifies the modulated signal and sends it to the antenna, where it goes over the reverse radio link to the receiver in the base station. This amplifier is the critical power component that limits call-duration battery life. It has to produce a radio signal with enough effective radiated power (ERP) to meet the signal-to-noise (S/N) demands of the base-station receiver and the air interface. While CDMA can demand high peak power levels in areas of poor reception or during busy periods, most of the time a CDMA channel is operating at a very low power level. This gives CDMA an advantage over AMPS and GSM in terms of power consumption and battery recharge interval.

One way to prolong battery life and to reduce radio interference as well is to shut down the transmitter when the user is not speaking. Normal human speech has short pauses between words and long pauses while the user is listening rather than speaking.[3] The air interfaces have protocols that allow for discontinuous transmission (DTx).[4]

The receiver receives the modulated signal coming from the base station. It also has an amplifier, but one that only needs enough power to produce an internal signal rather than a transmitted radio wave. Even that amplifier uses some battery power, and the receiver is always monitoring the paging channel while the mobile telephone is powered on. Thus GSM and CDMA proscribe a time schedule for paging so that the mobile telephone can conserve more power using discontinuous reception (DRx).

The transmitter and receiver share enough technology and components that it makes sense to form them into a single unit, a *transceiver*. AMPS and CDMA require the

---

[1]We have seen the SEND and END keys marked YES and NO on some portable user terminals.

[2]Telephones do this because speakers get nervous when they cannot hear their own voices. The round-trip delay of a telephone call, particularly a long-distance call, is long enough that it would be disconcerting to have the speaker's voice come back through the two-way link. The long-distance telephone network has *echo-cancelers* specifically to eliminate these round-trip echoes.

[3]Fax and modem signals do not have these pauses, and they present their own capacity engineering issues as a result.

[4]Tx is the radio engineering abbreviation for transmit.

transceiver to transmit and receive at the same time, whereas the GSM air interface maintains separate time slots for the forward and reverse signals. Since they do not have to transmit and receive at the same time, GSM mobile telephones can share such components as high-frequency oscillators. Not having to duplicate these components is a cost advantage for the GSM mobile telephone relative to AMPS and CDMA.

The demodulator is similar to the modulator. At the lowest level, the forward modulation is the same as the reverse modulation, but some of the other rules in the air interface are not symmetric in the two directions, as we will discuss in Secs. 8.8 and 8.9.

The speaker turns the demodulated forward link signal into sound for the user to hear. This requires an audio amplifier that consumes power, but it can operate at much lower power levels than the radio transmitter amplifier. After all, the signal from the audio speaker only has to travel about 1 cm to a user's ear, whereas the signal from the radio transmitter travels several kilometers to a base-station receiver.

3.1.2 Antennas and the air interface

Lastly, we come to the antenna, where electrical signals become radio waves. Unlike antenna arrays at base stations, the antenna of a mobile telephone is severely limited in size and directional control. In Chap. 4 we will go much further into the technical issues of antenna design, but it is essential to include some crucial points here.

All antennas have a *radiation pattern* that describes the antenna's gain in varying directions, essentially the fraction of the radio power going in each direction, as described in Sec. 1.8. The radiation pattern of an antenna is also its reception pattern when it is used to receive radio signals. In terms of antenna function, radio waves behave the same way in either direction.[5] Thus we refer to antenna radiation patterns for both transmitting and receiving.

This reciprocity of transmitter and receiver applies to other radio terms as well. One might expect that a beam antenna would transmit a beam, and it does. However, a beam antenna also receives on the same narrow beam—a fact that is not obvious from the term.[6]

In Chap. 4 we will discuss directional antennas, including phased arrays and smart adaptive antennas, more extensively. The mobile telephone antenna, however, is presently designed as an omnidirectional antenna, not just 360 degrees but a full 4π sphere with no antenna gain (0 dB gain). This is essential because a subscriber may rotate or tilt the mobile telephone and its antenna in any direction. If the mobile telephone could have significant antenna gain, then the more efficient radio link would improve coverage in poor reception areas and allow the battery to function longer without recharging. Also, if the mobile telephone could have some directional control of its radio signal path, then interference levels would be reduced and system capacity could be increased.

There is one technology, available as a third-party add-on to many cell phones. It is an antenna extender that fits over the back of the phone and is said to improve signal quality. If this technology is indeed beneficial, it might be included in future cell phone design.

---

[5]This same two-way reciprocity of signal path characteristics is used in engineering sound systems as well as radio systems.

[6]When the authors think of a beam, we think of a Star Trek phaser, the *transmission* of a narrow beam.

Looking into the future, we might ask if there are any other technologies that would increase antenna gain and improve the quality of wireless telephone service. If a smart array antenna could be built into a mobile telephone, then it could constantly direct the signal in the direction of the base station, even as the phone is moved around. It may seem far-fetched to imagine smart array antennas built into cell phones. However, the notion of an adaptive matched filter, the CDMA rake filter described in Sec. 8.4, seemed like science fiction rather than reality a few years ago. However, need drove the technology. Today, rake filters are standard in all CDMA telephones, and they manage delay spread instant by instant. What we have seen consistently in cellular engineering and other portable electronics is that demand drives engineering innovation, and therefore, it is a mistake to say that technology not available today will be impossible in 5 years.

3.1.3 Economics

What, then, are the driving forces in the technology for mobile telephones? And what lies ahead?

One driving factor from the past will continue to operate indefinitely: cost reduction. When we look at a mobile telephone system with its base-station towers, its elaborate switching systems, and its complex transport network, it is hard to realize that the mobile telephone is a major component of the cost of the cellular network. In fact, the mobile telephones may be the most expensive part of a wireless telephone system. It depends on usage patterns, of course, but a large system may have 10 switches, 1000 cells, and 1 million mobile telephones. If a switch costs $200,000, a base station costs $80,000, and each mobile telephone costs $175, then the mobile telephones would make up two-thirds of the equipment cost of the system.[7] As a result, reducing production cost for cellular telephones is very valuable to cellular providers, who often give away the telephone or discount it heavily to get new subscribers.

The production cost of electronics is often more dependent on the scale of production and the selection of appropriate and available parts than on the complexity of the design. Huge production quantities typically are required to make the use of application-specific integrated circuits (ASICs) economically viable in consumer electronics. The one-time cost of design and preparation for manufacture is huge, whereas the unit production cost is typically very small. Designers agonize over tiny part cost differences when those differences are multiplied by tens of millions of mobile telephones.

As we have said, battery power for a given size and weight has been a dominant driver of cell phone design until recently. It will continue to be a driver, but the trend of the use of that battery improvement is already shifting. Cell phones have become small and light enough to satisfy most users, and the keypad is now more a factor in limiting size reduction than the battery. As a result, the competition is now moving toward increasing battery life at the current size. Some phones can run 3 or 4 days of reasonably heavy use without recharging, and it would not be surprising if phones that only need to be recharged once a week become common in the next few years.

---

[7]That same large system may have fewer subscribers who use their cellular telephones more heavily. In that case, the mobile telephone fraction of the total cost would be much smaller than two-thirds.

One technology that may be used to increase battery life is rather remarkable. Philips Electronics has developed a pager that operates on an *asynchronous* processor chip, a chip without an internal clock. Nearly every commercially sold processor chip since the dawn of computing has had a quartz-crystal timer setting the chip speed. However, there is an alternate design: an asynchronous chip in which the clock function is replaced by a communications function that allows one part of the chip to tell another part when a calculation is complete. Asynchronous chips require radically different design and are harder to test than synchronous chips. As clock speeds grow faster, more and more time and energy are spent on the clock function, asynchronous chips may break into the processor market, especially for cellular telephones. Here are some of their advantages over standard synchronous chips:

- They use less power.
- They run faster.
- They generate less heat.
- They do not generate the radio frequency (RF) interference generated by quartz-crystal timers.
- Their encryption is harder to break because in synchronous chips the timer is a clue that can be used by decoding devices.

The Philips pager with an asynchronous processor had twice the powered-on operating life of its competitors. Philips created the initiative to make, test, and produce these chips so that it could continue to develop them for cellular telephones and other markets. At the time of publication, no other manufacturer has asynchronous chips in production, although IBM, Intel, and other companies have created them in their laboratories.

There is also a role for continued size reduction, but it depends on customer acceptance of the technology. At present, wrist cell phones, worn like a watch, are a novelty. However, if mobile telephones attached to the body become popular, we will see a further drive toward miniaturization. The first steps in that miniaturization are already occurring. Some users prefer an invisible headset with a phone attached to a small belt clip. In addition, voice-activated dialing is now a common feature. If the keypad is replaced by voice-activated controls, then battery size once again becomes the key driver in the move toward miniaturization. This demand would then drive battery technology even more rapidly in the direction of increased power at smaller sizes.[8]

Call quality is clearly a driver in the adoption of cellular systems and a significant reason why customers prefer CDMA to AMPS in the United States. In the development of mobile telephones, the quality of the radio is a factor in the quality and capacity of the network over the air interface. In current capacity planning, the relative benefit of improving the quality of mobile telephones versus improving the quality of base stations depends on the ratio of mobile telephones to base stations. If a system has 1000 subscribers per base station, it is probably more cost-effective to implement system performance improvements at the base stations because there are so many mobile

---

[8]Older readers will recall Dick Tracy and his two-way wrist radio from the comics. This innovation in cellular technology probably will be much more welcome in the market than the shoe phone made popular in the television show "Get Smart."

telephones served by each base station. However, if a system has only 100 subscribers per base station, it can be more cost-effective to improve the quality of the mobile units, even if this might mean offering customers a free trade-in for a new unit.[9]

In AMPS and GSM, a poor radio results in poor call quality for that user, making it easy for the customer to accept an upgrade to a new phone. However, in CDMA, each mobile unit transmitter is a source of interference for all the other callers on the cell. As a result, a poor radio or, for example, a radio with poor power control interferes with other users on the same cell. In these situations, the problem does not affect the cell phone owner; it affects other users. As a result, it may be more difficult to convince a customer to trade up to a new phone.

Overall, however, better radios in CDMA mobile telephones mean more calls per sector, requiring fewer base stations in a given area to provide the same quality of service to the same number of users. This is a factor in planning in that if phones are developed with better radios, it will be possible to slow the addition of base stations while meeting capacity requirements for growing demand. However, it is not an instantaneous solution to any existing problem on the wireless network because cellular providers cannot force customers to abandon their current phones immediately.

In fact, the need to support existing phones is often a major constraint in the development of cellular networks. Two particular examples come to mind. As AMPS use lessens, part of the AMPS frequency has been reallocated to support CDMA transmission. CDMA is much more effective in its use of bandwidth, so the preferred solution would be to abandon analog systems and convert everyone to CDMA, but this is simply not feasible. Instead, as AMPS users replace their older mobile telephones, providers are converting AMPS bandwidth to CDMA use and reducing the number of available AMPS channels, as described in Chap. 33.6.

The second example involves the design of cdma2000 and how it will replace cdmaOne in the United States. Although support for the higher capacity cdma2000 will arrive within the next few years, the systems will continue to support cdmaOne mobile units for perhaps a decade or more. This capability for legacy support was built into the cdma2000 specification.

However, many individual elements that may seem trivial to engineers drive the cellular market as well. In addition to personal preferences such as style and color, there are health concerns related to radiation, perceptions of brand quality, and other factors. These are nontrivial. Radiation shields that protect the brain from cell phone transmissions are already on the accessories market. It is fair to say that if a smart array antenna is developed for cell phones, it is likely to undergo evaluation as a potential health hazard and that no matter what the research findings, some customers will not want to put these units next to their heads.

Other user terminal features create engineering challenges and may change not only the cell phone but also other components of the wireless system. The voice messaging feature requires memory storage capacity on the cellular network available to users. Similar storage can be used to store user data, such as autodial lists and the sound clips needed for voice recognition on the cellular system, rather than in the cell phone. This could increase the demand for electronic memory on the cellular system and also increase usage of the network's signaling channels.

---

[9]The subscriber-to-base-station ratio depends on the subscribers and how much they use their mobile telephones. If people tend to spend a lot of time using their cell phones, then the system will have a lower ratio of user terminals to base stations.

3.2   Other User Terminals

As electronic technology becomes more pervasive, more and more different types of user terminals will interact with the cellular network. In this section we will discuss the possibility of cellular systems replacing land lines for the local loop to the customer's residence or place of business, various types of wireless data terminals, and terminals without a human interface. In addition to describing current technology, we also will risk a bit of prediction regarding what will come.

3.2.1   Wireless local loop (WLL)

The *local loop,* the landline link between each customer phone and the local branch exchange, is an expensive component of the public switched telephone network (PSTN). Engineers sometimes call the local loop just a *loop.* In the United States, Europe, and other developed areas, this infrastructure is in place and is unlikely to be changed any time soon. Over the last 20 years, there have been discussions of replacing copper twisted-pair cable with coaxial or fiberoptic cable to get higher transmission rates. The reality has been that, instead, we have increased the data capacity of copper twisted pair with technologies such as the Integrated Services Digital Network (ISDN) and digital subscriber line (DSL), so we could continue to use the existing infrastructure. Even as cable has been installed for television, we have not begun to use it for telephone.

There is a simple reason for this. Even if a newer technology is better and less expensive, if the old technology works, the least expensive choice is to keep using what we already have. As a result, in the first world, changes to local loop technology are gradual and consist primarily of installing fiberoptic cable to business customers in high-tech areas.

A few years ago an alternative local loop was proposed, particularly for developing nations. The idea was to use the air interface from base stations to fixed antennas instead of installing cable from a telephone central office to all the telephone subscribers in a neighborhood. For these subscribers, the wireless telephone system would be their primary home or business telephone system. This approach was called *wireless local loop* (WLL). On the face of it, the idea had many advantages, and for the last 5 years, it seemed to be on the verge of being adopted widely in the developing world. However, this has not happened, and in our opinion, it becomes increasingly unlikely as time goes on. To understand why, we need to look at a number of economic, social, and engineering factors.

The socioeconomic background of WLL is crucial in understanding why it has not yet been implemented. We tend to assume that the developing world will develop, and in looking over decades, it probably will. However, backing any particular technology to be deployed in the next decade or less is an extremely risky venture that depends on the political stability of each nation and other complex social factors. Backers of the Iridium satellite telephone network suffered heavy losses when a global customer base failed to materialize, and similar things may happen to those who expect WLL to be the primary technology for local communication in the developing world.

In these considerations, the People's Republic of China is a major factor. In fact, mainland China is often held out as the Holy Grail of global marketing for many products and technologies, including cellular. As the world's most populous nation, and one whose technological and economic development is managed by a central government, it appears to be a huge potential technology market. If the Chinese government committed to a single technology for developing telecommunications, it would mean hundreds of millions of customers and many years of committed sales. Production and sales at a volume that would satisfy demand in China would reduce the per-unit cost of WLL to the point where it would be affordable in other developing nations as well.

Many companies have hoped for, and perhaps even expected, such a commitment. Small WLL test systems have been developed and deployed in China. But no decision has been made, and no contract has been signed. Finding reasons for the actions of the Chinese government is a profession all its own. However, these factors should be considered: The government has a distinct ambivalence about Western economic and technological development and fears that it is a destabilizing factor. Political events, such as the worldwide live broadcast of the protest in Tiananmen Square, tend to exacerbate this hesitancy, especially with regard to enhanced communications. In addition, the government strongly prefers internal economic development, and mainland China has a growing capacity to manufacture electronics locally. This would be a reason for delay, since the later the initiative begins, the more equipment will be manufactured within the country, reducing the effect on mainland China's trade balance. Lastly, it is simply not possible to predict the behavior of any large bureaucracy as if its decisions were coming from an individual. Many individuals are needed to move any one decision forward, and as a result, a large bureaucracy is inherently conservative and slow.

Politics aside, technological issues and high customer expectations create challenges for WLL as well. Foremost among them is call availability. In the developed world, we view the mobile telephone as ancillary to the landline network and do not expect it to be as available as a landline telephone. A cellular system might block 2 or 3 percent of its busy-hour calls (and occasionally 5 or 10 percent) and still be providing excellent service. However, for home or business service, we consider 1 percent blocking to be the upper limit. In addition, higher call quality, with calls free of static and virtually no loss of service during a call, is expected. Another difference in quality between mobile service and WLL is that WLL really does have to provide good coverage throughout the entire cell. For mobile service, 90-by-90 coverage (90 percent of the region having acceptable service 90 percent of the time) is considered sufficient. However, this quality standard relies on the fact that customers will be mobile and that they will move through one area to another and try the call again. With WLL, however, the customer who lives in the area with poor coverage always has poor coverage.

The result of all these technical issues is that the higher signal quality requirement of WLL will require higher base-station towers and more tweaking than a primary mobile telephone system, so it may be more difficult to engineer and more costly per user. However, it would still be likely to be less costly than landline local loop.

Two other engineering components differentiate WLL from mobile cellular technology. The first is the WLL terminal equipment. This terminal equipment performs the same function as the transceiver, amplifier, and antenna of a mobile telephone. However, since it is not mobile, it can be powered by the local electrical system or a solar or other power generator. Instead of being an integral component of the cell phone, it terminates in a household jack, just as wireline service does in the United States. Customers can then attach any kind of phone or data terminal in the residence or workplace. The terminal equipment certainly will have a directional antenna pointed at the nearest base station, which will both improve call quality and reduce the interference it otherwise would generate for other cells.

One additional technical problem is created by the directional antenna of the terminal unit, as illustrated in Fig. 3.2. If a new base station is built closer to the customer,

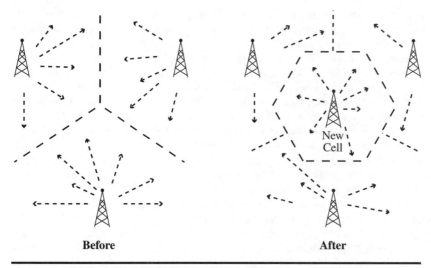

Before   After

Figure 3.2 Adding a cell with fixed, oriented antennas.

it is necessary to reorient the antenna toward the closest base station. This requires an expensive visit by a technician to every WLL unit with a directional antenna. The antennas would not have to be redirected immediately, but until it is done, the terminal unit will be interfering with a base station closer than the base station that provides it with service. In CDMA, more interference means fewer calls per cell, so the new configuration will not reach optimal capacity until all WLL terminal unit antennas are redirected to the nearest base station.[10] Adaptive phased arrays, smart antennas that can be reoriented electronically, might be used at WLL terminal stations. This might make it possible to reorient these antennas without visiting each location. However, this is an innovative idea, and it may turn out not to be either technically feasible or cost-effective.

The second special technology used in WLL is the creation of a single terminal equipment unit for multiple residences or offices. An apartment building, an office building, or even a development of multiple houses with local wiring within the development could all be served by a single multichannel terminal equipment unit. The air interface would have the functional role that a PSTN trunk group has from a private branch exchange (PBX) at that location to the local exchange on a traditional landline network as described in Sec. 11.1. It would provide access for all the calls from that location. Just as is true of local landline trunks, the total capacity of the air interface might well be less than the total number of users in the building or complex being served. Since not everyone will be calling at once, it might be possible, for example, for a link that can support 10 simultaneous calls to serve 20 residences. In the case of large customer facilities, a directional antenna could be installed at the base station pointing toward that one customer group. This terminal equipment unit is not a base station because it is on the remote end of the air interface. Functionally, it is the opposite of a base station because itreceives the forward signal and transmits the reverse signal. However, in terms of components, a complex terminal equipment unit is very similar to a base

---

[10]By nearest, we mean the one with the highest radio path gain.

station, and many of the design issues that apply to the base station discussed in Chap. 4 apply to these terminal equipment units as well.

Until recently, WLL seemed very promising. It seemed to be on the verge of exploding through the developing world. Its only competitor was landline service. In many developing nations, it can take months or years to get landline service installed. In others, copper is an expensive commodity, and wire is dug up and stolen shortly after it is installed. Overall, the existing landline network often has low availability and poor quality. WLL seemed ready to roll.

However, for various social, political, and economic reasons, WLL has not rolled yet. And now it has its first serious competitor. Oddly enough, that competitor is the mobile cellular phone system. And one of the key drivers is the comparatively low cost of cellular mobile units verses WLL terminal units.

As we mentioned earlier, a major factor in the per-unit cost of electronic equipment is the number of units being produced. Hundreds of millions of cell phones are in use worldwide, and sales remain high. Also, production facilities are already in place, whereas there is little or no mass production of WLL terminal units. As a result, the cost of deploying cellular continues to drop.

But what about quality? Earlier we argued that WLL has to be more reliable than cellular. However, that was based on the idea of meeting the quality standards of landline service in the first world. Realistically, cellular service is already better than the landline service in many nations.

One interesting example is Israel. In the last few years, cell phones have exploded across the country, outstripping the growth of the landline network. Most members of the middle class and above seem to be getting cell phones. There are a number of reasons for this. Although landline service is high quality, it does charge per minute even for local calls and is a monopoly, with only one service provider in any given city. There are at least two or three competing cellular companies, and as they compete, costs are driven down until they are similar to or better than the costs of landline services. Also, Israel is a very mobile society. There are many students and many young adults living at home after military service but planning to move out as soon as they save enough money. Having multiple generations share a single dwelling adds value to the privacy of a cell phone that allows you to walk outside and call someone. And in a nation often under siege, being able to call loved ones wherever they or you might be is a major incentive for having a mobile telephone.

Of course, Israel is much more prosperous than many developing nations. However, it would not be at all surprising if other nations followed a similar path. In this scenario, cellular providers would enter the market and compete. People would get cell phones when they move or when they could afford them, and usage would grow. In poorer villages, it could even happen that several families or even a whole village would share a cell phone the way they now share satellite phones or televisions.

Having talked about Israel, let us return to China. As of early 2002, China has chosen time division synchronized code division multiple access (TD-SCDMA) as the cellular technology it will deploy. It also has become the world's largest market for cellular telephones. Much of the equipment manufacturing is being done in China, and TD-SCDMA requires the use of adaptive phased arrays at the base stations. Economies of scale in manufacturing may bring down the cost of adaptive phased arrays so that they are affordable at the WLL user terminal equipment unit as well. It is too early to tell how much WLL China will develop, but they are clearly choosing a technology that is likely to make it feasible.

Given the growth of cellular, is there a future for WLL? In all probability, yes, because WLL has some other uses we have not mentioned. First of all, the infrastructure that supports cellular also can support WLL. As a result, in developing nations, WLL may be an ancillary service to cellular with better quality. Once the cellular networks are growing, multiunit residences and businesses could add WLL terminal units. Second, WLL has some specialized uses in the first world as well. WLL is an excellent way to provide high-quality phone service to temporary locations and in emergencies. In the construction industry, projects and costs can be managed much more effectively if there is reliable telephone service to construction sites. It can be cost-effective to bring either satellite or WLL phone service into a construction site that is in a remote area and not served by landline service yet. Also, if disaster damages or overloads the local PSTN, WLL can assist. After the collapse of the World Trade Center, WLL pay phones provided phone service in lower Manhattan for several months, as shown in Fig. 3.3.

3.2.2   Wireless data terminals

Wireless data terminals include simple devices such as pagers, as well as full-powered laptop computers and PDAs with wireless Internet access built in. Although these devices are the first that spring to mind, wireless terminals are extremely diverse in function and in design. In this section we will discuss the units that have a human

Figure 3.3 Wireless pay phone near World Trade Center, New York City.
(Yoav Levy.)

interface, and in the next section we will discuss fully automated units with no human interface.

Wireless data terminals can be divided into two major categories. Devices such as pagers receive and send data over the Short Message Service (SMS), which transmits packets of data up to 160 characters long. The SMS protocol, more fully covered in Chap. 19, uses free space on the signaling channel of cellular networks to transmit short packets of user data. Cell phones, pagers, and other devices can use SMS. For example, when any of us chooses the option to leave a callback number, we are using SMS, but not when we leave a voice message. As a result, the person we are calling or paging can see the callback number without dialing in to pick up messages. This works because the message is transmitted over the control channel and not over a voice channel.

SMS traffic is currently about 1,000,000,000 (an American billion) messages per day worldwide. However, this could increase significantly. Public institutions, such as elementary and secondary school systems, are starting to use SMS broadcasts to announce school closings for bad weather. As people come to expect to be notified of important events wherever they are, the cellular network will supplement and in some cases perhaps replace broadcast radio and television as media of public notification and communications. Once a service becomes common and inexpensive enough to be used by public institutions, it is likely to expand rapidly as many sectors of society find uses for it that were never predicted by the planners and developers.

The second category of wireless data terminals includes those devices which send large quantities of data over the cellular network traffic channels. Among these are PDAs and laptops for Internet access, as well as for access to proprietary or secure corporate and government networks using dedicated lines, encryption, or virtual private networks (VPNs). The demand for higher-speed data access is a major driver behind the 3G/UMTS standards initiative, as discussed in Part 9.

As more people send data over wireless for more reasons, and as new technologies develop, what can we expect to see? One possible growth area is in the use of the cell phone itself, especially if Bluetooth technology takes off. Bluetooth is a wireless network technology designed to allow digital devices to talk to one another up to a radius of 10 m (30 ft) using very inexpensive digital transceivers. One potential implementation of this is the *personal area network* (PAN). To get a sense of how a PAN would work, imagine walking down the street with a cell phone at your waist, a laptop in your briefcase in sleep mode, and an electronic wallet in your back pocket. You are wearing a headset and listening to MP3 music. If someone sends you an e-mail, it arrives over your cell phone. Your cell phone wakes up your computer and deposits the e-mail onto the hard drive. If the message is marked urgent, then it beeps your MP3 player and tells you that you have important e-mail. Then a new song you ordered is released. The song is downloaded over your cell phone to your MP3 player so that you can listen to it, and your electronic wallet is debited for the cost of the order. In this way, Bluetooth would allow all your electronic devices to talk to one another, to any other nearby Bluetooth device, and through your cell phone to any other electronic device on the Internet. The Bluetooth consortium certainly would like it if every third-generation (3G) telephone contained a Bluetooth chip, but it is too early to say if this will happen. However, with Ericsson, a major cell phone manufacturer, as a key member of the Bluetooth consortium, it certainly seems possible.

It is fair to say that given the cost of data service, demand for high speed digital data services will remain mostly in the private sector for some time. However, this demand may not be concentrated in the Fortune 1000 and other major companies. There are three reasons wireless data may have a broader subscriber base. First of all, cellular service providers and third-party providers relying on cellular networks are building turnkey prepackaged services and wooing small and medium-sized businesses. Second, it will benefit commercial vendors and service providers if their customers enter the high-speed wireless world as consumers, not just as employees. Third, the largest firms have the option of developing their own wireless infrastructure on a private band. At high volume, this can be cost-effective compared with leasing time on the cellular network. Doubtless some major companies will develop systems in-house and maintain control of them, whereas others will choose to outsource and rely on cellular data services. Some companies will create systems that rely on both, perhaps using their own network in their home cities and leasing wireless service elsewhere so that they get optimal cost performance in each region. Eventually, government and not-for-profit agencies will begin to use these services as well.

A variety of services will be provided. Beyond Internet access and access to organizational networks, we also can foresee the possibility that wireless networks will compete to provide a variety of specialized functions, especially if data capacity increases and costs come down. One possibility is that wireless networks will provide a lower-cost alternative to private radio networks. Some smaller police forces are already relying on cell phones rather than radio systems, although, of course, encryption technology will have to provide very strong security for such a service to grow in that field.

Many industries, such as gas and electric utility companies, taxicab services, and delivery services, already use private networks to carry wireless data. Demand is growing for increased data capacity as these customers are looking to provide real-time mapping information and other data-intensive functions. It is possible that the cellular networks will be able to compete to provide these services. In this arena, cellular has an advantage over satellite transmission, which has a very slow bit rate from the remote terminal back to the PSTN, as low as 8000 bits per second.[11] The general trend in our economy to outsource communications and information technology infrastructure will support more rapid growth in this market. In order to compete in a less regulated, more competitive market, companies want to own less and less equipment, and wireless service providers could get extra income from existing infrastructure by providing these services.

Lastly, we should expect the unexpected. Both crises and innovations drive adoption of new technology. The Candlestick Park earthquake in 1991 increased telecommuting in northern California, and the World Trade Center tragedy of September 11, 2001, which overwhelmed both the PSTN and the cellular network, triggered a boost in satellite phone sales. On a more positive note, very few people foresaw how the development of the World Wide Web standard would cause the Internet to grow so rapidly outside the ivory tower to transform our economy. Doubtless similar innovations will create completely unexpected uses for wireless data technology.

3.2.3 Terminals without a human interface

For people unfamiliar with the idea of automated devices communicating with each other, the idea of wireless communications without people may seem strange and unrealistic. However, such communications have been around for decades and are growing

---

[11]Some data services, including satellite and cable modem, have different throughput in different directions.

rapidly. For decades, computer systems have been designed that automatically call in to send or receive data, to report a malfunction, and even to be repaired or upgraded remotely. This usually takes place over a landline or, these days, the Internet, but it certainly can occur over a wireless network as well. And the technology is already spreading from computers to electronic and mechanical devices with computer controllers.

Key technologies that enable these services are the Global Positioning System (GPS) technology (which allows any device to calculate its own location on or above the earth's surface within a few meters), the satellite communications network, and the cellular network. Cars already have GPS-based mapping and driving direction services and will have Internet access soon. Cellular networks could compete with satellite services to provide this as GPS location reporting is incorporated into wireless telephone networks.

Utility companies are looking into gas and electric meters that report their status, eliminating the need for meter readers. Many campuses are going to smart-card vending machines for everything from soda machines to washers and dryers. Currently, charges are managed on the smart cards. However, the addition of cellular communications could provide these devices, which are already computerized and in locations that are costly to connect with a network, with the ability to report errors or even a need for routine service such as restocking of products for sale. Similar devices designed for households are already coming onto the market in Japan, which tends to be the first adopter of highly computerized technology. There are refrigerators that order groceries when you run low and microwaves that download recipes over the Internet. Of course, these do not have to be on cellular networks. However, the cellular network is already in place, and providers are seeking more customers, so it is reasonable that they will compete to provide service to such devices.

What will come next? No one can be sure, but there are some incredible uses for wireless data already. In Japan, a cell phone strapped around a cow's neck calls the cow home from the pasture for milking in the evening. Many of us have heard of the LoJack system that transmits a radio signal from a car if it is stolen, helping it to be recovered. A similar product is now available for pets, and it may be implanted, rather than worn on a collar, so that it cannot be detected. Such devices could transmit data over the cellular network as well. We are already willing to protect our cars and pets with locator technology. Will it be very long before we are willing to protect our children the same way? For some, the notion of hiding or implanting radio devices on ourselves or our children or requiring them for convicts on parole may seem anathema either because it seems to automate us and turn people into commodities or because of its implications related to freedom and privacy. However, increased concern for security as a result of the World Trade Center attack on September 11, 2001, and other events is causing many Americans to reevaluate their priorities. One of the authors (Kemp) attended a meeting of state police technology experts where the importance of being able to identify individuals correctly in real time was at issue, and such devices were suggested as likely innovations in the not-too-distant future.

3.3   Conclusion

The user terminal has grown a long way from the early mobile telephones installed in cars, or perhaps it would be better to say that it has shrunken and morphed. As many sizes, shapes, and uses as there are for wireless devices, there will be more in coming years. However, there are numerous engineering challenges to be overcome, especially with the advent of the 3G world of Enhanced Data rate for GSM evolution (EDGE), universal mobile telephony system (UMTS), cdma2000, and wideband code division multiple access (WCDMA). The ability to deliver high-speed data will require more sophisticated, more expensive radios with more exacting manufacturing tolerances. And these expensive precision piecesof equipment will need to be provided at lower and lower cost as they are included in existing equipment or given away at low cost or for free as incentives for choosing cellular service.

4

The Base Station

The base station is the critical component of a cellular network technically, economically, and oddly enough, aesthetically as well. In system terms, we can define the base station by its two interfaces: a noncellular pipe back to the mobile switching center (MSC) and public switched telephone network (PSTN) and an air-interface link using the system's cellular standard [Advanced Mobile Phone Service (AMPS), Global System for Mobility (GSM), code division multiple access (CDMA)] to the user terminals operating within the cell. On the air interface, the base station transmits the forward (downlink) radio signal and receives the reverse (uplink). As an engineering component, the base station is a unique technology that creates the cellular structure. Of course, every component of the cellular system has a crucial function, but base stations, due to their combination of telephony and radio components, are more specialized than, for example, MSCs, which are largely telephone switches adapted to cells. The engineering of the base station as a whole determines the reliability of the cell, and the engineering and management of the air interface are the most critical components in optimizing cell capacity and quality.

Economically, base stations are expensive to build, and a cellular system needs a lot of them. Also, due to their remote locations, they are often expensive to repair as well. Aesthetically, they are considered a nuisance in most neighborhoods, and this constrains their height and location and adds considerably to their cost.

In this chapter we will look at the internal components of the base station in detail. Those components are listed in Fig. 4.1.[1]

4.1 The Antennas

Starting at the top, a wireless base station has a set of radio antennas. Typically, there are three antennas per sector, one to transmit the forward signal and two to receive the reverse signal. The two receiving antennas provide microscopic spatial diversity to

---

[1]Where there are differences among AMPS, GSM, and CDMA, the CDMA components are illustrated. For example, some AMPS systems have multiple amplifiers sharing a combiner rather than one amplifier. These distinctions are discussed in the text.

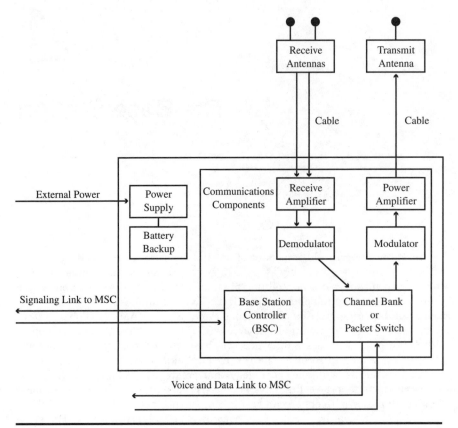

Figure 4.1 Base-station components.

combat the Rayleigh fading described in Sec. 2.1.4. At wireless frequencies, half a meter of spacing is enough.

In the case of the user terminal, both cost and size are major considerations. The
base-station antenna is high atop a tower, and its cost is spread out over hundreds of
users so that we can afford the luxury of putting the right piece of hardware up there.

In Sec. 1.8 we discussed how an antenna could have gain by concentrating its focus
over a small angle, the *radiation pattern* of the antenna.[2]

The horizontal radiation patterns of the base-station antennas match the sector plan
of the cell. We can have an omnidirectional 360-degree pattern, a three-sector 120-
degree pattern, or a six-sector 60-degree pattern. For conventional reuse [frequency division multiple access (FDMA) and time division multiple access (TDMA)] the three-
and six-sector plans make sense because the six important interferers are evenly
spaced around the serving cell. In CDMA, however, the important interferers are in the

---

[2]If the antenna is used for receiving rather than transmitting, then the term *reception pattern*
might be more appropriate, but we still call it the radiation pattern of the antenna. See Sec. 3.1.2
for a detailed explanation.

same cell, so it is possible to design other sector plans to optimize capacity, as discussed in Sec. 31.1.[3]

Due to the reciprocity of antenna transmission and reception patterns, the entire set of transmitting and receiving antennas is generally tweaked as a unit. The vertical radiation pattern of the base-station antennas determines the coverage range of the base station. We can extend the cell's radius by using high-gain antennas, which have narrow vertical range. The user terminals near and under the base-station antenna are out of the high-gain zone, but they are very close to the base station, so they still get more than adequate radio coverage.

The vertical radiation pattern can be used to shrink the coverage range as well. If we tilt a sector's antennas downward, then their maximum gain is closer to the base station. As a user terminal moves past the antenna gain peak, its radio path gets weaker both from increased distance and from reduced antenna gain. In this way we can use a high-gain, vertically thin-beam antenna to attenuate the signal path beyond a certain distance. It is not terribly hard to tilt antennas downward. This means that a large-radius coverage cell can be converted into a smaller-radius cell in a growth environment by reaiming the antennas already there, which is a lot easier and cheaper than replacing them.

The wireless telephony engineer can use the base-station antennas to do spot engineering. If a particular tunnel needs coverage, then the appropriate antenna can be selected and installed to reach subscribers there. Or service can be improved by aiming a base-station antenna down into a valley. If the spot engineering problem is too much for an antenna to fix, then we can employ microcells and repeaters (discussed in Sec. 4.6).

4.1.1 Smart antennas and phased arrays

Often radiation patterns are created by using multiple antennas in a *phased array*. By broadcasting the same signal with the appropriate radio wave phase angle delays from several closely spaced antennas, a signal beam can be created with a very specific radiation pattern. The same can be done by using time-delay filters on the receiver. Smart antenna technology electronically manipulates the phased array so that it can even be used on a call-by-call basis to focus each radio beam very narrowly.

Consider the array of six antennas shown in Fig. 4.2. If we want to send a radio wave toward the right side of the page, as is shown in the figure, then we can transmit the radio wave with a 0-degree phase shift at the left two antennas, a 45-degree phase shift at the middle two antennas, and a 90-degree phase shift at the right two antennas. The combination of these six radio sources coherently adds all six antenna outputs in a narrowly focused radio beam to the right.

If we want to send a radio wave up and to the right, as is also shown in the figure, then we can transmit the radio wave with a 0-degree phase shift at the lower left two antennas, a 45-degree phase shift at the middle two antennas, and a 90-degree phase shift at the two upper right antennas. These six sources coherently add up to a narrowly focused radio beam to the upper right.

---

[3]The primary source of interference in CDMA is traffic in the same *sector*, but after that, traffic in the other sectors in the same cell tend to contribute more interference than traffic in other cells.

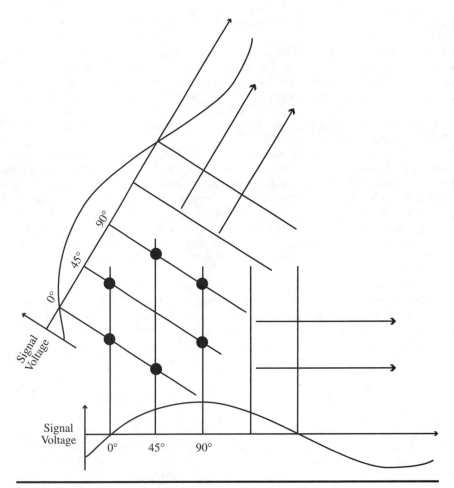

Figure 4.2 Phased array or smart antenna.

A phased array can transmit narrow beams in any direction with the correct choice of phase shifting. From the reciprocity of transmitting and receiving, we can deduce that the same phase shifts can be used in the receiver to select a narrow angle. When that narrow angle is dynamically adjusted by continuously optimizing the received signal, we call the configuration an *adaptive phased array*. Adaptive phased arrays are particularly useful for receiving a signal from a source that is moving.

4.2 The Tower and Cable

The big question for any base-station deployment is how high to build the tower. High towers have wider coverage, but they are more expensive, and they receive the most resistance from neighbors. It is far easier to get zoning approval to put up low towers or to *colocate* antennas on existing towers. Any kind of radio tower can be used for colocation. There are also a variety of techniques for making towers less obtrusive, including locating them on the tops of buildings and disguising them as trees or parking lot lighting poles. Antennas are placed on platforms at any height from the base to the top of the tower so that if an antenna needs to be lowered later, the tower is not cut shorter.

The initial cell layout of a new system is usually a grid of large coverage cells. We call them *coverage cells* because their mission is to create a seamless region of wireless telephone service. These towers need to be high to reach as far as possible. Coverage cells both meet initial customer needs and also provide a relatively low-cost method of meeting the Federal Communication Commission (FCC) regulatory requirements that a cellular provider must provide coverage to a certain percentage of the area and a certain percentage of the population covered by the provider's license.

Later on, when we have more customers, we add smaller *growth cells* and convert the original cells into smaller growth cells as well. Having the antennas lower helps the base station serve nearby subscribers in its own cell and not interfere with more distant subscribers in other cells. This means that the platforms on the original high towers may need to be lowered.

Each antenna has its own cable connecting the equipment on the ground to the antenna high up on the tower. Each cable has to have low enough electrical impedance not to have too much signal loss and has to be well enough shielded that it does not become an antenna itself and defeat the carefully engineered radiation pattern of a sectorized cell. If the tower is 50 m high (165 ft), then this means 50 m of radio frequency (RF) cable for each antenna. The transmit antenna cables have to be able to handle the voltages and currents of the transmitter.

Any receiving cable tens of meters long has a few decibels of loss. When a CDMA base station is serving a large-radius coverage cell, the tower is high, so the cable is long, and the noise level of the receiver is a significant factor when compared with the weak signals coming from distant user terminals. When this happens, we can put a *low noise amplifier* (LNA) at the top of the tower and connect it directly to the antenna. This boosts the received signal, creating a higher signal-to-noise (S/N) ratio at the receiver at the bottom of the base-station tower.

4.3   The Power Amplifier

Power amplifiers take low-level modulated radio signal and amplify it to sufficient power for an antenna to transmit it. There are many books on power amplifier design, both for audio frequencies and for radio frequencies, and we are only going to touch on the highlights that affect wireless telephony.

An amplifier has to deliver enough power over time. Consider the task of transmitting eight continuous radio signals of 6 W each. Normal intuition tells us that the total power output would be 48 W, eight times 6 W, and this is usually the case. A radio power amplifier called on to transmit these eight signals has to be able to deliver a continuous average power of 48 W without overheating. Sustained average power is a basic and important component of amplifier specification.

The amplifier also has to deliver the required amplitude, not just the average voltage, but also the peak voltage. Let's consider a simple case of two uncorrelated 1-W signals transmitted from the one antenna and powered by one amplifier. The total average power is 2 W. Into a typical antenna load, a 1-W signal might have a peak of 10 V. Once in a while, the two 10-V peaks of the two signals line up so that the amplifier is called upon to deliver 20 V. This is shown in Fig. 4.3. While a 20-V amplifier could handle a 4-W signal into the same antenna load, it takes a 20-V amplifier to send two 1-W signals. In this case we say that the amplifier requires a *peak* power

v (t)

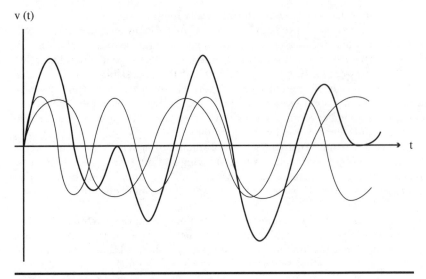

Figure 4.3   Peak versus average power.

of 4 W and an average power of 2 W. We also say that we have a *peak-to-average ratio* of two.[4]

As the number of signals k increases, the average power requirement increases in proportion to k, but the peak power requirement increases in proportion to k^2. Ten 1-W signals require an average power of 10 W and a peak power of 100 W.

Building a 100-W amplifier just to send 10 W seems like overkill. Another solution is to build 10 smaller amplifiers and use a radio *combiner* to add the signals together. There is significant loss in a combiner, about 6 dB, so three-quarters of the amplifier's output goes to heat the combiner. Still, ten 4-W amplifiers may be easier to come by than one 100-W monster.

There is another argument in favor of the combiner solution. FM, BPSK, QPSK, and m-ary modulation schemes maintain the same wave amplitude over time; we call these *constant-envelope modulations*. Building a constant-envelope amplifier is far simpler than building an amplifier that follows varying amplitude. To vary amplitude over time, the amplifier has to provide a proportional or *linear* amplification of its input signal, whereas a constant-envelope amplifier simply can swing from its minimum to its maximum voltage following the input phase instant by instant. Constant-envelope amplifiers are not only simpler; they also are more efficient electrically. Losing 6 dB in the combiner mitigates some of this efficiency, but the design efficiency is still a compelling argument in favor of using constant-envelope amplifiers with combiners instead of using linear amplifiers.

On the other hand, as linear amplifier technology has improved, we have moved away from the combiner solution. The combiner works best when the signals are narrow bands

---

[4]Since these radio waves typically look like sine waves, there is an extra factor of 2 in the peak-to-average ratio calculation. We are deliberately ignoring the sine wave factor of 2 to make the point that *combining* signals creates its own peak-to-average ratio problems in power amplifier design.

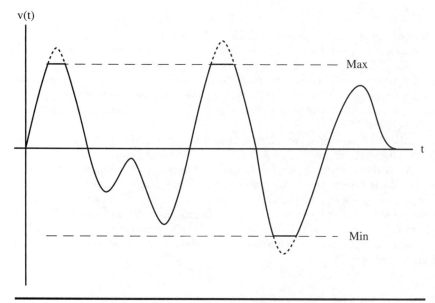

Figure 4.4 Minimal clipping causes minimal distortion.

of specified frequencies and does not work so well for broadband solutions such as CDMA or even TDMA.

There is another way to deal with the peak-to-average power requirement as the number of channels gets larger. The fraction of the time the instantaneous power is much larger than the average power is very small. No matter how many simultaneous channels we are trying to send, a peak-to-average value of 10:1 delivers a clean waveform 99.85 percent of the time.[5] Being willing to live with 0.15 percent clipping makes the design and cost of linear amplifiers more reasonable. Figure 4.4 illustrates the distortion caused by clipping.

We talk of the average power but not the average voltage. In almost any radio or audio system, the voltage swings high and low about equally, and the average voltage is zero. However, even the average absolute value of the voltage is not a meaningful measure of anything useful. Because voltage changes as the square root of power, we want the voltage that reflects the average power. This measure is the square root of its average squared value, which we shorten to *root mean square* (RMS). We speak of peak and average power, and we speak of peak and RMS voltage.[6]

---

[5]For the mathematically curious, we got this value by assuming that the sum of a large number of radio channels is a normal distribution in voltage and calculated that $\sqrt{10}$ standard deviations are exceeded 0.078 percent of the time on the high side and undercut 0.078 percent of the time on the low side.

[6]In the world of audio equipment, it is common to see a specification for amplifier power in RMS watts, which is technical sounding but wrong. What they are measuring is the *average* power in watts that could be derived from RMS *voltage*.

4.4 The Radio Receiver

The radio receiver is typically a separate piece of equipment for each voice channel in the cellular system. These are called *voice radios* or *modems* in digital radio. Often the term *modem* includes the individual voice channel equipment that supports the radio transmit function as well. These radio modems that modulate and demodulate the signal from the telephone link to the air interface should not be confused with the computer modems we use at home to modulate digital signals onto the analog landlines of the local loop of the PSTN.

FDMA and TDMA receivers have filters to find the specific frequency of their channels, and TDMA systems have demodulators and processing equipment to find their time slots. As we will discuss in Sec. 8.2, CDMA receivers are more complicated because each receiver demodulates the entire signal to find its own specific code.

In most cellular systems, the receiver equipment is connected to two antennas, and switched diversity is used to combat Rayleigh fading. Even in wideband CDMA systems, where fading is a minor problem, having two receivers combats multipath delay spread by using a rake filter to combine the two signals. We have more to say about rake filters in Sec. 8.4.

4.5 The Power Supply and Environmental Controls

Each piece of equipment in a base station uses electrical power. It is important to add up all the power consumption figures and build an ample power supply. If the current demand of the base-station equipment exceeds the power-supply capability, the voltage can drop, and the radio equipment can perform poorly or can fail altogether.

The base station must be protected from the weather by a small building or large cabinet, and it also may need climate control to operate in cold or warm conditions. Air conditioning in a hot climate is important because the base station will cease to provide service if it overheats.

The wireless subscribers expect their telephones to work, even when the power mains do not work. Even in the United States there are areas with frequent power outages. The wireless service provider has to decide how much backup power to build into a base station. In the early AMPS days, we had racks of huge batteries that would run a base station for many hours in the event of a power failure.

4.6 Microcells, Picocells, and Repeaters

We often expand coverage of a cell by using a higher base station tower, but sometimes there is a particular region that is hard to reach. Other times there is a particular community of interest, a lot of wireless telephone demand in a small space. We can deal with these issues locally using microcells, picocells, and repeaters.

A *microcell* is a cell smaller than 500 m in radius. This is a full base station connected to the wireless telephone system just as any other cell. Because of its small coverage area, a microcell can use smaller and simpler radio equipment. Microcells are sold to wireless service providers as a package, a single piece of equipment, sometimes with a built-in microwave link for its connection to the MSC.

When the service area is very small, perhaps a single building, then we use even smaller units called *picocells*.[7]

When the community of wireless subscribers is not large enough to justify a microcell but hard to reach with the usual base-station towers, we use repeater technology to extend coverage. The obvious examples are buildings, subways, and tunnels, but even a deep valley can be a coverage problem.[8]

As the name suggests, a repeater receives the signal from a sector of a base station and retransmits the signal into a hard-to-reach area such as a tunnel. Repeaters do require an external power source, and they have two sets of receiving and transmitting antennas. However, they do not have any direct link to the MSC or PSTN, the way base stations, microcells, and picocells do. They rely on their air interface to a base station to carry the signal back to the PSTN. The repeater can be a major source of multipath interference because it is another broadcast copy of the serving signal. In conventional reuse systems (but not in CDMA), the repeater can shift the frequency so that it communicates with the base station on one set of frequencies but communicates with user terminals on a different set. In CDMA systems, where the repeater is communicating with the base station on the same frequency used for communication with user terminals, the connection to the serving base station must be well isolated from the repeater serving antenna.

4.7   The Base-Station Controller (BSC)

The brain of the base station is the base-station controller (BSC) that directs the operation of the entire base station. Each sector of the cell has its own antenna faces and its own radio equipment, and they all come together at the BSC. As we will discuss in Chap. 7 GSM systems can have a single BSC control several cells.

The BSC coordinates pages and mobile originations, sets up and tears down wireless telephone calls, selects radio resources for each call, and manages the handoff process. When a handoff stays within the domain of one BSC, the BSC can perform the switching for the handoff.

The BSC also connects the base stations to the MSC through pipes that carry the digitally encoded voice and data from the subscriber calls.[9] The connection from BSC to MSC is called *backhaul,* and the network of base stations served by a single MSC has a backhaul network connecting the base stations to their MSC. We describe the backhaul network design in Chap. 35.

---

[7]Somewhere along the way they skipped nanocells.

[8]Before repeater technology was in use, the original AMPS installation in Pittsburgh, Pennsylvania, seamlessly covered a very hilly area and two difficult tunnels, Fort Pitt and Squirrel Hill. This was a major feat of engineering that took a lot of work to accomplish. Repeaters could have made this job a lot easier.

[9]In the old days, we had a set of dedicated trunks, described in Chap. 13. Today, the interfaces aggregate calls and packetize data so that the term *trunk,* and its association with a single line for a call, is no longer the right concept.

4.8   Component Reliability

Reliable base stations are key to service availability and quality. In this section we will look at both the mathematical modeling and the practical engineering and management issues related to designing, installing, and maintaining reliable base stations.

4.8.1   A simple reliability model

Base-station equipment can fail. Power supplies can go out, controllers can stop working, and amplifiers can burn out. Any one of these three failures will take a significant area out of service. Therefore, let us take a look at reliability as an issue of base-station design.

Let us use a simple model of a base station as three separate components, power supply, controller, and amplifier, components A, B, and C. The base station is working only when components A, B, and C are all working. Like an electric circuit, we call this a *series* system in reliability theory, as shown in Fig. 4.5. When the components themselves are reliable, we can simplify the mathematics and approximate the outage time for the series system as the sum of the outage times for the components. If components A, B, and C are each working 99 percent of the time, then we might say that components A, B, and C are 99 percent reliable and that the entire system is 97 percent reliable.

We can evaluate the reliability of each component by considering it to be in one of two states, WORKING or FAILED, as shown in Fig. 4.6. (These states are similar to the call states in Chap. 16.) We presume that the component has a failure rate λ and, once broken, some rate of repair μ. These are average rates rather than a specified schedule.

Consider a component that breaks down an average of once per month. Some months might have no failures, whereas other months might have several failures, but the failures over time average one per month. We would say that λ is one failure per month or about 0.032 failures per day. If the average time for repair is 1 day, then we would say μ is 1.0 repairs per day. Any given repair might take an hour or a week, but the *average* is 1 day.

In an equilibrium condition (over a long period of time), the probability flow from WORKING to FAILED will equal the probability flow from FAILED to WORKING, and the probabilities add up to 1. This gives us Eq. (4.1):

$$\mu p_0 = \lambda p_1$$

$$p_0 + p_1 = 1$$

(4.1)

where p_0 = probability of being in FAILED state
      p_1 = probability of being in WORKING state

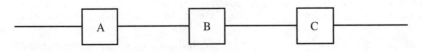

Figure 4.5   Three components in a series system.

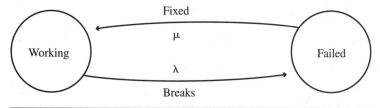

Figure 4.6 Two-state component reliability diagram.

We can solve these equations and see that this component will spend $\mu/(\lambda + \mu)$ of its time in the WORKING state and the other $\lambda/(\lambda + \mu)$ of its time in the FAILED state. Our example component will average 96.8 percent of its time WORKING and 3.2 percent of its time FAILED. If this is component A, then adding in the failure times for components B and C can only make it worse.

If this were our equipment, then we probably would decide that 3.2 percent is an unacceptable out-of-service time for a base station. There are two direct ways to improve the service level: Decrease λ by making the equipment more reliable or increase μ by having faster repair service. This may be difficult because more reliable equipment may be several times more expensive, and base stations are generally in remote locations that may be hard for a repair truck to reach quickly.

There is another approach. We can improve the reliability of a system by adding spare equipment. Now we have two each of components A, B, and C. So long as at least one component of each pair is working, the system is working. We call this a *parallel system*. The use of parallel subsystems for improved base-station reliability, is shown in Fig. 4.7.

We can model a two-element parallel system with three states, BOTH-WORKING, ONE-WORKING, and BOTH-FAILED, as shown in Fig. 4.8. We assume that each of the two components has the same λ failure rate as before and the same μ repair rate as before. The reason we have 2λ instead of λ as the rate from BOTH-WORKING to ONE-WORKING is that there are two components, either one of which could fail. We also assume that the repair in the FAILED state will fix both components. We can set up three equilibrium equations by insisting that the probability flows are equal in and out of the FAILED state and

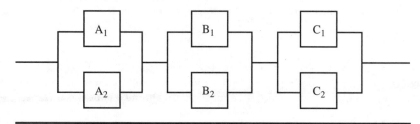

Figure 4.7 Six components in a series and parallel system.

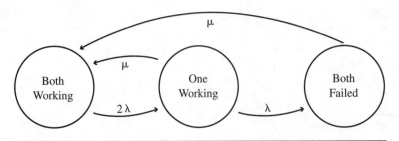

Figure 4.8 Three-state component reliability diagram.

the ONE-WORKING state and that the three probabilities add up to 1, as shown in Eq. (4.2)[10]:

$$\mu p_0 = \lambda p_1$$

$$(\lambda + \mu)\, p_1 = (2\lambda) p_2 \qquad\qquad (4.2)$$

$$p_0 + p_1 + p_2 = 1$$

where p_0 = probability of being in FAILED state
$\quad\quad p_1$ = probability of being in ONE-WORKING state
$\quad\quad p_2$ = probability of being in BOTH-WORKING state

The solution for the probability of being in the FAILED state is

$$p_0 = \frac{\lambda}{\lambda + \mu}\ \frac{2\lambda}{2\lambda + \mu}$$

This redundancy drops the outage time for our example component from 3.2 to 0.2 percent, a major improvement in reliability for a factor of 2 in equipment cost.

This example has both components of the parallel system able to fail in the BOTH-WORKING state. We call this a *hot spare* because both components are actually running. An alternative design has a primary unit running and a secondary unit that stays off until it is needed. When the primary unit fails, the secondary unit is powered up and goes into service until the primary unit is repaired. We call this a *cold spare* because the secondary unit is not running. In the cold-spare scenario, we presume that the secondary unit does not fail while turned off, the 2λ term in the reliability equation becomes just λ, and the outage time is about half that in the hot-spare case. This outage-time calculation does not count the time it takes to turn on and to warm up the secondary unit.

4.8.2 Real-world reliability issues

The analysis in the preceding subsection is the tip of the reliability science iceberg. The model makes a number of assumptions, some of which we will list here. In this key concepts section we will not go into a detailed mathematical analysis of these issues.

---

[10]The three equations for the three states are *redundant;* that is, any two are sufficient to define the problem. Thus we picked the easiest two equations to use in solving the system.

Rather, we will identify the most important real-world issues that the model did not address.

The model assumes that we can detect the redundant component failure even when there is no outage and that the repair in the ONE-WORKING case gets the same speed and enthusiasm as the repair of both components in the FAILED case. This requires monitoring systems and standard operating procedures for repair.

The mathematical model we used assumed that failures were independent of one another. However, when a single cause, such as a power surge or a flood, may disable multiple components, a different model is required.

Our models did not take all the components into account; for example, we did not address antenna failure or tower collapse. For relatively stable components, it is often best simply to engineer them well and to consider their failures as emergencies outside routine reliability studies. However, in supporting a large network, it is important both to reduce the likelihood of and to be ready for rare crises. We can reduce the likelihood of rare failures by adhering to a well-designed maintenance schedule. In fact, proper management and maintenance are crucial factors that generally do not fit into our neat equations.[11]

4.9 Conclusion

In the last two chapters we have discussed the systems on the two sides of the air interface: the user terminal and the base station that connects the cellular link to the PSTN. In the next chapter we will discuss the basics of wireless telephony. That discussion will lay the groundwork for descriptions of AMPS, GSM, and CDMA.

---

[11]Maintenance and management may not sound like capacity issues until somebody sends the bill to the wrong place, the electric company cuts off the service, and the batteries run out several hours later.

5

Basic Wireless Telephony

Interference used to be a bad thing in radio, a disease to be minimized rather than a phenomenon to be managed. The fundamental idea of managing, rather than avoiding, interference gave birth to the cellular phone networks that have sprouted up in the last two decades, growing to about 130,000,000 cell phones in the United States and 1,000,000,000 worldwide in the year 2002.

A key component in managing radio interference is the *cell,* a small transmission area where all mobile phones are served from one small radio tower. Having many small cells allows for reuse of frequencies and results in greater call capacity than a single radio service area with the same number of channels.

The model of a cellular transmission structure with managed interference is the basis of all cellular systems, including the original analog systems, Europe's time division multiple access (TDMA)–based Global System for Mobility (GSM) and the latest code division multiple access (CDMA) technologies used for Personal Communication Services (PCS) and universal mobile telephony system (UMTS).

5.1 The Wireless Signal Path

We follow the links in the chain of a wireless telephone call from the portable (or vehicular) telephone through the radio path and the wireless telephone system to wherever the call is connected. We will start with the full story from a user terminal to a regular landline telephone.

Figure 5.1 shows us the chain from wireless to landline telephone through the radio link, the base station, the base station to mobile switching center (MSC) pipe, the MSC, another pipe, the public switched telephone network (PSTN), and the local loop. After describing this chain in detail, we will consider the differences when the call is from one wireless telephone to another.

5.1.1 Wireless to landline

The user terminal is a complete two-way radio, as described back in Chap. 3. The user terminal can detect the presence of a wireless telephone environment through its paging and access channels. It synchronizes itself to the system and makes itself known to the system by registering itself. This tells the system that there is a user terminal in a

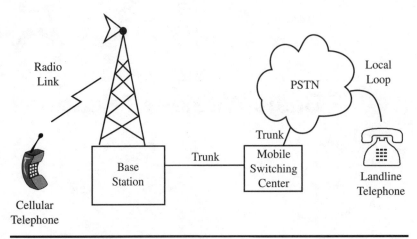

Figure 5.1 The wireless-to-landline telephone path.

specific area and provides its telephone number in case anybody is calling it. We will have a lot more to say about the wireless telephone number and incoming calls in Sec. 12.5.5.

Call initiation in wireless telephone systems is done through *preorigination dialing*. A subscriber enters the entire telephone number to be called and presses a special key usually marked SEND.[1] Preorigination dialing was conceived when the market for wireless was vehicular. The driver could enter two or three digits at a time with breaks to look up and check traffic. When the entire telephone number had been entered, the driver could press SEND to start the radio signaling that initiates a telephone call.

Once the telephone initiates or receives a call, it establishes a duplex radio channel supporting two-way voice communications with signaling. Whether this uses a pair of radio frequencies [frame division multiple access (FDMA), Advanced Mobile Phone Service (AMPS)], time slots in a larger channel (TDMA, GSM), or spread-spectrum components (CDMA) is not important here. The important thing is that the user terminal and the system are in continuous two-way, full-duplex voice-link communication. As each party talks, his or her voice is modulated, transmitted, received, demodulated, and reproduced in sound at the other end. Both speakers can talk at the same time, with both voices being reproduced at each other's telephones just like a regular landline-to-landline call, and there is some mechanism built in to carry signaling information. We call the duplex channel for voice plus the signaling for one call a *connection*.

The other end of the air interface from the user terminal is an antenna configuration atop a base station. In the AMPS trial system in Chicago in 1982, even the mobile telephones had two antennas for diversity. This was dropped in the commercial system for three reasons: First, the mobile telephone was already expensive enough without having an extra piece of wire. Second, most mobile customers were concerned with having their landline contacts have good sound even if their own sound had static from time to time. And third, a customer bringing an expensive car into a shop for cellular

---

[1]We have seen the SEND and END keys marked YES and NO on some portable user terminals.

telephone installation is annoyed enough at having *one* hole drilled in the roof without the extra indignity of having a *second* hole.[2]

The base station receives and demodulates the call as described in Chap. 4. The call is then routed onto a pipe connecting the base-station controller (BSC) to an MSC.[3] These pipes are almost always digital links carried over wires, fiber optic cable, microwave link, or whatever medium the wireless telephone carrier wants to use.

The MSC routes calls and reroutes them during handoffs. The MSC is a telephone switch that can direct calls instant by instant, although some BSCs can switch handoffs from sector to sector on the same cell.[4]

The MSC has yet more trunks going to the PSTN, where the call is routed through a series of switches and transport to a *local exchange office*. The local exchange office routes the calls onto the local loop to the called party's telephone.[5] Depending on the wireless switch architecture, there also may be trunks from the MSC directly to long-distance carriers. If load requires, it is also possible to add trunks to other specific destinations, including nearby MSCs, 911 emergency call centers, and possibly Internet service providers.

5.1.2 Landline to wireless

A call to a user terminal follows the same chain as a call from user terminal to landline in reverse, but it has the extra burden of *finding* the user terminal. The user terminal can be found because, whenever it is turned on, it registers with the local base station. If the user terminal is somewhere other than its home system, then this information is sent to the home system, a process called *roaming* discussed in Sec. 12.5.5. If the user terminal is turned off or out of communication range, the system detects this and, on PCS networks, routes the call to the customer's system-based voice mail.

Once the system knows where the user terminal is located, at least to the nearest cell, it sends a paging message to tell the user terminal that it has a call. The user terminal then initiates communication with the system the same way it does when the mobile subscriber is making the call.

There is another reason for the user terminal to detect and recognize the system. Forming a CDMA radio link takes time for the receiver to become synchronized with the transmitter. From a cold start with no prior timing information, the synchronization process can take half a minute, but from a warm start where the receiver already has some information about the CDMA timing sequence, the same synchronization process only takes about 1 second. A user pressing the SEND key expects to wait a second or two, but a 30-second wait to connect a phone call is out of the question. Instead,

---

[2]The early cellular telephone service was still a high-priced luxury item. Our AMPS trial customers in Chicago drove mostly Cadillacs, Lincolns, BMWs, and Porches.

[3]The MSC was called a mobile telephone switching office (MTSO) in the early cellular days.

[4]We are sure that somewhere out there are base stations that also do telephone switching. For our purposes, we would consider those to be both base station and MSC *colocated*. Also, in GSM the BSC administers several base stations, and it can execute a handoff (called a *handover* in Europe) between two of its own cells.

[5]In the Bell System days, this was often referred to as *plain old telephone service* (POTS), and the local switch was called a *class 5 switch*.

we do the cold-start synchronization when the user terminal is first turned on so that the individual call can do a warm start quickly.

5.1.3   Wireless to wireless

When a wireless customer calls a landline telephone, the call chain is from user terminal to base station to MSC to the PSTN to the landline telephone. If one wireless customer calls another wireless customer in another wireless network, then the call is routed through the PSTN from one wireless system to the other, two MSCs, two base stations, and two radio links. The chain is from user terminal to base station to MSC to PSTN to another MSC to another base station to another user terminal.

When the call is within one wireless network, on the other hand, the system designer may elect not to send it to the PSTN at all. Rather, it may be simpler to complete the call within the MSC or within the network of MSCs to send it directly to the other base station. The call then goes from user terminal to base station to MSCs to another base station to another user terminal. The two base stations are usually different, but they may be the same. Since BSCs have no call switching capability, the call is going to an MSC even if both user terminals are in the same cell coverage area. In the original AMPS systems, there were loopback trunks specifically designed to look like PSTN trunks while not going anywhere, and these loopback trunks were used for mobile-to-mobile calls.

Telephone network design is often about *signaling,* the communication among network components about calls. Any system that bypasses the PSTN has to make sure that these signaling issues are taken care of, both making the connection and billing the users appropriately for the call.

There is a specific reason for avoiding routing through the PSTN on mobile-to-mobile calls whenever possible. Wireless providers pay a lease fee to PSTN service providers for use of the PSTN. Mobile-to-mobile calls that do not go over the PSTN cost less for the wireless provider. Wireless providers can use this either to increase their net revenue or to offer special rate plans to attract customers.

5.2   Radio Architecture

We have looked at the basic concepts of the design of the cellular system. Now we will examine a cellular system from the vantage point of radio transmission and reception.

5.2.1   Cells and base stations

Cells are regions of radio coverage defined by having the same base station; base stations and cells go together in a one-to-one correspondence. Whatever antennas are on top and radio resources are under the roof and however many sectors the cell may be divided into, we think of the cell as a basic entity. There is a solid financial reason for this: Base stations are also the economic units of wireless telephone systems. With the added costs of building, tower, antennas, electrical power, telephone transport, and zoning issues for a radio tower, adding a new base station is a very expensive way to increase cellular capacity. If reengineering base stations can increase capacity without new construction, the service provider saves a lot of money. In addition, many customers want cellular service but do not want radio towers in their neighborhoods. A zoning permit increases the cost of a base station and also delays its construction. During the delay, customers in that area are doubly irritated: As cellular customers, they are experiencing poor service, and as residents, they are experiencing a vendor trying to reduce the quality of their neighborhood. The same people who clamor for cellular telephone service are not at all anxious for their neighborhoods to have a skyward spire prickly with radio antennas at the top.

Having more capacity in one base station is more economical than having multiple base stations because it aggregates transmission and reception for more users into one building. In addition, a base station also aggregates telephone trunking from the base station to the rest of the wireless system. This is a serious economy because getting any kind of digital link from one place to another is usually a much bigger problem than adding *more* capacity once a link is already in place. The paging and access channels are also aggregated at the cell level. While radio frames, antennas, cables, and trunks cost money, their costs are usually relatively small compared with the costs of the building, tower, power supply, telephone access, and zoning issues required for each base station.

5.2.2 Sectors and antenna faces

Cells are usually divided into sectors. As discussed in Sec. 4.1, omnidirectional, three-sector, and six-sector cells are the sector plans that make sense for conventional reuse systems (FDMA and TDMA). Other sector plans can make sense in CDMA systems, where the primary sources of interference are other calls in the same cell.

The cell sector is formed by having an *antenna face* at the base stations. Three antenna faces give some cellular towers their distinctive triangular platforms. The face contains directional antennas, both transmitting and receiving, to define the exclusive sector coverage. It is easy to use the terms *sector* and *antenna face* interchangeably, but the sector is the geographic region of coverage and the face is the antenna configuration defining the sector.

Each antenna face is connected to a collection of radio equipment. The configuration and cost of this equipment can influence base station design. If the traffic engineering algorithms described in Chap. 23 tell us that 17 radios serve the demand and if 16 radios fit on a radio frame, for example, then we might want to think about a small quality-of-service decrease to save the cost of an entire radio frame.

5.2.3 Server groups and extra cell radii

When we discuss cellular growth principles in Sec. 27.7, we deal with varying reuse distances in one channel set, as shown in Fig. 27.18. This is done by defining more than one cell radius, which we call a *server group.* When we draw the extra hexagonal lines in Fig. 27.18, the cellular engineer defines large-cell and small-cell server groups and then calculates the radio power level thresholds for handoffs between inner and outer cells.

The extra, smaller cell radius is called an *overlaid cell,* a smaller cell colocated with the larger cell. Each cell radius is a server group. When this is done to a sectorized cell, the intersection of a face and a server group is called a *logical face.* The cell shown in Fig. 5.2 has three faces, two server groups, and six logical faces.

By using server groups and logical faces, the cellular engineer can choose some radio channels for small-cell reuse and leave others for large-cell reuse. Also, since each

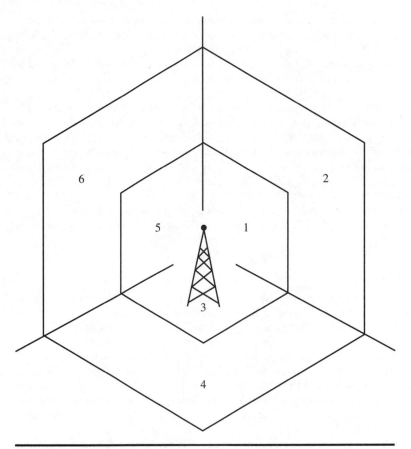

Figure 5.2 Overlaid cell with six logical faces.

radio can have its own independent radio threshold values, one radio frame can have radio channels from multiple server groups.

Multiple server groups create multiple headaches for neighbor list generation. Suppose that we have a cellular system laid out like Fig. 27.17. For each of the 33 cells in this picture, imagine calculating all the reasonable cells for a handoff. Now imagine taking the 57 logical faces in Fig. 27.18 and doing the same job. Remember that other logical faces on the same cell are handoff candidates as well.

Now take these same two pictures and imagine all the cells are three-sector cells so that most of these cells have 5 or 6 logical faces. A large cell may have two neighbors completely covering one of its large-cell sector areas so that the total would be 5 instead of 6 logical faces. Six-sector cells with two server groups can have 12 logical faces.

As you can see, the combination of faces, server groups, and logical faces can make life very difficult for a cellular engineer. This is only a 33-cell system, whereas big systems have hundreds of cells, and we have not considered radio propagation irregularities. Sometimes a cell's actual coverage has a pseudopod extending far enough so that the handoff neighbor list has some odd-looking entries.

The good news for the CDMA cellular planner is that CDMA has no need of multiple cell radii, server groups, or logical faces.

5.2.4 Individual radio channels

Each sector is served by a collection of radios for individual subscriber voice and data links. These radios may be separate units, individual transceivers in a row on a radio frame, or separate logical entities in one physical box. (The distinction between physical and logical models is described in Sec. 14.4.) These radios are the system's termination point for the air interface. In addition to the individual voice and data links for subscriber service, there are radio links for signaling, including paging and access channels.

5.3 Conclusion

Now that we have introduced the basics of the wireless signal path and the radio architecture of base stations, cells, and sectors, we can look at AMPS (FDMA), GSM (TDMA), and CDMA systems in greater depth in the following chapters.

6

Analog Wireless Telephony (AMPS)

Although many people think that Advanced Mobile Phone Service (AMPS) was the first mobile telephone system, it has a predecessor. Before AMPS, Improved Mobile Telephone Service (IMTS) was available, but it wasn't cellular. Each city was served by just one radio tower. AMPS was the first cellular service in the United States and a revolutionary technology in its day.

No matter how big and clumsy the old AMPS car phones look now, and no matter how primitive the analog FM technology appears to today's engineers, the fundamental principles of AMPS are the same as those of the latest Global System for Mobility (GSM) and code division multiple access (CDMA) technologies.

6.1 The Early Days of Cellular

In 1966, two members of the technical staff at Bell Telephone Laboratories, Richard Frenkiel and Philip Porter, realized that service capacity could grow enormously by reuse of voice radio channels.[1] A service area could be divided into *cells,* and interference could be managed by making sure that same-channel users were sufficiently far apart. The more aggressively we reuse channels, the more call capacity we get with the same cells within available bandwidth. On the other hand, the more aggressively we reuse channels, the more interference calls will experience. We must manage that interference to optimize the number of calls while ensuring satisfactory call quality.

The potential of this new technology got the Federal Communications Commission (FCC) of the U.S. government to authorize 60 MHz in the 900-MHz (upper UHF) band for cellular telephones, with 20 MHz given to wireline companies (the Bell System telephone companies in those days), 20 MHz assigned to nonwireline companies (other than telephone companies), and 20 MHz held in reserve for future use.

---

[1]Let us add a bit of history from the *Bell System Technical Journal* (January 1979): "The *cellular* concept and the realization that small cells with spectrum re-use could increase traffic capacity substantially seem to have materialized from nowhere, although both were verbalized in 1947 by D. H. Ring of Bell Laboratories in unpublished work." At Bell Telephone Laboratories, the spark that lit the cellular fire was the 1966 memorandum by Frenkiel and Porter.

In 1982, Bell Telephone Laboratories prepared a cellular business case for AT&T and Western Electric. Cost and revenue projections were based on a forecast of a million cellular telephones by the end of the century, the biggest believable number we could think of at the time.

The cellular principle is capacity growth in fixed radio spectrum through frequency reuse and managed interference. Regions of radio coverage, cells, can be made smaller and smaller as needed to accommodate higher density of mobile telephone demand.

By the end of 1983, there was commercial cellular service in Japan, Scandinavia, and the United States. As the capacity to serve mobile telephone customers increased, so the market grew as well. The new cellular telephone technology grew quickly from a novelty to a way of life; there were hundreds of millions of cellular telephones worldwide by the late 1990s. This growth beyond all early forecasts has made capacity management a key issue in the growth and financial success of cellular service providers.

6.2   The Cellular Principle

Prior to the deployment of cellular technology, mobile telephones for an entire city were served by a single tower with a single set of radio channels. The entire New York City area was served by just 14 radio channels and could only support 14 simultaneous mobile telephone calls. Adding one channel to the spectrum allocation added just one more telephone call for the entire city. Cellular offered the opportunity for unlimited service, as we saw it then, to as many as a million mobile telephones by the end of the twentieth century.

Advanced Mobile Phone Service (AMPS) was prepared for this rate of growth. With a hundred towers of 45 channels each, a large cellular system could handle thousands of calls. Because we could split cells repeatedly, there was no obvious limit to growth. We could double the number of base stations to double the system capacity and, if demand warranted, double it all again. Using four-to-one cell splitting, 8-mile cells could become 4-mile cells, 2-mile cells, and even 1-mile cells to build a network that would satisfy the market demand for cellular telephones. A library of sophisticated computer engineering tools would calculate channel assignments, handoff radio thresholds, neighbor lists, and a host of other complex parameters.

The primary market for cellular telephony was going to be the mobile worker and the business commuter because this was a *vehicular* telephone technology. There were portable telephone prototypes about the size of a single-lens reflex camera, but they were expensive and had a short battery life. Portable cellular telephones were a cute novelty, and that was all. The future of cellular was a car phone future.

In just a few years we knew that the market was larger and more based on handheld phones than our wildest visions. However, even back in 1983, when cellular telephone service was just starting, we knew that we had to use radio spectrum more efficiently than AMPS, and we were working on a narrowband frequency division multiple access (FDMA) digital channel, a 6-kHz channel for voice.[2] Analog AMPS did give way to digital systems, but these were time division multiple access (TDMA) and now CDMA rather than narrowband FDMA. The European GSM standard brought us a dig-

---

[2]We also explored higher-order digital constellations for voice privacy and cellular modems (data) within the AMPS channel.

ital channel more efficient than AMPS, and in the United States, CDMA systems brought a whole new approach to channel reuse with greater efficiency.

When a cellular telephone is activated, it is located within the system through radio signal strength measurements and assigned a particular channel for call setup. If the phone moves to another cell, the movement is detected by the change in relative signal strength coming from the phone, and the phone is *handed off* to a different cell in the middle of a call. The switching of calls while they are in progress was an innovation to telephony engineering required by the cellular network.

The important interference to control was cochannel interference from another cell other calls using the same frequency in other parts of the service area. Reuse patterns put the other transmitter far enough away from the serving cell to keep the interference level under control. Both AMPS (FDMA) and GSM (TDMA) use this method of managing cochannel interference. CDMA brings a whole new approach to managing interference and, with it, a very different set of issues and solutions.

The cellular environment today is portable telephones at traffic densities far beyond our 1983 forecasts using digital radio technology far beyond what we had then. The cellular environment tomorrow will add wireless local loop and high data rates. The cellular reality of 2002 and our visions of the cellular world of 2012 are nevertheless faithful to the cellular principles of 20 years ago.

6.3 Managed Interference

Since the dawn of radio technology, radio engineers have fought the signal-to-noise battle. More powerful transmitters, lower-noise receivers, and higher-gain antennas have been the focus of broadcast radio and television as well as dispatch services.

The concept of managed interference was the big breakthrough of cellular. We still have all the noise issues to contend with, but the history of cellular radio has been continuing refinement of the tradeoff between quality and quantity of calls. New technology has pushed the quality-quantity frontier further and further, but the cellular engineer's job remains the management of that tradeoff.

This is not a sellout. Many of the products we used to know and love have become more available and less expensive at the cost of lower quality. Things that used to be made of high-quality sheet metal are now stamped out of cheap plastic. Less expensive ingredients are substituted to bring a price-competitive product to market. That some wireless vendors have taken the low road in their service offerings is not an indictment of the cellular technology that made these choices possible.

The IMTS that cellular replaced was a terribly inadequate service. Only 14 simultaneous calls could be supported in the entire New York City area, the signal quality was lousy, and everybody in the world could listen in. Not only could everybody listen to your call, but also those who wanted to make a call *did* listen so that they could pounce on the call button to get a channel before the other waiting subscribers.

The failure of IMTS was not a technical failure, but rather a technology overwhelmed by demand for mobile telephone service. There is something amusing about seeing demand overwhelm our cellular technology that has capacity thousands of times greater than IMTS.

The cellular carrier has direct control of the cost-*versus*-quality tradeoff. Use a frequency a little more often, and the call quality is a little lower—and the cost is also a little lower. As we will discuss in Sec. 25.4, a radio channel used every seventh cell provides higher call quality than the same radio channel used every fourth cell. However, using the channel every fourth cell allows each cell to service seven-fourths (1.75) times as many simultaneous calls.

The cellular engineer can exercise this tradeoff on a small scale as well as system-wide. In a high-density region that needs a few extra channels, it may be worth a few noisy calls to serve many more subscribers.

There is no high end of mobile telephone call quality. Optimization is more about making more calls good than about making some calls great. Making a good call better is an insignificant improvement, whereas making bad calls good is of major importance. An AMPS call at a 10-dB signal-to-interference (S/I) ratio is awful, 15 dB is okay, 20 dB is good, and 25 dB is great. Going from 25 to 40 dB S/I is an almost imperceptible improvement. The important cellular engineering issues are at the low end of the SI range.

6.4   The AMPS FM Channel

The AMPS channel is a 30-kHz FM radio channel. The full AMPS channel is a two-way channel with 45 MHz between the two directions. The higher frequency, around 875 MHz, goes from the base station to the mobile telephone, the *forward* or *downlink* direction. The lower frequency, around 830 MHz, goes the other way, the *reverse* or *uplink* direction.

The audio bandwidth supported on the AMPS channel goes from 300 to 3300 Hz, the standard analog telephone bandwidth. The voice signal on an FM channel modulates the frequency up and down from the center of the band. The AMPS channel allows a maximum frequency deviation of 12 kHz, and the level is set so that the nominal talker averages 2 kHz from the center of the band.[3]

The voice transmission quality is improved by a two-to-one compandor system. The voice signal is compressed by two to one (in decibels) before it is sent out over the radio airwaves. The receiver expands the voice signal by the same two to one to restore the original voice signal. This effectively doubles the loud-to-soft dynamic range in the conversation.

The Dolby systems used in compact cassette recording and playback are similar to the AMPS compandor: They compress the high frequencies during record and expand them again in playback to suppress audible hiss. The AMPS compandor is used across the entire voice band rather than just high frequencies by compressing and expanding the entire AMPS audio bandwidth. When the voice signal gets 10 dB louder, for example, the compressed signal sent over the radio is only 5 dB higher. The receiver expands the 5-dB increase back to 10 dB so that the listener hears the voice as it was spoken. The effect of this compression and expansion is to reduce the perceived background noise dramatically. Speech coding algorithms do the same kind of processing in digital telephony.

The two FM carriers in the two directions form one full-duplex AMPS voice channel. The channel is continuously on in both directions. While some portable telephones can pause their transmission (sleep) for short time intervals to save battery life, the channel is not available for anybody else to use during the call.

An AMPS mobile telephone has two radios, one for transmitting and one for receiving. This means that during the call, there is no way for the mobile telephone to main-

---

[3]The average is the root mean square (RMS) described in Sec. 4.3.

tain a link on the control channel. Rather than add a third radio, the AMPS designers made this two-way link do the full job of maintaining a cellular telephone call. In addition to maintaining the audio voice, the channel handles link supervision, messages for cellular overhead, and subscriber call-processing messages. Call setup is done on separate paging and access channels, which are not used during a call.

6.4.1 Supervisory audio tone (SAT)

How does a telephone system make sure that it is still communicating? As telephone users, we have the same issue: Silent periods longer than a few seconds make us nervous. Both ends of the telephone call have to know when the call has ended. Also, the FM channel has brief outages due to the Rayleigh fading described in Sec. 2.1.4. We do not want to hear somebody else's conversation during those brief outages.

Maintaining and confirming continuity on a link is called *supervision*. In AMPS, we do this with a *supervisory audio tone* (SAT) that modulates the FM carrier along with the voice. The SAT is set out of the telephone audio band, at 5970, 6000, or 6030 Hz. This tone is sent by the base station and returned by the mobile telephone, as described later in Sec. 25.1, and it is filtered out so that the customer never hears it.

Should the SAT tone be absent or the wrong frequency of SAT be detected, the receiver mutes the audio so that the listener does not hear radio static or, worse, another telephone call.

6.4.2 Blank and burst

Once an AMPS call is on a voice link, the base station and mobile telephone communicate through a series of *blank and burst messages*. These messages are 100 ms (one-tenth of a second) of direct-binary frequency shift keying (FSK) data; a 1 is an increase of 8 kHz in the carrier, and 0 is a decrease of 8 kHz. The initial sequence of a blank and burst message is recognized by the audio receiver, and it mutes the audio so that the subscriber does not hear a "bzzzt" sound when these messages are sent.

Blank and burst messages are used for cellular overhead signaling such as handoff and power-control messages. The alert message to ring the mobile telephone is also a blank and burst message as AMPS establishes a voice link before the telephone rings.

Subscriber call processing during a call typically starts with a flash for call waiting or three-way calling. Three-way calling also sends its dialed digits via blank and burst. The blank and burst messages are designed to be robust because typically they are used in poor radio conditions, especially for handoffs. Some cellular systems use subband digital signaling, frequencies below 300 Hz, for their own digital messages, but this is not part of the AMPS standard.

6.4.3 Narrowband AMPS

Introduced by Motorola in 1991, narrowband AMPS (N-AMPS) replaces the 30-kHz FM channel with three 10-kHz FM channels. Compared with AMPS, the sound quality is diminished by the narrower channel.

However, N-AMPS has several advantages.

- *Greater capacity*. A three-for-one channel split actually increases capacity more than three times because of traffic engineering issues discussed in Chap. 21.

- *Compatibility.* N-AMPS can use the same control channels, and dual-mode mobile telephones can be used to operate under both standards.

- *Improved supervision.* The SAT tone is replaced by a 200-bit-per-second signaling stream.

- *Improved call monitoring.* Using a feature called *mobile reported interference* (MRI), the base station can interrogate the mobile for the forward channel signal strength and the bit error rate (BER) of the signaling stream.

- *Short message service (SMS) and paging.* Messages up to 14 characters long can be sent on the forward channel in a messaging or paging mode.

- *Voice-mail notification.* The system has the ability to activate a message-waiting light on the mobile telephone to tell a subscriber that voice mail is waiting.

6.5 AMPS Call Setup

AMPS calls are set up by the mobile telephone. The mobile subscriber may initiate a call by pressing the SEND button, or the mobile telephone may detect a page for an incoming call. However, it is the mobile telephone that selects the cell based on its radio measurements. As we will see in the next section, it is the cellular system that handles handoffs.

A mobile telephone has to register with the system. It may be in its home area or it may be roaming, but it must register and establish its identity before it can make or receive calls. Once identified, the mobile telephone stays tuned to the paging stream in case there is an incoming call.

In the case of an incoming call, the system makes two attempts to page the mobile telephone. If those pages fail, then the call attempt fails.

A call starts when the subscriber presses the SEND key or a page comes in for an incoming call. The mobile telephone searches all 21 access channels to find the strongest signal. If all the access channels are set to the same transmit power level, then this should be the cell with the highest path gain (least path loss), but Rayleigh fading and measurement error have a hand in the cell-selection process. Given that the mobile telephone has been monitoring the paging channel successfully, it is highly unlikely not to find a suitable access channel, but it could fail to find one at this point.

A mobile page response or call initiation is called a *seizure message*. The forward access channel bit stream has a busy/idle bit to tell the mobile when it is safe to transmit its seizure message. Two mobiles could see the same idle bit and transmit at the same time, causing a message collision. When this happens, the mobile telephone is required to wait a random amount of time before trying again. The mobile telephone makes 10 seizure attempts, and the call fails if it does not get through.

The cell may not be able to serve the call. The cell goes through all its logical faces (Sec. 5.2.3) and selects the best one where the mobile has adequate path gain based on signal strength measurement. The mobile telephone seizure message is sent with a power level directed by the system limited by its own maximum power level.

If no logical faces are available, then the mobile telephone is directed to retry a list of alternate cells. In the case of a smaller-growth cell where all the access threshold levels are set high, there may be regions where the mobile telephone chooses the cell and there is no adequate server. Such calls also are directed to retry alternative cells. There are directed retry counters to make sure that the mobile does not get into an endless directed retry loop.

Once the seizure message reaches a suitable cell, the cell sends a message with voice channel and initial mobile transmit power level.

If the mobile telephone does not acquire SAT within a few seconds, then the call is lost. If the base station receiver does not receive the SAT returned from the mobile telephone, then the call is lost.

Once the call has passed through all these stages, a two-way stable voice link is established, and the subscriber is ready to be involved. In the case of a mobile origination, the microphone and speaker are turned on, and the subscriber hears ringing, a busy signal, or whatever the telephone network sends. In the case of an incoming call, the base station sends an alert message via blank and burst, and the cellular telephone rings.

6.6   AMPS Call Maintenance

Once the call is set up with a voice link, the mobile telephone monitors SAT. So long as the mobile receives SAT of the correct frequency, it sends the same SAT back to the base station. Otherwise, the mobile telephone mutes the audio signal and waits. If too much time goes by without SAT, the mobile telephone drops the call.

The base station monitors the returned SAT. As in the mobile telephone case, momentary loss of the correct SAT cause audio muting, and sustained absence causes a lost call. During the call, the base station directs power levels and handoffs as needed.

6.6.1   AMPS power control

On call setup and every 5 seconds, the base station measures the signal strength of each AMPS call. Based on the mobile telephone transmit power, the system can infer the radio signal path gain of the call.

AMPS mobile telephones have eight power levels called *voice mobile attenuation codes* (VMACs). These range from full power (VMAC set to 0) to minimum power (VMAC set to 7), as shown in Table 6.1. Class 1 mobile telephones go up all the way to VMAC = 0, class 2 mobile telephones go up to VMAC = 1, and class 3 mobile telephones only go to VMAC = 2.

Once the base station determines the radio path, then it can set its own transmit power and tell the mobile telephone what power level to use. This power level is sent using a blank and burst message.

The technique of deducing the received power at the mobile telephone from the mobile telephone transmit power and the received power at the base station may sound trivial, but it was not obvious at the time. In fact, it was considered subtle enough for

TABLE 6.1   Voice Mobile Attenuation Codes

VMAC = 0	3.0 W	36 dBm
VMAC = 1	1.6 W	32 dBm
VMAC = 2	630 mW	28 dBm
VMAC = 3	250 mW	24 dBm
VMAC = 4	100 mW	20 dBm
VMAC = 5	40 mW	16 dBm
VMAC = 6	16 mW	12 dBm
VMAC = 7	6 mW	8 dBm

a U.S. patent to be issued for it. In Sec. 25.6 we discuss the strategy issues of power control on the AMPS channel.

6.6.2   AMPS handoff

When the signal goes below some level, called the *primary threshold* in the AMPS language, the base station decides to look for a handoff. There are three kinds of AMPS handoffs: to another server group on the same face, to another face on the same cell, and to another cell. Server groups are different cell sizes served by the same base station, as described in Sec. 5.2.3.

Once the call has been below primary threshold for a few measurements, all the antenna faces at the serving base station and some nearby neighbor base stations measure the signal strength of the reverse channel. If there is a logical face where the call is above primary threshold, the base station tries to put the call onto the strongest face (best radio path) and the innermost server group that can handle the call.

If the mobile telephone reverse signal is not above primary threshold on the primary neighbor list, then there is a secondary threshold and a larger secondary neighbor list. If the signal is below secondary threshold, then this wider survey of measurements is taken, and a more intense search is done for a suitable handoff.

When other cells measure the signal strength of a call, they can be instructed only to return a measurement when the signal at the requested frequency has the correct SAT. We want to avoid doing a handoff based on measuring the wrong call.

This is the basic outline of handoff search we designed at AT&T. Different cellular equipment vendors use different selection rules. And different cellular service providers set their system parameters differently.

Notice, please, that the mobile telephone has played no part in the handoff search process. While some of this may be attributed to a "we know best" attitude on the part of base station designers, there is also a more practical issue. The mobile telephone has a two-way radio completely tied up with the duplex transmission of the full-time FM signal. Maybe a third radio circuit could have been designed into the mobile telephone, or maybe it could have been designed to break the audio link for one-tenth of a second (the duration of a blank and burst message) to measure other channels. Whatever could have been done, however, the AMPS system design has the community of base stations making all the power-control and handoff decisions.

If there is nowhere to hand the call off and the call is below some yet-lower interference protection threshold (as we called it at AT&T), then the call is dropped. This threshold is not designed to protect subscribers against their own bad calls, since we have SAT to do that. The idea of the interference-protection threshold is to protect the rest of the system against a mobile telephone that has wandered so far away from its serving cell that it presents an unacceptable interference hazard to other cells.

After all the local measurements, the neighbor cell measurements, the determination that a handoff is appropriate, and the selection of a new radio is made, the base station finally directs the mobile telephone to change frequency. The new base station starts transmitting, the current base station sends a handoff message, and the new base station looks for SAT. The system tries its handoff message up to three times.

Handoffs can get lost in a variety of ways. The mobile telephone may change to a channel and find no signal, SAT can be lost, or the signal level of the new connection may be below the new cell's interference-protection threshold.

If the handoff does succeed, the base station starts making measurements and running its power control algorithm.

In this sequential paragraph form, the handoff process seems cut and dried. The steps of the handoff process fit neatly into the mental picture of a car driving smoothly out of one hexagon into a neighboring hexagon. The AMPS designers knew full well that the handoff procedure had to be robust enough to handle some oddball cases.

Even so, there are cases of strange handoff behavior. The measurements and threshold can do strange things. In the early Chicago days, we had a street corner in Elmhurst, Illinois, where we would sit at the red light and watch the channel indicator on our test unit cycle through three cells as A handed off to B, B handed off to C, and C handed off back to A.

There is an important disparity in the AMPS system design. In hindsight, it is easy to say somebody should have thought of this, but it certainly was not obvious at the time. The initial call setup is done by the mobile telephone, and the handoffs are done by the base station. This means there is a transfer of radio control from mobile telephone to base station just as the call is getting underway. There is a higher frequency of handoffs in the first 15 seconds of AMPS calls than after that.[4]

6.7 Conclusion

The AMPS system pioneered cellular technology in support of mobile telephony and defined many of the engineering issues and solutions still in use today. As we shall see in the next two chapters, many issues remained the same even as cellular engineering made the huge leap from analog to digital.

---

[4]Maybe back in 1978 somebody did think of it and decided it was okay. When we noticed the effect in our 1985 simulation work, we were concerned about the effect on system capacity of so many extra handoffs.

TDMA Wireless Telephony (GSM)

With effective cellular engineering already developed in Advanced Mobile Phone Service (AMPS), the designers of the Global System for Mobility (GSM) could focus on the next major step in the evolution of cellular. That step was the application of digital technology to voice telephony. The technology they chose was *time division multiple access* (TDMA), where each digitized voice signal is sent in a particular, periodic time slot measured in milliseconds.

GSM was developed as a pan-European standard with the hope that it would become a world standard. GSM is actually not a single standard but rather a set of standards specifying many differing interfaces. This allows different manufacturers in different nations to develop components of the GSM system. When the components are brought together on a single network, they are likely to work well together, although, of course, they should be tested thoroughly. The Europeans used an international standard to fulfill the same function that corporate consortiums often fulfill in the United States.

7.1 GSM Architecture

GSM is laid out somewhat differently from AMPS. Where AMPS has a controller in each base station, GSM concentrates the cell controller function in a single base station controller (BSC) for several cells. GSM refers to the controller-less base station as a *base transceiver station* (BTS). The entire network of base stations for one BSC is called a *base station system* (BSS). One mobile switching center (MSC) can control several BSCs. We can think of the BSC as another level in the GSM hierarchy.

The BSC is an important level in GSM because it can do its own handovers. (AMPS calls them *handoffs,* but GSM calls them *handovers.*) This means that the GSM BSC has to have some switching powers independent of the MSC.

7.2 The TDMA Channel Concept

In analog frequency division multiple access (FDMA), the concept of the radio channel is clear. A channel is a frequency range allocated to carrying a modulated signal from a transmitter to a receiver. The continuous time of the voice signal is represented by continuous time in the modulated radio carrier. This is the same for broadcast FM radio and AMPS cellular, except that AMPS cellular adds the idea of a duplex channel—two separate frequencies used for bidirectional conversation.

In GSM, the channel is more complicated than in AMPS in two important ways. First, it is a *digital* channel, where the voice signal is coded into a sequence of binary digits and the digits are sent over the radio link. Second, eight digital voice streams are combined into a single radio carrier. Each voice channel gets one of eight time slots. In TDMA, we maintain the distinction between the *carrier,* which includes the entire digital radio link, and the *channel,* which only includes one time slot of a carrier.[1]

Therefore, a voice channel, which in FDMA referred to a separate radio carrier with its own frequency range, in TDMA refers to a time slot within a multichannel carrier with a larger frequency range. The two directions are still two different frequencies, but their time slots are staggered so that the user terminal never has to transmit and receive at the same time.

The digital medium also changes the way control signals are managed in GSM. When a call is not in progress, in both AMPS and GSM control signals are sent on their own channel, a separate frequency in AMPS and a separate time slot in GSM. During an AMPS call, signals are carried on the voice channel by in-band blank and burst interruptions that interrupt the voice channel briefly and by out-of-band supervisory audio tones (SATs). During a GSM call, the signaling data are sent on the same digital channel as the digitally encoded voice bits, and the digital carrier has extra frames for signaling data.

7.3   The GSM TDMA Channel

The GSM radio carrier is a 200-kHz digital channel that carries eight conversations along with some signaling. This is a tougher channel than AMPS because it can tolerate more co-channel interference than AMPS. This difference will be discussed in Chap. 25.

The GSM radio carrier is a 200-kHz digital channel using gaussian minimum phase shift keying (GMSK) modulation, a variant of quadraphase phase shift keying (QPSK). This 200-kHz carrier is also called the *traffic channel* (TCH). The rate of the carrier is 270 kbps. The data stream is divided into 4.615-ms frames, and each frame is divided into eight time slots of 148 bits each. Some of these frames are for signaling and synchronization, so each of the eight voice channels has a raw data rate of 22.8 kbps.

The GSM voice channel uses error correction (as described in Sec. 17.2) to get higher accuracy at a lower bit rate. The resulting bit stream available for speech coding (as described in Chap. 18) is 13 kbps. The speech coder uses a mu-law coding and a speech-compression method called *regular pulse excitation—long-term prediction* (RPE-LTP).

The original European version of GSM has 50 MHz of spectrum in the 900-MHz band, just like AMPS. And the forward link was 45 MHz higher in frequency than the reverse link, just like AMPS. This means that the classic GSM configuration supports 62 radio carriers, or 496 time-slotted channels for voice or signaling compared with 416 for AMPS. Other versions of GSM use the same channel architecture at higher

---

[1]Our usage here is technically correct, but it is not the industry standard. GSM engineers call a particular frequency band a *channel,* whereas we are calling it a *carrier.* In GSM language, a single radio channel is time-divided to carry multiple voice channels. We chose to use the term *carrier* so that we would not have two different meanings for the term *channel.*

frequencies where more spectrum is available, so these systems can have more GSM carriers and channels, and there is more frequency separation between the forward and reverse directions.

The GSM channel has 30 μs between its time slots, which works out to 4.5 km for the round trip at the speed of light. If a mobile telephone is further from the base station than 4.5 km, then GSM allows for it to be resynchronized using a feature called *dynamic time alignment* up to 237 μs, about 40 km. Any range further than this requires the call to use *extended dynamic time alignment,* which takes two time slots for the call instead of just one. There is not much GSM traffic operating further than 40 km (25 miles) from the cell tower.

Pauses in speech can be matched by pauses in transmission using DTx. *DTx* is the radio buzzword for *discontinuous transmission.* (*Tx* is the abbreviation for *transmit.*) The speech coder uses voice-activity detection (VAD) to determine when to transmit and when to keep quiet. The GSM receiver turns silent frames into comfort noise so that the listener does not hear the clipped muting so common in less sophisticated voice-activated systems such as those often found on inexpensive tape recorders.

Rayleigh fading is less severe in a 200-kHz GSM carrier than it is in a single 30-kHz AMPS channel because Rayleigh fades are concentrated in a particular place at a particular frequency. Even with its broader carrier bandwidth, there is still some fading in a GSM radio path. To combat this remaining Rayleigh fading, GSM can employ a form of frequency diversity by using slow frequency hopping. The combination of frequency hopping and bit interleaving spreads the errors over enough time slots and reduces the number of consecutive bit errors. Strings of five or more errors in a row cause the GSM error-correction facility to lose its ability to correct errors.

There is a half-rate GSM in which a voice link uses only 1 time slot out of 16 instead of 1 in 8. Half-rate GSM may be twice as efficient, but it is not a commercial success. It does not sound nearly as good as full-rate GSM, and the processing equipment for its 6500-bps speech coding is larger and more expensive than the 13-kbps coding used in full-rate GSM.

In GSM, the transmit and receive time slots are deliberately staggered so that each voice channel transmits on one time slot and receives on a different time slot. This allows a tremendous cost saving in mobile telephone design, since the time separation of transmit and receive modes means that only one radio circuit is needed. Not only can a GSM phone use a single radio circuit for transmit and receive, the phone still has six other time slots available to make measurements of other GSM channels in the handover determination.[2]

7.4 GSM Signaling

Each GSM cell has two signaling channels. There is a *broadcast-control channel* (BCCH) that the mobile telephone tunes to when it is not on a call. The *cell broadcast-control channel* (CBCH) is another signaling stream that sends 80-character short messages and shares the same time slot as the BCCH.

---

[2]The base station does not have extra slots available, since the BTS is serving multiple user terminals simultaneously, so it has to be able to transmit and receive all eight time slots for each eight-channel GSM carrier.

The other cell-wide signaling channel is the *paging and access-grant channel* (PAGCH). Its reverse link is the *random-access channel* (RACH), where the mobile telephone requests access to the system.

There are two signaling modes on the *traffic channel* (TCH). The *slow associated control channel* (SACCH) uses parts of the TCH not associated with speech coding, and the *fast associated control channel* (FACCH) uses parts of the TCH normally used for speech. For this reason, the SACCH is called out-of-band signaling, and the FACCH is called in-band. Both of them are part of the same digital stream, but the FACCH takes bits from the voice channel stream and relies on the error-correction component of the GSM speech coder to fix it.

7.4.1   GSM call setup

In GSM call setup, we think of the base station system (BSS), all the base stations served by a single BSC, as a single entity. An important difference between AMPS and GSM is that GSM has much more comprehensive specification of interfaces other than the air interface. Any AMPS mobile telephone should work on any AMPS system, but the interfaces between base stations and switches are often proprietary, requiring that both components be made by the same manufacturer. In contrast, any GSM BSC should work with any GSM BTS and any GSM MSC, at least in theory. (We would still recommend testing equipment for compatibility before investing a lot of money based on the GSM compatibility claim.)

When the mobile telephone initiates a call, it uses the RACH to get the BSS to assign it a *stand-alone dedicated control channel* (SDCCH), which is sent over the access-grant channel (AGCH). The SDCCH is used until a voice traffic channel is assigned.

Once the mobile telephone is set with the BSS, a service request message is sent to the MSC, which informs its home location register (HLR) or its visitor location register (VLR) that the mobile telephone is requesting service. Through the BSS, the MSC requests and receives the international mobile equipment identity (IMEI), and it checks the IMEI for validity.

Now the mobile telephone sends its setup request along with its dialed digits, and the MSC uses the HLR or VLR parameters to handle the call. The BSS and mobile telephone establish a voice channel, they both switch to that channel, and the BSS informs the MSC of the change.

Finally, the public switched telephone network (PSTN) is involved in completing the call. Unlike AMPS, the mobile telephone is specifically informed, via signaling messages, when the landline telephone is ringing (network alerting) and when the telephone is picked up (connect).

When a call comes in from the PSTN, the MSC confirms the mobile telephone identity with the HLR and VLR and has the BSS page the mobile telephone on the paging channel (PCH). The mobile telephone responds on the RACH, and the signaling sequence from here is similar to the mobile telephone–initiated call, although there are some differences.

Once the voice link connection is made, the mobile telephone rings (alerts), sends a network alerting message and then sends a connect message when the subscriber answers the phone.

There is a similar exchange of messages when a call ends, all carefully prescribed in the GSM specifications.

GSM allows for a mobile telephone to power off its nonessential circuitry, putting it in sleep mode during periods where no paging messages will be sent. This feature, *dis-*

continuous reception (DRx), is supported by having the paging channel divide pages up into paging subchannel groups by the last digits of the mobile telephone number or the international mobile service identity (IMSI).

7.4.2 GSM power control

GSM defines five power classes of user terminal. There are eight power classes in the full specification, but only five seem to apply to mobile telephones. These are shown in Table 7.1. These power levels are the burst power for the one active time slot so the average transmit power is one-eighth of these values.

GSM power control is managed by the base station, the same as in AMPS. There are 15 power levels separated by 2 dB for a total range of 28 dB, the same as in AMPS. Each class of GSM mobile telephone can go down 15 steps from its own maximum power, the full range of 28 dB. The 2-dB changes can come every 60 ms, every 13 bursts at full rate. The GSM base station transmitters can bring their power all the way down to 13 dBm, 20 mW.

7.4.3 GSM handover

GSM handovers occur when a call has poor signal quality or low signal path gain. They also occur for mobile telephones that are too far away from their serving base stations and for calls where the system determines that the call can be served somewhere else at a lower power level.

After using one time slot to transmit and one to receive, the GSM mobile telephone makes use of the other six time slots to make power measurements of other cells. This information is sent on the SACCH to the serving BSC, where the handover determination is made. We call this a *mobile assisted handover procedure*.

There are four types of GSM handovers depending on how far up the chain the handover has to go.

- *Type 1, intra-BTS handover.* The call is changing from one face to another face in the same cell. The call keeps the same MSC, BSC, and BTS.

- *Type 2, intra-BSS, inter-BTS handover.* The call is changing from one cell to another cell covered by the same controller. The call keeps the same MSC and BSC but has a new BTS.

TABLE 7.1 Eight GSM Power Classes

Power class	Mobile telephone maximum power,	Base transceiver maximum power,
1	20.0 W	320 W
2	8.0 W	160 W
3	5.0 W	80 W
4	2.0 W	40 W
5	0.8 W	20 W
6	Not applicable	10 W
7	Not applicable	5 W
8	Not applicable	2.5 W

- *Type 3, intra-MSC, inter-BSS handover.* The call is changing from one controller to another within the same switch. The call keeps the same MSC but has a new BSC and BTS.

- *Type 4, inter-MSC, inter-BSS handover.* The call has a new MSC, BSC, and BTS.

When the mobile telephone is instructed to change to a new GSM carrier and time slot, it tries to synchronize to the new cell. A presynchronized change is called a *seamless handover*. In a seamless handover, the mobile telephone sends time-aligned bursts immediately, and there is no pause in the audio signal. This tends to happen when the cell sizes differ by less than 1 km.

7.4.4 GSM short message service

A paging capability is built into the GSM standard for short message service (SMS), which is described in Chap. 19. The messages can be up to 160 characters in length and are sent and received by a GSM mobile telephone that is not on a normal voice telephone call.

In the forward direction, the SMS messages are sent on the CBCH, the control channel, and they can be stored by the system if the mobile telephone is not receiving pages. Just as in the case of a page for a telephone call, the mobile telephone with the correct number picks up the message, and all the other mobile telephones ignore it. In the reverse direction, GSM SMS uses the SDCCH signaling channel.

On the GSM air interface, one can think of SMS as a call that stops at the very first signaling message. At current demand levels, SMS traffic has not been a capacity problem for GSM.

On the terrestrial interface, SMS is carried on the Signaling System No. 7 (SS7) network described in Chap. 14, where it competes with call signaling and *can* be a serious capacity issue. The SMS community may find a way to transport its messages over Transmission Control Protocol/Internet Protocol (TCP/IP) links, a cheaper and more plentiful pathway.

7.5 Conclusion

The GSM set of standards has proven highly successful in Europe, as well as in other nations around the world. The standardization of the entire system, rather than just the air interface, can be seen as part of the larger process of the socioeconomic unification of the European Union. However, GSM is roughly equivalent in the digital services it provides and in capacity with the North American cdmaOne standard, which we will turn to next. After that, we will explore the issue of the competition to support more digital traffic and become a world standard that is now developing between the latest version of GSM and the new cdma2000 and W-CDMA standards.

The CDMA Principle

With the basics of cellular worked out in Advanced Mobile Phone Service (AMPS) and the basics of digital radio telephony developed in the Global System for Mobility (GSM), cellular engineers turned to the task of using a new, complex technology that had potential to provide even better performance than time division multiple access (TDMA). That technology was *code division multiple access* (CDMA), an implementation of spread spectrum. In CDMA, multiple signals are sent from the same tower on the same frequency at the same time. Signals are distinguished not by time or frequency but by a code attached to each signal. At the receiving end, the signal is distinguished from all other signals by its code and is extracted from them.

This simple-sounding principle actually requires some very sophisticated mathematics, as well as sophisticated engineering, particularly in the realm of power control. In this chapter we will explore these details of the CDMA principle and look at its implementation in the cdmaOne standard (formerly called IS-95 and developed by Qualcomm) as it is deployed across North America and elsewhere.

8.1 Spread Spectrum

In contrast to conventional reuse, the spread-spectrum channel is no more efficient for one serving antenna but is enormously more resistant to interference. The effect of this is a more efficient *system.*

The origin of spread spectrum is military, where resistance to deliberate interference, or *jamming,* was the major consideration, with radio spectrum efficiency a nice side effect.

Picture a communication link in a hostile environment. If we send a signal with a traditional analog or digital modulation scheme, then a hostile party can broadcast a signal on the same carrier frequency, and our own receiver loses the message. Now suppose that we hop from one frequency to another in a prearranged pattern. The actress Heddy Lamar suggested this, and the patent for frequency hopping bears her name. So long as the hostile party does not know the frequency pattern, it would take an enormous amount of power over a broad band of frequencies to jam the signal.

More sophisticated spread-spectrum systems use a large frequency band continuously and typically at low power levels. Not only does this make it hard to interfere with the signal, but it also makes it hard even to *detect* the signal. The safest way to protect something is to have the hostile party not even know it exists. Ten watts of radio power in 10 kHz is easy prey, easily detected and easily jammed. The same 10 W spread over *10 MHz* is 1000 times less efficient in bandwidth usage but is a lot safer. Assuming that the enemy detects such a spread-out signal, the spread-spectrum system can be nearly 1000 times more resistant to interference.

The same technology that protects military users in battle against hostile interference can protect civilian users in shopping malls against friendly interference. The spread-spectrum system can protect its users against other users *of the same channel,* thus making capacity gains over conventional modulation.

There are two major differences in capacity planning for conventional reuse [frequency division multiple access (FDMA) and TDMA] and spread spectrum (CDMA).

First, the conventional reuse channel set is fixed and determined by the number of radios installed at each base station. Call quality is determined by channel reuse in other cells and not by how many calls are using the same-cell channels. In spread spectrum, capacity is determined not by allocation of channels but by radio conditions in the carrier used by all the calls.

Second, the primary interferers to a conventional reuse call are in other cells using the same radio carrier. In spread spectrum, the same-cell, same-carrier calls are the primary interferers, with other-cell users being of secondary importance. Each call has to have a certain share, a *percentage,* of the total power of the carrier at its receiver. If the call has a sufficiently high percentage of the received power, then it will be able to distinguish the signal from the other calls in the same frequency band.

This notion of a required fraction of the signal is an important and strange feature of spread spectrum. Making the pizza pie bigger does not serve more people if every diner insists on eating one-eighth of it. Similarly, if each call requires 3 percent of the total signal, then the thirty-fourth caller is going to have a tough time of it. Under noisy conditions, that same 3 percent limit may only support 15 or 20 calls, and the base station has to know when to turn away more calls *before* all calls run out of signal quality.

8.2 Processing Gain

The basis of CDMA is its *processing gain,* sometimes called *spreading gain.* This is the multiple-access part of CDMA. To see what processing gain is all about, consider a 2-bit digital channel with as much noise as signal. In our radio jargon, consider a binary phase shift keying (BPSK) channel plus additive white gaussian noise (AWGN) with a 0-dB signal-to-noise (S/N) ratio. The transmitter sends a +1.0 or −1.0 for each bit, and the noise adds a random bell-shaped normal distribution with a standard deviation of 1.0, as shown in Fig. 8.1. The solid bell-shaped curve on the right side is the distribution of received signal plus noise when the transmitter sends a +1 bit, and the dotted bell-shaped curve on the left side is the distribution for a −1 bit. The receiver considers any received signal on the right side of the center line to be a +1 and any received signal on the left side to be a −1. The receiver receives a bit error when the solid curve is on the left of the center line or when the dotted curve is on the right.

According to our college statistics course, a normal distribution is more than 1 standard deviation high 16 percent of the time, more than 1 standard deviation low 16 percent of the time, and in the fat middle part of the curve the other 68 percent of the time. This gives us a bit error rate (BER) of 16 percent.

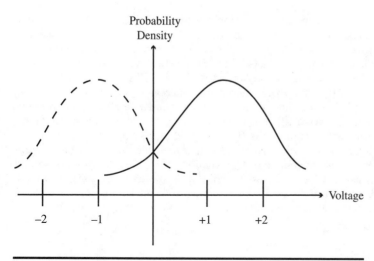

Figure 8.1 BPSK with a 0-dB S/N ratio.

If we do not want to live with a 16 percent BER, then we could boost the signal power by 7 dB (a factor of 5) so that we have an S/N ratio of 7 dB. A factor of 5 in power is a factor of $\sqrt{5}$ (2.236) in amplitude (voltage). This gives us the picture in Fig. 8.2, $+\sqrt{5}$ or $-\sqrt{5}$ for each bit with the same noise contribution. The peak of each curve is now further from the 0-V centerline. As before, we have a bit error when the solid curve is on the left of the centerline or when the dotted curve is on the right. The same college statistics course tells us that a normal distribution is more than 2.236 standard deviations high 1.3 percent of the time, more than 2.236 standard deviations low 1.3 percent of the time, and in the fat middle part of the curve the other 97.4 percent of the time. This gives us a BER of 1.3 percent.

There is an alternative to using more power. We could send each bit *five times* instead of only once and have the receiver add up all five signals. This gives us better BER performance, and we use the term *processing gain* to describe the increase in performance

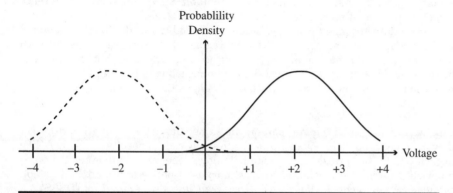

Figure 8.2 BPSK with a 7-dB S/N rato.

from repeated sending of the same bit. The magic of processing gain is that this five-fold repeating gives us the *same* statistical advantage as a fivefold power increase.

In both cases we have increased the energy per bit five times while keeping the noise constant. It is the ratio of bit energy to noise that is important here. We denote this ratio by E_b/N_0, and we will have more to say on E_b/N_0 in Sec. 27.1. The important concept for processing gain is that the E_b/N_0 increase by increasing power and the E_b/N_0 increase by sending multiple samples of each bit yield the same BER improvement; they are exactly the same, statistically speaking. In this example, the processing gain is 7 dB, a factor of 5, enough to change the BER from 16 to 1.3 percent.

We call the sub-bit samples *chips,* and we refer to the *chip transmission rate* or the *chipping rate.* The cdmaOne standard created by Qualcomm sends 1,228,800 chips per second to deliver 19,200 bits per second, a processing gain of 64 (64 chips per bit), or 18 dB.

For discussions about digital radio, it is customary to refer to bits as +1 and −1 rather than 1 and 0. It makes it much easier to talk about multiplying bit values, as shown in Eq. (8.1):

$$+1 \times +1 = +1$$

$$+1 \times -1 = -1$$

$$-1 \times +1 = -1 \tag{8.1}$$

$$-1 \times -1 = +1$$

If a digital radio engineer, in casual conversation, refers to −1 bits as zeroes, then it will not be the first time. There is a bit of schizophrenia about −1 and 0 depending on which notation makes the most sense in context. This arithmetic is being done with high-speed transistor gates that have two voltages for the two bit values and that have no idea what formulas we wrote on paper to describe them.

Simply sending the same bit over and over again improves the robustness of a digital channel, but if all we wanted was a more robust digital channel for a single user, forward error correction would do a better job. Instead, we take the sequence of redundant chips and multiply them by a sequence of random-looking +1 and −1 values. We call this multiplying sequence a *pseudorandom sequence* or a *pseudonoise* (PN) *code* in the next section.

Suppose that we have a digital signal with few enough bits and a radio channel with wide enough bandwidth that we can maintain acceptable E_b/N_0 with more interference than signal, perhaps a lot more interference than signal.

We can take a data stream of x bits per second and send it on a digital radio channel at x bits per second, or we can spread it k times with a PN code and send each bit with k chips on a digital radio channel at kx chips per second. Assuming that the transmit power is the same in both cases, the processing-gain principle tells us that the x-bit-kx-chip-per-second modulation will have the same E_b/N_0 as the plain old, ordinary x-bit-per-second modulation.

Why would we want to do this?

Because we can now send *another* call on the same channel, a call with a *different* PN code with no particular relationship to the first PN code. These two calls look like noise to each other. The spread kx-chip channel allows other calls with other PN codes to form noiselike interference, whereas the ordinary x-bit-per-second channel does not. This spread-spectrum technology is called *direct sequence spread spectrum* (DSSS).

This is the core of CDMA. Multiple calls can occupy the same channel at the same time divided not by frequency, not by time, but by their PN codes.

8.3   Pseudonoise (PN) Codes

Let us take a closer look at the PN-code concept in CDMA. Each bit a is transmitted in n chips c_1 through c_n. We derive each chip value by multiplying bit a by the PN code p_k so that the transmitted chip sequence is $c_k = ap_k$, for $k = 1 \ldots n$. The receiver adds up the n chips with the same PN code p_k as multipliers in Eq. (8.2) to get the received bit a_{rec}.

$$a_{\text{rec}} = \sum_{k=1}^{n} p_k \, c_k \qquad (8.2)$$

We can substitute $c_k = ap_k$ into Eq. (8.2) to get Eq. (8.3):

$$a_{\text{rec}} = \sum_{k=1}^{n} (p_k)^2 \, a \qquad (8.3)$$

Whether p_k is $+1$ or -1, the value $(p_k)^2$ is always $+1$. In the absence of noise or interference, this gives an exact answer $a_{\text{rec}} = na$, which will be positive when a is positive and negative when a is negative.

At the same time a is being spread with p_k, the transmitter[1] is processing b with q_k so that the total chip value is $c_k = ap_k + bq_k$. The receiver for a uses Eq. (8.2) to get the received bit a_{rec} in Eq. (8.4):

$$
\begin{aligned}
a_{\text{rec}} &= \sum_{k=1}^{n} p_k c_k \\
a_{\text{rec}} &= \sum_{k=1}^{n} p_k \, (ap_k + bq_k) \\
a_{\text{rec}} &= \sum_{k=1}^{n} (p_k)^2 a + \sum_{k=1}^{n} (p_k q_k) b \\
a_{\text{rec}} &= na + \sum_{k=1}^{n} (p_k q_k) b
\end{aligned}
\qquad (8.4)
$$

The receiver for b uses a similar calculation to get the received bit b_{rec} in Eq. (8.5):

$$
\begin{aligned}
b_{\text{rec}} &= \sum_{k=1}^{n} q_k c_k \\
b_{\text{rec}} &= \sum_{k=1}^{n} q_k \, (ap_k + bq_k) \\
b_{\text{rec}} &= \sum_{k=1}^{n} (p_k q_k) a + \sum_{k=1}^{n} (p_k)^2 b \\
b_{\text{rec}} &= nb + \sum_{k=1}^{n} (p_k q_k) a
\end{aligned}
\qquad (8.5)
$$

---

[1]This could be either the same transmitter or another transmitter adjusted so that both signals reach the receiver at the same power level.

The secret of CDMA is that p_k and q_k agree and disagree about the same amount, so the interfering cross term

$$\sum_{k=1}^{n} (p_k q_k) \tag{8.6}$$

is small compared to n.

The PN code of CDMA has to send the signal message with minimal interference with other PN codes. This can be done by generating a sequence of digits that does not repeat itself for a long time or by using codes tailored not to interfere with each other.

8.3.1 Pseudorandom sequences

Truly random sequences (without the *pseudo-* prefix) have three desirable statistical properties for CDMA over the long haul:

1. Half the bits are 0 and half are 1.

2. The numbers of same bits in a row, the *run lengths,* are just one in a row half the time, two in a row one-quarter of the time, three in a row one-eighth of the time, n in a row $1/2^n$ of the time.

3. Two sequences have half their bits equal and half not equal.

Generated sequences that follow these three rules are *pseudorandom sequences* (or codes). It is far easier to have the transmitter and receiver agree on a generating function than to have them agree beforehand on a truly random sequence long enough for a telephone call.[2]

A *maximal length linear shift register sequence* satisfies these three properties and is easy to generate. We generate a sequence of zero-one bits a_n from

$$a_n = c_1 a_{n-1} + a_2 a_{n-2} + \cdots + c_r a_{n-r} = \sum_{i=1}^{r} c_i a_{i-r} \tag{8.7}$$

The c_i are connections, and they are 1 when a_n is connected to a_{n-i} and 0 when there is no connection.

Multiplication is done the usual way:

$$0 \times 0 = 0$$

$$0 \times 1 = 0$$

$$1 \times 0 = 0 \tag{8.8}$$

$$1 \times 1 = 1$$

---

[2]A 10-minute cdmaOne telephone call is 750,000,000 chips long.

whereas addition is done *modulo 2:*

$$0 + 0 = 0$$

$$0 + 1 = 1$$

$$1 + 0 = 1$$

$$1 + 1 = 0$$

(8.9)

Mathematically speaking, these operations form an *algebraic field* with almost all the nice properties of the real numbers we use in regular algebra.

An example of a linear shift register sequence of four stages is to connect each bit to its four predecessors, $a_1 = a_2 = a_3 = a_4 = 1$. This means that $a_n = a_{n-1} + a_{n-2} + a_{n-3} + a_{n-4}$. Consider a starting seed of 0 0 0 1, and we get

$$0\,0\,0\,1\,1\,0\,0\,0\,1\,1\,0\,0\,0\,1\,1\,0\,0\,0\,1 \cdots$$

(8.10)

This linear shift register sequence repeats every 5 bits.

Another example of a linear shift register sequence of four stages is to connect each bit to its first and fourth predecessors, $c_1 = c_4 = 1$ and $c_2 = c_3 = 0$, so $a_n = a_{n-1} + a_{n-4}$ without the middle two terms. With the same starting seed of 0 0 0 1, we get

$$0\,0\,0\,1\,1\,1\,1\,0\,1\,0\,1\,1\,0\,0\,1\,0\,0\,0\,1 \cdots$$

(8.11)

Note that this linear shift register sequence repeats every 15 bits. This is the maximal sequence length for a four-stage sequence because there are 15 four-bit sequences available. The sixteenth four-bit sequence, 0 0 0 0, produces its own very dull sequence of all zeroes. This second example is therefore a maximal length linear four-stage shift register sequence, and its length is $2^4 - 1$.

In general, a linear shift register sequence with r stages repeats every $2^r - 1$ bits, and a maximal length sequence repeats no more often than that. For every r, there is at least one of maximum length, although it may be difficult to find.

A maximal length linear shift register sequence *almost* satisfies the three conditions at the beginning of this section.

1. There are 2^{r-1} ones and $2^{r-1} - 1$ zeroes in the $2^r - 1$ bit sequence. The fifty-fifty distribution of ones and zeroes is off by one part in 2^r. This is one part in a million for $r = 20$, a tiny deviation.

2. Run lengths of ones up to r in a row and run lengths of zeroes up to $(r - 1)$ in a row are in correct proportion, and there are no same-digit sequences longer than r. Thus the run lengths up to r are statistically overrepresented by one part in 2^r, again a tiny deviation.

3. Some mathematical algebra can be used to show that the difference between two offset sequences generated from the same r coefficients is another maximal length linear r-stage shift register sequence. Therefore, the difference between two offset sequences has half equal and half not equal within one part in 2^r.[3]

Using offset maximal length linear shift register sequences as PN codes in CDMA, an interfering signal looks like a noise contribution attenuated by the processing gain.

---

[3]The difference is the same as the sum using the two-element algebraic field in Eq. (8.9).

8.3.2 Orthogonal codes

Under the right circumstances, we can attenuate same-sector signals by more than the processing gain. We do this by selecting bit sequences specifically designed to cancel each other out. These are called *orthogonal sequences* or *orthogonal codes*.

Consider two signals A and B being sent on the same channel with two chips per bit. Let a_k and b_k be the A bits and B bits for time period k. If the two chips are $c_{k1} = a_k + b_k$ and $c_{k2} = a_k - b_k$, then the receiver can compute the sum $c_{k1} + c_{k2}$ for the A bits and the difference $c_{k1} - c_{k2}$ for the B-bits.[4] In each of these calculations, the cancellation of the other signal is not merely statistical; it is absolute. These two 2-bit codes agree exactly half the time and disagree exactly half the time.

The theme matrix of Hadamard-Walsh orthogonal codes is $\mathbf{H_2}$ in Eq. (8.12):

$$\mathbf{H_2} = \begin{array}{|cc|} \hline +1 & +1 \\ +1 & -1 \\ \hline \end{array} \tag{8.12}$$

Note that if we take any two rows of $\mathbf{H_2}$ and add their pairwise products, then we get zero.[5]

The sum of pairwise products

$$\vec{a} \cdot \vec{b} = \sum_{k=0} a_k b_k$$

is called the *inner product* of vectors $\vec{a}$ and $\vec{b}$. The inner product is also called the *dot product*. When two nonzero vectors $\vec{a}$ and $\vec{b}$ have an inner product of zero, they are said to be *orthogonal*.

We can take another step and build $\mathbf{H_4}$ from four copies of $\mathbf{H_2}$ in Eq. (8.13):

$$\mathbf{H_4} = \begin{array}{|cc|} \hline \mathbf{H_2} & \mathbf{H_2} \\ \mathbf{H_2} & -\mathbf{H_2} \\ \hline \end{array} \tag{8.13}$$

The four copies of $\mathbf{H_2}$ in $\mathbf{H_4}$ are the exact same structure as the four copies of $+1$ in $\mathbf{H_2}$. Filling in the $+1$ and -1 elements of $\mathbf{H_2}$ yields four orthogonal rows of $\mathbf{H_4}$ in Eq. (8.14):

$$\mathbf{H_4} = \begin{array}{|cc|cc|} \hline +1 & +1 & +1 & +1 \\ +1 & -1 & +1 & -1 \\ \hline +1 & +1 & -1 & -1 \\ +1 & -1 & -1 & +1 \\ \hline \end{array} \tag{8.14}$$

---

[4]This calculation gets $2a_k$ and $2b_k$. Since our receiver is using anything positive for $+1$ and anything negative for -1, the extra factor of 2 does not matter.

[5]Not a lot of choice here, but bigger matrices have the same property.

Any two of the four rows of $\mathbf{H_4}$ are orthogonal, so the entire $\mathbf{H_4}$ matrix is called an *orthogonal matrix*.

The next step builds $\mathbf{H_8}$ from four copies of $\mathbf{H_4}$ in Eq. (8.15):

$$\mathbf{H_8} = \begin{array}{|cc|} \hline \mathbf{H_4} & \mathbf{H_4} \\ \mathbf{H_4} & -\mathbf{H_4} \\ \hline \end{array} \qquad (8.15)$$

The resulting eight orthogonal codes of $\mathbf{H_8}$ (without the boxes) are shown in Eq. (8.16):

$$
\begin{array}{cccccccc}
+1 & +1 & +1 & +1 & +1 & +1 & +1 & +1 \\
+1 & -1 & +1 & -1 & +1 & -1 & +1 & -1 \\
+1 & +1 & -1 & -1 & +1 & +1 & -1 & -1 \\
+1 & -1 & -1 & +1 & +1 & -1 & -1 & +1 \\
+1 & +1 & +1 & +1 & -1 & -1 & -1 & -1 \\
+1 & -1 & +1 & -1 & -1 & +1 & -1 & +1 \\
+1 & +1 & -1 & -1 & -1 & -1 & +1 & +1 \\
+1 & -1 & -1 & +1 & -1 & +1 & +1 & -1 \\
\end{array}
\qquad (8.16)
$$

As in the previous two cases, $\mathbf{H_8}$ is an orthogonal matrix with any two of its eight rows having exactly four matches and four mismatches. Changing the sign of any combination of rows does not affect their orthogonal property.

Consider a transmitter sending eight CDMA signals using these eight rows as PN codes. Whatever the bits are, $+1$ or -1, the receiver sees *no net contribution* from any of the other seven codes. There is no room for a ninth signal in this scheme, but eight synchronized signals using $\mathbf{H_8}$ get a free ride with no interference among them.

We can mix orthogonal codes in powers of 2. We can use any row of Eq. (8.14) as an orthogonal spreading code at the expense of two rows of Eq. (8.16). For example, the first row of Eq. (8.14) used with a processing gain of 4 takes the first and fifth rows of Eq. (8.16), each of which otherwise could have been used with a spreading gain of 8. The first row of Eq. (8.16) sends $+1 + 1$ or $-1 -1$, and the fifth row sends $+1 - 1$ or $-1 + 1$. This technique is called *orthogonal variable spreading factor* (OVSF), and it is used in third-generation (3G) CDMA to mix varying data rates in the same orthogonal PN-code space.

There is another use for the $\mathbf{H_8}$ codes. These eight codes are maximally different. Consider 3 bits being sent on a noisy channel. The 3 bits can be used to select one of these eight 8-bit codes, and the receiver can translate the code back into the 3 bits. This is a processing gain of 8/3. The likelihood of selecting the wrong 3-bit combination is minimized by choosing codes so far apart in their eight-dimensional space.

This eight-to-three scheme, this maximal distance in eight-dimensional space scheme, is a noncoherent modulation scheme. To the demodulator, there is no difference between

$$+1 \quad +1 \quad -1 \quad -1 \quad -1 \quad -1 \quad +1 \quad +1 \quad \text{and} \quad -1 \quad -1 \quad +1 \quad +1 \quad +1 \quad +1 \quad -1 \quad -1$$

The eight-to-three processing gain strategy is less wonderful than eight noninterfering channels, but it does not require synchronized chips. The forward link is naturally synchronized with a common transmitter sending all the signals and can support eight simultaneous orthogonal codes in $\mathbf{H_8}$.[6] The reverse link is naturally unsynchronized, and it would be very difficult to stagger the user terminal transmit times for simultaneous *arrival* of the chips at the base station receiver. Thus the eight-to-three strategy may make more sense.

The logic described for $\mathbf{H_8}$ extends naturally to $\mathbf{H_{64}}$, where we have 64 orthogonal codes. This gives is 64 noninterfering channels forward and 64-to-6 processing gain in the reverse direction in cdmaOne systems.

8.4   Multipath and the Rake Filter

Life is simple when there is a single, direct path from transmitter to receiver. Everything that is not signal is noise. Radio is more complicated when there are multiple paths arriving at the receiver at multiple times, as shown in Fig. 8.3.

We can divide multipath into three rough categories, measured by the length of time between the arrival of the first signal and the last signal, which we call the *delay spread*. The delay spread can be roughly nanoseconds, microseconds, or milliseconds. In digital radio, these three work out to fading, chip confusion, and echo. It is not the actual *delay* that is the issue; rather, it is the difference in delay, which we call *delay spread,* that concerns us. The term *delay spread* is used both for the distribution of delay and for the time duration containing most of the distribution.

In the wireless telephone environment, echo is almost never a problem because radio paths rarely have such long delays. Echo is a major concern for communication links of 1000 km (600 miles) or longer. We'll let the long-haul microwave and satellite engineers worry about echo.

---

[6]In Sec. 8.8 we point out that this synchronization makes it economical to send a strong pilot signal that the receiver can use for coherent demodulation.

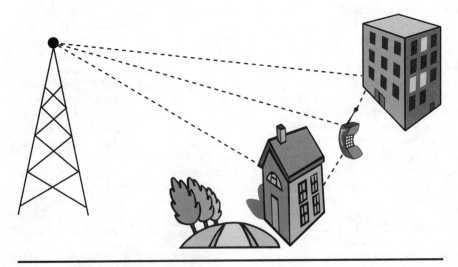

Figure 8.3 Multiple radio paths.

Figure 8.4 Delay spread graph.

The nanosecond differences in path length are on the order of meters, a few wavelengths of microwave radio. The effect of multiple paths this close together is Rayleigh fading, local dead spots about a wavelength in size. Thus there will be a bad call place, about human hand size, usually right where you want to make a call. Spread spectrum gives us significant protection against Rayleigh fading because the dead spots are different at different frequencies, and we are broadcasting our signal across a wide frequency band. When delay spread is in nanoseconds, our signal is unaffected by chip confusion. Even the highest chipping rates used in third-generation (3G) CDMA are no faster than 5 Mchips per second, and a 5 Mchip per second rate is 200 ns per chip, or 60 m. The cdmaOne CDMA standard uses chips five times this long.

It is the chip-sized delay differences that concern us in CDMA. Except for the Rayleigh fading, double images a microsecond apart are not a major issue in analog radio, but chip-sized delay spread in CDMA means that a receiver picks up two different chips at the exact same time.

Consider the delay spread graph shown in Fig. 8.4.[7] This is a plot of the radio power received as a function of time delay after it is transmitted. If we sent a single, instantaneous pulse of energy, then this would be the energy over time at the receiver. In Fig. 8.5 we show the same delay spread in the timeline of CDMA chips. If our CDMA receiver locks onto the signal at the moment of maximum power, then there is significant interference coming from the same signal at different times.

For a simplified example, suppose that the transmitter is sending a sequence of chips A, B, C, D, ... Consider three paths of equal strength, 2, 4, and 5 chip periods in length, as shown in Fig. 8.6. For example, when the E chip is transmitted, it arrives three separate times. The first time the E chip arrives at the same time as the second

---

[7]We would like to say, "Consider the *typical* delay spread graph," but radio paths vary so much that there really isn't a typical radio environment.

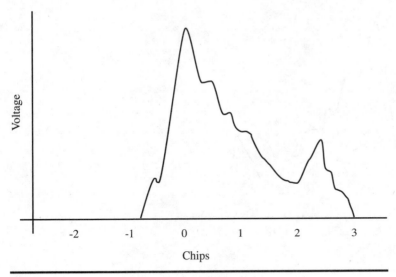

Figure 8.5 Delay spread graph with chips shown.

path of C and the third path of B, so each received chip is really the sum of three chips sent at three different times, $B + C + E$. The second time the E chip arrives, the three-chip total received is $D + E + G$. And the third time the E chip arrives, the three-chip total is $E + F + H$.

A single receiver synchronizes with one of the three paths and sees two-thirds noise no matter which of the three paths it chooses. Suppose that it picks the shortest path and looks for B, C, and D from $B + C + E$, $C + D + F$, and $D + E + G$. The other two paths are time-shifted so that the receiver treats them as uncorrelated noise; they may as well be other CDMA telephone calls. This comes across as a factor of 3 loss of S/N or E_b/N_0, a 5-dB loss.

The best solution is to add the three received signals in reverse order of their delay. We call this reverse-order summation a *rake filter* or, in more engineering-oriented dis-

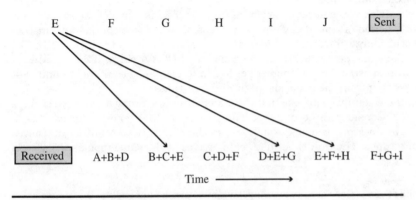

Figure 8.6 Transmit and received signals with three radio paths.

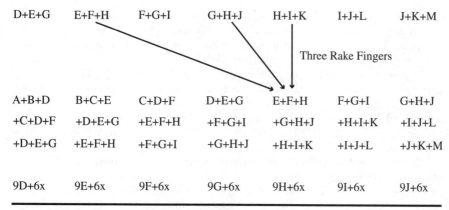

Figure 8.7 Rake filter reconstructing the signal.

cussion, an *adaptive matched filter*. Each of the representations of the signal in the rake filter is called a *finger* of the rake. It is a nice mathematical result that the time reverse of the delay spread is, in fact, the optimal receiver for the multipath radio channel.

In Fig. 8.7 we show a three-finger rake receiver recombining the three paths in Fig. 8.6. Consider the E chip leaving the transmitter. The three E-chip arrivals at the receiver are $B + C + E$, $D + E + G$, and $E + F + H$. The rake receiver combines these three triple-chip sums to get $(B + C + E) + (D + E + G) + (E + F + H)$, equal weight for equal signal paths in this example. This works out to $3E + B + C + D + G + F + H$.

The sum of the three received chips generates the desired chip with triple amplitude and six time sidelobe cross-terms. Triple amplitude is a ninefold power increase, so the total result of multipath and rake filter is nine signal units and six time-shifted units.[8] This is a 40 percent loss of S/N or E_b/N_0, a 2-dB loss. In this example, the rake filter recovers 3 dB of the damage caused by multipath.

In our example we knew that the delay was spread over three chips. How does an adaptive matched filter figure out the delay spread so that it can time-reverse it? The receiver measures the signal over many chips and does a mathematical maximization of signal strength. In the forward direction, the adaptive matched filter of the user terminal uses the known characteristics of the pilot signal. When the results from the adaptive filter match the pilot signal as closely as possible, they are also matching the data signal as closely as possible. In the reverse direction, however, there is no pilot. There is, however, an expected PN code in the signal. The adaptive matched filter at the base station receiver attempts to combine the signals from different time delays to reassemble the PN code. The maximum signal level of the PN code is the optimal receiver for the entire signal.

This is an interesting tradeoff in CDMA design. The data rate in chips per second and the frequency bandwidth in hertz are typically very close. For example, a 1.25-MHz

---

[8]Triple amplitude is nine times the power because power is proportional to the square of amplitude.

channel carries 1.2288 Mbps. If almost all the delay spread energy fits within one chip, then the rake filter and its time sidelobes have no damage to make up. However, if the delay spread energy is concentrated in a time that short, then the channel is narrowband in terms of its resistance to multipath Rayleigh fading. The result is that if delay spread is in the nanoseconds range, the signal suffers from Rayleigh fading but not from delay-spread interference. However, if the delay spread is in microsecond range, then Rayleigh fading is not a problem, but delay-spread interference is.

We have heard CDMA advocates tell us that one great advantage of CDMA is that "multipath is our friend." The idea seems to be that the adaptive matched filter creates a sort of processing gain of its own to discriminate between signal and interference. In actuality, the best possible rake filters achieve results almost as good as not having multipath at all.[9] As in the case of AMPS and GSM, more paths mean more power to combat receiver noise, but the advantage ends here for FDMA, TDMA, and CDMA. We see no particular advantage of having multipath present nor any particular advantage CDMA has in dealing with it.

We call the forward direction a *one-to-many broadcast environment*. In the one-to-many environment, the rake filter can reduce the damage of multipath, but each interferer is amplified by the same coherent addition plus time sidelobes as the signal.

The reverse direction is the opposite, a *many-to-one environment*. This gives us an advantage in the resolution of multipath because the interferers do not have the same delay spread as the signal.

The example of three equal paths in the reverse direction gives the same results for the signal path as in the forward direction. The receiver gets one triple-amplitude signal, nine power units, and six time sidelobes.

It is the interferers that have changed in this picture. In this example, the interferer also has three equal paths from the transmitter to the receiver, but those paths have spacing *different* from the signal paths. Thus, when the signal rake filter triples the interferer, each of its three paths appears three times for a total of nine occurrences of one power unit each. These nine power units of interference are a perfect match for the nine power units of the triple-amplitude signal. The time sidelobes of the signal are the only cost, just two-thirds of a single channel, no matter how many interferers there are.

The reverse multipath many-to-one picture is brighter than the forward one-to-many picture. With perfectly uncorrelated multipath and a perfect rake filter, the interference cost of each interferer is no worse than it is in the single-path case. The only destructive effects of multipath are the time sidelobes of the signal.

8.5   CDMA Power Control

All the analysis up to here has been based on setting CDMA power levels to be equal for each call, each PN code, with absolute certainty. Some calls are much closer to the serving antenna than others and have much higher radio path gains. We call this the

---

[9]The absence or presence of radio multipath is not a system design parameter but an attribute of nature to be dealt with. And once nature is unkind enough to create multipath, the rake filter is the best way to deal with it. We are objecting to the mythology that CDMA somehow turns this obstacle into a kind of virtue that makes the system *better*. The best we can hope for is to keep multipath from making the system worse.

near-far problem, that signal paths vary considerably, as much as 80 dB if one subscriber is 100 times closer to the base station than another. We use power control in CDMA to adjust transmitters so that the *received* power levels are balanced.

In AMPS and GSM, we are happy with power levels matching within a few decibels. This is nowhere near good enough for CDMA. A CDMA channel running at 85 percent of its full call capacity will overwhelm a call received at just 1 dB below optimal power.[10] Because CDMA requires each user to have a proscribed *fraction* of the signal pie, safety margins are not going to solve problems of inaccurate power control. Adding a 1.5-dB safety margin to the power-control algorithm only serves to reduce system capacity by 30 percent.[11] CDMA relies heavily on the power-control algorithm doing an effective job of managing power levels.

When the mobile telephone starts a call, it sets its transmit power based on measurements it makes of the forward channels. This initial choice of transmit power is done without base station participation, so we call it *open-loop power control.* Once the call radio link is established, within the two-way data stream is a flow of power-control messages going in both directions, power-up and power-down messages. This feedback-driven process is called *closed-loop power control.*

Sitting here, godlike, with full knowledge, we can solve the CDMA equations in Sec. 26.1 and determine optimal power levels. The live system has to try power levels, make measurements, and send corrections. This is a two-way street because the user terminal has to make its own measurements and send its own power instructions. In AMPS, the base station measures the received power of the mobile telephone, computes the radio path gain, and instructs its own radio and the mobile telephone what power levels to use. This kind of inference is not good enough for CDMA.

In CDMA, we determine the need for power-level adjustment by looking at the signal quality, as measured by the error rate. How do we measure signal quality in CDMA? The forward error-correction systems provide bit error rate (BER) estimates. The error correction figures out which bits are wrong and corrects them. It also can tell the power-control system how often it had to do this. If this number is higher than an optimal level, then the power-control system tells the transmitter to turn up the power. And conversely, when this is lower than the optimal level, it tells the transmitter to turn *down* the power. The radio link is therefore in a constant state of small corrections in its quest to maintain optimal power.

It may seem odd that we reduce power in order to increase the number of errors to an ideal level. However, in CDMA, a low error rate is an indicator that this signal is using more than its allotted percentage of the entire signal power. This call is clear, but at a cost of greater interference for all the other calls on the cell. Reducing the power on this one call will bring all the calls on the one cell closer to equal in received power.

Sometimes the raw bit stream has too many errors to resolve, but this happens when the signal is really lousy. When it does happen, there is no dispute that a power-up message is called for.

Another approach is to measure the E_b/N_0 directly rather than using the BER. For each bit, we take the total amplitude of all its chips—just add them all up. Then we do the PN-code arithmetic and recompute the total amplitude of all the chips. The first

---

[10]One decibel down is 80 percent.

[11]1.5 dB is 1.4 and −1.5 dB is 0.71, a 30 percent reduction.

sum is the estimate of signal plus interference, and the second sum estimates signal after processing gain. If the second divided by the first is low, then we increase transmitter power. When the ratio is high, we decrease power.

Both these techniques rely on statistical averaging for their success. Under normal conditions, they point in the right direction just over half the time. It might take a hundred measurements and messages to get power levels reliably within 1.0 dB of optimal. Whereas AMPS sends power-control messages every 5 seconds, cdmaOne sends almost 1000 power-control messages each second. Even at this high rate, a moving user terminal in a fading environment poses a power-control challenge.

Power control in CDMA uses all three methods. Open-loop power control is used to set initial power in the reverse direction, and closed-loop power control is used to maintain optimal levels. Closed-loop power control has an *inner loop* that measures the E_b/N_0 directly and sends power-change messages in 1-dB steps. There is an *outer loop* that measures the frame error rate (FER) and adjusts the target E_b/N_0 for the inner loop.[12]

There is an interesting tradeoff in power control. The energy efficiency of power control has a significant effect on total signal capacity. The power-control messages have to travel between the system and user terminal somehow, and this takes more power and costs CDMA capacity. We can send nearly perfect power-control messages by devoting a lot of power to power control. Or we can economize on power-control power, have a lot of power-control errors, and live with a system out of power balance. These are not the measurement errors in the last paragraph but rather the message errors in the power-control messages themselves. Again, large numbers average out the errors, but the power-control process becomes slower when the error rate is high.

If the measurement part is perfect—that is, if the receiver always makes the right power change determination—then a high rate of power-control messages is supposed to be more energy efficient. We could send a small number of large-change messages with enough energy per power-control bit to ensure a highly accurate power-control process. Or we could send a large number of small-change messages with less energy per bit. The power-control process would be less accurate bit by bit, but the larger number of attempts more than makes up the difference.

It is tempting to visualize a system with a thousand power-control messages per second maintaining exact power level and tracking the small boosts and fades of a telephone moving at 60 mi/h (100 km/h). The power control in CDMA does keep control of power, but this kind of accuracy is not part of the picture. There are significant capacity losses due to power-control errors in CDMA. If we found that power control were *not* a problem, then we would reduce the power allotment for power control and increase the capacity of the system in this way.

8.6 Varying Data Rates

The cdmaOne channel has voice activity detection (VAD) and can use lower data rates when there is less activity. The speech coder uses a full 9600 bits per second when required but can back off to 4800, 2400, or even 1200 bits per second when there is less

---

[12]All the 3G CDMA systems use this dual-loop fast power-control structure in both forward and reverse directions. cdmaOne uses it in the reverse direction and relies on its orthogonal codes and a slower power-control scheme in the forward direction.

voice activity. This is coordinated between the base station and the user terminal. Lower data rates require less radio power, which reduces overall interference levels and extends battery life.

In the forward direction, transmission of fewer bits per second is done by repeating the same bits over and over again at lower power, essentially increasing the processing gain by another factor of 2, 4, or 8. In the reverse direction, transmission of fewer bits per second allows the user terminal transmit only a fraction of the time. The base station does not expect transmission the other half, three-quarters, or seven-eighths of the time.

Why does cdmaOne user lower power in the forward direction and intermittent transmission in the reverse direction? In the absence of any other issues, we would prefer to have equal power over time with more spreading gain. The CDMA channel can carry more traffic that way. This is the primary factor in the forward direction, so we have the transmitter send repeating bits.

However, power control is more critical in the reverse direction because of the near-far problem, and the base station can make better radio measurements when the user is transmitting at a consistent power level rather than reducing its power level every time the VAD unit notices that the speaker has paused. The base station receiver measures each frame and makes a power-control determination, up or down, and this determination is much harder to calculate if the user terminal transmit power levels are changing because of voice activity. The reverse power-control measurement issue overrides the preference for smooth power levels over time. As a result, in the reverse direction, the best option is to have the user terminal either transmit at its full-bit-rate power level or not transmit at all. In this way, overall transmit power is reduced from the user terminal without interfering with power-control measurements.

8.7   Soft and Softer Handoff

CDMA uses a different kind of handoff, a *soft handoff,* where a user terminal is served by two base station antennas *at the same time.* The need for extremely accurate power control in CDMA requires a more sophisticated coordination among serving base stations than the break-and-make handoffs used in conventional reuse cellular systems.

As we will see in Sec. 26.1, it is important that a CDMA call be served by the right cell sector, the one with the highest path gain. When CDMA calls are served by the wrong cell sector, even momentarily, the overall capacity of the system is reduced dramatically. Radio signal paths vary enough from place to place that a mobile telephone near a cell boundary might change its choice of the right cell dozens of times per minute. Rather than execute dozens of handoffs (a ping-pong handoff situation) when radio signal paths are close, CDMA shares the call with *both* cell sectors.

Unlike AMPS and GSM, cdmaOne CDMA systems let the mobile telephone determine when a handoff should occur. When the mobile telephone detects a pilot from a new antenna face B, it informs its primary server A, and a link is established between A and B. The new antenna face begins transmitting, so A and B are both sending the same data. The base stations and mobile telephone coordinate their PN codes to make this work.

Let us take a moment and examine how a two-to-one link can work. Consider two base station antennas A and B serving a single mobile telephone. In the forward direction we have two separate copies of the same message coming from A and B. The mobile telephone resolves the two signals using the rake filter that it uses to resolve multiple radio paths, as described above in Sec. 8.4. When the mobile telephone evaluates the signal quality for power control, it sends power-up and power-down messages that both *A* and *B* obey.

In the reverse direction, we have two receivers picking up the same message and coordinating their signals. When the two receivers are two sectors at the same base station, the rake filter is used to combine them optimally. When the two receivers *A* and *B* are at different base stations, one of them remains the *primary base station,* while the others are called *secondary base stations.*

The signals from *A* and *B* are the same except, perhaps, for power control. When *A* and *B* send conflicting power-control messages, this means that the signal is good enough for one cell and not good enough for the other cell. The idea of soft handoff is to maintain just enough signal strength to keep one base station happy, so the mobile telephone powers *down* when it gets conflicting power-control messages.

When one of the two signals is no longer helping the call, that signal is dropped. If the signal being dropped is the primary base station, then a secondary base station becomes the primary base station for the call.

This concept of a shared radio channel is possible because CDMA uses the same radio carrier in the same frequency range for all cell sectors.

When *A* and *B* are two separate base stations, we have a *soft handoff* until one of them is dropped. When *A* and *B* are two sectors of the same cell, we have a *softer handoff,* as shown in Fig. 8.8.

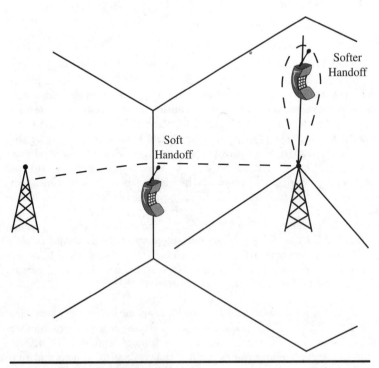

Figure 8.8 Soft and softer handoff.

The CDMA channel allows more than two signals in the soft and softer handoff process. cdmaOne supports up to six simultaneous signals.[13] Thus we have three-way soft handoffs, four-way soft handoffs, and various kinds of soft-softer handoffs.

Not all CDMA handoffs are soft handoffs. Sometimes the pilots of the two cells are too different for the mobile telephone to synchronize them. Some vendors permit soft handoffs from one MSC to another, but the ANSI-41 handoff standards in Sec. 15.1 do not require it. cdmaOne supports having multiple CDMA carrier frequencies to make the best use of more than 1.25 MHz of radio bandwidth. Handoffs under any of these conditions can be the old-fashioned AMPS-or-GSM-style handoffs that we call *hard* handoffs in CDMA. Handoff support requires signal coordination within the cellular network, as described in Sec. 11.4.

8.8   The Forward (Downlink) Direction

The forward link from base station to user terminal has one base station sending a common radio signal to many user terminals, a one-to-many radio environment. This offers the opportunity to share broadcast resources and to synchronize signals.

Having a one-to-many environment allows for a strong pilot signal for each receiver to use for coherent modulation and rake filter alignment. Carrier phase and multipath change rapidly for mobile and portable telephones and slowly for fixed wireless local loop.

The synchronization of all the signals allows orthogonal codes to be used to minimize interference. So long as the delay spread is not much longer than the chip duration, the same-sector interference can be reduced significantly.

On the negative side of the forward link, the common path for signal and interference means that the rake filter does as good a job coherently amplifying interferers as it does coherently amplifying the signal. The optimal receiver is matched to the delay spread of the signal. If the delay spread of the interferer is the same as the signal, then it is going to give each interferer the same advantage.

8.9   The Reverse (Uplink) Direction

The reverse direction of CDMA from user terminal to base station has many user terminals sending signals to a single receiving antenna at the base station, a many-to-one radio environment. While synchronization is nearly impossible, each receiver rake filter can take advantage of the different radio paths to distinguish its own signal from the other calls.

A reverse pilot has no economy of scale, so cdmaOne does not use one at all, and it uses a noncoherent modulation scheme. The lack of synchronization makes the virtual cancellation of interference using orthogonal codes not practical. However, the same orthogonal codes can be used as a noncoherent maximal-distance spreader.

On the positive side, each path has its own multipath characteristics, its own delay-spread graph. A rake receiver optimized for one path is not going to give interferers the same advantage. This relative advantage is only enough to combat the negative effects of multipath, but this is better than the forward direction.

---

[13]We wouldn't want to sit down and do the analysis of the power-control and rake-filter components for six simultaneous signals from four base stations.

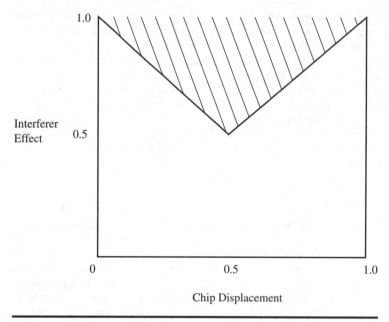

Figure 8.9 The advantage of nonsynchronized chips.

There is another advantage from the lack of synchronization in the reverse direction. The statistical advantage of processing gain is that nonorthogonal interfering PN codes look like noise. In the synchronized case, where the chips line up perfectly, the interfering PN code contributes some virtual noise level. Consider the case where the interferer is offset exactly half a chip. When two consecutive interfering chips are the same, the contribution is no less than the synchronized case, but when consecutive interfering chips are *different,* the contribution goes to zero. A perfectly staggered interferer loses half its impact, 3 dB, from this statistical effect.

Let us assume that this effect is linear in the phase offset so that a quarter-chip offset gives us half the benefit of a half-chip offset. Then the average interference attenuation of chip offset is one-quarter of the interference, 1.2 dB, as shown in Fig. 8.9.

If there is a 1.2-dB opportunity with nonsynchronized chips, then why not do it in the forward direction as well? Because the advantage of a common pilot and orthogonal codes far outweighs this 1.2-dB advantage.

8.10 Conclusion

The CDMA technology applies extremely sophisticated mathematics to electronic chipsets to obtain the advantages of spread-spectrum technologies. The engineers who worked to develop the original AMPS reasonably could be astounded by the sophisticated rake filters in CDMA user terminals, the encoding algorithms in Qualcomm chipsets, the complexity and precision of closed-loop power control, and the cooperative management of soft and softer handoffs.

Today's CDMA may seem to be the culmination of the integration of telephony, radio engineering, and digital technology. In some ways it is. In other ways, all that has been done merely lays the groundwork for what is to come. In the next two chapters we will take a closer look at how cdmaOne is currently deployed. After that, we will look at the new 3G/UMTS standards and their effort to increase vastly the data-carrying capacity of cellular systems.

Standards for Cellular Systems

This short part of the book, "Standards for Cellular Systems," addresses the crucial topic of standards. In radio and telephony, standards define interfaces, allowing different manufacturers to produce equipment and components that will work together on a single system.

In Chapter 9, "General Cellular Standards," we provide a background about the function and nature of standards and how standards change technology. We also describe all the standards that are specific to cellular telephony or have had a major effect on its development.

Standards are not purely technical; they define customer features as well as bit rates. We explore some of the deeper concepts underlying the standards that define cellular telephony, including the notions of vertical services and personal services. We conclude Chapter 9 with a discussion of the first three generations of cellular telephony, covering the evolution of the standards of today's systems.

In Chapter 10, "Worldwide CDMA Standards," we turn to today's worldwide code division multiple access (CDMA) standards for second-generation (2G) and third-generation (3G) systems. 3G may be exciting, but 2G is important. 2G systems still support about 900 million cellular users in 2002, so we provide in-depth technical details of cdmaOne, previously called IS-95. Moving to 3G, we describe the three CDMA systems that have been approved as 3G standards and are beginning to be deployed. cdma2000 adds voice capacity and high-speed user data support when overlaid on cdmaOne systems. Wideband CDMA (W-CDMA) will be used to upgrade European Global System for Mobility (GSM) time division multiple access (TDMA) networks and will compete with cdma2000 worldwide. And the newest member of the 3G CDMA family, time division synchronized CDMA (TD-SCDMA), due to be deployed in China, incorporates the most advanced develpments in several areas of cellular telephony and radio to create flexible high-capacity systems. We provide in-depth technical specifications for all these standards.

Standards are powerful engineering tools created through a process of human communication that is often politicized. Understanding the strengths and weaknesses of the standards underlying the systems we design and the equipment we install helps us do a better job of creating solutions that design around the weak spots and resolving problems on the systems we support.

General Cellular Standards

The whole notion of standards is rarely explained and can be quite confusing. This chapter offers the what, why, and how of standards. It also explains why standards run into trouble and sometimes fail to solve problems or even end up creating new problems. Engineers (those who write standards and those who work with them), executives, and customers all have different opinions about standards, sometimes very strong ones. However, once we move past the opinions and understand standards properly, they can be a valuable tool for specifying and developing high-quality systems and optimizing capacity. When we know the weak points and strong points of a standard, we know where to pay the most attention in planning new systems, and we know where to look when we need to solve problems.

9.1 The Role of Standards

Without standards, communication would be inherently unreliable. Prior to the standardization of English spelling, it was very difficult for a reader to be certain what word an author meant. We have very few signatures belonging to William Shakespeare, but we know he spelled his name three different ways. Worse, we are not sure what he wanted Hamlet to say at certain key moments. Was Hamlet talking about "this too, too solid flesh," as most scholars think, or was it sullied flesh or perhaps sallied (attacked) flesh? If Ben Johnson's dictionary had come out before Hamlet, we might be certain. But reading Shakespeare might be a lot less fun, too.

When we want to lay out instructions for how things should be done, we create a standard. Standards are rules, and people have odd reactions to rules. People who want to make sure that things work tend to like standards. People who like to have fun, do things their own way, or do things the best way they can do not like standards. The world of wireless engineers contains a lot of people with each of these attitudes.

In practice, however, standards are essential for radio and telephony. The reason for this is that standards specify what the two components on either side of an interface must do to communicate with one another. In defining what each component will send, we define what each component must be able to receive. If we then define the states of each component, the other relevant inputs, and the appropriate response of each component, we have defined how each device operates to fulfill its role in the system.

9.1.1 Types of standards

Standards are not laws. It is possible to develop systems without standards or with purely proprietary standards and succeed in business. One of the world's most ubiquitous and successful cases of standardization, the system of red, yellow, and green traffic lights, was never set as a standard. It was deployed, it worked, and it was accepted. However, given the cost of developing new technology, certification of a standard—and certification of a particular equipment model as meeting that standard—has a tremendous amount of power in the marketplace.

For practical purposes, we can put standards into six different categories. Here they are, with the most official ones first:

- *International standards* are approved by an official body recognized by multiple governments and many companies. For the cellular industry, the most important international standards organization is the International Telecommunications Union (ITU).

- *National standards* are approved by a quasi-governmental body such as the American National Standards Institute (ANSI) in the United States. Adherence to these standards also may be regulated by an official government body, such as the U.S. Federal Communications Commission (FCC).

- *Industry standards* are managed by industry groups, such as the Institute of Electrical and Electronics Engineers (IEEE). These bodies often develop standards and propose them to the more official standards organizations. Some industry groups are more technical, developing standards to solve problems, and others are more political or more business-oriented, promoting particular technologies as they develop standards for them.

- *Proprietary standards* are developed and maintained by a company who may license use of the standard to other companies so that they can develop compatible systems.

- *Open standards* arise when an organization makes its proprietary standard available for general use at no charge. UNIX is an open standard.

- *De facto standards* are accepted ways of doing things that are prevalent in society. The color of traffic lights that we discussed earlier is an example.

Of course, a single standard may operate at many of these levels and is likely to change status over time. For code division multiple access (CDMA), many of the standards were proposed by Qualcomm, which holds patents on many chip designs that implement CDMA standards. However, Qualcomm did not keep the CDMA standard proprietary. Only the patents on the chips are proprietary. ANSI, with the assistance of IEEE and other organizations, took the CDMA standard for cellular telephony through multiple revisions as Interim Standard 95 (IS-95) and then approved the standard for North American cdmaOne. The ITU defined the third generation (3G) of mobile telephony standards and requested proposals for standards to meet its requirements. The CDMA Development Group (CDG), an industry group more on the political and promotional side of things than IEEE is, proposed cdma2000, a standard based on cdmaOne but meeting the requirements of ITU's 3G. There is much more than this to the story of CDMA for 3G, and we will discuss it again later in this chapter.

Even when we know what the official status of a standard is, there may be confusion about what is being standardized. This is discussed in the next subsection.

9.1.2 Standards and interfaces

Standards can be confusing. One of the best ways to get clear about them is to ask the question, What interface is being standardized? We can think of an interface as the defined communication between two systems. Expanding on this idea, we can picture an interface as a gap between two systems that is being crossed and the standard as the agreed-on way to get information across the gap. At a detail level, each single standard specifies one interface. However, we often use the term *standard* to mean a set of standards. For example, we talk about the Global System for Mobility (GSM) standard. Actually, the set of standards in GSM defines many interfaces: the radio-air interface from base station to cell phone; the interface between the base station and the mobile switching center (MSC), the standards for power control, and others.

In contrast, the CDMA standards used in cellular telephony specify only the air interface. They define how voice signals, user data, and control signals will operate between the base station and the user terminal. The backhaul network from base station to MSC is managed by independent standards, such as ANSI-41.

For complicated systems, it is valuable to break a single interface standard down even further in several ways.

The best way to subdivide a standard sometimes depends on the nature of the technology. For example, in CDMA, we have subcomponents, including voice coding, the forward link, the reverse link, and power control, for each of the links. A section of any given CDMA standard (cdmaOne, cdma2000, W-CDMA, or TD-SCDMA) will be devoted to each of these subcomponents.

This makes standards flexible and easy to improve. For example, a change to the reverse link would not necessarily affect the forward link. As standards evolve, new services can be supported through the standardizing of new components. In the 3G initiative, the addition of standards that support high-speed data transmission is a major area of growth and change.

Another way of dividing a standard is to divide it into layers. The most common method of layering telecommunications and networking systems is the Open System Interconnect (OSI) model. The OSI model is, in a sense, a standard for creating standards. It allows the possibility of having parts of an interface specification independent from one another. For example, the Internet uses the Transmission Control Protocol (TCP) over the Internet Protocol (IP). Voice over IP (VoIP) systems transmit voice over the same Internet Protocol but do not use TCP.

9.1.3 The difference between standards and specifications

A standard is not, and should not be, a complete specification. There are two important differences between standards and specifications. First of all, a *standard* defines an interface, whereas a *specification* defines a component (also called as a *device*). Second, standards contain less detail than specifications.

Any component can be viewed as being surrounded by standards, each of which defines its interface to one other system component. The entire set of standards for a component's interface defines all the interfaces it must support to function well within the system. As a result, a component functions well within the system when it meets the standards for every interface. In such a case, the interface standards for all interfaces of a component define the functional standard for the entire component.

The challenge design engineers face is to understand the standard, to ensure that their understanding is correct, and then creatively to specify and to design equipment that meets or exceeds the standard in the best way. Problems arise when different engineers interpret standards differently. Then components created by different companies do not work together.

A full specification is far more detailed than a standard. Even for a given interface, standards answer some questions and define the remaining ones. One view of this is that a good standard defines what each component will do at an interface, but the question of how the component will do it is left to the design specification.

In addition, some devices do not have all interfaces defined by standards. A cellular telephone is a good example. The air interface is defined, but the human interface is not. Each vendor is free to design keypads, screen visuals, voice recognition, and bells and whistles for the user without following any particular standard.

As written here, it seems that there is a linear progression from standard to specification to design to production to deployment. This picture is far from reality. Sometimes a system is in production and being sold long before the formal standard is established, as was the case with IS-95, which later became cdmaOne. Even when the standard comes first, the process is iterative rather than linear. Standards themselves undergo revision. Specifications have both logical and physical levels, as discussed in Sec. 14.4. Specifications are verified to standards and, when prototypes are developed, they are tested for conformance to standards. The processes of testing by a manufacturer, standards organizations, and independent testing laboratories is too complex to go into in this book. However, it is important to note simply that any device may fail to pass a test and need rework.

Also, unfortunately, mistakes can be made at any step in this process, resulting in equipment that does not work or that works with lower capacity and quality than the standard seemed to imply it could.

9.1.4   The benefits of standards

When standards work well, components made by multiple vendors can be designed for each side of the interface, and all will work together on the same system. This enhances both competition and innovation. When a standard is made public, any company can produce equipment and seek to have it certified by the standards institution. With more companies trying to make similar equipment, prices are driven down. Also, if a company conceives of and develops equipment with new features, it will work on the system as long as it meets the standard at all interfaces. As a result, we have cellular phone wrist watches. They work on the network by meeting the air interface standard, and they work for the user who likes them because they are unique in an area of design that is not standardized, their shape.

Standards committees have been known to solve technical problems, particularly by bringing together experts with different specialties to create innovative solutions. Also, the job of specifying an entire system is huge. The work of standards committees, which often takes years, can be a first step, paving the way for accelerated resolution of technical issues within each company and thus for accelerated deployment.

9.1.5   Standards and politics

Standards also have a political side, much of which is not often seen as beneficial. In actuality, even this political dimension has neutral, problematic, and beneficial aspects.

A neutral aspect of the politics of standards important to our understanding of the world of CDMA is the way in which standards typically develop differently in North America and in Europe. In North America, standards are often proposed by one or more companies and then developed by a cooperative effort of businesses. Standards often first appear either in proposals from companies to relatively unbiased industry groups, such as the IEEE, or from industry consortia with an unabashed agenda, such as the CDG. From these organizations, standards often move up the hierarchy of officialdom to quasi-governmental or international organizations such as ANSI and ITU.

In Europe, standards are developed to simplify industrial and economic development throughout the European Union. Consortia have an international flavor. In fact, some might argue that creating, arbitrating, and managing standards is the European Union's biggest job. In any case, the ITU is the body based in Europe that is most relevant to cellular telephony. The ITU has two groups relevant to our work, the telephony group (ITU-T) and the radio group (ITU-R). Both groups create and certify standards relevant to cellular networks. Because the ITU is worldwide rather than European, standards often move from ANSI to ITU.

Another dimension of the neutral politics of standards for wireless telecommunications is the relationship between standards and government regulations. The largest governmental issue related to wireless is the allocation of specific frequencies. Available bandwidth is considered a limited natural resource and is managed by governments. They allocate particular frequencies in particular geographic regions for particular purposes, and the space fills up pretty quickly. This makes manufacturing of equipment for international sales difficult because the equipment must work in different frequencies for different countries. Fortunately, most countries follow either the U.S. pattern of spectrum allocation or the European pattern of spectrum allocation. The choice of frequencies of operation has been a major issue in the development of ITU's 3G standard. Initially it was hoped that governments might reallocate bandwidth or allocate unused bandwidth, allowing a single set of frequencies to be used for all terrestrial wireless communications worldwide. This did not happen, and the European and American bandwidths for their 2G systems (GSM and PCS) got carried over into 3G. In addition, Japan has allocated frequencies on its own unique program and needs to coordinate 3G deployment with its existing 2G Personal Digital Cellular (PDC) system.

When the 3G standards committee of the ITU failed to bring the world into one harmonious standard, it approved five different standards, saying that each of them met the requirements for certification as 3G. Each standard, of course, was proposed and developed by an industry group or consortium. Since ITU's decision, each of them has been developing and promoting its own standard in competition with the others, with billions of dollars at stake. This has led to significant clouding of technical issues as each group sometimes attacks the others and sometimes obscures real issues in complex verbiage to reduce its vulnerability to attack. We will discuss this unfortunate situation further in Sec. 9.2.3.

However, sometimes even the political side of standards clearly leads to good results. If we may illustrate with an example from the personal computer (PC) industry, consider the industry standard architecture (ISA) standard. When IBM developed the IBM PC in the early 1980s, it broke ground by publicizing the standard instead of keeping it proprietary and requiring expensive licenses. IBM did this because it was not sure that PCs would succeed. IBM thought that making the standard open would encourage innovation, and it did. Many small vendors invented PC hardware and software and launched the PC revolution.

However, this is not the end of the story. A few years later, IBM was upset about having to share such a rich pie with companies such as Compaq. (Perhaps IBM did not realize that when you open one side of a standard, you open the other side as well. IBM expected adapter cards, but not clones.) IBM tried to close off the PC clone market by coming out with a new, better PS/2 standard, which it kept proprietary.

The PC industry reacted first by getting together and declaring the original PC/AT standard the ISA. They could now produce and test equipment for compatibility and keep on selling even though IBM had left the fold it created. IBM went on to compete with the PS/2 interface by creating the extended ISA (EISA) standard. IBM's success kept innovation going, and prices dropping, in the face of a monopolistic effort to restrict technology.

The final chapter of this story is relevant to the cellular industry. The EISA standard outperformed the PS/2 standard, but it was less reliable. EISA told vendors what to build but not how to build it, and some of them got it wrong. It was all too easy to buy an EISA PC-compatible computer and an EISA card only to find that they did not work together.

A competitor named Intel joined the fray. Intel introduced the peripheral component interconnect (PCI) standard and provided computer chips for PCI controllers on the motherboard and PCI adapter cards. Cards from different manufacturers talked to motherboards from several other manufacturers, but at the actual interface, an Intel chip was talking to another Intel chip. The result was a highly reliable standard that still allowed for innovation. It was a bit more expensive, and you can guess where the money went, but it was worth the price because of its reliability.

Why is this story relevant to a book on CDMA? Many of the patents on CDMA technology were developed by Qualcomm. Qualcomm proposed the first CDMA standard for cellular telephony, which was accepted as Interim Standard 95 (IS-95) and then became cdmaOne. Qualcomm takes an approach similar to Intel. Qualcomm produces chips and licenses chip technologies that allow other vendors to create and sell reliable, innovative equipment.

Qualcomm's role in developing CDMA technology has been outstanding, but it has created a strong polarization within the industry. Some companies may choose GSM over CDMA even though CDMA is technologically superior because they do not want to pay a license fee to Qualcomm.

Until September 24, 2001, every CDMA telephone call ever made went through a cellular base station with a Qualcomm chipset or a chipset requiring a license from Qualcomm for use of its CDMA chipset patents. On that date, an innovative piece of equipment, a software-based radio base station developed by Advanced Communications Technologies, Inc., established a call to a standard cdmaOne cellular phone. If this technology succeeds, then it could create software-based radio base stations that cost perhaps 25 percent less than standard base station radio equipment and, very important, can be upgraded by software releases instead of requiring chip replacements.

Even though Qualcomm holds and licenses the patents currently required to create chipsets for CDMA, the international CDMA standard is independent of Qualcomm. This is an example of how standards support innovation and increase competition.

9.1.6   The role of standards organizations

Standards organizations are subject to a great deal of political lobbying by businesses with billions of dollars at stake. Some of the organizations, such as the ITU, ANSI, and

IEEE, seek to be unbiased. Others seek to promote one technology aggressively, mounting publicity campaigns and seeking to influence governments.

At the same time, standards committees are composed largely of engineers and technical experts from the industry. Even when there is no effort at pressuring the committee, each company's level of interest in the committee and its decisions has a significant effect on the quality of the committee. If a company feels that the problem the committee is addressing is real and important, it is likely to send its best engineers to solve the problem. If the solutions are sound, the company is likely to use and promote the standard. Conversely, if the issues seem trivial or ill-defined or so politicized that a good outcome is unlikely, then fewer expert engineers are sent, and the company's best experts stay at the office designing systems. This sometimes can lead to the unfortunate result that a company's best engineers are designing systems to standards developed by less talented people.

Standards organizations have a number of roles in relation to the development of new technology. A highly successful technology, such as Qualcomm's CDMA, can be deployed before it is accepted as an official standard. CDMA was an interim standard for many years even after it was supporting tens of millions of subscribers. In such cases, the market will draw vendors to produce products that meet the standard before—perhaps long before—the standard is final.

On the other hand, sometimes the work of standards committees can precede production and deployment of equipment and systems. This can happen in two ways. In one case, international, governmental, or independent industry standards committees, while solving one problem, look ahead to define and solve another problem. This is what happened in the evolution of 3G cellular standards. The ITU itself, having established 2G standards, saw the need for a single 3G standard for cellular voice and data, created a committee, and called for proposals. We'll have more to say about what happened to the 3G committee later in this chapter.

The second way in which committees precede standards is the case where an interest group representing a new industry or part of an existing industry is formed. These not-for-profit groups promote the standard, but they are biased toward the standard, and they expend effort marketing the ideas as well as defining them. This can lead to difficulties, as we shall see when we discuss the evolution and definition of Personal Communications Services (PCS).

9.1.7 The effects of standards on technology

In the case of a highly successful, rapidly accepted technology such as cdmaOne, success comes first, and standards are almost an afterthought. However, they are a valuable afterthought. They can solve problems, reduce costs, increase reliability through support of independent testing, and increase the life of a technology.

In the case of a good idea that does not get off the ground quickly, standards can elevate the idea to a long-term success. The imprimatur of a standards organization such as ITU or ANSI can cause both manufacturers and major corporate customers to invest in a technology when they have been hesitating, afraid that the technology might not be adopted. This can create a positive cycle of investment and interest, driving the technology to success.

On the other hand, a poorly designed standard or a standard with technical flaws can indemnify problems, creating long-term difficulties in the deployment of a technology. This increases costs and can reduce capacity or reliability.

The last major effect of standards is the case where two or more competing standards are approved, and a long-term battle in the marketplace ensues. This is the case with GSM (TDMA) and CDMA in the evolution toward 3G. The result is a mixed blessing. Each technology is better off because it is standardized. The European cellular system depends on the GSM standard, and the North American system depends on the cdmaOne standard and ANSI-41. However, the existing standards, as well as the existing installed base of technology, create a basis for long-term competition and conflict. The economic competition may not be a bad thing, but the conflict surrounding it often creates problems. The standards process works well when everyone agrees on terminology, concepts, and the definition of the issue being addressed. This is hard to achieve when the stakes are so high that people begin to attack one another's views and opinions. It can be very hard to find unbiased, intelligent explanations, and there seem to be even fewer unbiased, intelligent projections of what is likely to happen. When a standards process becomes highly politicized, the results are often costly technical problems and even costlier delays in deployment.

This is not to say that standards organizations should back away from politicized and controversial issues. Just the opposite. At their best, standards organizations can bring a rational process into defining a problem and developing solutions.

There are some smaller, but significant, effects of standards on technology in particular cases. Sometimes a small part of a standard becomes a significant driver in the marketplace. The short message service (SMS) is a case in point. It was a small part of the GSM system that allowed short text-paging-style user messages to travel over the cellular signal channel. It has become a major industry in its own right and is now spawning successors that can carry images and other attachments. Other times an effective standards committee quickly and quietly resolves a technical problem or a political conflict, allowing a technology to move forward more rapidly or more smoothly.

With all this said, let us take a look at some of the standards that have shaped the cellular industry.

9.2   Generations of Cellular Standards

The original North American Advanced Mobile Service (AMPS) systems were the proprietary AT&T and Motorola networks described in Chap. 6. From that point forward, however, cellular technology developed through industry standards, and there has been an effort to organize the standards into *generations,* where each generation has multiple standards, but the standards provide roughly equivalent services.

It is important to note that the notion of generations postdates the actual implementations of the systems and standards. Also, this introduction to the standards will be quite short because each one is described fully in other chapters.

9.2.1   First generation

The *first generation* (1G) label was applied retroactively to cellular networks with an analog air interface, primarily AMPS and narrowband AMPS (N-AMPS), which were described in Chap. 6. There was also a Nordic mobile telephone system developed by Ericsson and Nokia and deployed in two versions in Nordic countries and one version in France. Analog systems similar to or based on these technologies were deployed in other nations, including Japan, England, and South Africa. All first-generation systems used frequency division multiple access (FDMA) because other multiple access technologies require digital transmission.

9.2.2   Second generation (2G)

Digital transmission technologies were called *second generation* (2G). Services grew rapidly in the 1990s and also consolidated. Europe standardized on GSM using time division multiple access (TDMA), and many other nations adopted it as well. In North America, 1900-MHz PCS services grew, and local calling plans became first regional and then national. The North American services were a competitive mix of TDMA and CDMA. GSM TDMA was described in Chap. 7. cdmaOne, the final version of 2G CDMA in North America, is described in Sec. 10.1.

9.2.2.1   Worldwide usage profile.
The current CDMA technologies for cellular telephony got a later start than GSM. Throughout the 1990s, both services grew exponentially, but CDMA had only about 15 percent of the market share. In 2002 we have seen several industry reports and news articles suggesting that the European market for cellular phones is nearing saturation; that is, just about everyone in Europe who wants a cellular phone has one. Meanwhile, China has announced plans for rapid deployment of a CDMA-based cellular technology, time division synchronized CDMA (TD-SCDMA), in a market of 1,000,000,000 people.

As a result, market growth and market competition are likely to change. Rather than competing to enter a new market in the first world, cellular providers are competing to replace one another in the first world and deploy or expand in the second and third worlds.

Nations that have adopted digital cellular networks typically follow either the North American CDMA model or the European GSM model. Central and South America largely adopted CDMA, although some nations, including Chile and Brazil, have both CDMA and GSM. A few small countries with closer ties to Europe, such as the French West Indies, have only GSM. GSM is also the only 2G service throughout all of Europe, North Africa, the Mediterranean, and the Middle East all the way to Pakistan. The only exceptions to this are Israel and the former Soviet Republic of Kazakhstan, which use CDMA. GSM's dominance extends all the way south in Africa, although some central and southern African nations, including the Congo, Angola, Zambia, and South Africa, also have CDMA. Russia has a mix of CDMA, American TDMA, and GSM, whereas China has CDMA and GSM. India and Southeast Asia have a mix of both GSM and CDMA. The island nations of Indonesia, Polynesia, Taiwan, and New Zealand have a polyglot of standards with CDMA, American TDMA, and GSM, whereas Papua New Guinea, Fiji, and New Caledonia are solidly in the GSM camp. Australia has both GSM and CDMA. South Korea and Japan both have CDMA, but Japan also has its own standard, PDC.

Second-generation cellular services have been very successful and have grown rapidly. As of June 1999, there were 187 million GSM users worldwide, 42 million PDC users, 34 million CDMA users, and 25 million users of non-GSM TDMA systems. By summer 2001, GSM had 700 million users, cdmaOne had 100 million, and total worldwide cellular use was estimated at an American billion (a British thousand million) subscribers.

9.2.2.2   International intercompatibility.
There was some effort to create compatible systems internationally. The United States and Canada agreed to use the same plan for telephony (both landline and wireless), creating a single North American telephony system north of the Rio Grande. With the advent of PCS, the North American cellular network included personal communications numbers (PCNs), allowing a person to keep a mobile telephone number even when changing service provider or changing equipment.

GSM took a different, and very clever, approach to personal number assignment and made an effort to achieve truly worldwide interoperability. All the GSM user information is on a smart card that you can plug into any GSM phone equipment, making it your personal user terminal, even if you rent the equipment.

GSM services operate on five different frequency bands, GSM900, GSM1800, GSM1900, and more recently, GSM850, and GSM450. Each model of phone may be single band, dual band, or multiband. A phone that operates on both GSM900 (the primary European carrier) and also on either GSM1800 or GSM1900 (the North American frequencies) would offer seamless interoperability wherever there is coverage in Europe and North America without changing phones.[1]

However, 2G did not offer seamless global service. In part, this was due to a lack of demand. International travelers may be high-profile as well as high-profit customers, but the bread-and-butter customers that grew the 2G client base were the increasingly mobile citizens of each continent, not the globe-trotting executives.

This failure caused the ITU to propose an initiative for creating a single global standard. It was called the *3G initiative,* and it began in the late 1990s. In addition to unifying the world on one standard, the 3G initiative wanted to add high-speed data communications support to cellular networks. This would allow subscribers to check e-mail, browse the Web, log onto the corporate database, and generally be able to work all the time until commercial services are added, and we can work and play online, plugged in without wires wherever we are.

9.2.2.3 Additional data services (2.5G). It was reasonable to assume that it would take several years to define 3G standards and deploy 3G networks. As a result, the smaller goal of adding medium-speed services to existing 2G networks was defined. Standards that enhance 2G networks by adding data services were sometimes called *2.5G.*

Each of the major 2G services had some enhancements. GSM was enhanced with General Packet Radio Service (GPRS) and Enhanced Data Rate for GSM Evolution (EDGE). GPRS adds data capacity to GSM via on-demand packet switching. Under optimal conditions, rates are as high as 115 kbps are achieved by assigning the user terminal extra TDMA time slots for data when they are needed. EDGE is an enhancement of GPRS that has a theoretical maximum speed of 384 kbps. It achieves these higher rates through new algorithms that increase data per time slot.

When U.S. TDMA became an ANSI interim standard (IS-136A and B), it already supported data rates of up to 64 kbps and therefore could be classed as 2.5G. Its next evolution will be the addition of TDMA-EDGE (IS-136 Outdoor), which will be fast enough to be classified as 3G.

cdmaOne standards evolved from the 2G IS-95 standard and include data rates of 64 kbps, possibly up to 115 kbps. However, before these data services have come into wide use, cdma20001xEV-DO has arrived, moving CDMA into 3G.

9.2.3 Third generation (3G)

The 3G initiative had two goals: to add the capacity to carry user data at high speeds on cellular networks and to unify the world on a single standard. The ITU planned to

---

[1]However, we wouldn't want to claim that you could go all the way from Krakow to San Francisco without losing the call. There aren't enough cellular base stations in the Atlantic Ocean. And even if there were, you'd have to go by boat because, as we all know, cell phones must be turned off when an airplane is in flight.

have 3G cellular networks operational worldwide by the year 2000, but this did not happen. Significant large-scale deployment of 3G technology was just beginning in 2002. Let us look at the rate of deployment first and then turn to high-speed data services and the effort to create a unified world standard.

9.2.3.1 Slow deployment.
Some of the reasons for the delay had to do with general market conditions:

- Almost everyone already has a cellular telephone. The GSM market is reaching saturation, according to GSM telephone manufacturers and service providers.

- Not that many people want to use data on their cell phones. Cell phones are for talking. Wireless networking for personal digital assistants (PDAs) in public places is not that essential. Most heavy mobile data users are satisfied with plugging in at a hotel, and within an organization, wireless local area networks (LANs) make a lot more sense than cellular systems.

- Since 2001, the slump in the economy has slowed both demand and supply.

- Initially, prices were way too high. New pricing plans, including flat-rate plans, are making cost more reasonable as of summer 2002.

Among marketing people who promote technology, there is a general assumption that people will be as excited by the next enhancement to technology as they were by an earlier, truly revolutionary technology. This is unrealistic. Affordable portable cellular telephones were a real breakthrough with obvious, immediate uses. They were a genuine breakthrough for the cellular industry, just like the IBM PC with a spreadsheet and a word processor was for the world of computers. However, once people have a PC, they see that they do not necessarily need a better PC, at least not right away. The same is true with cellular phones. The biggest competition for 3G might be 2G because people already have it and it works.

In addition, the conflict that prevented the creation of a unified standard took time. The initial delays prevented deployment by 2000. The slow growth continued due to the bursting of the dot-com bubble, which slowed the demand for high-speed mobile data service and reduced available capital for major deployments in both information technology (IT) and communications. This was followed by the slowing of the general economy, which continues to limit the rate of growth of 3G system deployments.

Once the decision to accept multiple standards was made, each company or organization went on to develop standards and specifications for the technology it preferred. Instead of building a single worldwide group of experts developing a single technology, we now have five smaller groups competing against one another. It is impossible to know the net effect of the decision to have multiple standards on the rate of deployment. On the one hand, the experts are divided into separate camps, reducing the size of each brain trust. On the other, the competition to succeed first is pushing each group to accelerate its efforts.

CDG, the group that proposed and is promoting cdma2000, has developed a migration path from cdmaOne to cdma2000 and gotten its phase 1 standard approved as IMT-2000–compliant by the ITU. Both Verizon and Sprint PCS were upgrading their networks to cdma2000 in the summer of 2000.

Meanwhile, wideband CDMA (W-CDMA), the successor to GSM, is having difficulty implementing its standard. There are probably a number of reasons for this. One of them may be a subtle point related to the upgrade of technology from 2G to 3G. There is a significant difference in migrating from cdmaOne to cdma2000 on a given bandwidth versus migrating from TDMA (GSM or TIA-136) to W-CDMA on a given bandwidth.

When migrating from cdmaOne to cdma2000, both the old 2G technology and the new 3G technology are CDMA spread-spectrum technologies that are inherently resistant to mutual interference. The 3G carrier can be overlayed on the same frequency and appears as manageable interference to the 2G carrier. The ability of cdmaOne to manage interference from cdma2000 carriers is further enhanced by the use of codes that are orthogonal on both systems.

However, when migrating from TDMA to W-CDMA, the 2G TDMA system has no such protection from the interference generated by the W-CDMA carrier. As a result, there are only two ways to deploy W-CDMA. The service provider must either deploy the equipment in a separate frequency or drop TDMA and operate CDMA. Since the providers are already serving 700 million subscribers, dropping TDMA is not a realistic option. Instead, W-CDMA must either operate in a previously unused range of spectrum or slowly edge out TDMA in the existing spectrum.

By whatever methods, the cellular providers will continue to upgrade their networks to 3G. This will increase voice capacity and provide high-speed data capacity. At some point, some new use for high-speed data services will the catch corporate and public eyes, and we will see wide use of 3G data technologies.

9.2.3.2 High-speed data. The initiative is succeeding at reaching its goal of supporting enhanced data services. Standards for high-speed data have been defined and are being implemented in 2002, only a year or two after they were originally scheduled. The data rates are still on the low end of where 3G hopes to go, but they can move up.

In fact, for cdma2000, CDG has defined a migration path that we describe in Sec. 10.2.

9.2.3.3 Unification and harmonization. The effort to create a unified worldwide standard largely failed. There also was talk of harmonizing existing standards, but the competition between cdma2000 and W-CDMA can scarcely be called harmonious.

As the attempt to develop a truly global standard unfolded, the stakes were simply too high. Companies are already entrenched in different national markets, have hundreds of millions of customers, and have invested billions in the ability to manufacture current technologies. Understandably, each company wanted to continue to benefit from the manufacturing base it had developed. Long-standing rivalries and resentment over having to pay license fees (particularly to Qualcomm) did not make the situation any easier.

Even if all the companies had agreed on a technology, there was no way to sort out the bandwidth problem. Some nations follow the U.S. FCC in allocating frequency use, and other nations follow the European model. As a result, there is no single frequency available for terrestrial cellular communications worldwide. The ITU simply did not have enough influence with national governments around the world to sponsor a new, standardized allocation system.

In addition, although CDMA has distinct technical advantages, it is only used by about 15 percent of the world's cellular telephone users. Even if CDMA wins out in the long run, it will take a good deal of time.

Instead of insisting on a single world standard, the committee approved five different standards, saying that they met the overall standard of 3G communications. The overall standard was IMT-2000, which stood for International Mobile Telecommunications by the year 2000, the goal of the group. The five individual standards are as follows:

- *IMT-DS (direct spread),* also known as W-CDMA frequency division duplex (FDD)

- *IMT-TC (time code),* also known as W-CDMA time division duplex (TDD), which includes TD-SCDMA

- *IMT-MC (multicarrier),* also known as cdma2000 multicarrier

- *IMT-SC (single carrier),* also known as UWC-136 TDMA, the TDMA successor to GSM, North American TDMA, and GPRS

- *IMT-FT (frequency time),* also known as digital enhanced cordless telecommunications (DECT), which is for small cells with stationary or pedestrian users and which does not serve mobile terminals moving at vehicular speeds

We provide full specifications for the CDMA standards (2G and 3G) in Chap. 10. However, we organize the 3G CDMA standards a bit differently. Rather than grouping them by the type of technology used, as the ITU did, we organize them according to the systems as they actually will be deployed around the world. First, we discuss cdmaOne, the original North American 2G standard in a section of its own. We follow that with a discussion of the 3G CDMA standards: cdma2000, W-CDMA, and TD-SCDMA.

- cdma2000 is the successor to cdmaOne, which is being deployed across North America as this book is being written.

- European W-CDMA includes both IMT-DS and IMT-TC.

- TD-SCDMA is being covered in a separate section because it will be deployed in mainland China.

TD-SCDMA was included with IMT-TC for three reasons: It arrived too late to be given a spot of its own; it was designed to replace 2G GSM systems, and it uses a similar combination of time division and code division as does the original IMT-TC. However, it is a standard that will be deployed independently in China and may compete in the third-world market.

IMT-SC and IMT-FT are not included in this book because they are exclusively TDMA technologies, and they make no use of CDMA.

9.3   Conclusion

Standards are essential, and yet they often also create problems. This is natural. Standards committees face many complex issues. Even if we leave out the more political and irrational pressures, standards committees must try to define how untested technologies will be developed and will serve vaguely defined customer needs. The time spent creating and testing standards is essential, and yet it draws away from each company's own development efforts.

The most basic purpose of standards—to make it easier to create working systems faster—is usually achieved. Or, at least, not having a standard is worse than having one. When an interface is left undefined, the results are usually costly. One of the authors (Kemp) saw a particularly vivid example of this when he was a local area network

(LAN) manager. IBM PC compatibles had almost every essential standard defined. As a result, it was easy to swap out keyboards, adapter cards, modems, CD-ROMs, and monitors. However, there was no standard specified for a security device. Third-party products fit the niche. However, they required removing the cover from every PC and drilling a hole in it in order to install a lock bolt that could then be attached to a security cable. If the IBM PC had had a standard for physical security, installing a physical security system would have been a lot less time-consuming.[2]

ITU, ANSI, and the other standards committees have a remarkable track record at creating standards that enable communications systems. One essential fact should be highlighted: Once a standard is defined, it is possible to create independent test procedures that validate equipment models, each built to its own specification, as functioning according to the standard. If it functions according to the standard, it should be interoperable with other equipment in the system. This increases reliability while lowering the cost of manufacturing and helps products and services reach the marketplace at reasonable cost.

This technical face of standards is clear. The other face of standards, the more political one, is considerably murkier. Now that five different 3G standards have been approved, there is an ongoing battle among them that often obscures issues rather than resolving them.

Standards are a good tool for solving problems but not a good tool for comparing systems. And they are an even worse tool for predicting the future because the predictions will be based on the fallacy that the best technology wins.[3] How long it will take to move the existing 2G networks to 3G and to deploy 3G cellular and wireless local loop (WLL) worldwide is much more dependent on economics and politics than it is on the differences between CDMA and TDMA or the even smaller differences between W-CDMA and cdma2000.[4]

Now that we have defined the general background of standards and the history of cellular standards, let us turn our attention to the standards that define the topic of this book: CDMA systems. In Part 4, "Key Data Concepts," we will address information technology (IT) standards and the challenge of convergence. Convergence, or bringing together telephony, IT, and video standards, is a challenge because these three technologies have developed separately for decades. Each industry's standards were developed to meet different needs, provide different services, and solve different technological problems. As we shall see, convergence is definitely a work in progress.

---

[2]We did say at the beginning of this chapter that the lack of standards makes it easier to have fun. Drilling holes in 20 computers was certainly an unusual experience.

[3]If the best technology always won, we would be listening to music on reel-to-reel tape recorders and watching movies on Betamax tapes.

[4]As Ecclesiastes put it, "under the sun, the race is not to the swift . . . nor riches to the intelligent . . . nor favor to men of skill . . . but time and chance happen to them all."

Worldwide CDMA Standards

cdmaOne, originally American National Standards Institute (ANSI) Interim Standard 95 (IS-95), was the only second-generation (2G) code division multiple access (CDMA) standard for cellular telephony. With the advent of third generation (3G), CDMA has come into its own. cdma2000 was designed as the successor to cdmaOne, adding high-speed data capacity and improving a number of technical features. It is being deployed in North America in 2002.

Wideband CDMA (W-CDMA) is the CDMA standard designed to be a successor to the Global System for Mobility (GSM) for the European market and other countries where GSM is already in place. There is also WCDMA without the hyphen, a general term for any wideband CDMA standard with 3.75 MHz of bandwidth or higher per carrier, three times the bandwidth of cdmaOne. With this terminology, both cdma2000 and W-CDMA would be types of WCDMA.[1]

CDMA is constantly evolving. In fact, efforts to define a fourth-generation (4G) set of standards are already underway, as described in Part 9. The latest fully developed CDMA standard is, however, time division synchronized CDMA (TD-SCMA). The TD-SCDMA standard was developed too late to be considered for the United States and Europe, but the People's Republic of China, now the world's largest market for cellular telephones, has chosen to deploy a TD-SCDMA system. The effort behind the TD-SCDMA concept seems to be to create a successor to GSM that combines several multiple-access methods to achieve optimal capacity and optimal bandwidth flexibility.

These standards are summarized in Table 10.1. Please note that some of the summary information is provided in more precise detail later in the chapter. For example, when cdma2000 and W-CDMA shift from CDMA transmission to time division multiple access (TDMA) transmission to support higher data rates, they use additional forms of modulation, not only quadraphase phase shift keying (QPSK).

10.1   cdmaOne (IS-95)

Qualcomm developed the chipset and proposed the standard that became first ANSI IS-95 and then cdmaOne. ANSI IS-95A was a circuit-switched technology, and it was

---

[1]The authors did not invent this terminology. We just have to live with it.

TABLE 10.1   Comparison of CDMA Standards

	cdmaOne	cdma2000	W-CDMA	TD-SCDMA
Assigned spectrum, MHz	869–894	869–894	2110–2170	2110–2170
Rx	1930–1990	1930–1990		
Tx	824–849	824–849	1920–1980	1920–1980
	1850–1910	1850–1910		
TDD not paired			1900–1920	
TDD not paired			2010–2025	
Voice users/carrier	15–50	15–100	196 (max.)	48
Speech rate	8/13 k	8/13 k	up to 13 k	8 k
Voice activity	4 levels	4 levels	on/off	on/off
Radio conditions	No change	No change	8 rate AMR	8 rate AMR
Data services	2G (few)	3G	3G	3G
Normal data rate	14.4 k	307.2 k (1x)	936 k	384 k
Ideal conditions	114 k	2 M (1x)	4 M	2 M
Typical mode	FDD/CDMA	FDD/CDMA	FDD/CDMA	TDD/TDMA/CDMA
Typical carrier	1.25 MHz	1x 1.25 Mhz	5.0 MHz	1.6 MHz
		3x 3.75 MHz		
Chip rate (Mchip/s)	1.2288	1x 1.2288	3.84	1.28
		3x 3.6864		
Modulation	QPSK	QPSK/OQPSK	QPSK	QPSK
Reverse pilot	No	Separate	In channel	Midamble
Forward power control	Slow	800/s	1500/s	Slow
Reverse power control	800/s	800/s	1500/s	Slow
Smart antennas	Maybe	Maybe	Maybe	Yes
Joint detection	No	No	No	Yes
Cell identification	Timing	Timing	PN code	PN code

followed by IS-95B, which added support for high-speed packet data. This technology was an innovation of the 1990s, bringing spread spectrum into civilian use for the cellular network. Major North American networks such as Sprint PCS and Verizon Wireless, as well as smaller regional companies and some third-world nations, have deployed cdmaOne. There are about 100 million subscribers in North America in 2002.

10.1.1   cdmaOne services

The service of cdmaOne is primarily voice telephone calls, but there are three categories of data service in cdmaOne: short message service (SMS), asynchronous data and facsimile (fax), and packet data services.

Voice service in cdmaOne has evolved beyond the original IS-95 standard of IS-96A code excited linear prediction (CELP) speech coding using rate set 1 (RS1), 8.55 kbps. In the next section we point out the enhanced variable-rate codec (EVRC) and a higher-bit-rate codec using rate set 2 (RS2), 13 kbps.

SMS is the service described in Chap. 20 that allows subscribers to send messages up to 160 bytes in length. Typically, recipients of short messages are other mobile subscribers, but there are also SMS subscribers outside the wireless telephone network. These short text messages have a function similar to text paging. SMS user messages are data carried on the signal channel without requiring call setup and without using a dedicated channel.

Asynchronous data and facsimile are circuit-switched data services with a dedicated data path for the duration of a call. Data rates of up to 14.4 kbps are supported by the Radio Link Protocol (RLP) with automatic repeat request (ARQ), forward error correction (FEC), and flow control. ARQ is an error-control protocol in data transmission. The ARQ receiver automatically requests the transmitter to resend the packet when it detects an error in the packet.[2] Efficient ARQ and error correction can bring the error rates down from typical cellular channel bit error rate (BER) values of 10^{-2} to acceptable data transmission values of 10^{-6}.[3] With IS-95B, up to eight RS1 or RS2 channels can be aggregated into a single digital stream so that the subscriber can see data rates as high as 114 kbps.

Packet services provide direct digital links through a service called *cellular digital packet data* (CDPD) from the subscriber to a public packet data network (PPDN) such as the Internet. The wireless telephone system conserves its air-interface resources by putting inactive packet users into *dormant* states where they temporarily cannot send or receive packets. Voice or circuit-switched data can be sent while the packet service is dormant. Simultaneous voice and data on the same traffic channel are supported for subscribers with user terminals equipped for this capability. These services are supported even when the user terminal is handed off to another mobile switching center (MSC).

10.1.2   cdmaOne speech coding

The speech coder used in cdmaOne is the code excited linear prediction (CELP) described in Sec. 18.2.1. The standard IS-96A speech coding for cdmaOne is RS1, 8.55 kbps. Two enhanced modes are offered, the enhanced variable-rate codec (EVRC) using RS1 and the 13-kbps speech codec developed by the CDMA Development Group (CDG) using RS2. A *codec* is a speech coder and decoder the same way that a modem is a data modulator and demodulator.[4]

EVRC is a more sophisticated version of CELP that reduces the number of bits required for its linear predictor coefficients and pitch synthesis, so it has higher voice quality at the same 8.55 kbps as the original IS-96A coding, as shown in Table 18.2.

RS2 is a higher-quality, higher-bit-rate CELP speech coding system. While it sounds even better than EVRC, its higher bit rate of 13 kbps reduces the capacity of a cdmaOne system by 40 percent compared with IS-96A or EVRC.

All these speech coders have multiple rates depending on the subscriber's speech patterns. When the speech pattern is less complex, they go to rate one-half. IS-96A and

---

[2]This process is repeated until the packet is error-free or the error continues beyond a predetermined number of transmissions. ARQ is also sometimes used to guarantee data integrity with GSM communications.

[3]Fax data are more error-tolerant than computer data. In computer data, flipping a single bit can corrupt a program or render a data file unreadable. In fax data, flipping a single bit will turn one dot from black to white, or vice versa, and the fax will still be readable. A low error rate improves the appearance of the fax and also reduces the likelihood that crucial control characters will be transmitted incorrectly.

[4]Codecs are used for music-quality audio and for video as well. The term *codec* refers to the component that encodes and decodes the audio or video source into a digital signal. In cellular telephones, the component is a hardware chip or chipset. On a computer, a codec may be implemented in software, perhaps as a device driver. Codecs code and decode a signal according to one or more standards. The standards often, but not always, perform compression. The term *codec* is also sometimes used to refer to the standard rather than to the device.

RS2 have a rate one-quarter mode for even simpler speech patterns. And they all have a rate one-eighth mode that is used when the subscriber is not speaking.

10.1.3   cdmaOne radio bandwidth

cdmaOne operates using pairs of carriers for two-way communication, employing frequency division duplex (FDD) with a 1.25-MHz carrier in the forward direction and another 1.25-MHz carrier in the reverse direction. Each cdmaOne carrier uses a total of 2.5 MHz for bidirectional communication. Each pair of carriers theoretically could support 64 channels, each with its own Walsh code, but installed equipment usually sets a maximum of about 30 channels, and practical limits created by interference typically limit the number of calls per carrier to about 25. The cdmaOne system architecture supports multiple carriers in one system, 11 in the North American PCS band.

The multiple carriers of a large cdmaOne system are treated as a single resource in channel assignment. The wireless telephone system selects an available carrier for each channel at the start of a call, and the system can instruct the user terminal to change to a different carrier during handoff, performing a carrier-to-carrier handoff. For both voice and data service, one large single resource is more efficient for traffic engineering than several small resources, as discussed in Chap. 21.

10.1.4   cdmaOne forward channel coding

The cdmaOne forward channel takes the RS1 voice signal, 0.8 to 8.55 kbps, and adds frame, quality, and overhead information to bring it to 1.2, 2.4, 4.8, or 9.6 kbps. This is doubled by convolutional forward error correction (FEC) to 2.4, 4.8, 9.6, or 19.2 kbps. The lower bit rates are simply repeated to generate a speech-coded and forward-error-corrected stream of 19.2 kbps.

RS2 uses a less aggressive forward error correction scheme, 4 to 3 instead of 2 to 1, so its full 13-kbps rate still comes out to 19.2 kbps.

This stream goes through a block interleaver so that bursty errors in the channel do not become adjacent errors in the bit stream. The bits are multiplied by a long pseudo-noise (PN) code specific to the wireless call. Eight hundred power-control bits per second are sent on a subchannel created by puncturing the 19.2-kbps data stream. Each 24-bit power-control group has 2 bits replaced with a power-control bit. The position of the power-control bit in the 24 bits is one of the first 16 bits based on the last 4 bits of the last power-control group.

Each voice channel is assigned one of 64 orthogonal Walsh codes so that the 19,200 bits per second are spread to 1.2288 Mbps. These bits are then spread once more by the short PN code into in-phase and quadrature signals.

The pilot is sent at a high power level, transmitting all zeroes using Walsh code zero, so its only coding is the short PN code. A synchronization channel of 1200 bits per second is sent using Walsh code 32. The paging channel is sent on another Walsh code (between one and seven, inclusive) and has its own long PN code.

10.1.5   cdmaOne reverse channel coding

The reverse cdmaOne channel takes the RS1 voice signal, 0.8 to 8.55 kbps, and adds frame, quality, and overhead information to bring it to 1.2, 2.4, 4.8, or 9.6 kbps. This is tripled by convolutional forward error correction to 28.8 kbps or less. When the speech coder is running at lower rates, we use the data burst randomizer below rather than repeating the bits at this stage. The reason why the forward direction repeats the bits and the reverse direction sends them in bursts is discussed in Sec. 8.6.

As in the forward direction, RS2 uses a less aggressive forward error correction scheme, 2 to 1 instead of 3 to 1, so its full 13-kbps rate still comes out to 28.8 kbps.

Each 6 bits is spread using the Walsh functions to 64 bits for a rate of 307.2 kbps. These Walsh codes are not orthogonal in the reverse direction, but they do present a good noncoherent spreading scheme with a ratio of 64 to 6. When the speech coder is not operating at full rate, there is a data burst randomizer that sends rate one-half, rate one-quarter, and rate one-eighth data at full rate one-half, one-quarter, or one-eighth of the time. The bursts are not genuinely random, of course, since the receiver has to know when to detect them, but they are well distributed over time.

The 307.2-kbps stream is spread by another factor of 4 using the long PN code to get a 1.2288-Mbps data stream. This goes through an offset quadrature phase shift keying (OQPSK) noncoherent modulator before going out over the air.

10.1.6 cdmaOne power control

Reverse power control is the three-step process described in Sec. 8.5. Power control begins when the user terminal requests a channel for an incoming or outgoing call. The user terminal sends a message called an *access probe* at the open-loop power level and awaits a response. The entire process of sending access probes to set up a call is called an *access attempt*. If the access probe is unsuccessful, then the user terminal waits a random period of time and sends another access probe with 1 dB more power. If 16 access probes fail, then the access attempt has failed. In this first step, power control is handled based on the power level of the signal channel.

If the call channel is established, step two begins. The user terminal sets its initial power level for the reverse channel using open-loop power control, measuring the pilot signal power on the forward channel and adjusting its transmit power accordingly. There are parameters telling the user terminal how much to add to its received power. Since this power level is based on forward power statistics and the forward channel is on a different frequency with different fading characteristics, this initial reverse power level is just an estimate to be refined using the closed-loop power control.

Once a call is in progress, the third step begins, and the closed-loop power control takes over. The closed-loop power-control system operates with feedback at two levels. The inner loop is constantly measuring E_b/N_0 800 times per second and sending power-control messages to optimize E_b/N_0. The outer loop is adjusting the target E_b/N_0 for each call by measuring frame error rate (FER).[5]

The outer loop is where the system pays for the higher-bit-rate speech coder, RS2. While both speech coders, RS1 and RS2, send 28.8 kbps in the reverse direction with forward error correction, the higher bit rate requires a higher E_b/N_0 to maintain the same BER after the error-correction stage at the receiver.

Increased vehicle speed also affects the target E_b/N_0 required for a satisfactory BER. The base station does not have to know if the user terminal is in a moving vehicle because the outer loop will increase E_b/N_0 to compensate for declining BER performance.

---

[5]Please recall that in CDMA, an E_b/N_0 or an FER that is too low indicates that this channel is too powerful, generating excessive interference for other channels.

Forward power control is more casual in cdmaOne because the orthogonal PN codes remove most of the same-cell interference. The user terminal keeps a running total of frame errors and reports any sequence of frames over the required FER using a power-measurement report message (PMRM). The base station reduces its power if it receives no PMRM over a period of time. When a PMRM message does arrive, the base station changes its power according to the message. This is typically a power increase, but it can be a decrease as well. If the user terminal reporting threshold is lower than the required FER, then the base station could receive a PMRM message telling it that FER is higher than necessary. While RS1 timers can be as long as one full second, the RS2 forward power is monitored and adjusted every two 20-ms frames.

10.1.7 cdmaOne cell identification

The forward pilot channel is the reference for all user terminals in the demodulation process. It is also how the user terminal selects and identifies the serving cell at the beginning of a call. Pilots for all base stations are set 4 to 6 dB higher than traffic channels so that the user terminals can detect them.

The pilot channels have a PN code of 32,768 chips, a duration of 26.66 ms, and 75 repetitions per second. All antenna faces use the same pilot PN code, and each cell sector is identified by a unique time offset of its pilot PN code. The timing is coordinated through global positioning system (GPS) radios at the base stations.[6] These offsets are in increments of 64 chips, so there are 512 unique offsets. At the speed of light, about 300,000 km/s, these offsets work out to 15.6 km. Only the busiest systems have so many cells that there is confusion among pilots at the user terminal. This confusion occurs in regions where a user terminal detects so many pilots that it does not reliably identify the best signal. We will have more to say about this in Sec. 30.1.

10.2 cdma2000

cdma2000 standards development is being promoted by the CDG. CDG is attempting to deliver a series of incremental standards that provide a smooth, quick migration from cdmaOne to cdma2000. The first versions are referred to as *cdma2000 1x*. The *1x* describes the carrier bandwidth as a ratio to the 1.25-MHz bandwidth of cdmaOne. cdma2000 systems now being deployed are all 1x, with 3x expected later. cdma2000 adds data services, starting at about 144 kbps, and also has the potential to double the voice capacity of cdmaOne networks.

In deploying cdma2000 1x systems, vendors hope to achieve the following objectives:

- Increased voice capacity

- Added or enhanced data capacity

- Installation of software-upgradeable cdma2000 equipment, which will reduce the cost of upgrading to later cdma2000 standards

- Support for cdma2000 and cdmaOne user terminals on the same carrier frequency

---

[6]GPS systems, in addition to providing timed triangulation allowing a receiver to determine its exact location, also provide a readily available synchronized clock service.

The cdma2000 standard will doubtless go through a number of modifications, as all standards do. As of the time of publication, the International Telecommunications Union (ITU) has approved cdma20001xEV-DO as a 3G standard. This is a current implementation of cdma2000 with 1x (1.25 MHz) bandwidth using EVolution Data Only (EV-DO) as a packet carrier.

As of this writing, we are aware of these cdma2000 detail standards[7]:

- 1x data services up to 144 kbps.

- 1x EV-DO, called 1X-EV phase one (also known as high-rate packet data air interface), provides a completely separate carrier for data, increasing the data rate to 2.4 Mbps.

- 1x EV-DV, called 1X-EV phase two, is expected to support data rates ranging from 3 to 5 Mbps. Several proposals have been submitted to the standards committee.

- 3x is approved by the ITU as a 3G standard. It uses the space of three cdmaOne 1.25-MHz bands plus outside guard bands for 3.75 MHz of bandwidth transmission in a 5-MHz space for each direction or 10 MHz for full-duplex service. It can overlap its carrier on the same frequencies with cdma2000 1x or cdmaOne carriers.

cdma2000 is being deployed in North America by both Sprint PCS and Verizon Wireless in 2002 as this book is being written. Also, Cingular, the eighth largest cellular service provider in the United States and one of three major U.S. GSM (TDMA) providers, has announced its plans to convert to cdma2000 with rollout beginning in November 2002. cdma2000 networks are also being deployed in Korea.

10.2.1 cdma2000 services

cdma2000 claims to be backward-compatible with the current cdmaOne standards. With 100 million cdmaOne subscribers out there and 100 million cdmaOne user terminals in circulation, this should be an important criterion. Some existing cdmaOne handsets, however, seem to have problems working with cdma2000.

While cdma2000 has its own voice calls, the significant new service offering of cdma2000 is the 3G suite of voice and data services. These services, based on the 1x and 3x standards described earlier, are expected to provide data rates of 144 kbps for high-mobility traffic, 384 kbps for pedestrian traffic, and up to 2 Mbps for stationary traffic in picocells where user terminals are near their base stations. These data services can be circuit-switched or packet-based.

There is a wide range of circuit-switched data rates:

- 32- to 64-kbps computer connections analogous to dial-up modem connections

- Integrated Services Digital Network (ISDN) access up to 144 kbps

- 64- to 384-kbps connections for videoconferencing

Packet services are looking even brighter. Plans include multimedia and other data services to a public packet data network (PPDN) such as the Internet with peak rates ranging from 64 kbps to 2 Mbps. This includes Internet Protocol (IP) packet-based videoconferencing. Traffic data services are carried in RLP.

---

[7]We wanted to call them substandards, but they may not be substandard.

TABLE 10.2   cdma2000 Bandwidth

Bandwidth	Suffix
1.25 MHz	1x
3.75 MHz	3x
7.5 MHz	6x
11.25 MHz	9x
15.0 MHz	12x

10.2.2   cdma2000 speech coding

The speech coding standards for cdma2000 are the two optional codecs from cdmaOne, the 8.55-kbps EVRC codec using RS1 and the 13-kbps codec using RS2.

10.2.3   cdma2000 radio bandwidth

The radio spectrum of cdma2000 is based on the cdmaOne FDD 1.25-MHz channel as its quantum level of radio spectrum. The cdma2000 standard offers five bandwidths in each direction, as shown in Table 10.2 with their multiplier suffixes. Add an extra 1.25 MHz for guard bands in each direction, and double these bandwidth figures for a full two-way FDD system, so a 3x system uses 10 MHz of radio spectrum.

There is more flexibility than just a selection of five bandwidth choices, however. First, the wireless service provider may elect to use different bandwidths for forward and reverse. For a community that surfs the World Wide Web, a 3x forward link and a 1x reverse link may be ideal. Second, the bands can be overlaid and combined.[8] There is a multicarrier mode for cdma2000 3x that not only sends its data on three carriers at 1.2288 Mchips per second but also coordinates its Walsh codes to be orthogonal to the cdmaOne channels in use.

10.2.4   cdma2000 forward channel coding

The forward modulation scheme in cdma2000 has separate in-phase and quadrature bit streams as opposed to cdmaOne, where they are combined in the final stage of processing gain. This gives cdma2000 an automatic factor of 2 in bit rate capacity at the expense of a factor of 2 in processing gain.

cdma2000 supports two kinds of traffic channels, fundamental and supplemental. The fundamental channel is used for voice calls and for low-bit-rate signaling during a data call, the same 9600-bit-per-second RS1 and 14.4-kbps RS2 as cdmaOne.

Supplemental channels can be dedicated to a single user terminal or shared by multiple packet mode users. Every user of a supplemental channel must have an active fundamental channel that may do nothing more than power control and low-bit-rate signaling at one-eighth rate and one-eighth power. Forward supplemental channels are controlled by the wireless service provider, so subscribers can share fixed supplemental channels in packet mode or can have their own supplemental channels, one or more

---

[8]Even if there were no specific provision for carrier coordination, the direct-sequence spread-spectrum (DSSS) principle makes the carrier resistant to narrowband jamming, even if the narrowband jamming is from another DSSS carrier.

on the same forward link. In the forward direction, the base station does the packet scheduling and tells the user terminals which time slots are assigned to them.

To support the wide range of channel data rates, the Walsh codes use the orthogonal variable spreading factor (OVSF) described in Sec. 8.3.2, so different channels on the same carrier can use different spreading gains. cdma2000 3x supports 256 Walsh codes with spreading gains as high as 256 and as low as 4. With 2 to 1 turbo coding FER, 3x can support data rates over 1 Mbps. The cdma2000 pilot, synchronization, and paging channels are similar to cdmaOne.

Recent work has developed the potential of the cdma2000 1x carrier. The 1xEV-DO (1x EVolution Data Only) carrier is specifically designed for packet data. For user terminals close to the base station (picocell service), 1xEV-DO achieves a peak forward direction rate of 2.45 Mbps in only 1.25 MHz of radio spectrum. In its highest data rate configuration, 1xEV-DO abandons QPSK and CDMA and goes into TDMA with higher-order constellations in its modulation, 8-ary and 16QAM as demand and radio conditions allow. It also uses turbo forward error correction, described in Sec. 17.2.2, which is designed for large data blocks. Small changes in data rate are accomplished by repeating sequences of bits to stretch them out or by adding a few extra bits into the stream, not enough to break the forward error correction scheme. This technique is called *puncturing* the bit stream.

Other recent work has developed even higher-order modulation schemes in the cdma2000 1.25-MHz camp. 1Xtreme uses 64QAM, 6 bits per symbol, as the key to even higher speeds, with claims of packet data rates of up to 5.2 Mbps and CDMA data throughput of 1.2 Mbps.

10.2.5 cdma2000 reverse channel coding

In contrast to cdmaOne, the cdma2000 reverse link supports multiple channels for a single user terminal. The reverse channel set includes a pilot that allows coherent demodulation at the base station and more accurate measurement for power control and the rake filter.[9] The multiple reverse channels of a cdma2000 link are spread with orthogonal Walsh codes to keep them separate. The pilot uses a Walsh code of all zeroes, the dedicated control channel uses an eight-chip Walsh code, the fundamental channel uses a four-chip Walsh code, and the supplemental channels use 2-bit Walsh codes to allow higher bit rates in favorable environments. Since the user terminal is sending fewer channels than the base station, the reverse link can use shorter Walsh codes to maintain orthogonality among its separate channels.

The fundamental channel is used for voice calls and for low-bit-rate signaling during a data call, the same 9600-bit-per-second RS1 and 14.4-kbps RS2 as cdmaOne. Unlike cdmaOne, the power-control bits are in the pilot channel rather than inserted into the fundamental bit stream.

Supplemental channels can be dedicated or shared, the same as in the forward direction. A single link can have only a fundamental channel, or it can have one or more supplemental channels along with a low-bit-rate fundamental channel. When a single reverse supplemental channel is shared among many user terminals, scheduling is

---

[9]The potential of higher data rates makes the extra capacity overhead of a pilot worthwhile. In cdmaOne, the highest reverse data rate is 14.4 kbps, which is not worth the extra overhead and design expense of a separate pilot.

handled by a medium access control (MAC) protocol. User terminals send MAC messages across their own dedicated control channels and get confirmation before they use a shared supplemental packet data channel.

The reverse cdma2000 link puts the pilot and control channels on its in-phase bits, puts the fundamental and supplemental channels on its quadrature bits, and balances them using a technique called *complex spreading*. cdma2000 uses orthogonal complex QPSK (OCQPSK), a technique to reduce out-of-band radio emissions. Because OCQPSK came out of international committees from the United States, European Telecommunications Standards Institute (ETSI), Japan, and Korea, it is also called *harmonized PSK*.

The 1xEV-DO reverse link is CDMA with quick connection and teardown to facilitate sleep modes to save battery power and fast reconnect so that the subscriber sees little delay in packet service.

10.2.6   cdma2000 power control

Power control in cdma2000 reflects a concern for the need for greater accuracy in user data transmission than in user voice transmission. If a cdmaOne subscriber walks or drives into a fade area and the system does not respond for 20 ms, then the gap in voice performance is small and not likely to be noticed. The same 20 ms in a high-speed download could lose hundreds of bytes of user data. As a result, cdma2000 supports high-speed power control in both forward and reverse directions, 800 times per second. The reverse pilot contains the 800-bit-per-second power-control stream.

The power-control architecture is the same as cdmaOne. The open-loop method is used to set initial reverse power based on received forward pilot power. Once a call is in progress, the closed-loop power control takes over. The inner loop is constantly measuring E_b/N_0 800 times per second and sending power-control messages. The outer loop is adjusting the target E_b/N_0 for each call by measuring FER.

All the channels in cdma2000 use a fixed chip rate that depends on the carrier bandwidth: 1.2288 Mchips per second for 1x and 3.68 Mchips per second for 3x. However, they differ greatly in their BER requirements. A fundamental channel using RS1 at one-eighth rate is sending 1200 bits per second over the same carrier as a supplemental channel carrying a live video signal or downloading a large database. The low-bit-rate channel has far more spreading gain built into it and can transmit at far lower power than high-bit-rate services, creating far less interference. And even among high-rate services, there are dramatically different requirements because lost bits in a video signal are far less consequential than lost bits in an accounting spreadsheet or an executable program. The outer loop controls the amount of radio resource, in the form of interfering power, that each channel uses.

10.2.7   cdma2000 cell identification

The forward pilot channel is the reference for all user terminals in the demodulation process, just as in cdmaOne. Also, the user terminal uses the forward pilot to select and identify the serving cell at the beginning of a call. Pilots for all base stations are set 4 to 6 dB higher than traffic channels so that the user terminals can detect them. All pilots share the same PN code staggered with different time delays coordinated with GPS receivers in the base stations, the same as in cdmaOne.

10.3   W-CDMA (UMTS)

Wideband CDMA (W-CDMA) was designed as a 3G upgrade path for GSM networks and is being deployed in Europe and Japan. The Universal Mobile Telephony System (UMTS) is the European catch phrase for systems using W-CDMA.[10] In addition to specifying the use of W-CDMA, UMTS also specifies a frequency band of 1885 to 2025 MHz for terrestrial cellular networks.

As a standard, though, W-CDMA does have a significant innovation. The incorporation of different classes of service, offering quality-of-service definitions appropriate to different functions (live personal voice/video communications, interaction with computer system, e-mail, and so forth) into the design standard of a cellular network is likely to speed development of worthwhile products and services for subscribers. These classes of service are described below along with other W-CDMA specifications.

10.3.1   W-CDMA services

On registration with the system, the W-CDMA user terminal tells the network its radio capabilities in one of six categories:

- 32 kbps for basic speech service and limited data rates up to 32 kbps

- 64 kbps for speech and data, with the possibility of simultaneous speech and data

- 144 kbps for videotelephony and other data services

- 384 kbps for more data services, including advanced packet data

- 768 kbps as an intermediate step between 384 kbps and 2 Mbps

- 2 Mbps is the state of the art and is defined for the forward direction only

We will discuss the four quality-of-service (QoS) classes in W-CDMA: conversational, streaming, interactive, and background. Delay sensitivity is the main distinction among these W-CDMA QoS levels.

10.3.1.1   Conversational class. Conversational class is mostly circuit-switched voice, but voice over IP (VoIP) and videotelephony are expected to require this type of service. This is meant for communication between live human beings.

The traffic is symmetric in character, and the end-to-end delay must remain low. Human perceptions define the requirements of the W-CDMA conversational class of service, and people seem to allow 400 ms of end-to-end delay before the channel starts to annoy them. End-to-end delay is from the caller's mouth to the called party's ear, so delays created during encoding and decoding in the user terminals must be included, as well as propagation and circuit delays.

Circuit-switched voice is old-fashioned telephone service, with the same voice quality as 2G GSM when radio conditions permit. Unlike cdmaOne and cdma2000, the

---

[10]Note that although the United States has a reputation for arrogance, it is the European telephony community that has declared its systems first global and now universal. GSM is operating on several continents but is far from global. The authors' research indicates that no cellular system is likely to go beyond global, extending past earth's biosphere to the larger universe, any time soon.

W-CDMA speech codec responds to varying radio conditions, as we will describe in Sec. 10.3.2.

Video has delay requirements similar to speech, about 400 ms end to end. Compressed video has a more stringent BER requirement than speech; that is, it requires a lower error rate. W-CDMA has specified the use of ITU-T Recommendation H.324M as the standard for videotelephony. ITU-T Recommendation H.324 is a standard for multiplexing audio, video, user data applications, and system control onto a data stream suitable for transport in the public switched telephone network (PSTN) or a wireless circuit-switched network.

10.3.1.2   Streaming class.   Streaming data are processed as a steady and continuous stream. Typically, this approach is used for large multimedia files downloaded from the Internet. Using this class of service, the receiving browser can start displaying the data before the entire file is sent. The receiver has to collect data and send them to the application in a steady stream.

Streaming applications are very asymmetric, with one direction sending almost all the data. Picture a multimedia application where the subscriber watches an audio/video presentation and clicks a mouse once or twice a minute to direct the transmission. These applications tolerate more delay, and they can handle more jitter, that is, more variation in delay (described in Sec. 17.1), than symmetric conversational services.

Some of these services are designed and targeted for lower bit rates, 28.8 kbps, whereas others are aimed at a high-end market, offering over 100 kbps.

10.3.1.3   Interactive class.   Interactive class is designed for a person or a machine online requesting information from remote equipment. This includes World Wide Web browsing, database retrieval, and server access. The key quality element in this class of service is the duration of the round-trip time from the subscriber stimulus to the data response.

Location-based services fall into this category. Finding the nearest gas station or hotel, or even a local map, depends on knowing the location of the user terminal. This can be obtained through cell coverage-based positioning using radio propagation delays or through network-assisted GPS methods.[11]

Computer games generally fall into the interactive class of service. However, if the game is highly interactive, then it would be better to play it in the conversational class with its even stricter delay requirements.

10.3.1.4   Background class.   E-mail, SMS, and database downloads are less delay-sensitive because the subscriber does not need to make an immediate response. A delay of several seconds or even a few minutes may make little difference in these services. Typically, these data have high accuracy requirements and must be delivered error-free. If the forward error correction is insufficient, it may be supplemented by error detection and retransmission because the lack of delay sensitivity allows this.

---

[11]In network-assisted GPS, the base station forwards GPS timing information to the user terminal, but the user terminal does not have its own GPS. A third option would be to have a GPS in the user terminal. In that case, the user terminal would identify its own location and inform the base station.

10.3.2 W-CDMA speech coding

W-CDMA uses adaptive multirate (AMR) speech coding with multiple rates that depend on radio conditions. This is not the same as cdmaOne and cdma2000 lowering their bit rates during times of reduced voice activity. The change in coding is triggered by changing radio conditions, not by changing speech patterns. The W-CDMA codec has eight rates from 12.2 kbps down to 4.75 kbps. The 12.2-kbps AMR speech codec is the same as GSM, 7.4 kbps is the same as the US-TDMA codec, and 6.7 kbps is the same as the Japanese PDC codec. Every 20 ms the system evaluates the load on the air interface and the radio link quality of the speech signal. Depending on the results of these measurements, the system may change the speech coding rate. When the transmit power is getting close to its maximum, the AMR bit rate is reduced.

The coding scheme is algebraic code excited linear prediction (ACELP), and the multirate version is referred to as MR-ACELP. The bits are arranged according to their subjective importance into three classes, A, B, and C, where class A is the most sensitive. The forward error correction is designed to provide the most protection for class A bits.

AMR has voice activity detection (VAD) for discontinuous transmission (DTx) to extend battery life and to reduce CDMA interference. The AMR transmitter sends a silence descriptor (SID) frame at regular intervals so that the receiver can present comfort noise during the silent intervals.

The bit rate is controlled by the wireless network, so lower bit rates can be used during high system loads. The lower bit rates are also used during periods of poor radio connection to extend cell coverage. The AMR system allows a smooth tradeoff between system capacity and coverage on one side and speech quality on the other.

10.3.3 W-CDMA radio bandwidth

The carrier spacing of W-CDMA is 5 MHz.[12] There is some debate about the need for guard bands in W-CDMA. In the standard, the proposal is to use complex filters so that guard bands are not needed, and 5-MHz bands can be 5 MHz apart, center to center. However, the CDG (which promotes cdma2000) has doubts about the feasibility of this approach. The CDG claims that it remains to be seen if a system can be specified and developed that will meet power emission requirements of the U.S. FCC. The CDG also says that in many nations there may be difficulty placing 5-MHz W-CDMA adjacent to other wireless services without adding guard bands. The potential for unmanageable problems is higher if the adjacent service is not CDMA because that service will not be able to manage interference as well as a CDMA service would. To date, all wireless services of different types have been separated by guard bands.

If operated in the FDD mode, probably the most common duplexing method, then separate forward and reverse carriers add up to 10 MHz. In time division duplex (TDD) mode, a single W-CDMA carrier serves both the forward and the reverse channels in 5 MHz of bandwidth.

---

[12]There are references to other W-CDMA bandwidths, 1.25, 10, and 20 MHz, but most of the literature we have seen on W-CDMA refers only to the 5-MHz carrier width. Later versions of the W-CDMA standard include the multicarrier mode of cdma2000 described in Sec. 10.2.3 as part of the W-CDMA standard.

10.3.4   W-CDMA forward channel coding

The chip rate for W-CDMA, in either FDD or TDD mode, is 3.84 Mchips per second. All channels in W-CDMA have pilot bits included at 1500 per second. By including pilot bits in each channel, we avoid having to use a separate pilot channel the way cdmaOne and cdma2000 do. In either case, the pilot is overhead—radio power dissipated into the CDMA carrier spectrum to lock its receiver onto a known signal that does not carry any subscriber data.

The W-CDMA carrier supports up to 512 forward channels using orthogonal variable spreading factor (OVSF) to vary spreading gain. Voice channels use the longest codes, 512 bits, whereas higher-speed data channels use shorter codes down to 4 bits.

Speech and data can be multiplexed onto a single channel with 1500 time slots per second. These are not TDMA time slots apportioned to separate channels but time slots to allow a single radio channel to offer multiple services to one subscriber.[13]

Packet mode uses a shared channel, with TDMA over the radio link to multiplex data to different users. The base station can send data to a variety of users by assigning different time slots to different users.

Both dedicated circuit and packet modes use 3 to 1 convolutional coding for low rates and 2 to 1 turbo coding for high rates. For any required data rate, the system uses a spreading gain to expand the result to near 3.84 Mchips per second and then repeats some bits or punctures the bit stream to get the desired data rate to within 100 bits per second of the target rate.

The TDD mode has no continuous pilot because there is no continuous transmission. Instead, each data burst has a set of bits for coherent demodulation. These bits are called a *midamble* because they are placed in the middle of the burst. The data bursts are 2560 chips at 1500 per second, just a little bit longer than a GSM time slot. The TDD mode also does not support soft handoff. Instead, all handoffs are mobile-assisted hard handoffs. Also, the spreading gains used are lower, ranging from a maximum of 16 to a minimum of 1 (QPSK with no spreading gain).

10.3.5   W-CDMA reverse channel coding

The reverse channel chip rate is 3.84 Mchips per second, the same as the forward direction. All the W-CDMA channels have pilot bits included at 1500 per second, also the same as the forward direction. This ensures coherent detection in both forward and reverse directions.

The reverse carrier uses OVSF to allow spreading gains from 256 for voice and low data rates down to 4 for high data rates. Speech and data can be multiplexed onto a single channel with 1500 time slots per second.

Packet mode uses a shared channel. Time slots are assigned using MAC messages sent on the same shared channel using a slotted ALOHA protocol.[14] High-priority

---

[13]We say radio channel because engineers with a data-processing background would describe it differently. We would see the time division supporting multiple voice and data channels through a single device for a single user. This would allow, for example, two different applications on a personal digital assistant (PDA) to use time division duplexing to carry both a voice call and e-mail over the wireless channel. We have a single radio channel, a single subscriber, and voice/data multiplexing over that channel.

[14]This is different from cdma2000, where the MAC messages are sent on a separate control channel.

subscribers are allowed to use shorter random times for retrying their attempts to improve their access to the shared channel.

W-CDMA reverse channels use 3 to 1 convolutional coding for low rates and 2 to 1 turbo coding for high rates. For any required data rate, the system uses a spreading gain to expand the result to near 3.84 Mchips per second and then repeats some bits or punctures the bit stream to get the desired data rate to within 100 bits per second, the same as the forward direction.

The reverse TDD mode uses data bursts of 2560 chips, 1500 per second, with a midamble, and the spreading gain can vary from 16 down to 1.

The reverse W-CDMA link balances its in-phase and quadrature components with complex spreading using OCQPSK, also known as harmonized PSK, the same as cdma2000.

10.3.6   W-CDMA power control

Open-loop power control is used to set initial reverse power levels. Closed-loop power control in W-CDMA, both forward and reverse, is sent 1500 times per second in the same cycle as the pilot. The inner loop adjusts the power to meet the E_b/N_0 target, while the outer loop adjusts the E_b/N_0 target based on BER. As in other CDMA technologies, this dual loop is important because W-CDMA anticipates that some of its subscribers will be stationary while others are moving and will require a higher E_b/N_0 to achieve the same BER performance.

In TDD mode, the power control is slower, from 100 to 750 times per second, using a closed loop in the forward direction with the same inner and outer loops. The reverse uses only open-loop power control. We can get away with this because TDD uses the same frequency for both directions, so the fast-fading characteristics will be exactly the same for forward and reverse radio paths. In FDD systems, the different frequencies for forward and reverse carriers make the Rayleigh fading characteristics different enough that separate power control is necessary.

10.3.7   W-CDMA cell identification

Cell sectors are differentiated by 512 distinct scrambling codes rather than through the use of timing differences. The user terminal acquires the cell signal by a three-stage search:

- Every sector has a primary synchronization channel (SCH) with the same 256-chip Golay code transmitted 1500 times per second. (Golay codes are a set of PN codes with special correlation properties.) The user terminal locks onto the strongest primary SCH.

- The user terminal looks to the secondary SCH, synchronized with the primary SCH, for one of 16 groups of scrambling codes.

- Each of the 16 groups has 32 codes, and the user terminal tries these in sequence to find a match.

This procedure is much faster than searching through 512 PN codes in sequence.

Having distinct codes for each cell sector avoids any problem of timing errors due to radio propagation delays at the speed of light.

10.4   TD-SCDMA

Time division synchronized CDMA (TD-SCDMA) is the newcomer among 3G CDMA standards, and it has a different approach. cdma2000 was built from North American cdmaOne, and W-CDMA was built from European GSM. As a result, they tended to reinforce, rather than resolve, the economic competition and rivalry between their 2G predecessors. TD-SCDMA took a different approach. Although its network components are optimized as a GSM upgrade path, its air-interface standard appears to be an effort to optimize capacity and flexibility using almost every available technology.

TD-SCDMA flexibly allocates air-interface resources using four different technologies for dividing the available capacity of the air interface: time division, code division, frequency division, and space division.[15] In addition, it is optimized for practical data applications. In addition to supporting variable bandwidth in each direction, it has special features to support a variable bandwidth ratio, a key support factor for Internet browsing, graphical downloads, and other data functions.

In this section we will explain how multiple technologies were brought together to optimize the potential for the air interface in the standard that the People's Republic of China has chosen for its cellular deployment to the world's largest subscriber market.

10.4.1   TD-SCDMA services

TD-SCDMA offers the entire suite of 3G wireless voice and data services. Symmetric voice and data service are supported, and they can migrate smoothly to asymmetric data service where most of the flow is in one direction. Reverse data rates go up to 384 kbps, whereas forward data rates can go as high as 2 Mbps under the most favorable radio conditions. All these data services are available in circuit or packet mode engineered for efficient IP data service.

10.4.2   TD-SCDMA voice channel

The speech data rate for TD-SCDMA is 8 kbps. A single carrier can handle as many as 48 simultaneous voice calls with a spectrum efficiency three times greater than GSM. The technology is designed for smooth migration from GSM while increasing spectral efficiency by a factor of 3. Simulations show the 1.6-MHz TD-SCDMA carrier handling 28 calls per cell at 60 km/h, 24 calls per cell at 120 km/h in the city, and 22 calls per cell at 120 km/h in a rural setting.

10.4.3   TD-SCDMA radio bandwidth

TD-SCDMA uses a TDD carrier of 1.6 MHz. There is no FDD mode because the TD-SCDMA research is clearly oriented toward serving a voice and data market that can be balanced in its forward and reverse usage and can be highly asymmetric with large

---

[15]But TDCDFDSDMA would be a bit of a mouthful.

downloads of data. It is easier to find narrower radio spectrum allocations of 1.6 MHz than the broadband 5-MHz carriers require, especially in their preferred FDD mode, where they require 10 MHz.

The seven time slots of TD-SCDMA allow the system to vary its forward to reverse ratio from 6-to-1 to 1-to-6 dynamically as demands change.

10.4.4   TD-SCDMA channel coding

The TD-SCDMA channel is both time division and code division multiple access and divides its resources as required by its user community. With time slots of 10 ms, it has 100 opportunities per second to adapt to its service requirements. The TDMA mode is the dominant mode because it can operate at full rate. The CDMA mode of TD-SCDMA can handle spreading gains of up to 16.

Like the TDD mode of W-CDMA, TD-SCDMA uses a midamble in its data bursts for coherent demodulation because a steady pilot is impractical without a steady carrier in each direction. TD-SCDMA user terminals compensate for the distance from their serving base stations by advancing their timing. This eliminates collisions between adjacent time slots that would occur due to propagation delay at the speed of light. The timing advance allows the system to use very short time gaps between adjacent time slots so that air-interface time is not wasted.

10.4.5   TD-SCDMA space division with smart antennas

A central design point of TD-SCDMA is the use of adaptive antenna arrays, also called *smart antennas*. Since the radio frequency is the same in both forward and reverse directions, we can rely on measurement of the reverse signal to determine the phase relationships in the forward direction.

An array of eight antennas measures each channel 200 times per second and adapts its phase and amplitude relations to achieve an improvement of 8 dB in signal to interference (S/I) over omnidirectional antennas. This is referred to as *space division multiple access* (SDMA) because it takes advantage of the distance in space among user terminals to reduce interference.

10.4.6   TD-SCDMA dynamic channel allocation

While the time slots and code division control intracell interference, TD-SCDMA minimizes its intercell interference using four multiple access technologies:

- *TDMA* is used by allocating traffic to the least-interfered time slots.

- *FDMA* is used by allocating traffic to the least-interfered radio carrier because multiple 1.6-MHz carriers can fit in a radio spectrum of several megahertz.

- *Space division multiple access* (SDMA) is enabled through the use of adaptive phased arrays, that is, smart antennas.

- *CDMA* is used by allocating traffic to the least-interfered code, with 16 codes per time slot.

These allocations are done dynamically to optimize radio resources according to the interference at the moment.

10.4.7   TD-SCDMA joint detection and power control

Joint detection (JD), also known as *multiuser detection* (MUD), is a mathematically complex technology that filters out same-sector CDMA interference more effectively than the PN-code multiplication used in other current CDMA standards. JD uses a training sequence within each time slot so that the receiver can estimate the parameters of the radio channel. In TD-SCDMA, the emphasis on time division multiplexing allows the CDMA component to operate at a lower spreading gain. Having only 16 possible codes allows the receiver to do the calculations. The high number of codes used in other 3G CDMA systems would require huge computer processors to perform the calculations necessary for optimal multiuser reception.

The claims are that JD offers significant gains in CDMA capacity and that it alleviates the near-far problem by allowing differences as large as 20 dB in reverse channels signals. The result of this is that TD-SCDMA does not require high-speed power-control algorithms. Eliminating sophisticated and rapid power control reduces the CDMA capacity overhead for power control. This allows the resource that otherwise would be dedicated to power control to carry subscriber voice or data, increasing capacity. This change also makes the equipment simpler and cheaper.

10.5   Conclusion

We have now explored the standards that allow multiple vendors around the world to produce interoperable equipment for CDMA systems. Standards are intended to support the rapid deployment of reliable, compatible, high-quality systems. Implementation in the real world always falls below the ideal of the standards to some degree. As we move forward, we will first provide additional background information in data systems and telephony and then move on to discuss the real-world capacity and quality issues, helping you to improve your real-world systems so that they approach the optimal design implied in these standards.

Key Telephony Concepts

The primary purpose of the cellular system is to carry telephone calls. However, readers with a radio engineering or data systems background will not have an in-depth knowledge of the fascinating history and deep concepts that underlie telephony. The public switched telephone network (PSTN) is a worldwide technological marvel, and much of its technology is adopted and adapted into the wireless network.

Part 3, "Key Telephony Concepts," provides knowledge of telephony engineering and technology and its application to wireless networks.

In Chapter 11, "The PSTN and Telephone Switching," we look at the basic structure of the PSTN, including the local loop and local access provisioning, PSTN trunks and switches, and private branch exchanges (PBXs). We then take an in-depth look at telephone switches and their functions. With this background, we can present the switching functions that are key to cellular telephone networks, including roaming, paging and registration, all the different types of handoff, and the call statistics that allow us to trace problems and to bill our customers. Chapter 11 provides the tools you need to design and troubleshoot the switches at the mobile switching center (MSC), the nerve center of our cellular network.

In Chapter 12, "Telephony Engineering Concepts," we describe the steps of a telephone call and define the issues that determine quality of service. We also describe the architecture and engineering issues significant to wireless telephone networks.

In Chapter 13, "Telephone Transport," we introduce the technology that carries telephone calls over distance. After a discussion of the protocols of telephony, we take an in-depth look at wireless system transport, which is the basis of the backhaul network that connects base stations to the MSC.

In Chapter 14, "Signaling with SS7," we turn our attention to the system that provides signaling control for the entire worldwide telephone network. We describe the underlying principles and features of the Signaling System 7 (SS7) network and its operational functions.

In Chapter 15, "ANSI-41," we introduce and explore ANSI-41, the standard that defines the signaling operations for all code division multiple access (CDMA) networks, both second generation (2G) and third generation (3G). ANSI-41 defines the communications that must be used to support inter-MSC handoffs, roaming, and the short message service (SMS). It also contains functions for operations, administration, and maintenance (OA&M) and over-the air service provisioning. ANSI-41 is crucial to 3G network design and operations because it defines how CDMA networks will communicate with the Internet and other data networks around the world. Although ANSI-41 officially sets the standard only for interoperability between networks, in reality, it functionally defines the core aspects of every CDMA cellular system.

In Chapter 16, "Call States," we introduce a practical design and troubleshooting tool, the state diagram. As we introduce the tool, we also provide models of landline and cellular call states for traditional telephone calls, call waiting, and three-way calling. If you are not already familiar with state diagrams, you will find that they help engineers define problems clearly, speeding the way to a solution.

The capacity of any system is limited by the capacity of its weakest link, its bottleneck. While CDMA is the technology of the air interface between our subscriber's cellular telephones and our network, an optimized CDMA radio link will not solve all our capacity or quality problems. With a deep understanding of telephony, we can design and implement optimal cellular networks not only at the base station but also at the MSC and in the backhaul network that connects the base stations to the MSC.

Chapter 4

The PSTN and Telephone Switching

Telephone switches are the core of the public switched telephone network (PSTN). It is hard to imagine a telephone system without switches, although there was one on Wall Street about 100 years ago. In some very old movies you will see a businessman's desk with over a dozen phones. One rings, and the executive fumbles, picking up phone after phone until he knows who called. Each phone had only one wire leading to one destination, with one phone at the other end. A person needed a separate telephone with a direct line to each person he or she might ever want to call.

Switches eliminated that limitation, allowing any one telephone to reach any other telephone. The earliest switches were manual patch-bays, where an operator would sit in front of a board and make connections as subscribers requested them. These connections may have started with people's names, but the telephone company was soon large enough that individual lines were referred to by number: "Operator, please connect me to 7283." These gave rise to another Hollywood image, that of the operator plugging and unplugging cords, one after the other, throughout a busy day.

With the advent of automated switches, people were not part of the switch any more, and Hollywood lost interest. But this is where our interest begins. In this chapter we will explore the development and function of the automated telephone switch from the early mechanical days to the latest electronic technology. This chapter's final section will look at the additional functionality added to switches to support mobile telephony services.

11.1   The Public Switched Telephone Network (PSTN)

The core of the PSTN is the network of switches that are used to carry a telephone call from the switch that connects to the calling party to the switch that connects the called party of a telephone call. The cabling and equipment that connect telephones to the network of switches are often referred to as the *local loop*.

11.1.1   The local loop

The vast majority of residential subscribers access the network using analog devices such as two-wire analog telephones, fax machines, and modems. The PSTN services these customers by providing a twisted pair of wires to connect the telephone to the local exchange office.[1] A novel feature of the analog loop is that it uses a single twisted pair for both the talk path and the listening path of the call, as well as the signaling of digits, sensing of on- and off-hook, and ringing the telephone. Of particular interest to us is the ability to carry both the talk and listening paths. This is accomplished by using a device called a *hybrid circuit* at both ends of the loop. The hybrid circuit essentially channels the incoming signal to the earpiece while channeling the signal from the microphone or mouthpiece out to the line without excessive energy going to the earpiece. Ideally, it also ensures that no energy is reflected back to the other end of the line, a phenomenon called *echo*.

Modern local exchange offices may use specialized equipment located close to the telephone to minimize the length of the twisted pair of wires. For the purpose of this brief introduction, the scenario used will be that the twisted pairs extend all the way back to the local exchange office. It is important to note that each telephone number is assigned a physical presence—think of it as a pair of terminals—on the local exchange office. A call to a telephone number will result in the switch sending the call to that physical location on the switch. All telephones connected to the other end of the twisted pair are assumed to have an assigned telephone number.

The actual procedures used between the telephone and the telephone switch to communicate the user's desire to place a call or to alert the user that the telephone number is being called are the topic of Sec. 12.1.

A useful concept is that of a *line*, a transmission path that can be switched only at one end. The twisted pair that connects the telephone to the switch is a line, since it is static at the telephone end and only switched at the local exchange office end. If the twisted pair were connecting a pair of switches such that the twisted pair could be switched at both ends in order to complete a circuit, then the twisted pair would be called a *trunk*, a transmission path that can be switched at both ends. If the trunk is virtual, that is, digitally multiplexed with other trunks onto a digital facility, then it is still referred to as a *trunk*. The physical facility on which the multiplexed digital signals are being carried is often referred to as a *line*, but the trunks that are being carried remain trunks. A collection of several trunks between two switches is referred to as a *trunk group*.

11.1.2 PSTN equipment

The term *local exchange office* is often used to describe the telephone switch used to connect customer equipment to the PSTN (via the local loop). Two other common terms for this switch are *central office* and *class 5 office*. The latter term has its origins in the early architecture of the PSTN, which used a hierarchy of switches, class 1 through class 5, to perform the function of routing calls across long distances.

Unless the calling and called parties are connected to the same local exchange office, the call will be routed through additional switches to get from the calling to the called party. The calls will be routed between the switches over trunks. The trunks may, in rare instances, still be a pair of analog wires. A more common form of analog trunk uses

---

[1]Twisting of the wires reduces the susceptibility of the wires to interference from magnetic fields.

a pair of wires in each direction so that the two voice paths remain separate. The vast majority of trunks in use today are virtual in nature rather than physical trunks because they are digitized and multiplexed onto some form of digital facility capable of handling multiple simultaneous trunks. (In telephony, we are using the term *virtual* for these trunks because they are not separate wires but multiplexed into a larger facility. These trunks have a well-defined presence on the specified facility, with specific bits reserved for them. In data communications, when we refer to a *virtual circuit* the connection is defined only by its end points and not by any specific facility or route.) The basic digital building block in North America is the DS-1, capable of supporting 24 trunks. These trunks may traverse many other types of equipment in order to get from one switch to another, such as multiplexers capable of routing the trunk over both copper and optical fiber, cross-connects, and equipment designed to cancel any echo from the other end of the call.

The most common form of digital encoding used by these digital facilities is referred to as *pulse code modulation* (PCM) and has been standardized to be a 64-kbps bit stream consisting of 8-bit samples taken at 8000 samples per second. This data encoding technique (described in Sec. 18.1) is assumed for the design of the various digital transmission facilities and equipment used to interconnect telephone switches and assumed in the design of the switches themselves.

The DS-1 facility, capable of carrying 24 simultaneous calls, is formed by concatenating 24 eight-bit samples into a frame and then adding a framing bit used for frame synchronization. Therefore a DS-1 frame is 193 bits long, carrying 192 bits of information. The repetition rate is 8000 frames per second, as required by the PCM coding method, resulting in a bit rate of 1.544 Mbps, of which 1.536 Mbps is information, that is, the 24 samples. The position of the sample within the frame is commonly referred to as a *time slot*.

A DS-1 facility actually consists of two DS-1 framed transmission paths, one for the forward direction and one for the reverse direction, so that the voice path is completed in both directions. The 24 time-slot pairs that result when used to connect switching systems are still referred to as *trunks*. A DS-1 facility can be used for purposes other than carrying trunks, so a more general term for a time-slot pair is a *channel*.[2] Chapter 13 describes in greater detail the hierarchy of digital transmission facilities.

There are two standards for interpreting the analog value of the voice signal and creating an 8-bit digital representation. These two standards are commonly referred to as *μ-law* and *A-law*. μ-Law is used in North America and Japan, and A-law is used in Europe. Many switching systems are capable of handling both, as well as converting between them for international traffic. This subject is treated further in Sec. 18.1.2.

11.1.3 The private branch exchange (PBX)

The private branch exchange (PBX) was developed originally to be a small switch with many of the functions of a local exchange office but located within the customer's offices and owned by the customer. Modern PBXs can be used in this manner for a few

---

[2]DS-1s can be set up without predefined channels. The subdivision of a DS-1 into 24 time slots is commonly referred to as *D4 channelization*.

telephones up to tens of thousands of telephones and can be equipped not only with the features of a modern local exchange office but also with many advanced features required for the business customer, such as call centers and computer-telephony integration. Some PBXs such as those manufactured by Avaya, Inc., actually have been used as small local exchange offices and for enhanced services.

Originally, businesses would purchase separate telephone numbers from the PSTN, such as AT&T in the days of the Bell System, and the PSTN would run a separate pair of wires to the business for each telephone number purchased. Over time, as business services matured, a business service called *Centrex* emerged. With Centrex, users within the customer's community of telephones could use abbreviated dialing (less than seven digits) to get to other telephones within the community. Special features were added to attract business customers. The customer's telephones were still connected by dedicated wires to the local exchange office. If the customer purchased a telephone number for each telephone, outside callers could reach the customer's telephones directly, a service called *direct inward dialing* (DID). The customer also could purchase a limited number of telephone numbers and use an operator or second dial tone to reach individual telephones. Each telephone in this arrangement had an extension number, a number not visible to the PSTN.

Some business customers chose to purchase only a few numbers from the PSTN and to install a switchboard from which an operator would connect the numbers to telephones within the business. This approach used the telephone lines more efficiently, but a human operator was required to connect calls.

Jumping forward in time, business customers today still have the option of purchasing Centrex services, but many choose to own their own switching equipment in the form of a PBX. PBXs and Centrex remain in heated competition, but the PBX offers as advantages the economies of being able to use fewer connections to the PSTN, the ability to set up private networks, and advanced features not generally available from Centrex.

The modern PBX connects to the local exchange office and sometimes directly to interexchange carriers as well using groups of trunks. The actual number of trunks used is selected as appropriate to the number and length of calls expected and the customer's tolerance for blocking, that is, getting a busy signal. The selection of the correct number of trunks involves a discipline known as *traffic engineering,* discussed at length in Chap. 23.

Most PBXs support DID, the ability for callers to use a PSTN telephone number to dial the telephones directly behind the PBX rather than to call a main number and then to give an extension number. DID requires that the business purchase from the PSTN a range of telephone numbers, which the business then assigns to the telephones behind the PBX. When a call comes to the local exchange office for a telephone in that range of numbers, it will seize an idle trunk from the trunk group going to the PBX and pass the call to the PBX, much as it would to another local exchange office. The PBX is responsible for associating the telephone number with the line leading to the correct telephone and completing the call.

A recent development in the PBX marketplace is the Internet Protocol (IP) PBX. In its most basic form, it is a PBX that places calls through an IP network and internally routes calls to the correct telephone, which itself is an IP telephone. Advanced IP PBXs are capable of routing calls from IP or traditional telephones either via an IP network or the PSTN.

11.1.4 The organization of the PSTN

In 1985, the North American PSTN was organized as a clear hierarchy of levels from the class 5 local exchange office to the most central class 1 tandem switches of the long-distance network. Many of the concepts associated with telephony today draw their heritage from the Bell System, which was broken up in the 1980s. The Bell System of the early 1980s provides a comparatively straightforward example of the architecture of a telephone network and for this reason will be used as a starting point for our exploration of the organization of the PSTN.

Prior to the breakup of the Bell System, all telephones were connected to the local exchange office. The local exchange office was responsible for monitoring each line to detect when a customer went off-hook, then to connect the appropriate resources to detect any dialed digits, and then to signal the customer that dialing could commence by providing dial tone. The digits collected would then be used to determine how to connect the call. The local exchange office had the additional responsibility of monitoring the length of the call and creating a record of the call start time, length, and called number [called an *automatic message accounting* (AMA) record] for use in rendering a bill for the call.

This local exchange office occupied level 5 in a hierarchy of switches in the network and alternately was called a *class 5 office*. Each class 5 office was identifiable by a three-digit number called the *office code*.[3] The lines within the office were identifiable by a four-digit number. This combination of three and four digits was the seven-digit telephone number that customers would dial (when within the area code) to reach another telephone.

The three-digit office code was cleverly designed so that the digits 1 and 0 never appeared in the second position of the code. When a class 5 office saw the three-digit code with neither a 0 nor 1 in the second position, it knew to expect only 7 digits. Once the digits had been collected, the switch would use its internal parameter database, called *translations,* to determine what to do with the call. If the office code belonged to this class 5 office, then it would terminate the call locally. If the office code did not belong to this class 5 office, then it would examine its translations to determine what trunk to route the call out over. If the call were to another class 5 office to which it was connected directly, then it would select an interconnecting trunk. If there was no direct connection available or all direct connections were busy, then it would then route the call to a class 4 office.

If the first three digits dialed by the caller contained a 1 or a 0 in the middle-digit position, then the class 5 office knew to expect 10 digits and to treat the first three digits as an area code. As a general rule, all calls with an area code were routed directly to the class 4 office.[4]

The class 4 office would then use its translations to determine if the call should be sent directly to a class 5 office, to another class 4 office, or up to a class 3 office.

Toll switches, classes 1 through 4, prior to the breakup of the Bell System had no responsibility other than routing calls and passing signaling information back and forth. The AMA messages were created by the class 5 office where the call originated.

---

[3]A single office code supports up to 10,000 numbers. Therefore, larger offices would have more than one office code. There also were instances where very small offices would share an office code.

[4]Other mechanisms are used today, although not consistently, to identify if the office is to expect 7 or 10 digits. Hence the restrictions on the use of 1 and 0 largely have been dropped in the North American numbering plan.

The breakup of the Bell System created major architectural changes and new responsibilities for the class 4 office. First, all class 5 offices were given to the local exchange carriers (LECs). AT&T, in its new role as an interexchange carrier (IXC), retained the class 4 offices and above. Since there were many small IXCs, as well as many class 5 offices, it was deemed unreasonable for each IXC to have to connect to every class 5 office. Therefore, the concept of an access tandem office was created.[5] The access tandem office was essentially a class 4 office owned by the LEC. Connection to an access tandem would permit an IXC access to all the subtending class 5 offices. Some carriers, most notably AT&T, would continue to connect directly to class 5 offices. Whichever access arrangement was chosen, the access point itself was referred to as the *point of presence* (POP).

The class 4 offices owned by the IXCs now had the responsibility of tracking for billing purposes the calls that were routed to the IXC's network. For a long time after the AT&T breakup, this capability was not available, and carriers, particularly AT&T, paid the LEC to render its bills.

Section 11.1.1 described how the telephone communicated the desired telephone number to the local exchange office. If the local exchange office needed to send the call to another office, it also had to communicate the desired telephone number to that office. A method still in use today to accomplish this is to signal directly over the trunk selected for the purpose. A detailed description of the technique used will be left to Chap. 13. This method does not provide, however, the rich feature set available in the network today and largely has been replaced by an out-of-band signaling approach, commonly called *common channel interoffice signaling* (CCIS). CCIS uses a data network to communicate the desire to set up a call through the various switches in the call path. It is possible to reserve all the trunks needed to complete the call before switching occurs.

This data network uses an internationally standardized protocol called Signaling System 7, commonly abbreviated *SS7*.[6] This signaling system is essentially a datagram service designed with redundancy for high reliability. When the Bell System was broken up, each operating company operated its own signaling system. Calls placed between LEC and IXC traversed the POP using an access protocol (called *Feature Group D*) that was based on signaling over the interconnecting trunks. Today, SS7 is used commonly to communicate between LECs and IXCs for a more seamless and efficient network.

The European PSTN is functionally and architecturally similar to the American PSTN. It is a highly reliable network using SS7 switching. However, the digital facilities for voice and signaling are E lines rather than DS lines, as described in Sec. 13.1. The European network's system of telephone numbers is quite different because of its different history. The European system is developing through the increasing unification of what were separate national phone systems, whereas the North American system is developing largely through the continuing diversification of what was once a corporate monopoly.

Other nations around the world have a variety of PSTN networks of greater or lesser reliability. Most are based on either European or North American models. The same is true with the allocation of frequency for wireless telephony: Most nations follow either the European model or the North American model of radio bandwidth allocation.

---

[5]A tandem office is used to switch a call received over a trunk to another trunk.

[6]Sometimes SS7 is referred to as *Common Channel Signaling 7* (CCS7).

11.2   Telephone Switch Technology

Once switches were automated, the process of providing a number to the switch needed to be automated as well. The pulse dialing method was adopted, and the telephone company developed the rotary dial that could send electrical pulses in the form of momentary line disconnections.[7] The telephone central office switches detected these pulses and completed calls using the dialed digits. These were step-by-step switches in which each pulse would move a motor one step at a time and, later, crossbar switches in which combinations of dialed digits made connections. The crossbar was named for its relay-driven rows and columns of connecting rods. These were magnificent machines offering machine-operated telephone service in a time before computers, when few other things were automated.

In 1947, Bell Telephone Laboratories helped open the age of digital electronics with its invention of the transistor. Less than 20 years later, the digital computer brought a new era to telephone switching with introduction of the electronic switching system (ESS). Electronic switching systems use computer software to control the electronic signal pathways. In 1965, Western Electric, the technology and manufacturing arm of the Bell Telephone System, came out with the 1ESS, which could handle 60,000 lines. In 1976, the 1AESS more than doubled the 1ESS capacity with its more advanced 1A processor. By 1985, the 1AESS switch was able to handle 250,000 subscriber lines.

The 1AESS is an *analog switch*. While the 1A processor is a digital computer, the telephone wires going in and out of the 1AESS are analog wires, and the connections inside the 1AESS are mechanical switches. These reed switches are controlled by tiny electromagnets that open and close circuits.[8]

The early Advanced Mobile Phone Service (AMPS) cellular systems used the 1AESS as their mobile telephone switching offices (MTSOs), so AMPS was a fully analog system, from the mobile telephone along the analog FM radio link, through analog base station electronics, along analog facilities to the MTSO, through an analog telephone switch, and out to the PSTN. Today's AMPS calls go through digital facilities and digital equipment, but the air interface remains analog FM.

A digital switch not only has a digital computer but also has the ability to interface directly to digital facilities, which we will discuss in Chap. 13. The Northern Telecom DMS-10 switch was introduced into small local exchange offices in 1981, and the DMS-100 was introduced for larger offices. AT&T's Western Electric Division also introduced a digital local exchange office switch, the 5ESS. Because of the popularity of digital switches, implementations of analog switches quickly declined.

The 4ESS is a digital switch using the 1A processor originally used in the analog 1AESS. The 4ESS is a *tandem switch* designed to connect to trunks rather than lines. Trunks are shared facilities that can be used by any call received by the switch rather than dedicated, as in the case of lines. Therefore, trunks can be engineered for higher utilization. The result is that the number of trunks terminated on a tandem switch is a fraction of the number of lines terminated at a local exchange office.

---

[7]In many places one can still dial a call by pressing the switchhook rapidly up and down on a regular telephone, even a Touch Tone telephone.

[8]Diehard fans of analog switches miss the *sound* of a telephone office. Other than for the whirring of cooling fans, a telephone office with a digital switch is quiet. Step-by-step, crossbar, and reed switches have their own distinctive audible character, and a telephone switching office used to be a noisy place.

11.3   Telephone Switch Functions

Telephone switches are specialized according to the functions they are to perform. Some of this specialization is achieved by optimizing the architecture of the switch, and for some functions, specialization is achieved by the activation of software features.

11.3.1   Connections

Local exchange offices are optimized for the functions that are needed to terminate individual lines (telephones). They are highly modular, capable of terminating a large number of individual lines. Each line is physically associated with a telephone number. The average utilization of a line is very low, so local exchange offices are designed to terminate a large number of lines but typically have a somewhat limited ability to connect those lines either to other lines or to other switches. Finding the right balance between the number of lines and the internal switching resources is another problem for the art of traffic engineering.

It is impractical to bring every telephone number's twisted pair all the way back to the local exchange office, especially for very large offices. In practice, the connections to the local exchange office are distributed by using remote switching modules and subscriber-loop carrier systems, essentially a form of multiplexer.

The vast majority of lines will be installed and administered, provisioned in telephony jargon, as two-wire analog lines. The LEC also may offer two- or four-wire digital service, called *ISDN Basic Rate*. This service provides two 64-kbps time slots and a separate 16-kbps time slot used for signaling. ISDN Basic Rate is widely deployed in Europe. It is not widely available, or used, in North America.

Local exchange offices are also used to provide service to PBXs, and this is accomplished with trunks. Trunks may be either analog or digital, as required by the customer contract. Digital trunks typically are provisioned as DS-1s. The number of trunks will be engineered for the traffic level and quality of service required by the customer.

The trunks provisioned on digital facilities may emulate the trunk signaling used with analog trunks by embedding the signaling within the 8-kbps samples, a technique referred to as *robbed-bit signaling*. The trunks also may be provisioned with out-of-band signaling. This is accomplished by taking the twenty-fourth time slot of the DS-1 and dedicating it to a signaling channel for the remaining 23 time slots. DS-1 facilities provisioned in this manner are referred to as *ISDN primary rate interface* (PRI) facilities.

The DS-1 facility is primarily available in North America. Europe and much of the rest of the world have standardized on a similar digital transmission technology, called an *E-1*. E-1 has 30 channels as compared with DS-1's 24, for a data rate of 2.048 Mbps.

The trunk side of local exchange offices is connected to other switches almost exclusively by using digital facilities. The basic building block for these digital facilities is the DS-1 (or E-1, where deployed). Since the actual number of trunks required in LEC and IXC offices is large, the carrier facilities used typically operate at higher rates than DS-1 or E-1. This is accomplished by multiplexing multiple DS-1s or E-1s onto a larger facility.

A digital signal hierarchy has been established for these higher rates. For instance, four DS-1s are multiplexed to form a DS-2, and seven DS-2s are multiplexed to form a DS-3. The DS-2 is, in practice, an intermediate step in the hierarchy, and the transmission facility actually used will carry a DS-3, consisting of 28 DS-1s. Optical facilities may be used to carry even higher rates. Internationally, E-1 facilities are also multiplexed up to higher rates.

11.3.2 Switching

Modern digital switches are designed to switch individual calls in 8-bit samples at a rate of 8000 sample per second. Analog signals, such as those common on two-wire lines, are converted to four-wire signals using hybrids and digitized to the 64-kbps rate. Digital channels derived from the time slots present on digital facilities are presented directly to the switch.

Two channels, one for each direction of the call, must be switched to have a complete call. Since the incoming and outgoing legs of the call may not be in synchronization with the internals of the switch, buffers are used by the switch to store the 8-bit samples until the switch is ready to accept them for switching and also prior to placing the sample in the outgoing time slot. The buffering capability of the switch is limited, requiring timing of the incoming and outgoing DS-1 signals to come from a reliable source. LECs and IXCs have strict synchronization plans to ensure that disruptions in the signal path caused by timing errors are minimized.

11.3.3 Managing calls

The jobs we usually associate with a telephone switch are call processing and accounting. These tasks are easier in a wired system than in a wireless system because for land telephones, the telephone numbers are related directly to physical presence on a specific switch, and all accounting is associated with that switch. Wireless switches need to do more work to process calls and handle accounting functions because of the mobility of the wireless terminals.

11.3.3.1 Call processing. The local exchange office has the task of strobing each line to detect an off-hook condition, applying the internal resources needed to collect the dialed digits from the caller, and then interpreting the caller's intent. The local exchange office also must locate and ring the called telephone and send tones that sound like ringing to the calling party's telephone. The local exchange office must detect if and when the called party accepts the call. Once the called party accepts, the originating local exchange office must be notified so that the originating office can remove the locally generated ringing tone from the calling party's line and cut through the voice path. At this time, the local exchange office that connects the calling party initiates a billing record for the calling party.

When either party hangs up, the on-hook condition must be sensed and communicated to the other local exchange office participating in the call so that a disconnect can be forced at that end and billing can be stopped.

The switch also deals with call states other than normal call completion, such as busy signals, blocked trunks, and various telephone network failures. Failures of telephone equipment are rare, but the telephone company does not throw its arms in the air and give up just because a piece of equipment stops working. When a call needs a piece of equipment that is not available, the switch has to know what to try next, a shift to alternate equipment, a fast busy signal, or a recording. The software in the telephone switch has this knowledge programmed into it. Most of the software in a telephone switch is for situations that never or seldom occur.

Call processing also includes the internal activity of trunk selection for each call. The local loops at each end are determined by the calling party or called party, respectively, but the intermediate links are selected and managed by telephone switches.

11.3.3.2 Accounting.   The local office switch also has the billing function, not always as popular among customers as call processing. Calls have to be tracked and monitored not only to be paid for but also for planning the growth and maintenance of the telephone network. The switch generates the data for these business functions and stores them as AMA records on magnetic tape or in a computer data file.

The other accounting function is account verification, which ensures that subscribers who do not pay their bills are not allowed to make calls. This function requires that the system read and evaluate the subscriber's identification data from the system at the beginning of the call. This is considerably more complicated than storage of data about a call, which requires only writing data, not reading them.

11.4   Mobile Switch Functions

The two big differences between landline switching and mobile switching are the need to switch calls while they are in progress and the need to support a radio link across the air interface rather than a landline local loop. These are not cosmetic or administrative differences. A mobile switching center (MSC) spends much of its time managing radio link assignment and handoff. An MSC also must convert the compressed voice channel optimized for the radio link into a pulse code modulation (PCM) signal acceptable to the PSTN. We will now look at several of the management functions specific to mobile switches.

11.4.1   Registration

Each wireless user terminal registers with the system to announce its presence. The system therefore has a complete picture of all the wireless telephones in its service area. This picture is not perfect because wireless telephones can leave the service area or lose power. In Sec. 12.5.2 we describe technologies designed to identify user terminals that were registered but which are no longer accessible on the system. When a user terminal recognizes a new system, it registers again.

Each MSC keeps a list of the user terminals that are active in its coverage area, and it pages only those terminals for incoming calls.

11.4.2   Paging

When a call comes in for a registered user, the wireless system pages the telephone. Only in the earliest days of AMPS were there few enough cellular telephones that we could afford the luxury of paging the entire system for every incoming call. In the huge wireless market of today, the relationship between registration and paging is an important capacity design issue.

While unanswered pages are wasteful of system resources, not paging a subscriber expecting an incoming call is far worse. The system design has to be aggressive in tracking and hunting down the intended recipient of any incoming call attempt. Many subscribers are familiar with a typical result of a network's failure to find a subscriber's phone. The subscriber never hears the cellular telephone ring but then does receive notification of voice mail from someone who called just minutes before. The network is designed to prevent this kind of frustration whenever possible, but momentary failure of the link across the air interface or unrecognized movement of the mobile terminal from one serving area to another can create these failures to connect a call coming into a mobile telephone.

In the Internet/ethernet world of today, the notion of a broadcast page to a specific user does not seem strange. In the world of telephony, however, the concept of calling a telephone that might be anywhere or might be nowhere was a new and strange concept.

The switch also has to respond to calls initiated by subscribers. When the user terminal starts a call, it sends a radio signal to the base station. This signal is similar to the signal sent in response to a page.[9]

11.4.3   Roaming

Since subscribers can roam from system to system, the MSC finds itself part of a network that keeps track of all its telephones all over the world. The home and visiting systems have to communicate on several levels.

11.4.3.1   Defining home and roam.   Even the terms *home* and *roaming* have different meanings at different levels of the cellular system. In the ANSI-41 standard, a user terminal is on its *home system* when it is being serviced by any base station served by its home MSC, and it is *roaming* when it is registered at any other MSC.

There is an intermediate level of the network that is a system (containing one or more MSCs and their base stations) all identified by one system identification (SID). (See Chap. 15 for details.) For some purposes, roaming may refer to being on any system with a SID other than the SID of the home system. Call management is particularly difficult when the equipment on the system on which the user terminal is registered is different from the equipment on the home system. For example, hypothetically, a subscriber might subscribe to voice-mail service but be unable to reach his or her voice-mail messages if the voice-mail function is not supported on the system where the user is currently registered. A user terminal's ROAM indicator will go on when the SID of the MSC where it is registered is different from the home system's SID.

At the largest level, one could consider the entire continental network owned by one service provider to be the home system and roaming to be the state where a user terminal is being served by a different service provider. From the customer's perspective, this is the case for Sprint PCS customers in North America. All home rates apply whenever the user terminal is using the Sprint PCS North American Network. If a Sprint PCS base station is not within range, the cellular telephone attempts to find first an alternate digital provider and then an analog provider. If it succeeds, the subscriber is notified of the digital roam or analog roam state and warned of roaming charges that will apply, which may exceed 40 cents per minute.

---

[9]Here is an example of a potential glare condition (described in Sec. 12.3): The subscriber can initiate a call while the user terminal is being paged for an incoming call. The MSC software has to resolve this.

In the following technical discussion we will be referring to the ANSI-41 definition of roaming as what happens when the user terminal is registered at any MSC other than the home MSC unless we specify otherwise.

11.4.3.2 Roaming registration (HLR and VLR). When a subscriber terminal registers with the system, it sends its *mobile identification number* (MIN) and its *electronic serial number* (ESN) to the network. The MIN is a unique 10 decimal digits (40 bits).[10] The ESN is a 32-bit binary number encoded during the manufacturing process. The serving MSC uses this information to retrieve the subscriber's service profile record stored in its *home location register* (HLR). If the subscriber is roaming, then the serving MSC is not the home MSC, and it sets up a record in the *visitor location register* (VLR). The VLR record is temporary and only remains active while the user terminal is registered at that particular serving MSC. The HLR is informed of the location of the serving MSC and stores the current registered location and status of the user terminal inside the subscriber's record.

In addition to establishing identity and location, the system must verify the subscriber's right to use the system. Are this subscriber's wireless telephone bills paid? If the subscriber is cut off from service at home, then we want to cut off service everywhere else as well. The system must allow a roaming subscriber to use only services that are in the subscriber's contract and must track the usage of each service and bill it at the appropriate rates.

Incoming calls may come to the home MSC and may then be routed to the MSC serving the user terminal.[11] Even in landline telephony, it is not unusual to involve multiple telephone switches in one call, but it is strange to have two telephone switches both involved in the local-loop termination of a call. However, this is a normal situation for incoming calls to roamers.

11.4.3.3 Roam dialing. A call from a roaming phone is an administrative task unlike any faced by a landline central office switch. Since the days of AMPS, the rule has been "When you roam, dial like home." This has become harder in today's international roaming world, and the wireless telephone systems seem to be keeping up, at least some of the time. In the Sprint PCS system, where the entire Sprint PCS North American Network is the home system from the perspective of the subscriber and for billing and service purposes, the original dialing rule is not fully functional. As long as the subscriber dials from his or her home area code, there is no problem. Dialing either a 10-digit number or a 7-digit number will work for a local call. However, if the subscriber leaves his or her home area code, then the subscriber must use 10-digit dialing for all calls, including both calls to the home area code and calls to the area code where the subscriber is currently registered as a visitor. A subscriber can work around this limitation by using the cellular telephone's capacity to use 10-digit dialing even in the home area code. If all numbers are dialed with 10 digits, then the subscriber dials all numbers the same way anywhere on the North American network.

---

[10]The MIN originally was designed to contain a North American directory number, the subscriber's mobile telephone number. In the worldwide wireless telephone environment, the MIN and the mobile directory number (MDN) are separate entities that are often set to the same number.

[11]The telephone world is moving toward more sophisticated signaling, and calls may be routed directly to the serving MSC.

11.4.3.4 Roaming support for PCS.

The PCS standard intends to change the goal of telephony from being point-to-point communications to being person-to-person communications. PCS services, and the equivalent custom calling landline services, demand new and different management functions in switching systems.[12]

Call forwarding requires the switch to store an alternate phone number entered by the user and to provide that number as the destination number for incoming calls. Three-way calling requires that the single radio link be connected simultaneously with two other PSTN-to-caller connections and that the addition and departure of callers be managed. Call waiting requires the handling of two simultaneous calls and the switching between those calls. Caller ID and last-number-dialed functions require the capture and temporary storage of the phone numbers of calling parties. Call answering requires storage and management of voice messages. All these functions require additional capacity in a variety of switch components, as discussed in Chap. 24.

Traditionally, switches had only a current state, plus data (stored, perhaps, on a computer tape) recorded—but not retrieved—by the switch. The new requirements of PCS not only demand complex software based on sophisticated call state models, they also need storage *and retrieval* of transitory information by the switch in real time. This requires the addition of memory or storage capacity to the switch, a significant cost item, especially if voice messages are included. Signaling data, such as the number of the calling party, do not require a great deal of storage, but voice messages do. Also, few signal data items are stored beyond the duration of the call, and when they are, usually only one item is stored. In contrast, subscribers can store many long voice-mail messages for a period of weeks.

11.4.4 Handoff

Handoffs are the biggest difference between landline and wireless switching. The traditional PSTN does not switch active calls and does not support any function equivalent to handoff. Call forwarding does redirect a call from one number to another, but it does this before the call is set up, not during the call. Some PBX systems do support host call transfer, where a user requests an active call to be switched to a different extension. However, this is a call-switching action initiated by request from the subscriber, not an automatic process that is invisible to the subscriber.

In the original AMPS cellular trial system, the switch did almost all the work for handoffs; later designs have moved this function into the base stations, where radio measurements and administration are easier to do. Also, moving the workload into the base station relieves the computational load at the MSC. Today the base station controllers are sophisticated enough to handle handoffs within their domain without any participation from the MSC. This means that sector-to-sector handoffs (or handoffs within a cell between server groups described in Sec. 5.2.3) are done completely in the base station.

---

[12]Aside from custom calling, the landline support of local number portability (LNP) may require similar kinds of switch-to-switch coordination as wireless roaming. LNP allows people to keep their phone numbers even when they change address and therefore is part of the general trend away from location-based and toward person-to-person telephony.

Even with the base stations measuring call quality and determining handoffs, both deciding when to look for handoff and where to go, the MSC is still busy with handoff work. When a base station considers a call for handoff, the MSC is used as a base-station-to-base-station message switch to request and receive radio measurements. As a handoff is being considered from one MSC to another, there is communication between the two base stations and between the serving and target MSCs.

When a handoff goes from one MSC to another, we have two telephone switches both involved in the same local-loop end of the same call, an event unheard of in landline telephony. The call goes from the anchor MSC to the serving MSC, as described in Chap. 15. If the call is an incoming roaming call, then *three* MSCs can be involved (home, anchor, and serving), all terminating the same call, and even more MSCs can be involved if this call is handed off to yet another MSC.

In code division multiple access (CDMA), a soft handoff, with a transition process from one serving site to another, is the normal form of handoff. Hard handoff was the only kind of handoff available in the AMPS days, but in CDMA, hard handoff is used only when soft handoff will not work. Therefore, we will discuss soft handoff first and then hard handoff.

11.4.5 Soft handoff

Soft handoff (see Sec. 8.7) requires close coordination between or among cells whose only connection is through the MSC or the MSC network.[13] The MSC receives multiple copies of speech data from multiple base stations for the same call. For each speech data frame coming from the user terminal, the MSC analyzes all the copies it receives and selects the version with best speech quality.

Each base station receiving a signal from a particular user terminal sends power-control signals to the user terminal and also forwards the frames it receives to the MSC. This performs two important functions:

- It allows the user terminal to adjust to the *lowest* power level acceptable to any receiver, reducing interference for all receivers.

- It allows for higher call quality because the MSC can choose the best voice signal frame by frame.

The soft handoff process allows the system to manage user terminals as both sources of calls and as sources of interference.

Any time a call is being served by two or more base stations, it is in the soft handoff state. Soft handoff is a *state* of a call rather than a transition from one state to another. The serving base station receives radio signal information from the user terminal and decides when a call should go into the soft handoff state. The call remains in soft handoff as long as the radio conditions warrant. When a base station's signal for the call becomes too weak, that base station is dropped.

If a call should start being served by base station A and then be in the soft handoff state, served by both base station A and base station B, and then move so that it is out

---

[13]Most soft handoffs are *between* two cells, but some are *among* three cells, a subtle point for the English grammarians reading this book.

of range of base station A and is served only by base station B, we can say, in casual usage, that a soft handoff has occurred.

In soft handoff, there is no change of radio frequency; multiple base stations are serving the call on the same CDMA carrier throughout the process.

11.4.6   Softer handoff

Softer handoff is a handoff between faces of a single base station. It is managed entirely at the base station, without the MSC participating at all. The antennas on the two faces deliver their signals to the base station controller (BSC), which puts the two signals through a rake filter to create a combined signal. The rake filtering at the base station does its calculations at the chip level, which is what makes softer handoff different from soft handoff, where the MSC makes its selection frame by frame, not chip by chip.[14] Like soft handoff, softer handoff involves no change of CDMA carrier frequency.

11.4.7   Hard handoff and semisoft handoff

In CDMA, hard handoff is used in three situations where soft handoff is impossible. They are

- When a call leaves an area serviced by CDMA, and the call is switched to another service.

- When a call goes to a new cell, and the call's carrier frequency is already busy on the new serving base station.

- When the timing of the two base stations cannot be coordinated.

Some CDMA systems use a form of hard handoff called *semisoft handoff*.[15] In hard handoff, the second link to the MSC is set up after the second receiving base station establishes communication with the user terminal. This creates dead space in the middle of the call. In semisoft handoff, the link between the MSC and the new serving base station is established before the switchover of the radio link service to the new base station. The switchover occurs when the user terminal receives and acts on an instruction to change its frequency or timing. As soon as the user terminal switches over, the new receiving base station picks up the signal. The new base-station-to-MSC link is already established, and there is no interruption in the call. Semisoft handoff makes use of the MSC's ability to maintain two links for the same mobile telephone call to eliminate the dead space of a hard handoff. However, semisoft handoff is a type of hard handoff in that the user terminal must make a change (of transmission type, of frequency, or of timing) to establish the new radio link and accomplish the handoff. The switchover occurs at the moment the user terminal makes that change and the semisoft handoff function has already put the new link across the landline portion of the wireless network in place.

---

[14]Another way to make the distinction between soft handoff and softer handoff would be to say that soft handoff is a telephony switching function at the MSC, whereas softer handoff is a radio filtering function at the base station.

[15]The authors also consider *gelato,* a semisoft ice cream, to be a type of hard ice cream.

11.4.8 Call statistics

On the PSTN, as well as on wireless networks, call statistics are kept for the purposes of traffic engineering and system troubleshooting, as well as to support customer billing. The call-tracking issues are, of course, much more complicated in the wireless world. Engineers need data to track call activity by time of day, by base station, and by sector. Ineffective attempts and lost calls need to be tracked. And handoff activity is a critical capacity issue that needs to be measured.

In addition, wireless billing is much more complicated than landline billing. On the traditional PSTN, there was an operating assumption referred to as *calling party pays*. A party receiving a call was never charged for the call or any services related to the call unless he or she accepted a collect call. With the advent of wireless, all that changed. Wireless subscribers pay for air time when they receive calls. They also may pay roaming charges and other fees as well.

From the perspective of call statistics, the switch must record all relevant data in the correct category, and all switches on the network must record the information in a consistent fashion. If this is done, the service provider's billing application can generate accurate telephone bills from the records. The subscriber uses his or her mobile telephone to make calls and receive calls just about anywhere, the switches record the data, and the bills come out straight. This topic is covered in more detail in Sec. 16.4. Billing errors not only frustrate the subscriber, they also increase customer service costs for the provider and may cause a loss of customers or a loss of revenue.

11.4.9 Speech coding

The wireless MSC has to deal with speech coding of varying rates during a wireless telephone call. The normal digital landline telephone switches in PBXs and the PSTN have DS-0 or E-0 circuits coming into and going out of them, and each switch changes the paths of data flow. In wireless telephone systems, the MSC must communicate with the PSTN in some form of pulse code modulation (PCM) over standard DS or E connections, but the MSC also communicates with the base stations of the cellular network in more compressed speech-coded forms appropriate to the lower available bandwidth on the air interface. As a result, the cellular provider must add a translation function that translates the landline-to-wireless voice signal from PCM to the wireless data compression and also translates the compressed voice signal from the wireless network into PCM for delivery to the PSTN. This speech coding translation function sits between the MSC switch and the PSTN in a conceptual model and usually resides physically at the MSC.

The cdmaOne and cdma2000 systems use Qualcomm code excited linear prediction (QCELP) speech coding, whereas Wideband CDMA (W-CDMA) uses regular pulse excitation–long-term prediction (RPE-LTP) from the Global System for Mobility (GSM) and lower bit rates when it can. Both these speech coding systems are aggressive in conserving bits over the air interface. The backhaul communication link between the base station and MSC enjoys that lower bit rate efficiency as well by using the low bit rate of the speech coding.

Some systems use IP-based packet communications to send the speech coded data between base station and MSC. A packet carries a set of bits, a fragment of the entire transmission. In this way, packets are similar to the frames transmitted over digital radio links or in the time slots of time division multiple access (TDMA) systems. However, packets are different from frames in an important way. Frames do not necessarily contain information that identifies the call of which they are a part. Frames are carried in sequence over a particular fixed channel, and as a result, their location in the system identifies the conversation of which they are a part and the frame's place in the sequence in the call. Packets contain a header with information defining the source location, target location, position within the transmission, and other signal information along with their user data. As a result, packets from a single conversation can travel over different pathways or out of order and still be reassembled correctly at their destination. This form of packet transmission is the basis of the Internet. In telephony, however, most of the time, packets are carried over single direct link from point to point. These links are called *packet pipes.*

Packet pipes are more flexible than the fixed pipes of traditional PSTN telephony and are potentially more efficient, as discussed briefly in Sec. 13.2.1 and in more detail in Chap. 35 and Sec. 44.1.

While packet switching using packet pipes of flexible size may be the future of landline telephony, the present world of landline telephony is steady-state circuit switching using PCM links of fixed size. Current CDMA technology adds speech coding from MSC to base station and from base station to user terminal over the air link, achieving greater compression, but for the most part, it still uses fixed pipes. Packet transport is a new concept in the telephony world, and it will require new equipment and new engineering methods. The equipment is being developed and deployed, but it will be some time before cellular systems move to the voice over IP (VoIP) packetized technology described in Sec. 19.4 and realize the benefits of flexible pipes (and other benefits of VoIP as well). This conversion will be accelerated where there is demand for greater data capacity and providers deploy 3G cellular solutions such as cdma2000 to meet that demand. However, it will be a fair number of years before the legacy systems of the PSTN are converted to VoIP.

11.5   Conclusion

Cellular networks make new demands of switching systems beyond the requirements of the traditional PSTN. These come from specific technical requirements of mobile technology, including the need to support roaming, the need to track many new system performance parameters, and the need to bill customers for use of specific facilities. Each of these functions makes demands for particular types of capacity within switches. In this chapter we have discussed the functions and the switches. In the next chapter we will take a closer look at the telephony engineering principles that describe how the switches and other system components function to support and manage calls.

Chapter

12

Telephony Engineering Concepts

This chapter provides an introduction to the basics of telephony for those who come from a radio engineering or data systems background, as well as for those who wish to review these fundamental concepts. The first four sections describe the key concepts of landline telephony, whereas the latter two sections introduce ideas specific to the world of cellular telephony.

12.1 Telephone Call Sequence

Before delving into the components and organization of a telephone network, it is worth a few paragraphs to examine the sequence of a normal landline telephone call. A landline call starts when somebody picks up the receiver. We call this person the *callING party,*[1] and we say the telephone *goes off-hook*. The telephone network detects the off-hook condition and connects an internal resource to the line for the purpose of collecting the digits dialed by the caller. The telephone network then sends a dial tone to tell the callING party that the telephone network is waiting for instructions in the form of *dialed digits*. We call them dialed digits even though they are often dual-tone multi-frequency (DTMF) tones generated by the familiar buttons on the telephone keypad.[2]

Once the callING party completes dialing a telephone number, the telephone network tries to establish a connection to the callED party through the network. If the attempt is successful, the network will both ring the phone of the callED party and locally generate a ring tone that is heard by the callING party. If the callED party picks up the phone, the network senses the off-hook, removes the ringing voltage at the callED party's end, and signals through the network for the office at the callING party's end to remove the artificial ring tone. We call the person at the other end the *callED party.*[3]

---

[1]We use the uppercase *ING* to differentiate between the callING party and the callED party, and we emphasize the *ING* syllable when speaking about telephone calls.

[2]Telephone subscribers in rural areas are loathe to pay the monthly charge for Touch Tone service. Rotary phones are alive and well.

[3]The uppercase *ED* is to avoid confusing the callING party and the callED party, and we pronounce it in two syllables, "call-ED," with the accent on the second to make the point.

The call is then connected and is now said to be a *stable call*. The call stays in the stable state until one party, either callING or callED, hangs up the receiver or, in telephone talk, *goes on-hook*. Once this happens, the call is complete, and any trunks used to complete the call go away.

We are using this traditional call sequence to illustrate the structure of a telephone call. If setting up this call required that the call be routed between offices, then one or more interoffice trunks would have been used. Traditionally, the trunk would have been selected by examining the trunks in the appropriate trunk group until an idle trunk was found, identified by the signaling bits (assuming a digital facility). If the call had to be routed through to another trunk, then the process would be repeated at the next switch in the call path. The method used today by many carriers is to negotiate for available trunks over the entire transmission path using Signaling System 7 (SS7) before actually switching the trunks. This is discussed in Chap. 14.

Every step of this process has alternative outcomes. The callING party may not complete a full telephone number, the line may be busy, or the callED party may not pick up the receiver. There are alternatives internal to the telephone network, but they happen far less often. There can be equipment failures, or all the circuits can be busy.

The addition of custom calling features, call waiting and three-way calling, adds a number of other possibilities to the sequence of events during a call. However, it is important to remember, in all the complexity and sophistication of telephone network design, that the primary mission of the telephone network has been to complete voice calls from one person to another.

12.2   Quality of Service

Telephone service can be evaluated in several ways, but there are three basic areas where a call can go wrong. It can fail to complete, it can fail while stable, or it can sound bad.

12.2.1   Ineffective attempts

Users make calls, and some of them do not get through. The simplest case is when the callED party does not answer or the line is busy. These events, however, are not failures of the telephone system. If the call does not go through due to a failure of the telephone system, it is called an *ineffective attempt* on the part of the telephone system. And when the telephone system fails to complete a call because it has no more resources to devote to calls, then we say the call is *blocked*. Blocked calls are discussed in some detail in Chap. 23.

A wireless telephone system may have the resources to complete a call, but the radio link is too poor to set up the call. The radio link can fail because its call setup signaling fails or because the voice quality is below the minimum standard. Code division multiple access (CDMA) technology designers face a tradeoff: The system can deny service to a caller when the system is getting close to its maximum level, or it can serve the caller at the risk that radio conditions may change and drive out one of the users in the sector.

One of the frustrations of cellular engineering is that even something as simple as blocking rate is hard to measure. It seems like the simplest thing in the world: Count the number of calls that are blocked, divide by the total call attempts, and multiply by

100 for a percentage. The problem is that people whose calls are blocked try again and again rather than going away quietly. If the average caller tries three times before giving up, then the measured blocking rate could be three times the true blocking rate.

Measuring ineffective attempts in a wireless system is particularly difficult when some of those are due to poor radio signal. If the original call attempt message is lost, then the system does not even detect a call attempt. Also, a cellular telephone with a defective radio or a broken antenna wire will perform poorly, but this is indistinguishable from a user being almost out of range. Thus it is difficult to distinguish failures of user equipment from errors in wireless system planning or engineering.

12.2.2 Lost calls

Once a connection is formed, it stays until one party ends the call. In wireless networks, a changing radio environment or a moving radio telephone can cause signal quality to drop and a wireless call to be lost.

As in the case of ineffective attempts, measurement is vague for lost calls. Long calls are more likely to be dropped not only because of their greater time exposure but also because their radio environment changes more. We would expect 8-minute calls to experience more user motion and more radio environment change and so to be lost more than twice as often as 4-minute calls, but we have no data on this. And a wireless telephone system has no obvious way of telling a true lost call from a user terminal shutting down during a call, perhaps from a weak battery, or a user becoming frustrated with lousy sound and ending a call prematurely.

12.2.3 Sound quality

Call sound quality may be measured subjectively, but most of us agree on what constitutes a good-sounding call.[4]

- The sound should be *pleasant,* a faithful reproduction of the source without being strident or muffled.

- Speech should be *intelligible* so that we know what the person is saying.

- The background should be *quiet.*

- The connection should be *continuous,* with dropouts kept short and infrequent.

- The call should have *minimum delay* so that the conversing parties do not get confused.

Unpleasant sound reproduction makes subscribers less anxious to use their telephones and makes less money for telephone service providers. In the world of high-fidelity audio reproduction, we measure *frequency response* to ensure that each frequency in the musical audio band is played back in proportion to how it was recorded. In telephones, we compromise this objective by emphasizing higher frequencies to make speech clearer, but listeners will find severe frequency-response errors objectionable. By emphasizing certain frequencies, we can improve the telephone's ability to

---

[4]The first step in measuring quality is to define the multiple aspects of quality from the user's perspective. Once these elements are defined, engineers can find ways to capture the right data and make appropriate measurements.

communicate human speech. Research on the sound shapes (formants) of speech goes back to the 1930s at Bell Telephone Laboratories, and telephones are designed with deliberate frequency-response deviations so that the sounds of speech, especially consonants, are heard more clearly. Old radio broadcasts without this sound-shaping technology lost their high frequencies and were difficult to understand. Much of the consonant sound spectrum is outside the 300- to 3300-Hz telephone bandwidth, and yet we have no trouble understanding people in a clear and quiet telephone call setting.

In addition to having a pleasant frequency response, the reproduction should have low enough *distortion* that the voice sounds clear enough. Distortion is the component of the reproduction that does not sound like the signal source.[5]

Hearing something other than the telephone call is distracting and annoying. There is one exception: A slight background hiss is actually a good thing because it reassures the listener that the call is still in progress. Very quiet lines get an occasional "Hello? Are you still there? Hello?" because one party or the other is concerned that the call may be quiet because the line is dead. As the noise contour changes from a wide-frequency hiss to a narrow band of frequencies, or as the noise gets louder, it becomes annoying rather than reassuring.

Our senses respond to change. As a pulsing light seems far brighter than a steady light, a changing noise level calls attention to itself. A hissing or rumbling sound that comes and goes or changes its frequency content is much more of a problem than steady, unchanging noise.[6] The worst case is when the noise mimics the rhythms of human speech; we call such interference *syllabic*. The most likely cause of syllabic interference on a telephone call is leakage from another call. Listeners feel that they can almost understand what is being said on the interfering conversation, even when it is synthesized by a machine and has no human speech content other than its rhythm.

Short dropouts are not even noticed by human listeners. A gap of 100 ms (one-tenth of a second) is seldom perceived, and 200-ms gaps are a problem only when they are frequent. Advanced Mobile Phone Service (AMPS) handoffs are about 100 ms long, and Rayleigh fades typically are much shorter. As we discuss in Sec. 27.1, most test listeners reported that 2 percent signal outage was still good call quality.

Signal delay is the last quality issue. Very long distance calls from the Americas to Asia often use two satellite links and take over 1 second for the round trip. Humans are not accustomed to conversations with delay, and they get confused when what they hear is 1 or 2 seconds behind what they are saying.

Radio conditions for good speech quality are the same for analog and digital radio: strong signal, weak interference, high signal-to-interference (S/I) ratio, and high E_b/N_0 as described in Sec. 8.2. The nature of the interference is important, too.

Analog interference is best when it is smooth. FM with its constant radio envelope, described in Sec. 4.3, eliminates the syllabic nature of AM interference. Digital interference causes bit errors, and these bit errors tend to be *bursty,* occurring in the same time interval. Statistically speaking, the likelihood of a bit error is greater when other bit errors occur at nearly the same time. Also, voice reproduction is more sensitive to

---

[5]While this definition of distortion is not rigorous enough to satisfy audiophiles or electrical engineers, it does describe how a system can change a signal for the worse with only a small change in its frequency response.

[6]Not only does the pulsing light seem brighter when the average power is the same, it even seems brighter when the *peak* power is the same.

some bits than others. Speech coding and forward error correction can work together to produce the best voice reproduction in a noisy radio environment.

12.3 Reliability and Redundancy

Networks have to be smart enough to know what to do when something fails. One approach to network recovery is to have all the routing decisions made at one central place by either a technician or a cleverly programmed computer. When a link fails, the network operator or facilities program figures out how to route its traffic and executes the change. This is a difficult environment to govern, and of course, one has to plan for what will happen when it is the administration center itself that fails.

The other approach is to have individual network nodes take over their routing functions, with each node programmed with the proper response to a link failure. These self-healing networks are hard to design and often difficult to provision, but they have no central processor that acts as a single point of failure for the entire network.

One of the difficult points of designing decentralized self-healing networks is that each node working independently still has to contribute to a whole network working well. Usually, the first response of a decentralized network to a link failure is fine: Each end point does the right thing and reroutes its calls over the route the designers planned for the failure. Later on, when some of the links are overloaded from the overflow and there is another failure, the system may not fail so gracefully. Figuring out all the multiple states of a network in transition after a failure is a daunting task. As a result, when a second failure does occur, circumstances may arise that the designers did not foresee at all. Switches, cross-connects, and wireless base stations all have software systems that also can be overworked during a link or component failure.

One problem that can arise due to an error in network design is called a *glare condition*. A glare condition occurs when two parts of a telephone system are both waiting for the other component to do something, A simple example of a glare condition in a setting familiar to many telephone users is a clash between call waiting and three-way calling. A landline subscriber who has both features signals to add another call using three-way calling. At the same time, a new incoming call arrives from someone and is connected through call waiting. The subscriber is waiting for a dial tone, whereas the new caller is waiting for a conversation. In this example, both the components in the glare condition are people, but human beings are part of the telephone system, too, and human beings also can get caught in glare conditions.

A common form of glare in the network occurs when two switches attempt to seize the same trunk. The trunk looks idle at both ends, based on the signaling bits, and both will attempt to seize. The signaling protocol is sufficiently clever that both switches are able to detect the glare condition. One of the switches, however, will have been programmed to accept that the other end has seized the trunk successfully and to wait to receive the dialed digits.

12.4 Wireless Telephone System Architecture

From the perspective of the public switched telephone network (PSTN), the wireless system is an *access* technology. The role of the wireless network is clear: A mobile switching center (MSC) looks like a private branch exchange (PBX) to the PSTN, and the base stations and air interface are analogous to cross-connects and local loops. The wireless system looks like a large community of end users with a set of telephone numbers and a collection of trunks. When wireless systems communicate with the PSTN, they must send signals according to the SS7 standard and calls packaged in pulse code modulation (PCM).

12.5   Wireless Telephony Engineering Issues

There are some telephony issues specific to wireless telephone systems. The *great big difference* is that wireless telephones are typically *mobile* telephones. They can initiate calls from anywhere, they can receive calls anywhere, and they move around during a call.

When they initiate calls, subscribers expect prompt service with their full feature capabilities. And subscribers expect to receive calls wherever the wireless telephone systems are compatible with their home service.

12.5.1   Tromboning

Most mobile users probably have a story similar to this one: Meeting a New York friend in Krakow, one of us (Rosenberg) called a mutual acquaintance back in Long Island. We dialed his local New York telephone number, and the wireless system in Krakow routed the call to New York for handling. It turned out that the mutual acquaintance was in Italy that day, so we need not have worried about calling too early and waking him up. The New York system routed the call to Italy, and we enjoyed our conversation. The call, however, was a bit excessive in its use of facilities because it had two superfluous transoceanic links. This kind of routing, out to some far away place and almost back again, is called *tromboning* because the route looks like a trombone. Tromboning is expensive, and newer switching and signaling technologies are being developed to minimize it.

A landline number with call forwarding would have operated in the same way. This is the case where the shift from location-based telephony to personal telephony has much the same effect as the addition of mobility to the telephone system provided by wireless networks. The difference is that in the case of wireless telephony, subscribers are using their own telephone numbers at distant locations. Mobile telephone engineers face the problem of tromboning frequently because mobile telephones encourage their users to be mobile in their telephone usage.

12.5.2   User terminal registration

The establishment of location and identity of user terminals on wireless networks is a continuing process that consumes radio and transport resources and which has no analogue in the wireline network. The system has to broadcast its own identity clearly enough for the user terminal to synchronize itself to the access channel, to identify the system, and to respond in a narrow time window so that the system can detect the response.

The user terminal *registers* with the wireless telephone network by sending a message on the signaling channel. The registration message identifies the user terminal to the serving system. The user terminal registers when it is powered on and recognizes the wireless system paging channel. The user terminal registers again when it recognizes a new wireless telephone system. The user terminal may move to another cell and detect another base station's paging channel, but it only reregisters when it detects a different system identification.

A user terminal also will reregister after a specific period of time. There is a timing cycle dictated by the system and communicated on the paging channel. When the appropriate time interval has passed, the user terminal registers again. We call this *autonomous registration*. If we did not have some kind of autonomous registration, then the wireless system would have to keep registrations current for a long time. A user terminal might leave the service area or be powered off, and it has no way of knowing that this will happen. As a result, the system cannot rely on the user terminal notifying the system that it is no longer connected and available to receive calls. Autonomous registration allows the system to drop a phone's registration if it fails to reregister at the specified interval. In this way, the system's record of registered phones is reasonably close to the reality at any given moment.

The registration timeout cycle is a parameter that affects both the currency of the paging list and the amount of signaling traffic. Having a short timeout for registration keeps the paging list current but increases signaling channel traffic with more mobile telephone registrations. Having a longer timeout, on the other hand, reduces signaling channel congestion at the expense of paging more subscribers who are not there to receive their calls.

12.5.3 Call setup

The entire process of establishing a link to the subscriber's phone on a call-by-call basis is a traffic issue specific to wireless. A landline telephone has a local loop, a link that is always there, managed by the local exchange office, perhaps with the help of a PBX. The local loop can always be there because the subscriber's phone is always in one place—it is not mobile.

A call attempt is made after the user terminal is registered via the paging and access channels. As a result, the user terminal and the system have identified one another. When the user terminal makes a call attempt, the system already knows who the subscriber is and where the mobile telephone is. In response to the call attempt, again using paging and access capacity, the terminal and system send several messages back and forth to establish the call itself and its radio environment.

12.5.4 Handoff

As an active wireless call moves from one cell sector to another, the call gives up its radio link with the old sector and is connected to the new sector. We call this process a *handoff* (or *handover* in GSM terminology). CDMA allows a soft handoff where the call is served simultaneously by two or more cell sectors.

Handoff was a new concept in telephony, and it still has few analogues in the landline network.[7] A landline call can be viewed as a point-to-point connection, but a wireless call is something more than its connection. The call is maintained while the connection changes, following the mobile user terminal.

---

[7]Answering the telephone in the bedroom and running downstairs to the kitchen to talk there is not quite the same thing.

The connection not only can go from sector to sector and from cell to cell, it also can go from one MSC to another. When this happens, we have one MSC setting up the call, connected to the PSTN, and a connection to another MSC that actually serves the call. We call the setup switch the *anchor MSC* to distinguish it from the *serving MSC*.

The call can even go from one wireless system to another so that the anchor and serving MSCs are in different systems. This requires close signaling coordination and a connection between the two MSCs.

When a handoff to a new MSC is being considered, we call the not-yet-handed-off switch a *target MSC*. The details of inter-MSC handoff are specified by the ANSI-41 standard, as described in Chap. 15.

12.5.5 Roaming

Roaming is unique to cellular systems, although the recent service of local number portability (LNP) is bringing similar issues to the landline network. The roaming subscriber expects services similar to home and expects to receive calls dialed to the home telephone number.

This is effected by close coordination in the signaling network and maintaining subscriber records in the home system. We call the home system repository of subscriber information the *home location register* (HLR). Anytime and anywhere a user terminal changes status, its HLR record is updated. The HLR keeps track, minute by minute, of the current location of the user terminal. It also knows whether the user terminal is available for call delivery. This allows the home MSC to make the appropriate choice in response to an incoming call: sending the call to the user terminal if it is registered on the home network, sending it to an alternate MSC if it is available through roaming, or directing the callING party to voice mail (or sending a busy signal) if the callED party's mobile telephone is not available.

A registration in a system other than the home system creates a record for the subscriber in the new system's *visitor location register* (VLR) and updates the location and status information in its HLR record. At this time, the serving system receives verification of the user terminal's identity and the service capabilities the subscriber should receive. Authorization can be denied to mobile identification numbers known to be fraudulent and to subscribers not paying their cellular telephone bills.

12.5.6 Small numbers of calls

Much of the engineering design of the PSTN is oriented toward supporting many subscribers and many calls through a relatively small number of local exchange offices and other facilities. This model reduces cost by achieving economies of scale.

In contrast, wireless telephone systems are limited by the capacity of the air interface, and they grow by adding base stations. Cellular networks grow by adding new locations and facilities rather than by expanding the capacity of existing facilities. Wireless network providers add mobile switching centers as well, but the primary element of growth is new cells. Since a typical cell handles 10 to 100 simultaneous calls, the base-station-to-MSC transport in a wireless telephone system never reaches the economy possible in systems where one transport pipe has hundreds of lines. In landline telephony, small trunk groups usually grow into large trunk groups as systems expand to serve more subscribers. In wireless telephony, small pipes grow into more small pipes as more cells are added to meet increasing subscriber demand.

The number of user terminals is dictated by the number of subscribers. After user terminals, the most numerous and expensive part of a wireless telephone system is the collection of base stations. At a typical busy-hour occupancy of 70 percent, this radio equipment is idle three-tenths of the time. There are ways to design the equipment to handle this inefficiency more economically, but this is the reality of small numbers, the reality of wireless traffic engineering.

The financial planners are usually not telephony engineers. They have to understand that they cannot divide the radio count by the length of busy-hour calls to get the busy-hour capacity of their systems. There are a two major reasons for this.

First, in conventional cellular systems with mixed cell sizes, the frequency-allocation pattern actually reduces the number of radios available on some faces. In CDMA, we cannot count the number of calls by counting the number of radios, but increased use in surrounding cells does add interference and reduce capacity at each cell.

Second, we cannot use all the voice links all the time and maintain an acceptable blocking rate. This adds a complex statistical element to our calculations. The mathematical models for this are discussed in Chap. 23, but here are some of the essential issues:

- The moment-by-moment demand during the busy hour follows a statistical Poisson distribution that depends on the average call demand level. The number of calls at any given instant can vary considerably from the average.

- Keeping the system available for subscriber calls is the same as maintaining a low blocking rate. Two percent blocking is excellent service, and 10 percent may still be acceptable.

- The number of available voice links required to maintain a low blocking rate can be significantly more than the average demand. As a result, the links are occupied only some of the time even during the busy hour.

- Lower engineered blocking rates require lower occupancy, which means less efficient use of voice links.

- While smaller demands require smaller numbers of voice links, the occupancy required for a given blocking rate goes down, so smaller quantities of call demand use voice links less efficiently.

- At the 2 to 10 percent blocking levels typically desired in cellular systems and at the numbers of radios in cellular systems, the busy-hour occupancy ranges from 50 to 80 percent. Offering reasonably low blocking rates, under the best of circumstances, requires at least one-fifth of the voice links to be idle even during the peak-usage period.

Within a cellular service provider, there is a danger that financial planners working on a cellular growth plan will set budgets based on an oversimplified model of demand and capacity. It is essential that cellular engineers communicate the complexity of these issues early and often and that they explain the results of their statistical modeling in a way that the financial executives can understand. It is also essential that the organization foster good communication and mutual respect between these groups so that the financial plan is sufficient to support a realistic engineering growth plan.

12.5.7 Changing channel conditions

A single CDMA carrier varies in capacity over time. The notion of a carrier or set of carriers with varying capacity is a new concept to most telephone engineers. Even those of us familiar with conventional reuse (FDMA and TDMA) are not used to a carrier set whose capacity changes over time. In CDMA, calls with poorer radio links (coming from inside buildings, for example) use a larger slice of the CDMA pie than other calls, add more interference to the other calls, and reduce the carrier's total capacity. The number of calls that can be served by a single carrier depends on the local conditions where the calls are calling from.

Cell capacity also depends on the quality of the user terminals. Different subscriber equipment may have different levels of radio waveform distortion, adding different amounts of noise to the CDMA channel. This variation is not only call by call but also moment by moment. A subscriber walks into a building, and the CDMA channel goes from barely good enough to not good enough for that call or perhaps for other calls on the same carrier.

The wireless service provider has to make some decisions about service quality in anticipation of changing radio channel conditions. Consider this example for planning capacity for a particular sector. If the average call duration is 2 minutes (120 seconds) and the expected capacity is 24 calls, then the expected time until somebody hangs up normally is 5 seconds. In this case, if worse than usual conditions do not allow high-quality calls when the twenty-fifth call comes in, then the subscribers experience about 5 seconds of excessive interference before someone hangs up. This may be acceptable if it happens occasionally, but if it happens too often, subscribers will be dissatisfied with the service. If this network continues to allow a twenty-fifth caller onto the system, then service may be unsatisfactory. The network can anticipate the increased error rate and block calls, maintaining a maximum of 24 calls in the sector under these conditions. The tradeoff between capacity and quality manifests itself in the decision whether to hold the line at 24 calls of good quality or to allow 25 calls and to tolerate the occasional 5 seconds of substandard voice quality.

12.5.8 Varying overflow characteristics

Overflow is the capacity of a network to provide alternate routing if demand exceeds the capacity of the normal or primary route of a call. On the traditional PSTN, if the connection from switch A to switch B is full, it might be possible to carry the overflow from switch A to switch C and then from switch C to switch B. However, in the traditional PSTN, there is no alternate route to replace the local loop.

Overflow in a wireless network is far less straightforward than it is in landline telephony, but the big difference is in the radio link. As far as transport is concerned, the wireless telephone system is still basically a telephone transport network. The overflow issues are a bit more complex due to handoffs and due to the mobility of call setup. We must take them into account when planning capacity and overflow, but we do not need to introduce any radically new ideas. We can use the same thinking we would use in planning PSTN capacity.

In conventional reuse, a busy cell sector can direct extra traffic to other cell sectors, providing traffic overflow for the air interface, equivalent to local-loop overflow on the PSTN. Knowing traffic can overflow allows us to be more aggressive in our traffic engineering, as discussed in Sec. 23.4 than we would be without overflow. We can block more calls from each sector when they have a significant chance of having somewhere else to be served.

In CDMA, however, sending extra traffic to another cell sector is generally *not* a good idea. When we solve the CDMA equations in Sec. 28.1, we see that adding traffic to the wrong sector is even more damaging to the right sector than simply overloading the right sector. Because all the CDMA sectors use the same frequency band, the interference impact of a call is minimized by serving it on the cell sector with the best radio path (highest path gain), where it can operate at what probably will be the lowest acceptable power level.

There is another, more subtle kind of overflow in CDMA. In conventional reuse and landline telephony, we have some number of telephone lines, and we can keep serving calls until they run out. This number remains fixed, regardless of activity on neighboring cells. In conventional reuse, interference from neighboring cells reduces call quality but not call capacity. In contrast, the CDMA channel capacity is variable and depends on the call volumes in neighboring cell sectors, a kind of global capacity even without call overflow. In Sec. 30.6 on CDMA soft blocking we discuss how we can use this global capacity in CDMA engineering to mitigate some of the capacity loss from other-cell and other-sector interference.

12.5.9 Nonlocal service quality

The last notion in traffic engineering we want to address, because it is new in wireless engineering, is the concept that service quality in one area depends on what is happening somewhere else, sometimes moment by moment. We have changing conditions in the landline network, and even some strange stories where conditions in one area changed and another area had a service problem. However, these are considered strange. In wireless, however, it is normal day-to-day reality that radio channels on one side of town affect cellular performance and capacity on the other side. The ball game lets out, and all other cells sharing the same channel set have more interference. This is not a new or deep concept in radio, of course, but it is a little bit strange to those who think in telephony terms.

12.6 Conclusion

This introduction to telephony engineering issues lays the groundwork for understanding the CDMA capacity and engineering issues we will discuss in Parts 5, 6, and 7.

Chapter

13

Telephone Transport

The methods used to transport telephone communications have evolved over time from the use of uninsulated open-wire pairs strung on poles to fiber optic systems capable of transporting over 100,000 simultaneous calls. The early open-wire systems gave way to the use of insulated twisted pairs, often bundled together in cables. It is common today for the cable drop to a residence to consist of six twisted pairs in a cable, and cables containing as may as 2700 pairs are in use.

Today there are many methods in use for transporting information through the network. Some of the more common methods will be described in this section. We prefer to use the term *information* because today's public switched telephone network (PSTN) transports not only voice but also various forms of data. It even carries portions of the Internet itself. First, we will introduce several general concepts that will aid in the understanding of later sections of this chapter.

Traditional telephony has as its roots the concept of a *circuit*. A circuit can be thought of as a dedicated pair of wires or a channel in a digital facility such as the DS-1 (described in Chap. 11). Traditional telephony creates connections between telephones by connecting circuits end to end until the entire transmission path is complete from caller to called party. This approach is commonly referred to as *circuit switching*. The term *circuit* applies most precisely to a two-way point-to-point link. However, the term is also used to refer to a series of circuits connected through switches connecting two end points.

Data communication uses a statistical approach instead. Communications between many devices on a data network may pass over the same transmission path, each identified uniquely by address information imbedded in the communication. This sharing of a transmission path is referred to as *statistical multiplexing*.

Many transmission technologies can transport far more bandwidth than is either required, desired, or affordable for a particular data communications application, even the Internet. The result is that that data communications often share transmission facilities and therefore are assigned a circuit on that transmission facility. The circuits provided by the various telephone transport methods can be applied to many purposes. In Chap. 11 we saw the use of circuits as lines for connecting telephones to local exchange offices. We also saw the use of circuits as trunks to connect switching offices. Circuits may be used to connect private branch exchanges (PBXs) to form a private network. These circuits are sometimes referred to as *leased lines*. The use of the term *lines* in this context may appear confusing because the leased lines connect switches rather than terminating at a fixed point. However, in the telephony context, they are not being switched by the PSTN. Finally, we saw the use of circuits, again called *leased lines,* to connect data communications equipment.

Wireless networks depend heavily on a combination of digital trunks and leased lines for connections between base stations and mobile switching centers (MSCs) and connections to the PSTN.

13.1   Telephone Transport Protocols

Analog and digital facilities alike rely on protocols, whether it be for the separation of channels through the use of frequency division techniques or the use of framing techniques such as DS-1 framing and Asynchronous Transfer Mode (ATM). For those familiar with the Open Systems Interconnection (OSI) model, this raises the question of how the apparent framing and addressing techniques used in transport relate to the OSI model. It can be stated generally that all transport technologies, including ATM, are considered to be OSI layer 1 protocols.[1]

13.1.1   Analog transport

The simplest protocol is that for the analog telephone call. It transports a human voice from 300 to 3300 Hz, 3 kHz of analog bandwidth. This bandwidth is enforced by the telephone receiver and by the local exchange office and is not a characteristic of the analog line itself. That human voice can be replaced by any other analog signal that fits between 300 and 3300 Hz, such as a modem or fax signal.

Analog trunks in the United States are bundled together in *L-carrier.* Each call has 3 kHz of bandwidth, and analog circuits modify the sound by shifting its frequencies into the ultrasonic range for transport. Table 13.1 shows the L-carrier range as of 1972.[2] As recently as 1990, the telephone network in the United States was mostly analog L-carrier. Competition forced the rapid conversion of the PSTN to all-digital trunks during the 1990s, and L-carrier is, for all intents and purposes, an obsolete technology.

13.1.2   Protocols for digital transport

The basic building block for digital transport protocols, both in North America and in Europe, is 64 kbps. This is referred to as the *DS-0* in North America and the *E-0* in Europe. This 64-kbps building block is transmitted in 8-bit blocks at a transmission rate of 8000 blocks per second. When used for voice transmission, each 8-bit block contains an 8-bit sample of the analog voice signal. The method used for taking the sample is different in North America from that used in Europe. The procedure used in North

---

[1]The reader will observe that framing, addressing, error detection, and the application (transported information) are attributes of most transport protocols, and the natural tendency is to assign these attributes to upper layers of the OSI model.

[2]There was an L-2 system used briefly before World War II.

TABLE 13.1   The North American Analog L-Carrier Transport Protocol

System	Bandwidth	Capacity
L-1	3 MHz	1800 voice circuits
L-3	8 MHz	9300 voice circuits
L-4	17 MHz	32,400 voice circuits
L-5	57 MHz	108,000 voice circuits

America is referred to as μ-*law,* and the procedure used in Europe is referred to as *A-law.* The μ-law and A-law are described in Sec. 18.1.2.

The DS-0 channels are grouped together to form a DS-1 frame. The DS-1 frame contains 24 DS-0s and one framing bit, for 193 bits. DS-1 frames are transmitted at a rate of 8000 frames per second, as required to deliver the DS-0 samples at a rate of 8000 sample per second. The DS-1 frames are further grouped into groups of either 12 frames (a *superframe*) or 24 frames (an *extended superframe*). The sequence of 1s and 0s assumed by the framing bit identifies both whether a superframe or extended superframe format is being used and where the first frame is in the transmission stream. The equipment receiving a DS-1 will examine the bits in each of the 193 bit positions until it detects the proper sequence of 1s and 0s. Once this is detected, the receiving equipment knows where the framing bits are in the bit pattern and proceeds to count off the frames in 193-bit chunks.

Signaling information can be transported within each DS-0 channel by using a technique called *robbed-bit signaling.* This technique is accomplished by overwriting the least significant bit in the DS-0 sample. It is not necessary to overwrite the least significant bit in every sample. Since the DS-0 is being transported within a DS-1 and the DS-1 has grouped the DS-1 frames into groups of 12 or 24, it is sufficient only to rob bits from a subset of the DS-0s. The practice used is to rob only the bits in the DS-0s transported in the sixth and twelfth frames of a superframe or, in the case of an extended superframe, the sixth, twelfth, eighteenth, and twenty-fourth frames.

The approach used in Europe with E-0s is similar to that used for DS-0s. E-0s are grouped together to form an E-1. The data rate of the E-1 provides for thirty-two 64-kbps channels, but only 30 of them are used for the transport of E-0s. The remaining 2, occupying the sixteenth and thirty-second time-slot positions, are used for the transport of signaling and for synchronization of the E-1 frame, respectively.

Both DS-1s and E-1s can be further multiplexed up to higher bit rates. The DS protocol hierarchy is shown in Table 13.2.[3] The European E-0 links are similarly grouped into their higher-rate packages, as shown in Table 13.3.[4] Alas, these are not compatible with the North American standards.

---

[3]The DS-0 channel can be one voice channel, one 64-kbps data channel, or even a collection of subrate channels.

[4]Both DS-2 and E-2 links exist in the world of telephony specifications, but like the L-2, neither is used in practice. In all three cases, the leap is from 1 to 3.

TABLE 13.2 North American Digital Service Transport Protocols

Signal	Speed	Channels
DS-0	64 kbps	1
DS-1	1.544 Mbps	24
DS-3	44.736 Mbps	672
DS-4	274.176 Mbps	4032

13.2 Wireless System Transport

Wireless systems have their own specific transport needs both within a single mobile switching center (MSC) service area and between or among MSC regions. Now that we have explained the fundamentals of telephone transport, let us look at the four kinds of telephone transport in a wireless telephone system: from the base station to the MSC, signaling using ANSI-41, MSC to MSC, and from the MSC to the PSTN.

13.2.1 Base station to MSC (backhaul)

Base-station-to-MSC transport is outside the realm of conventional telephony for several reasons:

- The voice channels use low-bit-rate speech coding.

- The voice channels use nonstandard data rates.

- A single pipe carries voice and data connections with differing rates.

- This transport network has many thin pipes.

- The network grows by adding nodes and pipes to those nodes, not by thickening pipes.

The engineering of the base-station-to-MSC transport network will be addressed in detail in Chap. 35 and Sec. 44.1, but we can start to look at the issues and challenges here. Base-station-to-MSC transport is outside the ANSI-41 standard.

In traditional telephony traffic engineering, we look at a random distribution with an average number of calls. A link might have a busy-hour average of 25 calls and a requirement of no more than 1 percent blocking. It is the traffic engineer's job to make sure that there are enough lines to serve that traffic distribution so that 99 percent of the calls get through. We discuss traffic engineering in Chap. 23. In this usual telephony traffic engineering world, there is a clear-cut notion of one telephone call, one unit of demand, and one unit of transport.

Wireless voice channels are designed to conserve radio spectrum rather than to make telephone transport easier, but we can take advantage of the economy of wireless voice

TABLE 13.3 European Digital Transport Protocols

Signal	Speed	Channels
E-0	64 kbps	1
E-1	2.048 Mbps	30
E-3	34 Mbps	480
E-4	144 Mbps	30,720

channel design in our transport networks. We can take the simple view that each base station has a certain amount of radio capacity, some number of bits per second, and that those bits have to get to and from the MSC through a transport network.[5] The low bit rate of code division multiple access (CDMA) and the Global System for Mobility (GSM) is maintained from the base station to the MSC.

As a result, each base-station-to-MSC link is a data link with varying numbers of varying-bandwidth users over time. The total usage of the link is limited by the total capacity of the air interface between the base station and the user terminal. So long as this link has more bit capacity than the air interface, there should be no limitation of service. As a result, planning capacity for a single-base-station-to-MSC link is a simple matter of adding up the total capacity of the base station's air interface and making sure that the link is a little bigger. We can use traffic engineering principles to determine just how much bigger it should be. The challenge in this arena is not planning for capacity to a single base station but rather planning the transport and estimating transport costs for a network growing by the addition of new base stations. Normal telephone networks grow by expanding capacity from point to point. Wireless base stations, on the other hand, do not gain capacity as a system grows. Radio spectrum capacity at a single base station is typically fixed, and we grow the total network capacity by splitting cells and adding new base stations. We have more links with the same demand distribution rather than the same links with growing demand.

Of course, changing technology, services, or facilities on the radio link or in the wireless network are likely to require changes to transport facilities. In planning transport facilities, it would be appropriate to ask if conditions such as these are likely to arise:

- Acquisition of more radio spectrum will increase capacity at existing base stations, requiring additional transport.

- If a cdmaOne wireless network is being upgraded to cdma2000, then it will be necessary to evaluate base-station-to-MSC transport capacity and upgrade that capacity wherever the new air-interface capacity exceeds the capacity of the existing pipe.

- If a CDMA system is being modified to support wireless local loop (WLL), particularly at high data rates or for large groups of subscribers, additional base-station-to-MSC transport might be needed.

- If the base-station-to-MSC pipe might be converted to an Internet Protocol (IP) packet pipe, then greater efficiency could reduce the size of the pipe needed after the conversion.

13.2.2 Signaling with ANSI-41

ANSI-41 is a signaling standard for wireless telephony that tells various parts of the wireless world how to communicate with each other. These signaling links almost universally use Signaling System 7 (SS7) for telephone signaling. Chapters 14 and 15 describe the architecture and function of SS7 and ANSI-41. ANSI-41 links are *signaling* links only; subscriber voice and data go on other facilities. There are two exceptions to this, the short message service (SMS) and its successor, enhanced message service (EMS), which send small packets of user data over ANSI-41, as described in Chap. 20.

---

[5]The reality of CDMA is that increasing traffic in one cell decreases the capacity of its neighbors, so there is some notion of community capacity we are not addressing here.

ANSI-41 allows the MSC to communicate with

- Other MSCs for inter-MSC handoff signal management.

- The home location register (HLR) for roamer registration.

- The visitor location register (VLR) for roamer calls.

- A message center for short message service (SMS).

- The over-the-air activation function (OTAF) for over-the-air service provisioning (OTASP).

This list doubtless will grow as wireless telephone systems add capabilities and features to meet customer demand.

ANSI-41 operates on SS7 networks, which are designed so that each switch and each link run at no more than 40 percent capacity, allowing for one of a pair of redundant facilities to carry the load if one piece of equipment should fail. As a result, transport facilities carrying ANSI-41 systems must be designed and expanded to maintain sufficient excess capacity to meet the requirements of SS7.

13.2.3   MSC to MSC

Voice and data links are required between MSCs for inter-MSC handoffs. These facilities are dedicated to handoff traffic. Other MSC-to-MSC communication is separate from handoff communication.

Consider the handoff process at its most basic level: A call changes from one cell sector to another. If the two sectors are in the same cell, then the MSC may not even need to know about it. If the two cells are served by the same MSC, then the one serving MSC handles the circuit switching, and the handoff has no need for MSC-to-MSC communication. When the call crosses an MSC boundary, however, one MSC needs to communicate with another. For this section we need only consider the capacity requirement of handoffs that cross MSC boundaries.

Let us consider a handoff in more detail. Several cells and the user terminal make measurements and determine whether there should be a handoff and where it should go, and not all of these cells are served by the same MSC. ANSI-41 specifies how MSCs communicate with each other to make measurements for inter-MSC handoffs, how they complete the handoffs, and how they maintain efficient paths through multiple inter-MSC handoffs.

Once two MSCs use the signaling network to decide to do an inter-MSC handoff, they use dedicated facilities between them to carry the subscriber's voice and data traffic. While these links may follow the same *physical* paths as the PSTN links used for roaming or call forwarding, they are not the same *logical* paths because they perform a different function. Logical and physical models are discussed in Sec. 14.4.

Multiple-handoff path minimization is important in today's wireless world. If we picture a user moving from system to system to system in a linear fashion, say, Boston to Hartford to New York, then the notion of multiple inter-MSC handoffs may seem like a rare event.[6] Increasing demand for wireless service has increased traffic density and

---

[6]Back in the early days of cellular, when we were working on abutting systems, we wanted to claim that one could start a call in Boston and drive all the way to Denver without losing the call. None of us could think of any reason why anybody would want to *do* that, however.

made MSC regions smaller, so their boundaries are more numerous and more frequent.[7] Also, there are triple points where three MSC regions come together, and traffic in those areas easily may have many handoffs among all three systems. It is not unusual to have a user terminal hand off from A to B to A to C to A and then to B, and it would be silly to transport that call over five inter-MSC voice links. We discussed this out-and-back transport issue, called *tromboning,* in Sec. 12.5.1. Path minimization reduces tromboning.

There is another function of inter-MSC pipes. Some wireless carriers may wish to support roamer terminations and mobile-to-mobile calls without having to use expensive capacity leased from the PSTN. If the wireless carrier establishes additional voice capacity among MSCs, then that carrier can serve roamer terminations and mobile-to-mobile calls end-to-end on its own transport facilities, possibly at reduced cost.[8] If a vendor is doing this, then the total capacity for the pipe between two MSCs should be the sum of what is needed for handoffs and roaming plus what is needed for mobile-to-mobile communications. Although the two functions may be supported on the same physical channel, their logical functions are separate. In addition to the separation of the two data links between two MSCs, the ANSI-41 channel capacity must be dedicated to ANSI-41 and not used for other purposes.

13.2.4   MSC to PSTN

If you are an old-fashioned telephony engineer, then you can relax for a while as we discuss MSC-to-PSTN transport. The connection between MSC and PSTN, at least at the time we are writing this book, is ordinary circuit-switched pulse code modulation (PCM) on DS-0 trunks bundled 24 to a DS-1 or E-0 trunks bundled 30 to an E-1. To the PSTN, the MSC looks like a local exchange office or a PBX.

The MSC-to-PSTN trunks do not carry handoff traffic because that is handled by dedicated MSC-to-MSC links. Incoming calls to roamers come into the home MSC and are forwarded to the serving MSC just like any other forwarded call. The signaling system may be smart enough to redirect the call to the serving MSC without it having to make a detour into the home MSC and back out again. As we will discuss in the next chapter, signaling eliminates much inefficiency in telephone transport.

The MSC-to-PSTN trunk group is ordinary not only in its format but also in its size. An MSC with 100 base stations with a capacity of 500 calls each, for example, will require tens of thousands of trunks, several DS-4 or E-4 links. Traffic engineering, described in Chap. 23, allows us to use less than 50,000 trunks, perhaps a lot less, depending on the subscriber demand usage patterns. These are the kinds of trunk groups that traditional telephone transport engineers are used to working with.

---

[7]This is a legitimate mathematical concept. Whatever measure we want to use for coverage area and boundary, the ratio of boundary to coverage increases as the service area decreases. If radio signal path variation makes inter-MSC handoff likely over a 1-km-wide region around the MSC service region, then there will be more calls in those 1-km regions as the regions themselves become smaller in geographic area.

[8]A vendor will choose the amount of capacity to allocate for mobile-to-mobile calls based on a complex optimization with routing through the PSTN as an alternate route.

13.3 Conclusion

In this chapter we have introduced the principles and standards for telephony transport. We have discussed the potential effect of a shift from traditional telephony to voice over IP. We also have introduced a number of issues of cost management and capacity planning, with a focus on topics most relevant to wireless networks.

14

Signaling with SS7

14.1 Introduction

Signaling System 7 (SS7) is both a network architecture and a signaling protocol that has been adopted internationally as a method of signaling for calls in the public switched telephone network (PSTN) and for providing advanced features such as 800 service. It is the most recent implementation of a concept called *common channel interoffice signaling* (CCIS).

Previous chapters described how signaling for a call could be carried on the pair or pairs of wires that transported the call. We also briefly touched on the concept of Integrated Services Digital Network (ISDN), in which all signaling for a group of calls could be carried in a separate channel on the same DS-1, a technique referred to as *associated signaling*. More generally, associated signaling refers to the use of a signaling path that parallels the trunks between two switches and which signals for those trunks.

CCIS offers a more powerful approach. Rather than using signaling paths that paralleled the trunks, CCIS sets up a separate data network for signaling. The SS7 network makes use of the 64-kbps DS-0 for the physical layer connections between network elements. These connections are called *signaling links*. The SS7 network does *not* replace the trunks that are used to carry voice and data traffic between switches. It is a separate data network established for the purpose of signaling for those trunks. The architecture of the SS7 network provides for high reliability through the use of redundant signaling links and redundant nodes in the signaling network.

14.2 SS7 Network Architecture

The SS7 network is designed to be highly physically redundant for reliability. It is maintained as a physical network separate from the facilities and switches that carry voice and data traffic. The two networks touch only at the switches themselves, as required to communicate the signaling functions for which SS7 was designed from the SS7 signaling network to the switches it controls.

The local exchange office at the telephone company is a service switching point (SSP) for SS7. The SSP is a separate logical entity connected to the local telephone switch, but the entire local exchange office is sometimes referred to as the SSP. The

SSP converts voice-switch signaling into SS7 messages. Most of the SSP SS7 traffic is related to voice circuits.

In addition to voice-circuit signaling, the SSP uses SS7 for database access. This started with toll-free 1-800 number lookups and added 1-900 number lookups, but local number portability is changing the SS7 traffic mix. Now that landline subscribers can take their telephone numbers with them when they move to a new location, almost every call will require checking the local number portability (LNP) database to determine which network provides services to the called number. Once the routing number that identifies the call's actual destination has been retrieved, then the SSP can begin circuit connections for a call.

The entry point to the SS7 network is a signal transfer point (STP). Physically, the STP can be attached to a voice switch, so tandem switches provide voice switching and SS7 STP services using a colocated computer. Standalone STP equipment allows companies to centralize their SS7 operations. STPs come in redundant pairs because all SS7 network components must be implemented in redundant pairs.

The service control point (SCP) is the SS7 interface to telephone company databases. These databases have routing numbers for toll-free, area code 900, and LNP, as well as credit-card validation data, fraud protection, and Advanced Intelligent Network (AIN) services used for creating subscriber services. The SCP itself may not store the data but merely provides SS7 access to a computer database.

The basic flow of SS7 signaling messages from one SSP to another SSP is shown in Fig. 14.1. There are six types of SS7 links (A through F):

Access links (A) connect the SSP to the STP. There are at least two for redundancy in case one link or one STP fails. The maximum number of A links to one STP pair is 16 to each STP.

Bridge links (B) connect mated STPs to other mated STPs. Each STP-to-STP pair connection requires four separate B links, often referred to as *quad links*. Some SS7 networks in Europe do not use all four B links as shown here. B links are used for STP

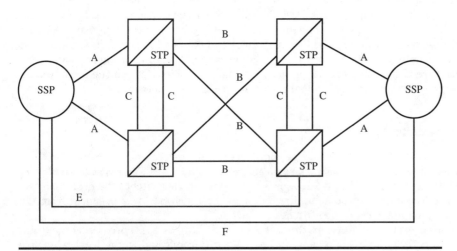

Figure 14.1 SS7 flow diagram.

connections at the same hierarchical level in the SS7 network, whereas D links (see below) are used when the STPs being connected are at different levels.

Cross links (C) connect the redundant STPs to each other. Even C links come in pairs to maintain redundancy in the SS7 network. These links are used for SS7 network management messages only except when the network is so congested that the C links are needed for SS7 traffic or when an equipment failure requires SS7 traffic to use the C links.

Diagonal links (D) are used to connect STPs at different hierarchical levels in the SS7 network. SS7 networks are not required to have a hierarchy, but it can be useful when there are centralized functions, such as SCP database access, or large numbers of central offices that benefit from having concentrator STP nodes. D links come in quad arrangements just like B links. The full SS7 picture with hierarchy is shown in Fig. 14.2.

Extended links (E) connect SSPs to remote STPs. This is normally used for traffic overflow when the home STPs become congested. E links are also a diversity backup in case of equipment failure.

Fully associated links (F) are direct CCIS links between two SSPs. F links typically are used when there is a large amount of signaling traffic between two SSPs, enough to justify a dedicated signaling link. They also can be used when a remote SSP cannot be connected directly to an STP for some reason. The F links then allow the remote SSP to access SS7 databases without direct A links to an STP.

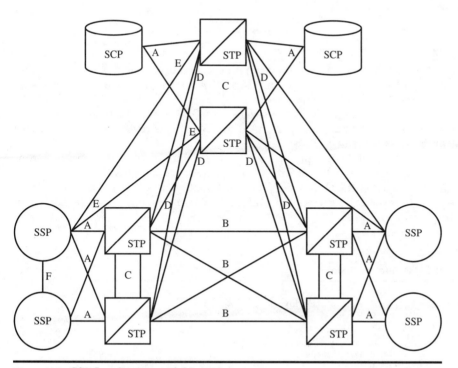

Figure 14.2 SS7 flow diagram with hierarchy.

Figure 14.3 How we usually visualize SS7 networks.

The transport used on these A through F links is entirely at the discretion of the network designer; SS7 does not care how its packet messages get from source to destination. In terms of the OSI model, we would say that SS7 specifies levels 4, 5, and 6 and leaves the implementation of levels 1 through 3 open. The links do have to be dedicated, that is, available for SS7 traffic at all times, and not used for anything else. When a link fails, the other links in the same set have to take over the traffic load. If an STP (or some other piece of SS7 equipment) should fail, then the remaining STP of the pair has to do the work of both of them. SS7 links and equipment are designed to use no more than 40 percent of their capacity so that the network will perform well in the event of a single failure. Because the telephone engineers did their job so well, we usually can think of SS7 as shown in Fig. 14.3.

14.3 SS7 Protocol

The SS7 protocol suite is an evolutionary product, and over time, many changes and additions have been made to it. The purpose of this section is to give a brief overview of its major features and functions.

The SS7 protocol suite predates the seven-layer OSI model, but this model is very useful in describing the functions performed by SS7. Let us review the OSI reference model outlined in Table 14.1.

The physical layer of the SS7 protocol is the 64-kbps links described in Sec. 14.2. Riding on this physical layer is the SS7 message transfer part (MTP), which can be looked at as layers 2 and 3 of the OSI model. The functionality of the MTP that most closely corresponds to layer 2 is referred to as *MTP-2,* and correspondingly, the layer 3 functionality is called *MTP-3.*

TABLE 14.1 Seven-Layer OSI Reference Model

7. Application layer
6. Presentation layer
5. Session layer
4. Transport layer
3. Network layer
2. Data link layer
1. Physical layer

14.3.1   Message transfer part (MTP), level 2

The layer 2 functionality of the MTP is to frame the message, to check the frame for errors, and to provide sequence numbering, which can be used both to ensure that all frames are received in sequence and to provide flow control to avoid congestion. Framing is accomplished by separating messages with a *flag,* which is an 8-bit pattern `01111110` that is never repeated within the message and therefore can be used reliably to detect the start of a frame.

14.3.2   Message transfer part (MTP), level 3

The layer 3 functionality of the MTP includes carrying the network addresses of the sending and receiving nodes and information useful for routing and congestion control. The MTP can be loosely compared to the functionality of the IP in that it creates the addressed envelope needed by upper layers of the protocol to get the message to the destination. Once at the destination, it is the job of the upper layers to get the message to the correct application. Standards efforts are presently under way to define how to do the work of SS7 over the IP network.

Now that we have an envelope to get messages across the SS7 network, we can define how to get the information in the messages to the correct destination.

14.3.3   Signaling connection control part (SCCP)

The SCCP of the SS7 protocol suite provides a way to address the individual applications within a node of the SS7 network. Such an application might be the conversion of 1-800 numbers into PSTN numbers. It also provides the ability to provide both connectionless and connection-oriented services to these applications.[1] SCCP can be thought of as filling in the OSI layer 3 services that are missing from MTP-3 and adding some layer 4 services.

14.3.4   ISDN user part (ISUP)

A peer of the SCCP is the ISDN user part (ISUP). ISUP is used to signal the voice and data calls carried over the PSTN. ISUP does not use SCCP, so ISUP messages are addressed directly to the switch itself instead of to other applications. ISUP has functions that straddle layers 4 through 7 of the OSI model.

14.3.5   Transaction capabilities application part (TCAP)

TCAP defines the messages and protocol used to communicate between applications. It is necessary to get the TCAP messages out of the MTP envelope and to direct them to the correct application. TCAP uses SCCP for this routing function. Among the services supported by TCAP are database lookup services such as needed to provide 1-800 service and calling-card service. TCAP can be thought of as providing a subset of OSI layers 5 and 6 services.

---

[1]A connectionless service is a datagram service such as provided by User Datagram Protocol (UDP) over IP and a connection-oriented service is a virtual circuit such as provided by the Transmission Control Protocol (TCP) over IP.

14.3.6 Mobile application part (MAP)

MAP provides services to mobile switching centers (MSCs). MAP messages are carried within TCAP messages. MAP is used to communicate between the databases in a mobile telephone network and facilitates the authentication of mobile subscribers and the support of roaming. MAP can be thought of as an OSI layer 7 service. MAP is used to support ANSI-41, which is the topic of Chap. 15.

14.3.7 Base station system application part (BSSAP)

BSSAP provides similar functionality to MAP, but for GSM systems. BSSAP and its subprotocols are sometimes referred to as *GSM-MAP*. BSSAP does not use TCAP but rather uses the services of SCCP directly.

14.4 Logical and Physical Models

When we delve into the world of communications systems and message protocols, we run into the distinction between *logical models* and *physical models*.[2] Both the logical and physical models of wireless system connectivity fall into the category of *network reference models* that show the interfaces among the basic functional components of a network.

A logical model describes the components for how a system functions and the interfaces between relevant pairs of components, whereas a physical model shows how the actual parts are put together. A simple example is the user terminal described in Sec. 3.1.1, where we consider the radio transmitter and receiver as logically separate items, even though they are physically designed as one transceiver unit.

Consider three MSCs, each connected to the other two. Each MSC has an associated home location register (HLR) and visitor location register (VLR). We will explain what HLRs and VLRs are in more detail in Chap. 15, but for now, let us look at the logical network. Each MSC is connected to all three HLRs and to its own VLR. The logical model of this network is shown in Fig. 14.4.

The physical layout may look nothing like Fig. 14.4, however. In real life, the MSC, HLR, and VLR may all live in the same cabinet, as shown in Fig. 14.5. The logical links connecting each MSC to its own HLR and VLR are software subroutines because they share the same computer memory and disk drives. Between each pair of boxes is a DS-1 pipe with 2 DS-0s used for each MSC-to-HLR connection and the other 20 DS-0s used for MSC-to-MSC data. If we were to walk in and ask to see the 9 MSC-to-HLR links shown in Fig. 14.4, then we would be told that 6 of them are time slots inside the three DS-1 links. Three of them do not physically exist as links at all. In fact, the left two MSCs in Fig. 14.5 may be colocated with landline telephone local offices, and their DS-1 link is actually one time slot on a DS-4.

The physical layout might just as well look like Fig. 14.6. In this case, one machine performs the HLR and VLR functions for all three MSCs, as well as a digital cross-connect function to manage the MSC-to-MSC links. In case one breaks down, this network has two HLR/VLR machines for redundancy, and these machines share the load when both are running. Each MSC is connected to both HLR/VLR machines with an

---

[2]This distinction is completely separate from the distinction of layers in the OSI reference model. The OSI reference model, SS7, and ANSI-41 are all logical models with varying physical models used in implementation.

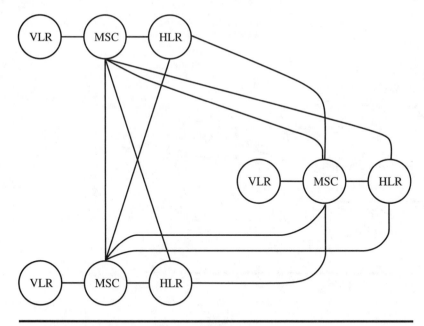

Figure 14.4 Logical diagram of three MSCs, HLRs, and VLRs.

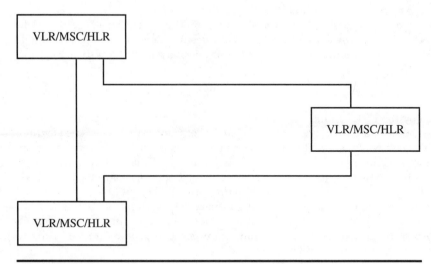

Figure 14.5 A physical diagram of three MSCs, HLRs, and VLRs.

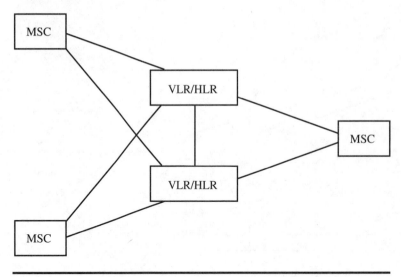

Figure 14.6   Another physical diagram of three MSCs, HLRs, and VLRs.

E-1 link, with 2 E-0 links dedicated to HLR and VLR traffic and the other 30 E-0s used for MSC-to-MSC data. The individual MSC-to-HLR and MSC-to-VLR links are four packet streams on the 2 E-0 links. The two HLR/VLR machines have a dedicated E-1 link between them to keep their databases synchronized.

These are two very different physical models that implement the same logical model.[3] The ANSI-41 standard provides a logical model of some functions of a wireless telephone network that may be hard to recognize from the physical layout of the system.

14.5   Conclusion

SS7 is used by the PSTN to increase the speed and efficiency of the network by transporting the network's signaling needs over a separate data network. It is designed for high reliability through redundancy. The use of a separate signaling network provides the opportunity to look at and manipulate the signaling information. This capability makes possible number translation services, such as 1-800 and 1-900 services, and support for LNP.

Users of the PSTN can subscribe to services that provide the ability to connect to the SS7 network and to make local decisions as to where calls such as 1-800 numbers are to be routed on a call-by-call basis. This service is particularly useful for balancing call volumes across PBX-based call centers and, more generally, for redirecting calls to where the party being called can be found.

---

[3]One of the authors (Rosenberg) tends to think like an engineer from the bottom up. Rosenberg would say that the two very different physical models are represented by the same logical model. Kemp, thinking as a planner from the top down, would say that one logical model has two very different physical implementations. The use of the distinction between logical and physical models can help team members who think very differently create networks (or books) as a team.

15

ANSI-41

The second-generation (2G) wireless world is divided into two camps, the Global System for Mobility (GSM), which does not use ANSI-41, and everybody else, who does. GSM has its own mobile application part (GSM MAP) for its signaling protocol. Other companies proposed Interim Standard 41 (IS-41), which then became the American National Standards Institute's ANSI-41 standard in use today.

It may seem odd that an American standard is used globally. The development and formal acceptance of standards run parallel to the deployment of equipment according to those standards, but the two are not synchronized. When a standard is useful, it is often deployed before it is formally approved. [This was certainly the case with Qualcomm's IS-95, which was used widely on code division multiple access (CDMA) systems before it was accepted as cdmaOne.] ANSI-41 is a valuable and useful standard that allows the mobile switching center (MSC) and other signaling-related equipment to interact on and across cellular networks. As a result, it is being built and deployed worldwide, even though, at the time of publication, it is officially only an American standard. The American National Standards Institute is a member of the International Telecommunications Union (ITU), and it is quite likely that the ITU-T (the Telephony Division of the ITU) is considering or will consider adopting ANSI-41 as a formal international standard.

ANSI-41 is a moving target as it evolves to meet increasing customer demand for features and mobility management and increasing system operator demand for operations, administration, and maintenance (OA&M) support. The major third-generation (3G) systems, cdma2000, wideband CDMA (W-CDMA), and UWC-136 (time division multiple access, or TDMA) all use ANSI-41 signaling.

ANSI-41 can use one of two signaling environments, X.25 packet protocol or Signaling System 7 (SS7). Today X.25-based systems are legacy systems because SS7 is the signaling service of choice, particularly for large carriers that already have SS7 networks in place. Such ANSI-41 entities as MSCs, home location registers (HLRs), visitor location registers (VLRs), and so on connect using SS7 A links to signal transfer points (STPs).

Figure 15.1[1] shows the first version of the ANSI-41 logical network reference model. We know most of the cast of characters already, but there are a couple of new faces.

---

[1]Reproduced under written permission of the Telecommunications Industry Association.

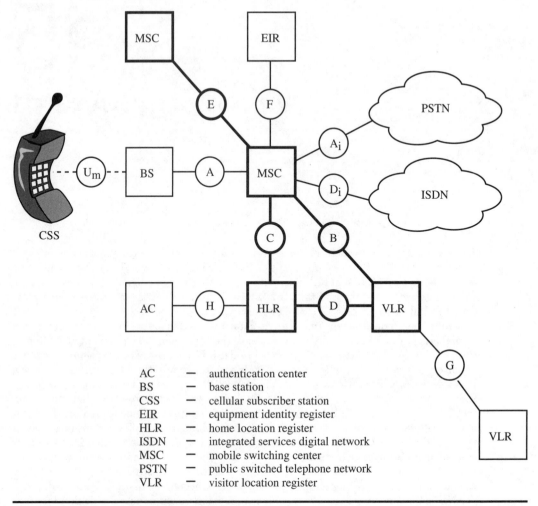

Figure 15.1 Original ANSI-41 network reference model.

The MSC is the mobile switching center, connected to a collection of base stations (BS) by an *A* interface not specified in ANSI-41. The base stations are connected to *cellular subscriber stations* (CSSs) by an air interface U_m, also not specified in ANSI-41. The MSC is also connected to the public switched telephone network (PSTN) and the Integrated Services Digital Network (ISDN) by A_i and D_i interfaces, also not specified in ANSI-41. (*A* and *D* originally stood for *analog* and *digital*.) Thus a typical mobile telephone call goes from cell phone to base station to MSC to PSTN to landline telephone without using an ANSI-41 link. (ANSI-41 signaling is still used for subscriber account verification and establishing call processing features, however.)

It would seem that the entire wireless telephone system is not specified by ANSI-41—that its designers have ignored every essential part of the system. However, ANSI-41 is not there to support the essential cellular concept; rather, ANSI-41 is there to enhance wireless telephony by allowing its users to roam freely supported by seamless handoffs from one MSC to another. ANSI-41 is also there to support short message service (SMS) and over-the-air service provisioning (OTASP) for the subscribers and to support operations, administration, and maintenance (OA&M) for the wireless service provider.

In ANSI-41, the notion of a wireless *system* is the MSC and its associated base stations. Elsewhere we have referred to a collection of MSCs and base stations as a single system because it is owned by one vendor and it serves one subscriber community. For the rest of this chapter, however, a *system* is an MSC, and *intersystem activity* is anything going from one MSC to another, whether the two MSCs are operated by the same provider or not.

The HLR is the home location register, a system that keeps track of the status of each user terminal at its home base. Even the simple mobile telephone call described above may involve some communication between MSC and the HLR, so ANSI-41 may yet be involved in a typical call. The interface between MSC and HLR is the C interface defined in ANSI-41. The VLR is the visitor location register. Its B interface to the MSC and its D interface directly to the HLR are both defined in ANSI-41. VLRs are allowed to communicate with each other using a G interface not defined in ANSI-41. HLRs can communicate with authentication centers using an H interface not defined in ANSI-41. MSCs communicate directly with other MSCs over ANSI-41–defined E interfaces.

The ANSI-41 network reference model has grown to look like Fig. 15.2.[2] The basic picture in Fig. 15.1 has changed a little, but a lot has been added. The CSS has been renamed the mobile station (MS), and the H interface between the HLR and the authentication center (AC) is now part of ANSI-41. Thus we have a fully specified signaling network within ANSI-41 connecting MSCs, HLRs, VLRs, and ACs to authenticate mobile subscribers and to let them roam and hand off where they please.

The SMS message center (MC) is connected to the MSC by the Q interface and to the HLR by the N interface, both ANSI-41–specified. The M_1 interface from the MC to the short message entity (SME) is specified. Also specified in ANSI-41 are the MC-to-MC M_2 interface and the SME-to-SME M_3 interface. SMS is designed into ANSI-41 as an integral part of today's wireless service.

The Internet has been added in the form of the public packet data network (PPDN) with an unspecified P_j interface. The interworking function (IWF) provides protocol conversions between packet-switched and circuit-switched entities. An example is converting circuit-based voice data streams from subscriber calls to packet-based voice over Internet Protocol (VoIP) to be sent over the Internet.[3]

Portable telephone numbers have been added in the unspecified Z interface to the number portability database (NPDB), which is standardized in TIA/EIA/IS-756. This is for telephone numbers formerly owned and administered by a wireless service provider and now belonging to subscribers served by another wireless or a landline provider.

The OTASP function uses ANSI-41 SMS operations to transfer customer information between the serving VLR and the HLR to support a new subscription. This allows customers to buy a cellular telephone and have it automatically activated when it is first

---

[2]Reproduced under written permission of the Telecommunications Industry Association (TIA).

[3]There is a separate interworking function, discussed in Sec. 15.6, that translates between ANSI-41 networks and GSM networks.

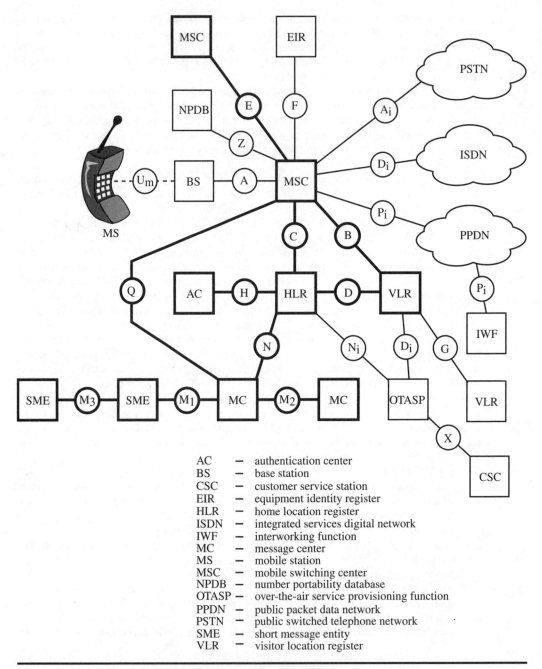

AC	—	authentication center
BS	—	base station
CSC	—	customer service station
EIR	—	equipment identity register
HLR	—	home location register
ISDN	—	integrated services digital network
IWF	—	interworking function
MC	—	message center
MS	—	mobile station
MSC	—	mobile switching center
NPDB	—	number portability database
OTASP	—	over-the-air service provisioning function
PPDN	—	public packet data network
PSTN	—	public switched telephone network
SME	—	short message entity
VLR	—	visitor location register

Figure 15.2 The most recent ANSI-41E network reference model.

turned on. This makes sales much easier because it makes it easy for untrained sales staff at third-party vendors to sell the cellular telephone with service without having to know how to set up the service. The OTASP function communicates with the MSC over the unspecified N_1 interface, with the VLR over the unspecified D_1 interface, and with a customer service center (CSC) over the unspecified X interface.

There are seven roles defined in ANSI-41 for the MSC during the call process. An MSC will have one or more of these roles in relation to each call during each moment of call setup, call operation, and call takedown:

- *Home* is the MSC with the subscriber's HLR.
- *Serving* is the MSC with the base station and radio link.
- *Anchor* is the MSC that was serving at the beginning of the call.
- *Tandem* is an MSC, not the first or last, in the handoff chain.
- *Target* is the MSC where we want to hand off.
- *Originating* is where the subscriber's telephone number is.
- *Gateway* is an originating MSC that is not the home.

The gateway MSC function occurs with mobile-originated calls if the call is made while the cellular telephone is roaming.

Now we have all the circles and arrow in place for ANSI-41, but we have not talked about what it does for us. ANSI-41 defines the processes that allow for handoffs, path minimization, short message service (SMS), and operations, administration, and maintenance (OA&M). We will discuss each of these in turn.

15.1 Inter-MSC Handoffs

Intersystem handoff allows a wireless telephone call to hand off from one MSC to another. There are five distinct functions in ANSI-41 handoff support:

- Handoff measurement
- Handoff forward
- Handoff back
- Path minimization
- Call release

For any of this to work, the MSCs must have their cell identification scheme coordinated so that each MSC knows where the neighbors are for its own cells. Intersystem handoff requires dedicated transport facilities from MSC to MSC, as well as the ANSI-41 E interface links. The length of the handoff chain on these dedicated facilities is reduced by using the handoff-back operation and by path minimization and ultimately is limited by parameters set by the wireless service providers.

In this section we speak about MSCs signaling to user terminals and MSCs detecting user terminal behavior. Of course, this signaling and detection go through an air interface, a base station, and a base-station-to-MSC signaling link. For the purpose of this discussion, we can treat these three steps as intermediate links in a communication chain. Also, we refer to a user terminal as changing to a new channel in the handoff process. In CDMA, it may be a different pseudonoise (PN) code on the same radio frequency.

According to the latest papers available at the time of publication, ANSI-41 does not appear to specify inter-MSC soft handoff, but some vendors, including Lucent Technologies, do support inter-MSC soft handoff within their own equipment.

15.1.1 Handoff measurement

In an intersystem handoff, we have a serving MSC and a target MSC where we are considering a handoff. As in the single-system handoff case, the CDMA user terminal provides measurements to the serving base station. For base station measurements, the serving MSC designates candidate MSCs and sends handoff measurement request messages to them.

The candidate MSC may not respond if it cannot support the radio channel characteristics of the call, if it does not detect a strong enough radio signal to measure, or if its own traffic conditions render it unavailable for a handoff. This is an implicit response that no handoff is possible here. If a candidate MSC does respond, it sends radio signal strength measurements to the serving MSC.

Depending on the signal strength measurements received from the candidates, the serving MSC may select a target MSC for handoff. The target MSC has to be compatible with the call mode, Advanced Mobile Phone Service (AMPS), CDMA, TDMA, and so on; it has to support the user terminal's power class and discontinuous transmission (DTx) modes; and it has to support whatever encryption algorithms the subscriber is using. If the target MSC is already involved in this call, then a handoff back is required. Otherwise, the serving MSC can try a path minimization or simply can choose to perform a handoff forward.

15.1.2 Handoff forward

Once the serving and target MSCs have established that a handoff forward is appropriate for this call, the procedure begins. The handoff chain is going to be one link longer after the handoff than before, but it must stay within the limit set by the system parameters.

Here is a typical handoff-forward scenario:

- The serving MSC requests a handoff forward and identifies a link.

- The target MSC accepts the handoff forward.

- The source-to-target inter-MSC link is set up.

- The serving MSC tells the user terminal to change channel.

- The user terminal changes channel.

- The target MSC detects the user terminal on its channel.

- The target MSC notifies the serving MSC.

- The serving MSC makes the connection to the target MSC.

- The handoff forward is complete.

Once the handoff forward is complete, the voice channel of the user terminal is supported by the new serving MSC, and the call is routed on a dedicated pipe back to the prior serving MSC, which may be the anchor MSC or a tandem MSC.

15.1.3 Handoff back

During a call, the serving MSC may discover that the best handoff is a handoff back to an MSC that is already involved in the call (either the anchor MSC or a tandem MSC). In this case, a handoff back is required. A handoff back is like any other inter-MSC handoff in that it is a change of serving MSC, but it has the advantage that the handoff back should shorten rather than lengthen the handoff chain. A handoff back prevents tromboning.

Here is a typical handoff-back scenario:

- The serving MSC requests a handoff back.

- The target MSC accepts the handoff back.

- The serving MSC tells the user terminal to change channel.

- The user terminal changes channel.

- The target MSC detects the user terminal on its channel.

- The target MSC requests release of the extra link.

- The serving MSC accepts the release.

- The link is released.

- The handoff is complete.

15.1.4 Path minimization

Path minimization is more complex than handoff back. If the serving MSC or the anchor MSC performs path minimization, then the handoff link will go as directly as possible from the anchor MSC to the target MSC. A third alternative is to have a tandem MSC perform the path-minimization process, which creates a path from that tandem MSC as directly as possible to the target MSC.

Here is a typical path minimization scenario:

- The serving MSC sends a path-minimization message to the anchor MSC.

- The anchor MSC sets up a direct path to the target MSC.

- The anchor MSC sends a handoff message to the target MSC.

- The serving MSC tells the user terminal to change channel.

- The user terminal changes channel.

- The target MSC detects the user terminal on its channel.

- The target MSC notifies the anchor MSC.

- The anchor MSC connects the call to the target MSC.

- The old circuits are released.

- The handoff is complete.

Path minimization reduces the number of facilities being used to maintain a call by creating a more direct route between the serving MSC and the anchor MSC. Here is a scenario where path minimization would be of value. Let us say that there are three MSCs, *A*, *B*, and *C*, with each MSC having a direct connection to the other two.

- Initially, the call is served by MSC *A*, the anchor and serving MSC, with no other MSCs involved.

- The subscriber moves into the area covered by MSC *B*, and a handoff forward is completed. Now, MSC *A* is the anchor MSC linking the call to the public switched telephone network (PSTN), and MSC *B* is the serving MSC. The call is being carried from the subscriber to MSC *B*, over the link between MSC *B* and MSC *A*, and on to the PSTN.

- The call continues, and the subscriber moves to the area best served by MSC *C*. If MSC *B* were to establish a handoff forward to MSC *C*, then MSC *B* would become a tandem MSC. Switching resources at MSC *B* and two sets of pipes (from MSC *C* to MSC *B* and then from MSC *B* to MSC *A*) would be used for the duration of the call. This is not the most efficient option.

- Instead, MSC *B* requests a path minimization, and MSC *C* informs MSC *B* that a pipe directly from MSC *C* to MSC *A* is available. MSC *A* sets up a link on this pipe. The call is then handed off from MSC *B* to MSC *C* with path minimization.

- After the handoff, MSC *C* is the serving MSC. MSC *A* is the anchor MSC. The call is using resources on one pipe, MSC *C* to MSC *A*. The path is minimized; MSC *B* and pipes connected to it are not participating in the call.

15.1.5   Call release

ANSI-41 also tells us how to *release* a call in intersystem handoff. The ANSI-41 functions tell the system how to tear down the handoff links in an orderly fashion once the call has ended.

Here is a typical call release scenario:

- Somebody hangs up (or presses the END key).
- The anchor MSC releases the link to the tandem MSC.
- The tandem MSC releases the link to the serving MSC.
- The serving MSC accepts the release and sends billing information.
- The tandem MSC accepts the release and sends billing information.
- The handoff circuits are all released.

15.2   Automatic Roaming

Management of roaming is the bulk of ANSI-41 activity. Roaming management is needed every time a user terminal is activated anywhere on the system other than at the subscriber's home MSC. Intersystem handoffs are much rarer because they are only needed if the subscriber moves from one MSC serving area to another during an active call. Also, roaming is more complicated because the subscriber has to be authenticated and the full array of subscriber features has to be sent from the HLR to the serving system.

15.2.1   The HLR and VLR

The home location register (HLR) is the functional entity defined in ANSI-41 that maintains subscriber information and status. In addition to being a large and powerful database, the HLR provides control and processing-center functions. The HLR is the home database for wireless subscribers.

The relatively stable information in the HLR on each subscriber includes the directory number, subscribed services and features, and whether the subscriber should be offered or denied service. The more transient information is the location and call status of the user terminal.

While each MSC has a separate logical HLR, a wireless service provider may opt to put several HLRs in one centralized computer. It is often more efficient to administer one large computer in one place than a collection of smaller machines at separate physical locations.

The visitor location register (VLR) is a database of valid roaming subscribers in an MSC region. These are subscribers who have registered, though they may never make an actual wireless call. The VLR keeps a record so that the MSC is ready to serve them for incoming or outgoing calls. The VLR is a separate logical entity, but it is usually physically part of the MSC itself. This physical colocation has the advantage that it streamlines the MSC-to-VLR signaling at the beginning of a call.

15.2.2   Service qualification

When a user terminal registers, the serving MSC determines whether it is a home subscriber or a roamer. In either case, the MSC communicates with the HLR to verify that this subscriber pays the bills promptly enough to get wireless telephone service. In the roaming case, the serving MSC contacts the subscriber's home HLR and its own VLR to figure out the status of the user terminal; it already may have registered recently in this system. The serving MSC contacts the HLR when the user terminal registers, when it moves out of the qualified area, when its allocated time authorization runs out, or when the HLR asks for an update.

The HLR can revoke service privileges for a user terminal by sending a message to the serving MSC and VLR. ANSI-41 does not tell the serving system *how* to revoke service; it might play a prerecorded message when the subscriber tries to make a call.

15.2.3   Location management

The HLR keeps track of where its user terminals are located. This location is at least as precise as knowing what system the user terminal was in when it most recently registered, but it may have more precise information. The HLR is notified whenever the user terminal registers in a new system. This is the location-update process.

There is also a location-cancellation process. This occurs when the user terminal sends a power-down signal to the serving MSC or when the user terminal fails to respond when a registration request is sent to it.

The roamer location management system has specific protection in it for intersystem confusion at call setup. When a user terminal tries to access the serving system, a neighboring system can pick up the signal and think it is being accessed for a call on one of its own base stations. We call a cell that can pick up a false access attempt a *border cell*. The HLR knows that something is going wrong when it gets two information requests from two separate systems for the same call. One of the signals is from the intended serving MSC, and the other is from an unintended MSC. Part of the call-attempt message from a serving MSC to the HLR is a radio signal path measurement, at least for potential border cells. When it gets the two conflicting messages, the HLR uses the two radio signal strength measurements to determine which of the two systems should be serving the call. It then tells the other system to abandon it.

15.2.4 User terminal state management

The user terminal state is either active or inactive. An active mobile telephone is one that is available for call delivery. While an inactive mobile telephone cannot receive telephone calls, it may still be able to accept SMS deliveries.

A user terminal is inactive under these circumstances:

- It has not registered anywhere.

- There is no valid location.

- It goes out of radio contact when it misses a registration, whether an autonomous registration or one scheduled by the system.

- It is in *sleep mode,* a subscriber-designated setting where it can place calls but not receive them.

- The serving system has designated the user terminal inactive according to the server's own internal rules.

An inactive user terminal remains inactive until a serving MSC sends the HLR a registration notification message.

15.2.5 HLR and VLR fault recovery

The ANSI-41 specification allows for the case where an HLR or VLR can lose its current status data for its subscribers. While we would hope that wireless service providers would do frequent backups on their HLR databases so that the account records would be safe from mishap, the HLR and VLR keep a moment-by-moment record of the location and registration state of a large number of mobile stations. The HLR and VLR have procedures to follow if these data are lost.

The HLR is required to keep a nonvolatile storage area with a list of all possible serving systems and another list of all active serving systems, systems actually serving one or more of its subscribers at any particular time. (A disk file is nonvolatile enough for these purposes. The ANSI-41 requirement for nonvolatile is simply that the file is not lost when the power goes off or when a computer is shut down and restarted.) In case of data failure at the HLR, when the HLR recovers, it sends an unreliable roamer data directive message to every serving system on its list so that the serving MSCs can send messages updating the HLR data files. In this way, the HLR can start fresh with a clean and correct representation of its own subscribers.

The VLR has similar protection against data failure. It keeps a nonvolatile storage area with a list of all possible HLRs and a list of all HLRs with active visiting subscribers. Should the VLR have a failure, when it recovers, it sends a bulk deregistration message requesting an update to all the HLRs on its list.

15.2.6 Roamer call delivery

When a roamer makes a call, the serving MSC and VLR coordinate with the HLR to authenticate the subscriber's validity and level of service. This should be done when the user terminal registers. From this point, the subscriber's call looks like an ordinary mobile telephone call.

Incoming calls require more coordination between home and serving systems because the call comes into the originating system rather than the serving system. ANSI-41 allows that the originating system where the call comes in may not be the same as the subscriber's home system. The originating system is called the *gateway system* when it is not the home system.

Automatic routing of roamer incoming calls relies on the user terminal location data in the HLR and that the user terminal is in the active state. For the duration of call completion, the user terminal is assigned a temporary local directory number (TLDN). A subscriber can disable the automatic call delivery feature.

Here is a typical call delivery scenario:

- The originating MSC queries the HLR about the subscriber status.
- The HLR sends a route request to the serving MSC.
- The serving MSC allocates a TLDN to the user terminal.
- The serving MSC tells the HLR the TLDN.
- The HLR tells the originating MSC the TLDN.
- The originating MSC routes the call to the TLDN.
- The serving MSC associates the TLDN with the user terminal.
- The serving MSC pages the user terminal.
- The user terminal responds to the page.
- The radio link is created for the call.
- The serving MSC sends an alert message.
- The call proceeds normally.

15.3 Short Message Service (SMS)

Wireless telephone subscribers and PSTN subscribers sending pages and similar messages use short message service (SMS) to send messages of one or two sentences to another SMS subscriber. The SMS subscribers do not have to be wireless, but the service certainly seems to be concentrated in the wireless community.

15.3.1 Short message entities

A short message entity (SME) is anything capable of originating and receiving a short message using SMS. An SME may be in a fixed network outside the ANSI-41 network, or an SME may be within the ANSI-41 network, typically a mobile terminal. It should be able to

- Compose short messages.

- Send short messages.

- Receive short messages.

- Store received short messages.

- Manage stored short messages.

- Display short messages.

- Request supplementary services.

ANSI-41 does not specify how these tasks are done. These functions are defined in the SMS standard (see Chap. 20). Many SMEs, including broadcast paging services, are on data networks. ANSI-41 only defines the functional components and processes that allow SMS messages to move through the cellular network.

15.3.2 Message centers

Message centers (MCs) are the ANSI-41 store-and-forward hubs for most mobile-originated short messages and for all mobile-terminated short messages. These are usually physically separate machines but may be combined with other functional entities. Each SMS subscriber is associated with a home MC in the subscriber's home system. The MC should be able to forward short messages to the user terminal SME, store short messages for unavailable user terminal SMEs, and perform other administrative services as required.

For roamers, the serving MC coordinates with the HLR for location update and user terminal state. To receive a message, the user terminal must be located, be in an active state, and be authorized to receive short messages in the serving system.

15.3.3 Short message processing

Here is a typical short message scenario from mobile *A* to mobile *B:*

- Mobile subscriber *A* sends a short message.

- Serving MSC *A* uses ANSI-41 and SS7 to route the message to MC *A.*

- MC *A* sends acknowledgment to serving MSC *A.*

- Serving MSC *A* sends acknowledgment to user terminal *A.*

- MC *A* sends the message using the Transmission Control Protocol/Internet Protocol (TCP/IP) to MC *B.*

- MC *B* uses ANSI-41 and SS7 to route the message to serving MSC *B.*

- Serving MSC *B* sends the message to user terminal *B.*

- User terminal *B* sends acknowledgment to serving MSC *B.*

- Serving MSC *B* sends acknowledgment to serving MC *B.*

- The message transfer is complete.

- It is up to subscriber B to respond to subscriber A.

15.4   Operations, Administration, and Maintenance (OA&M)

ANSI-41 allows for control of the MSC-to-MSC trunks used for intersystem handoffs.[4] The trunks can be removed and reinstated, and they can be configured for loop-back testing. This capability only affects intersystem handoff, not roaming and not ordinary call completion. An MSC-to-MSC link has four states:

- Active

- Locally blocked

- Remotely blocked

- Locally and remotely blocked

The trunk is locally blocked if it is removed from service at the local end, and it is remotely blocked if it is removed from service at the other end.

Loop-back testing has one end of a normal two-way trunk looped back at one end, as shown in Fig. 15.3, so that the signal received is immediately transmitted back the other way. Once the loop-back is in place, the trunk can be tested by sending signals along the trunk and making sure that the same signal comes back the other way.

15.5   Over-the-Air Service Provisioning (OTASP)

OTASP allows a subscriber to get wireless telephone service without having to visit a retail establishment. When OTASP is working properly, it makes service activation easy, it avoids programming errors from manual data entry, and it reduces the costs associated with operating retail wireless phone-center stores.

Using only the radio link, the base station transmits encrypted data to the user terminal in order to program its number-assignment module (NAM). Typically, the potential subscriber who already has a mobile telephone speaks to a highly trained service representative to obtain service. Financial information is provided by the subscriber and confirmed by the sales representative, and then OTASP is used to activate service for the user terminal.

Over-the-air parameter administration (OTAPA) makes it possible to adjust preferred roaming lists (PRLs), to update intelligent roaming databases (IRDBs), and to change the mobile identification number due to area-code changes. OTAPA works whenever the user terminal is powered on and requires no input from the subscriber.

OTASP and OTAPA require good radio signal because we do not want any mistakes to happen here. We also do not want anybody to overhear these transactions, so a very sophisticated encryption scheme is used for OTASP and OTAPA, the Diffie-Hoffman Key Agreement Standard. Diffie-Hoffman has no public keys to publish and no private

---

[4]Although we avoid the term *trunks* when referring to the wireless network, in the case of inter-MSC pipes, these are dedicated transport channels.

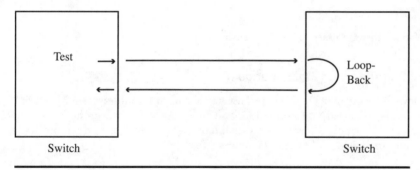

Figure 15.3 Loop-back trunk testing.

keys to store securely, but it is computationally complex and typically takes a user terminal several minutes to calculate. It is worth this kind of wait to ensure ultrasecure transactions for service provisioning and parameter adjustment. CDMA-based ANSI-41 networks use the SMS capability to send their OTASP data.

Proprietary services can extend the functionality of OTASP and OTAPA. For example, some cellular services, including Sprint PCS, are able to reconfigure telephones to correct technical problems or update user features across the air interface. The service provider may maintain a database of cellular telephone equipment and its internal system control codes, making it possible to modify specific equipment in specific ways.

15.6   Interaction with Other Networks

Most of the internetwork issues for ANSI-41 networks are solved by SS7. We do not have to worry about network addressing for packet messages, internetwork communication issues, or even international roaming because the worldwide SS7 network has already dealt with most of these issues. There are issues with international numbering plans because the mobile identification number (MIN) is usually the mobile *directory number* in the United States and Canada. This creates some confusion because international country codes look a lot like U.S. area codes.[5] New numbering plans are in the works as the international standards committees work on the mobile numbering problem.

cdmaOne, cdma2000, and wideband CDMA (W-CDMA) all use ANSI-41 networks, but there are 700,000,000 GSM cell phones in service that use the GSM signaling protocols. Since W-CDMA is an outgrowth of GSM, we expect W-CDMA user terminals to be dual-mode mobile telephones supporting both GSM and W-CDMA. This means that the same user terminal may find itself in a GSM system one day and in an ANSI-41 system the next, so we need some form of signaling communication between GSM and ANSI-41.

While the two signaling languages are not the same, they have much vocabulary in common. Since telephone signaling systems do not have the universal translators used on the "Star Trek" television show, we have to employ a more down-to-earth solution.

---

[5]For example, Mexico's country code is 52 and Arizona has a 520 area code, so an Arizona system could refuse service to a Mexican roamer with a confusing number.

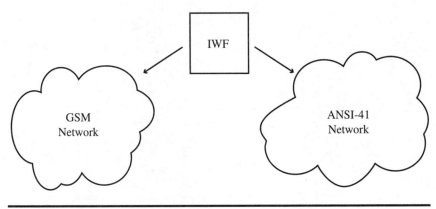

Figure 15.4 Interworking function (IWF).

The term *interworking* is used for the communication between GSM and ANSI-41 signaling systems. There are enough differences that there is not going to be full compatibility between the two systems, but there is enough in common that a subscriber can have a single subscription that operates in both environments. The *interworking function* (IWF) is the functional entity that does the translations. The IWF could be physically located in either system or somewhere between the two systems, but the basic picture is shown in Fig. 15.4.[6]

The most basic form of the IWF is a separate machine that connects to the GSM and ANSI-41 signaling networks just as the figure shows. Another implementation of the IWF is a dual-mode HLR, a single HLR machine that signals in both GSM and ANSI-41 protocols. In any case, neither GSM nor ANSI-41 is going away any time soon, so internetwork communication is going to be a part of the wireless signaling system for many years to come.

15.7 Conclusion

The ANSI-41 standard governs inter-MSC communications in support of advanced cellular services. This standardization of the cellular network specification simplifies management of cellular networks and reduces the cost of equipment design. ANSI-41 continues to grow, meeting new demands of cellular subscribers and service providers.

---

[6]There is a separate interworking function that translates between circuit-switched and packet-switched (IP) transport networks described at the beginning of this chapter.

16

Call States

The call state diagramming method is a useful tool for describing the operations of a telephone network. Modeling with call states allows system designers to make sure that the system is conceptually robust. If the physical model and physical implementation deliver the conceptual model successfully, and the conceptual model is robust, then the system should work as specified. State diagrams are also useful to engineers who are tracing network problems.

16.1 Defining Call States

A telephone call is a sequential process with a beginning, an end, and a logical progression of *call states* along the way. We model sequential processes as states and stimuli that can move the component or system from state to state. Making state diagrams and enumerating the various stimuli are major components of almost any system design.[1]

A state diagram is only a model and is not required to have every level of detail. Just as a map can be high level or close up, state diagrams can show varying levels of detail depending on what we are using them for. In creating the state diagram, we decide which states are important to model. Once the states are determined, we need to find each relevant stimulus that changes the state of the telephone. The notion of a stimulus being relevant gives us latitude in describing the call process. Depending on what we want to accomplish, we can choose a more general diagram to give a simple big picture, or we can work through all the possible stimuli and get the details right.

Allow us to illustrate the process with a basic description of the states of a landline telephone. At the highest level, a normal residential telephone has two basic states, sitting idle or being used. When the telephone is sitting idle with the receiver in the

---

[1]A call-state diagram is not quite the same as the flowchart that our teachers told us to make before writing a computer program, but the two are similar. Making call–state diagrams forces system designers to think about all the relevant states and stimuli in an organized way. For people with an information technology background, call-state diagrams are deceptively similar to data-flow diagrams. However, the circles in a call state represent states, not processes, and the arrows represent transitions between states, not data items that move from one process to another.

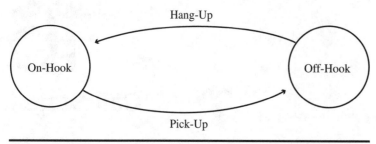

Hang-Up

On-Hook

Off-Hook

Pick-Up

Figure 16.1   Call-state diagram of the two basic states of a landline telephone.

cradle, we say that it is in the ON-HOOK state. And any time the receiver is not in the cradle, the telephone is in the OFF-HOOK state. These states are depicted in Fig. 16.1. We use circles for the states and arrows for the state-to-state transitions. The labels on the arrows name the stimulus that causes the transition from one state to another.

In the case of these two states, there are only two stimuli that can cause the telephone to change from one state to another. When the telephone is idle, in the ON-HOOK state, somebody can come along and pick up the receiver and put the telephone into the OFF-HOOK state. This is the *PICK-UP* stimulus.

The telephone will act differently depending on both what event occurs and what state it is in when this event occurs. In the ON-HOOK state, the telephone rings when a 90-V, 20-Hz signal is applied. In the OFF-HOOK state, the telephone sends digits and engages in two-way voice communication when someone makes a call. So long as the telephone does these things, the telephone is working properly, at least at this level of modeling. It is important to understand that the telephone behaves differently in each of the two states of this model.

Another important behavioral difference between the two states is the response to an incoming call. This two-state model may not be sophisticated enough to represent the difference between incoming and outgoing calls, but the rest of the telephone network still cares. A telephone in the ON-HOOK state is receptive to an incoming call. When a telephone is in the OFF-HOOK state, the switch returns a busy signal to the caller.[2]

There is a level of abstraction in analyzing states. Terms such as *PICK-UP* are general terms that can represent a large number of actual events. For example, the *PICK-UP* stimulus that sends the telephone from ON-HOOK to OFF-HOOK can be picking up a telephone receiver, pressing the SPEAKER button on a speakerphone, pressing the TALK button on a cordless telephone, or having a fax machine automatically answer. So far as this model of a telephone is concerned, all these are the same *PICK-UP* stimulus. In the case of modeling the fax machine this way, the telephone component is being picked up by the fax-modem component.

State diagrams have many uses. In Sec. 4.8 we used equipment-state diagrams to determine reliability, that is, the likelihood of the equipment being in a failed state. In call processing, state diagrams are used as a kind of checklist. The diagrams help the designers to make sure nothing unexpected is going to happen, for example, to ensure that there are no glare conditions, as described in Sec. 23.6. In call processing, we do not like surprises.

---

[2]Call waiting and three-way calling are discussed later.

16.2   PSTN Call States

The public switched telephone network (PSTN) provides many services besides basic call setup and teardown and handles these extra capabilities by cleverly using the call states we discussed earlier and extending them to provide these services. Many of these extra capabilities were added in support of the idea that the telephone service should switch from being location-based to being personal, or subscriber-based. We call these extra capabilities *vertical services* that started with custom calling (see Sec. 12.1) and have expanded into voice-mail messaging services and beyond. Let us examine the basic PSTN call first and then delve into call waiting and three-way calling.

The other two custom calling features, speed dialing and call forwarding, are less interesting to us here. Speed dialing determines how a call is routed from its abbreviated dialed digits, and call forwarding affects how the network routes an incoming call, a landline function similar to roaming.

16.2.1   Basic PSTN call states

Consider regular landline telephone calls being made on a regular landline telephone, with no call waiting and no three-way calling. This model will be more detailed than the two-state model examined earlier. Let's examine the four states QUIET, RINGING, DIALING, and TALKING shown in Fig. 16.2.

The QUIET state has the telephone just sitting there. Restricting our attention to PSTN call processing, we ask ourselves, "What can happen next?" One stimulus is that somebody calls, and the phone starts ringing. A ringing telephone is in a different state than an idle telephone; it responds differently when somebody picks up the receiver and goes off-hook. An *INCOMING-CALL* stimulus sends the telephone from the QUIET state to the RINGING state. The other stimulus is somebody picking up the receiver to

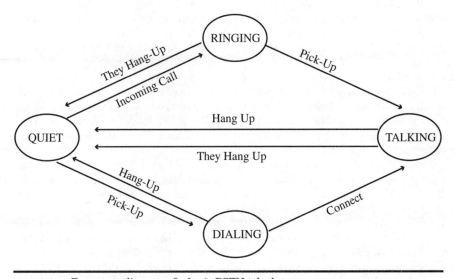

Figure 16.2   Four-state diagram of a basic PSTN telephone.

make an outgoing telephone call. A *PICK-UP* stimulus sends the telephone from the QUIET state to the DIALING state.

From the RINGING state, two things can happen. One stimulus is that the caller gives up, and the telephone stops ringing. The *THEY-HANG-UP* stimulus sends the telephone from the RINGING state to the QUIET state. The other stimulus is that the telephone is answered, in which case the *PICK-UP* stimulus sends the telephone from the RINGING state to the TALKING state, the stable call we described in Sec. 12.1.

Is the *PICK-UP* stimulus that sends the telephone from QUIET to DIALING the *same* stimulus as the *PICK-UP* stimulus that sends the same telephone from RINGING to TALKING? This model has two stimuli from two different states with the same name. From the state-diagram point of view, it does not matter: It is the state and stimulus *together* that determine the flow. We could be more precise and call them *PICK-UP-WHILE-QUIET* and *PICK-UP-WHILE-RINGING,* but this does not change the state model.

From the DIALING state, the *HANG-UP* stimulus sends the telephone back into the QUIET state. And the *CONNECT* event sends the DIALING telephone into the TALKING state. Finally, the *HANG-UP* stimulus or the *THEY-HANG-UP* stimulus sends the telephone from the TALKING state back into its QUIET state.

This four-state picture shows the progression of a telephone in a simple call. A busy signal is a call in the DIALING state that does not connect.

16.2.2 Advanced PSTN call states

In Fig. 16.2 the only meaningful stimuli in the TALKING state are *HANG-UP* and *THEY-HANG-UP*. Somebody else trying to call gets a busy signal, and the current call status remains unchanged.

Call waiting is a custom calling feature that allows an incoming call to interrupt a call in progress. A subscriber with this feature hears a tone during a call to signify a second caller. The second caller has the usual ringing sound, no signal that the call attempt is interrupting anything, The subscriber with call waiting can switch between the two callers using a *flash* (also called a *switchhook flash*). On a regular, plain-Jane telephone, a flash is executed by holding the switchhook down for about half a second, not long enough to end the call. Some fancy telephones have a special FLASH key, usually with a little lighting bolt on it, and this key is programmed to break the link for the right amount of time for a flash. Cordless telephones flash by pressing a button, often the TALK button.

This flash-to-switch arrangement continues until one of the calling parties hangs up and the call reverts to a normal two-way call. If the subscriber with call waiting hangs up to end a conversation with one caller and the other caller stays on the line, then the telephone rings back with the other caller on the line. This callback feature works even if the subscriber has not flashed yet.

Adding call waiting adds more states to the diagram shown in Fig. 16.3. In this case, a TALKING telephone can have an *INCOMING-CALL* stimulus and change to a state we will call TALK1. The telephone user is informed about the change of state from TALKING to TALK1 by the beep interrupting the voice telephone call.

Unlike the TALKING state, the TALK1 state responds to a flash to talk to the other party. From the TALK1 state the *FLASH* stimulus switches to the other call-waiting state, TALK2. Another *FLASH* stimulus sends the telephone back to TALK1.

From either call-waiting state, TALK1 or TALK2, having *ONE-OF-THEM-HANGS-UP* changes the telephone call back to the regular TALKING state. If the TALK1 or TALK2 tele-

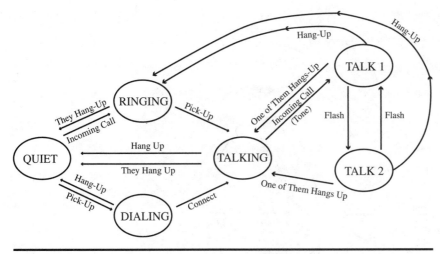

Figure 16.3 PSTN call states with call waiting included.

phone has a *HANG-UP* stimulus, then it goes into a RINGING state with the other party on the line. This is a feature of call waiting where the subscriber who hangs up gets called back with the other party on the line.

Three-way calling is a custom calling feature that allows a subscriber to talk to two other parties at once. The subscriber flashes to make a second call, the first call is put on hold, and the subscriber gets a dial tone. This dial tone usually starts with some tone pulses to make it sound distinctive, and this is called a *broken dial tone*. The subscriber can dial another telephone number and wait for it to connect while the first call remains on hold. This second call can continue for a while until the subscriber with three-way calling flashes again and all three parties are on the same call. This is different from call waiting, where flashing alternates the two calls and the three telephones never share the same line. Often we say the second flash *conferences in* the other party from the original call. Another flash from the three-way calling subscriber disconnects the second call and leaves the original call intact.

If the three-way calling subscriber hangs up, then both calls end, but if either other party hangs up, then the call continues with the two remaining parties as an ordinary two-way call.

In some office systems run by a private branch exchange (PBX), when the original three-way calling subscriber hangs up, the other two parties remain connected in their own call. This feature is often called *call transfer* because it is used to transfer incoming calls.[3]

Adding three-way calling to a diagram with call waiting makes the call-state picture more complicated. The interactions between the two features may differ on different systems. In most, if a single call is in progress and the originating caller initiates

---

[3]The typical scenario is I get a call from somebody who really should be talking to you, so I flash and three-way call you; I tell you who is calling; I flash again to make the call three-way; and then I hang up.

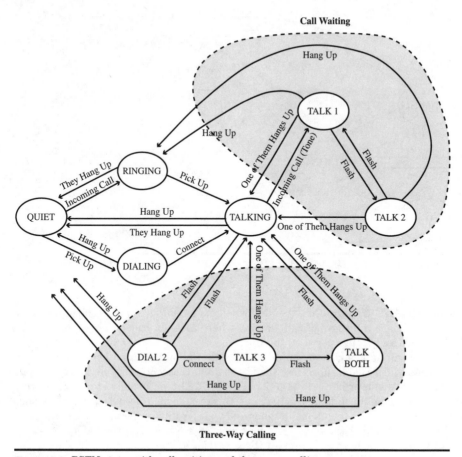

Figure 16.4   PSTN states with call waiting and three-way calling.

another call with a flash, the call will become three-way when the originating caller signals the system with a second flash. However, if the call is interrupted by a third party, then the call-waiting mode will be active, and the flash key will alternate between the two calls. A fairly complete state diagram with call waiting and three-way calling is shown in Fig. 16.4.[4]

Call state diagrams can be used to describe the state of a single piece of equipment, the states of an entire network, or the states of multiple components interacting on a network. When designing a messaging system using a protocol such as Integrated Services Digital Network (ISDN), it is important that *every* state of *every* node have some

---

[4]We say fairly complete for two reasons. First, not all telephone systems do the exact same thing when call waiting and three-way calling are both involved. We have seen systems that do not allow both call waiting and three-way calling to be used at the same time, and we have seen other systems that do. Second, as discussed earlier, call state diagrams operate at a level of abstraction that may not include all details.

response to *every* message in the protocol. This response may be not to do anything, not to change state, or not to send any messages anywhere, but freezing up and not knowing what to do is not a healthy design option. The entity sending the message has its own state space and is responding to its own stimuli, including other protocol messages. The design challenge of a multinode messaging system is in resolving the confusing combinations of states without creating glare conditions.

One example of this is in the ANSI-41 protocol (see Chap. 15). A serving mobile switching center (MSC) looking for handoff opportunities asks candidate MSCs to respond. If the candidate MSC is unable to take the call, it simply does not respond within a specified time period. This is a valid action. The serving MSCs call state diagram would include an appropriate action to take when not receiving a response. That action would be to drop the switch that did not respond from the candidate list.

16.3   Wireless Call States

With its preorigination dialing, the wireless telephone has no DIALING state. Instead, dialed digits are kept in the user terminal until the SEND key is pressed. The digits are dialed and kept in the cellular telephone until the subscriber presses SEND, and then the system sends a message requesting dialed digits.

The simple four-state wireless diagram is in Fig. 16.5. Dialing digits, the *DIGITS* stimulus, puts digits into the internal telephone registers but does not change the state of the telephone. While dialing digits does not change the state on this diagram, we feel that it is subscriber behavior that is important enough to put on the state diagram.

The QUIET, RINGING, and TALKING states for incoming calls are the same as landline states. Instead of a DIALING state, wireless calls have a CALLING state, where the system is trying to complete the dialed call. This state is distinguished by having a voice link to the subscriber with no connection to another telephone.

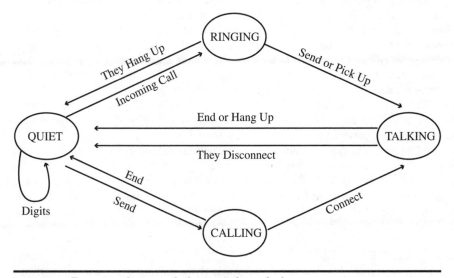

Figure 16.5   Four-state diagram of a basic wireless telephone.

Custom calling had to be reengineered for wireless to take preorigination dialing into account. Using the SEND key for a flash was natural, but the notion of dialing digits for three-way calling has some confusion.

This means that wireless call waiting is almost exactly the same as regular landline call waiting with the flash key being the SEND key.

In 1981, the Advanced Mobile Phone Service (AMPS) system trial telephones had no tones associated with dialed digits because they were only used for preorigination dialing. In fact, they were deliberately made silent so that a subscriber could key in a telephone number as it was spoken (by directory assistance, for example), press END, and press SEND to make a new call. The notion of pressing buttons to send Touch Tone digits on a telephone call required the telephone design to change, and the buttons on wireless telephones have been dual-purpose ever since. The numbered buttons on a wireless telephone put digits into a register to be sent the next time the SEND key is pressed, and those same numbered buttons send Touch Tone signals when a call is in progress.

Wireless three-way calling uses the same digit registers as initiating a wireless telephone call, so the Touch Tone sounds serve no purpose. A three-way caller dialing a second call almost certainly does not want spurious tones on the first call, so an extra step was added to the wireless three-way calling process.

The first flash sends the call into a quiet mode so that the subscriber can send digits for the second call.[5] Once those digits are entered to the subscriber's satisfaction, SEND is pressed again, and the system attempts to complete the second call. From this point on, a wireless three-way call is the same as the landline version, with the SEND key being used for a flash.

Figure 16.6 shows a call-state diagram for wireless with call waiting and three-way calling. As in the landline case, the interaction between these two custom calling features can vary a little from system to system.

The voice-mail service (sometimes called *call messaging* or *call notes*) provides the services of an answering machine, with the telephone network storing messages for the subscriber. It has an additional advantage in that if a new call comes in while you are on the line (and you do not have call waiting or have disabled it), the call will go to the automated answering service. This is not possible with a traditional landline answering machine. Voice mail is available both on the PSTN and on cellular systems. The voice-mail system does not require any new states for the telephone itself, but one could create a state diagram to represent the system states.

On the PSTN, the voice-mail service notifies the subscriber of new voice mail with a broken dial tone. The customer calls a special phone number to reach an automated message center and enters a passcode to listen to and to manage messages.

On cellular systems, the network pages the cellular telephone over the signaling channel, and the telephone signals the user with a distinctive ring, a flashing light, a text message, or even all three. On some phones, the user can call the automated message center to listen to and manage messages using a single keystroke. A passcode is not needed if the caller is picking up messages from his or her own cellular phone.

---

[5]Digits also can be entered during the nonquiet part of the call, but then there will be Touch Tone sounds on the line.

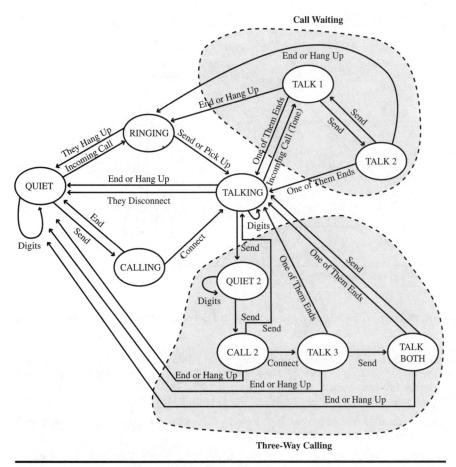

Figure 16.6 Wireless states with call waiting and three-way calling.

16.4 Wireless Radio States

In the same way that call feature designers use state diagrams to work out the flow of a telephone call, radio technology designers use similar state diagrams to work out the air interface. A high-level view of the wireless call setup is shown in Fig. 16.7. The user terminal is turned on and sends its registration message. The wireless system accepts the message and acknowledges the registration. The user terminal is now ready to make and receive calls.

Three things can happen to a registered wireless telephone, a call can come in, the subscriber can make a call, or the subscriber can turn it off. Once the user terminal receives a page, it establishes a connection with the system the same way it does when initiating a call. Radio link parameters are established, and a voice link is established.

This state diagram is a picture of what happens when things go right. It does not show all the alternative pathways where things go wrong. Some of those alternatives are explained clearly in technology specifications, whereas others are left up to the

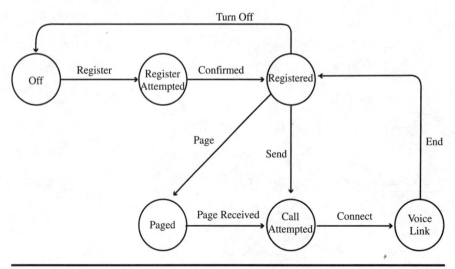

Figure 16.7 High-level wireless call setup state diagram.

equipment vendor. Without going too far into the details, let us go over some of these cases:

- A mobile telephone may not detect the system paging data stream. Perhaps the radio environment is too hostile, or perhaps the system's radio interface is not a technology the mobile telephone can support. A Global System for Mobility (GSM) telephone should not recognize a code division multiple access (CDMA) paging stream.

- The user terminal's registration can be rejected because the telephone is not working correctly or because it is corrupted by radio noise.

- The system may determine that the telephone is not to be served. This could be the result of a customer not paying the bill or a cellular telephone reported stolen. However, the registration may still be completed because wireless systems should still be able to give emergency service (911 in the United States) to telephones even when they are otherwise out of service.

- The system acknowledgment of the registration also can fail, and there is no two-way communication between the system and the mobile telephone.

- There is another curious kind of registration error created by the type of wireless telephone fraud called *cloning*. A telephone may be cloned, and the copy will have the same registration information as the original user terminal. Thus a registration attempt may appear with the same serial number and telephone number as another registration. The system operator may decide to serve one, both, or neither, but the system does have to deal with duplicate registrations.[6]

- Registration has to be renewed periodically. Otherwise, a user terminal would remain in the system database long after the subscriber shut it off or moved too far outside the

---

[6]Aside from fraudulent usage, there are also combinations of two cellular telephones with the same number so that an incoming call pages both and both telephones ring.

service area. The repeat registration can be lost or garbled, or it may be received and its acknowledgment may not be.

- Incoming call pages may not be received. When a system does not get a response for an incoming call, it tries again and it may try a third time, but it has to give up eventually. The caller gets a recording about the cellular subscriber being unavailable. At this point, a subscriber's voice-mail service may come in and take a message.

When a page is received or when the subscriber presses the SEND key, the user terminal establishes a link with the system. Several messages go back and forth, including dialed digits for wireless originations, and any of these can be missed or garbled. These messages have been designed aggressively to have enough redundancy not to be misinterpreted. It is pretty safe to say that they are received correctly or not received at all. It is also pretty safe to say that the same designers made sure that such an unlikely event is handled to avoid glare conditions, as described in Sec. 23.6.[7]

The voice link can fail while the call is in progress. Not only may the user terminal be moved into a poor coverage zone, but it may be shut off midcall. Back in the carphone days before user terminals had battery power, it was common to park the car while talking on the telephone, to follow the usual routine, to shut off the ignition, and oops, to lose the call. Even today, however, batteries can go dead during a call, or users can drop the telephone, causing the battery to pop off.

This is just the highest-level overview of wireless call setup; each of these states can be divided into a more detailed view of the process. Much of this detail is described in specifications, but quite a bit is left to the individual equipment designer.

We also can create call state diagrams for each aspect of cellular technology: for call maintenance, for power control for handoffs, and so forth. Soft handoff improves CDMA call performance, at some cost of capacity, and it increases the radio state space considerably. Not only are soft and softer handoffs (described in Sec. 8.7) themselves more involved than the older hard handoffs, but soft-handoff systems also still have to support hard handoffs between CDMA carriers, between CDMA and analog, when the pilots are not compatible for soft handoff (disjoint active sets), or when there are not enough radio resources (frame offsets) to support soft handoff.

There is one piece of good news in wireless call state analysis. The call-processing states and the radio states are almost completely independent. From the call processing point of view, establishing the radio link is part of a single step, and the call-maintenance tasks of power control and handoff occur completely behind the scenes. And from the radio link point of view, the call processing is almost as invisible. Custom calling features require an occasional SEND key, but otherwise the radio link does not even know if the call was completed. A completed call and a busy signal are both voiceband signals.

There are two exceptions to this rule, and both are relatively benign. The first is that user terminals have restrictions and conditions of service more complex than landline telephones. Landline telephones are active or they are not active, whereas wireless telephones, even fraudulent or in default, can still make emergency calls according to

---

[7]One glare condition observed by one of the authors (Kemp) appears to be the result of a design error in message handling. This is a case where a CDMA cellular telephone is turned on by the subscriber and, as the user is dialing digits, a signal comes in informing the user of new voice mail. The telephone enters a state where the user is unable to make a call out until the subscriber either checks the voice mail or waits a considerable time before trying the call again.

the Federal Communications Commission's (FCC's) enhanced 911 (E911) standard. Thus the call processing has to have a state where the telephone is able to call 911 and not much else.[8]

The second is an accounting issue for three-way calls. Wireless has a separate billing concept called *air time*. We pay not only for how much telephone time we use but also for how much radio time we use. Some wireless providers waive the air-time charges for incomplete, busy or no-answer calls, and it is tempting to use the call-processing billing records. After all, each complete call appears, from beginning to end, in the automatic message accounting (AMA) records.

The rub is that three-way calls are counted twice in the AMA records. Consider a 10-minute call using 10 minutes of air time. Suppose that the subscriber spent 6 of those minutes on a three-way call. The total call time is 16 minutes, but the total *air time* is still only 10 minutes. Just using AMA records will double-bill the customer for those 6 minutes.

This billing issue means that the system has to have some record of radio-link starts and stops as well as telephone-call starts and stops. For those wireless service providers not billing customers for busy and no-answer calls, the accounting system has to have some combined understanding of radio-link and call-processing states.

16.5 Conclusion

Call-state diagrams are valuable tools for component design, network planning, and network troubleshooting. Any cellular engineer will benefit from being able to read and interpret state diagrams. In addition, it is valuable to practice creating them and explaining them to others. Walking through plans and problems together with diagrams allows cellular planners and engineers to validate designs and resolve problems quickly.

---

[8]A cellular telephone also can be *locked* to prevent others from making calls, but it can still receive incoming calls. This is the electronic equivalent of the lock and key we used to put on a rotary-dial phone, and it does not really change the call-state space.

Key Data Concepts

Today's and tomorrow's cellular networks rely on digital technology and offer services that carry user data. However, engineers with a telephony or radio engineering background may not have a deep familiarity with the data standards and protocols that have developed over the last 30 years.

"Key Data Concepts" provides the background in information technology (IT) networking needed to plan and grow cellular systems as the convergence of voice and data systems continues and as third-generation (3G) cellular systems are being deployed.

In Chap. 17, "Quality of Service (QoS)," we address the most fundamental challenge facing engineers who are developing systems that will carry both voice and data: the different quality requirements of each service and how to manage those different requirements on a single network.

In Chap. 18, "Speech Coding," we provide the details of a key application of data technology to telephony and cellular systems: the process of digitizing voice signals.

Starting in Chap. 19, we turn to the world of IT. In Chap. 19, "Hybrid Voice-Data Networks," we take a look at existing and emerging solutions for carrying both voice and data on the same transport. After an extensive discussion of voice over Internet Protocol (VoIP), we look at the use of VoIP on cellular networks and its implementation in the process of upgrading the backhaul network from base stations to mobile switching centers (MSC).

The last two chapters in Part 4 discuss data services for cellular subscribers. Chapter 20, "Short Message Service (SMS)," discusses the SMS system, which carries small user messages over the cellular signal channel without establishing a circuit, and its descendants, enhanced message service (EMS) and multimedia service (MMS), which add small voice and video attachments. In Chap. 21, "Wireless Data Services," we discuss the high-bandwidth data services being deployed in 2.5G and 3G cellular systems.

Quality of Service (QoS)

The convergence of voice and data is a key issue for code division multiple access (CDMA) quality and capacity, particularly with the deployment of third-generation (3G) systems. QoS is the most crucial challenge for this convergence. Overall, *convergence* is the term used for the integration of the services, industries, and networks that provide voice, video, and data services around the world. Television-quality video has not yet become a major issue for wireless communications, so our primary concern is with voice and data convergence. The public switched telephone network (PSTN) and the Internet developed separately. Each of them is designed to provide high-quality service for the services it offers: The PSTN offers real-time audio with good voice quality for telephone calls, and the Internet offers slow but very reliable transmission of data.

At the same time, each system performs relatively poorly for the type of service it was not designed to provide. A traditional analog telephone line can support, at most, a 56-kbps data stream, and yet the same copper twisted-pair wire, when not specifically configured for voice, can support a data capacity 24 times higher in the form of a digital subscriber line (DSL). Certain design choices on the PSTN, such as the addition of filters that limit range of frequencies carried by local loops, are actually specifically disadvantageous for data service. The Internet, meanwhile, offers free long-distance Internet telephony, but when it comes to call quality, you get what you pay for.

The challenge in growing CDMA systems is to develop and implement standards and designs that provide satisfactory QoS for both voice and data, ideally on the same carrier. Meeting QoS requirements for voice and data but doing so by using data-only (DO) carriers is not an ideal solution. Using DO carriers alongside voice and data carriers limits the flexible allocation of carriers to meet subscriber demand for either voice or data.

17.1 The Customer Experience of Quality

The market for cellular telephones is quickly reaching the point of saturation in North America and Europe. At this point, pretty much everyone who wants cellular service has it, and the market shifts from a phase of rapid growth into a mature phase of competition. Wireless service companies will not grow so much by entering new territory first. Instead, they will grow by offering better services to their customers and increasing market share by taking it from their competitors.

In the mature competition phase of an industry, customer perception of quality and value is the key to success. However, much of the customer's perception involves things for which the cellular engineer is not primarily responsible. Some customers get a new cellular telephone because they like the color or style, and others switch services because they like the speed, politeness, and effectiveness of the customer-service department. Although engineering does not affect either of these directly, it does have an indirect effect. If cost-effective engineering projects are delivering reliable service on schedule, the whole business benefits. The customer-service department receives fewer problem calls, so it can do its job better, and more money is available for marketing and design.

Our engineering affects the customer's perception of quality in direct ways as well. While engineers may separate quality and capacity, the customers just want things to work right. Excessive call blocking is a capacity issue to an engineer, but it is a QoS issue to the customer. High static causing subscribers to hang up in frustration appears at first to be a quality issue. However, in CDMA, that static may be the result of a cell approaching its capacity limit so that too much interference is being generated by other calls. To satisfy customers and increase market share, we must work with the interrelationships of capacity and quality issues.

17.2 Capacity

As we deploy third generation (3G) systems, we need to understand the key principles governing CDMA capacity for data, for voice, and for a combination of voice and data.

17.2.1 Acceptable delay

Practical data transmission requirements are very different from voice transmission requirements. As long as a voice call is in progress, there is some level of communication in each direction. 3G systems can take into account the reduced amount of speech and lessen the number of bits transmitted to reduce interference to other calls. However, at any moment, the person may begin speaking again, and voice transmission should resume at the appropriate quality level immediately.

Data transmission does not have this requirement. A subscriber might open a data transmission to receive e-mail, reply to it, and send out the replies. After the download, the data session can move into a dormant state, reducing the demand on the capacity of the cell. When the user finishes creating replies and sends them, the system can reactivate the data transmission link. This response does not have to happen within a fraction of a second. If it takes a few seconds for the e-mail to start moving, then that is no problem. And a few seconds is a few thousand milliseconds. This flexibility in time creates opportunities for optimizing capacity for multiple users. For example, a system could be designed so that if capacity is near peak at a particular moment, a new request for data services encounters a brief delay. This allows one or more of the other transmissions to complete and go dormant. Rather than getting excessive interference, each transmission takes its turn, and all subscribers are satisfied.

17.2.2 Error recovery

In voice transmission, forward error correction (FEC) is used. As a result, the transmission of each person's voice relies on only one-half of the duplex channel. This is familiar to us in the case of static on only one channel. One party is saying, "I can't hear you," while the other party is saying, "I can hear you just fine."

With data transmission, the situation is quite different. Data transmission relies on error recovery, where the receiving system checks the packets it receives and requests retransmission of any corrupted or missing packets. As a result, even if a data file is being sent in only one direction, it relies on an operating two-way communications channel. This can be observed by watching the number of bytes moving in each direction while downloading a file from the Internet. Although most of the bytes are coming on the downlink, there is a simultaneous use of the uplink as the local computer confirms receipt of the packets. Capacity calculations for data systems need to take this two-way flow of data used for verification or error correction into account.[1]

17.2.3   Data-only carriers

All three of the 3G CDMA standards (cdma2000, W-CDMA, and TD-SCDMA) are developing detail standards for DO frequencies that are designed to provide high data bandwidth to support high-speed data transmission.[2] For example, the cdma2000 1x EV-DO standard, also known as *EV phase 1,* supports data rates of up to 2.4 Mbps, whereas the standard for a mixed voice and data carrier, cdma2000 1x, supports a data rate of 144 kbps. The DO carrier can only offer its highest rate by giving the entire carrier to one subscriber for a period of time. As a result, it would be impractical to support real-time voice on the same carrier. At the present time, creating a DO carrier allows data transmission capacity per call to increase by a factor of more than 16.

Wideband CDMA (W-CDMA) takes a different approach to managing DO channels. This approach has several elements that are described in Sec. 10.3. The elements include the ability of the user terminal to define its voice and data capacity, the definition of four QoS classes, the use of time division carriers to support high-speed data services, use of the Aloha protocol to allow user terminals to request supplemental data bandwidth, and the availability of high-speed data transmission on the downlink to support services such as Web browsing that require higher download speeds than upload speeds.

Time division synchronized CDMA (TD-SCDMA) provides the most flexible service, allowing variable data rates on demand in both directions, with rates equivalent to those found in other 3G services. TD-SCDMA has the ability to assign a data channel to the least-interfered time slot and then to select the optimal CDMA code within that time slot.

As these approaches to data-only carriers evolve, it will be up to system planners to decide which carriers to allocate to DO services.

---

[1]In data communications on systems with very low error rates, the data system may be designed not to use error recovery. In this case, the data transfer relies on FEC alone, just as voice calls do. Application designers can choose depending on the criticality of the information being delivered and the consequences of an error. For example, a Web browser reading HyperText Markup Language (HTML) will, at worst be unable to read the page if it contains an error. The user will be notified of the problem and press or click on Refresh to solve it. However, if a file is being downloaded for later use, using the File Transfer Protocol (FTP) rather than HTTP, error recovery is essential to ensure that the file, which may not be used until much later, is received correctly.

[2]As we discussed in Chap. 9, we use the term *detail standard* to refer to a standard that is part of a larger standard.

17.2.4   Mixed voice and data carriers

Existing carriers already support low data rates, and improvements to the channel will increase these rates. However, mixing voice and data on a single CDMA channel is likely to be less efficient than using separate carriers most of the time.

When a DO carrier is in place, it can offer a very high data rate to a single subscriber by temporarily reducing service to other subscribers on the same cell or even by locking them out for brief periods. This is acceptable to most users in data communications. We are all familiar with clicking to receive new e-mail and seeing that sometimes we get an immediate response, but sometimes we have to wait several seconds before anything comes through. Current DO channels rely on this willingness to wait so that they can provide high-speed services to individual subscribers. This is incompatible with carrying voice calls on the same channel because it is not acceptable for a voice call to drop out or experience extremely high interference for a few seconds. As a result, it would be inappropriate for a single subscriber to dominate a carrier supporting voice calls.

There are some other issues that will need to be addressed in the design and implementation of 3G standards. With CDMA, high-speed power control is a crucial tool in managing interference, and this power control uses capacity on the carrier. Also, a channel must be maintained at an acceptable error rate rather than be set for the fewest possible errors because having fewer errors requires a higher power level for that one channel, and this produces more interference for other channels on the same CDMA carrier.

For cdma2000, support for cmdaOne on the same channel requires backward compatibility. This limits the flexibility of design because any system must continue to support existing user terminals.

As a general rule, the wider the bandwidth a carrier has, the more bits it can carry on each channel. There are two ways of giving an individual subscriber a higher data rate. One approach uses supplemental channels, and the other uses carriers with wider bandwidth.

In the first alternative, supplemental channels are provided when they are needed. This is handled differently depending on whether the request is for high-speed data on the reverse channel or on the forward channel. For the reverse direction, the user terminal requests additional capacity for a higher data rate, and the base station sets up the supplemental channel. In the forward direction, the base station sets up the supplemental channel and informs the user terminal as to what channels to access. As long as the user terminal has only one receiver, the maximum data rate is limited by the maximum capacity of the carrier. The technique of using supplemental channels can only increase the data rate to the maximum data rate of the carrier. To achieve higher data rates, carriers with greater capacity are needed.

This problem is resolved in 3G CDMA implementations by offering carriers with wider bandwidth.[3] These carriers have higher chip rates and bit rates than the narrower 1.25-MHz cdmaOne carriers. This increases the maximum data rate that can be offered on a single carrier. It is important to note that while the maximum data rate does increase, spectral efficiency does not. Using 1 percent of a 3x carrier is no more efficient than using 3 percent of a 1x carrier.

---

[3]This is the W in W-CDMA.

The carrier bandwidth for 3G systems will migrate from the second-generation (2G) standard of 1.25 MHz up to 5 MHz in the near future. cdma2000 has two ways of offering 5-MHz service, 3x is a single 3.75-MHz carrier (5 MHz with guard bands), and multicarrier is three 1x carriers coordinated to offer data rates similar to 3x. The 3G standards for cdma2000 allow the possibility of even wider carriers, each a multiple of a 3x carrier, 6x with 7.5 MHz of bandwidth, 9x with 11.25 MHz, and 12x with 15 MHz plus whatever guard bands are required.

17.2.5 Allocating voice and data carriers

So far we have been focusing our discussion on the data capacity of individual channels and the methods for increasing that capacity. There is a separate capacity challenge for high-speed data services looming in the near future. How will the system perform when many subscribers all want high-speed data services on the same cell sector? All the advanced planning and careful design of standards and specifications done thus far will not answer this question. The proof is in the pudding, and sometime in the next few years, as subscribers use more and more wireless data service, the pudding will gel.

One manageable element in optimizing CDMA cells is the distribution of carriers assigned for DO versus those assigned to a mix of voice and data. This optimization is potentially quite complex.

Here are two approaches worth exploring:

- Should allocation of DO carriers on a cell change with time of day? Perhaps more bandwidth should be given to DO services during business hours and more to voice in the evenings.

- It seems highly likely that certain cells, such as those which serve downtown business districts, will need more channels devoted to data capacity. What measurement tools, analytic tools, and statistical tools are best for estimating the total demand for data capacity?

These questions will need to be resolved in the coming years as demand for high speed wireless data services increases.

Another variable that may be significant in overall system data capacity but which is not as easy to manage is the amount of capacity reserved for 1x channels that support cdmaOne user terminals. These terminals operate at somewhat lower efficiency, especially on the reverse channel, which has a pilot in cdma2000 but is pilotless in cdmaOne. As described in Chap. 28, cdmaOne has no reverse pilot, which keeps its reverse voice link simple, whereas 3G systems have a reverse pilot that makes them use the air interface more efficiently at the cost of extra complexity.

When it comes to improving the air interface, changing the base stations is only half the story. Overall cell capacity and data capacity might be affected by the number of cdmaOne user terminals, versus the number of cdma2000 user terminals, active on a cell at a given time.[4] Some sources predict an increase in capacity by a factor of 1.5 to 2 when cdma2000 1x service replaces cdmaOne service as long as cdma2000 user terminals are used. It remains to be seen if these improvements actually appear in a frequency-limited real-world radio environment. If the improvements are real, then

---

[4]It is also possible that future 3G terminals will improve the capacity of the air interface in ways we have not foreseen. The same management issue would be relevant in that case as well.

service providers will realize a significant cost savings if they can entice subscribers to upgrade to cdma2000 user terminals.[5] However, the service provider has no way to control those numbers directly. If it turns out to be the case that needing to support large numbers of cdmaOne users reduces QoS or capacity significantly, cellular service providers will have some options. The service providers could provide free upgrades or even upgrades with bonuses to subscribers with cdmaOne equipment. Or the service providers could lobby governments for the allocation of new bandwidth to be used for 3G services only.

17.3 Latency, Jitter, and Loss

There are three kinds of impairments in a data system: latency, jitter, and loss. Like the noise, interference, and distortion of an analog signal, these three interact with each other in contributing to a level of service for a subscriber.

The *latency* of data transport, or the *delay* in more casual usage, is simply how long it takes the data to get somewhere. We can think of a data connection as having an end-to-end latency performance, and we can think of each stage as having its own latency. The transport links used in the data connection have some latency because it takes some nonzero amount of time for the bits to get from one end to the other. A transmitter is in one place, a receiver is in another place, and information does not travel faster than the speed of light.

However, there is another source of latency in the tandem nodes along the way. While circuit-switched data take some time to get from input to output inside a circuit switch or cross-connect, packet data are much more susceptible to delays inside a packet switch. Packets are received, checked for errors, stored until transport space is available on an appropriate route, and then sent. This notion of keeping the packets inside the machine until there are transport resources to send them is called *store-and-forward technology*. This gives packet links extraordinary efficiency compared with circuit-switched data services, especially for bursty data such as Web surfing, but it does mean that a packet can wait for a long time to get out of a packet switch. Since a data connection can have several packet switches in its path, the store-and-forward delays can add up to significant latency. The relationship between the congestion of a packet switch and the latency of its pipes is discussed in Sec. 23.3.

The other time issue in data connections is that all the data packets in a single transmission do not take the same amount of time to get to their destination. Packets are routed based on available transport, so some packets may follow a less direct route than others. The physical length of the packet pathways varies, and the number and congestion of the packet switches vary as well. Not only might packets arrive with different delays, they also may arrive in a different order. Thus the packet receiver has the responsibility not only of receiving all the packets of a transmission but also of putting them back together in order to form the same data message that was sent. The variability of packet delay is called *jitter*.[6]

---

[5]Increased capacity per cell can reduce the need to deploy new base stations as demand grows.

[6]Jitter is primarily a packet issue, but it does come up in high-performance data circuits as well. We hear a great deal about jitter, equally spaced data bits coming at unequal time intervals, in audio compact disk playback technology.

We can think of latency and jitter as statistical moments, average and variability. There is some mean delay in a data connection, its latency, and some standard deviation or variance, its jitter component. This is a simplification because the distribution of packet times is often a far cry from a bell-shaped statistically normal distribution. However, it gives us a visualization of the issues in maintaining QoS under time constraints.

The third impairment in a data system is loss of information. A packet can be lost because it simply does not get there, or it can be lost because it gets there with errors, as we discuss in the next section. In either case, a data connection loses some fraction of its packets. Since a subscriber pays good money to have his or her data arrive safely, it sounds negligent to admit that we lose some of them. However, it is not as terrible as it sounds because voice and video services are adequate with losses or errors as high as 1 percent of the bits, and end-to-end data subscribers have their own software applications to ensure that their data losses are detected and either repaired or resent.[7]

The three impairments, latency, jitter, and loss, interact with each other in contributing to service impairment. Consider a voice telephone call where a total delay of over 200 ms from speaker to listener starts to become annoying. A wireless voice-coded data stream goes from one user terminal through a radio link to a base station, through a backhaul pipe to a mobile switching center (MSC), from the MSC through the public switched telephone network (PSTN) to a telephone, a landline telephone in this example. The voice data stream might tolerate a bit error rate (BER) of 1 percent. This means that 99 percent of the packets have to be both correct and timely. If 0.5 percent of the bits have errors and 0.5 percent of the packets take too long, then we have our 1 percent BER. If we design a packet link with less latency and jitter, then we can afford higher loss in the radio link and still offer adequate voice quality.

Asynchronous Transfer Mode (ATM) deals with the different requirements by having four classes of service, with class D traffic split into two subclasses. In these descriptions, a service is *time-sensitive* if it requires low delay and latency. If the class of service is sensitive to loss, then it requires a low error rate.

- Class A, constant bit rate (CBR), serves a low bit rate at a steady stream and is time-sensitive and also sensitive to loss. It is suitable for circuit-switched data service.

- Class B, variable bit rate (VBR), serves a steady or bursty data stream and is time-sensitive. It is suitable for real-time voice or video applications.

- Class C, non-real-time (VBR-NRT), serves a steady or bursty data stream and is tolerant of some delay. It is suitable for video or audio streaming.[8]

---

[7]We can think of these losses like airlines losing baggage. Loss of luggage is rare enough that many passengers check their luggage routinely and maintain a backup plan for the few times they arrive and their suitcases do not. Fortunately, packets in subscriber data services typically arrive with a bit error rate (BER) of 10^{-6}, one in a million, considerably better than airlines' rate of lost luggage.

[8]Streaming is a technology in which a relatively high quality voice or video file is received and is placed into a buffer. The file can begin to play for the user before it is finished downloading. Increases in delay or problems with jitter will reduce the buffer size during play. However, if the service is robust enough, the playback will complete without interruption. Streaming video and audio are used commonly on the Internet.

TABLE 17.1   Data Service Requirements

Service	Rate	BER	Delay
Circuit-switched	Steady	Low	Low
Voice	Steady low	High	Low
Video	Steady high	Medium	Low
Web access	Bursty high	Low	Medium
LAN	Variable	Low	High
Data downloads	Medium	Low	High
E-mail	Low	Low	High

- Class D, available bit rate (ABR), serves a bursty data stream, is tolerant of delays and some cell (packet) loss, and guarantees a level of data throughput. It is suitable for local area network (LAN) traffic.[9]

- Class D, unspecified bit rate (UBR), is a bursty data stream that is tolerant of delays and more cell (packet) loss. It makes no guarantees other than best effort.[10]

We picture an array of data performance requirements as shown in Table 17.1. The circuit-switched data subscriber requires a steady stream of highly accurate bits with low delay, the voice call permits higher error rates but still requires low delay, and a nightly data download requires high accuracy but can wait a few minutes or even hours.

17.4   Error Detection and Correction

There are several techniques for dealing with bit errors. To deal with errors, we have to detect them, and this requires enough redundancy so that errors appear wrong. If every combination of letters in English spelled a word, then proofreading would be very difficult, and there would be no spell checkers.

In general, we can divide error-correction schemes into two types, *forward error correction* and *error recovery,* sometimes called *backward error correction.* For forward error correction, we transmit enough redundant error-correcting information to allow the receiving system to reassemble lost data (most of the time) without having to request retransmission. In the older error-recovery technology, only enough information is sent to allow the receiving system to determine that an error has occurred. The receiving system must request retransmission of the packet containing the information that was not received correctly.[11]

---

[9]The packet loss is tolerated because in the LAN environment error recovery using the Transmission Control Protocol (TCP) or a similar protocol will ensure that the lost or corrupted packets are resent.

[10]This class of service is not really of much value for commercial purposes.

[11]There really is no specific term for the older send-and-retransmit method because it was the only method available, so it did not need a unique term. Allow us to use the general term *error recovery.*

17.4.1   Error recovery

Error recovery is generally used for data services. True data service requires a much higher guarantee of accuracy than digital voice or video. A single bit error can, for example, corrupt an entire computer program if it is not detected. However, since packets do not need to arrive in order, the receiving computer system can store all received packets as they are received and request retransmission of missing or failed packets. When the missing packet is received successfully, it is included in the message, which is then assembled into a computer data or program file or other large unit of data. Data transmission methods such as the Transmission Control Protocol/Internet Protocol (TCP/IP) were designed to use data received out of order. As a result, error recovery, which is more accurate but does not guarantee that packets are received quickly, is usually better for transmission of subscriber data.

The TCP, as part of TCP/IP, is a data-recovery system. When a packet has an error in it, the receiver sends back a request to retransmit the packet. The detection of error is done using a cyclic redundancy check (CRC).

The CRC function performs a binary calculation on a frame and produces a frame check sequence or *frame checksum* (FCS), which is sent along with the frame.[12] This means that the total transmission is the original frame and the CRC-calculated FCS. The same calculation is performed at the receiving end, and the packet is accepted only if all the frames have the same FCS received and calculated.

We add bits to the frames to produce redundancy. The amount of redundancy we design into the system, the relative lengths of the frame and FCS, depends on the error conditions we are trying to protect our data against. A frame with 16 bits of FCS added to it has 1 chance in 2^{16}, 1 in 65,536, of getting a wrong answer that passes the test.[13]

A simpler system of error recovery is the *parity bit,* where a bit is added and set to 0 if the original bit stream has an even number of 1s and 1 otherwise. We call them *parity bits* because they measure the even or odd parity of a collection of bits. The effect of this is that the entire bit set, including the parity bit, has an even number of 1s. This is typically not used in data transport, but it was used in older modem systems and is used in memory systems.[14] This is done because parity bits require so much less computation to check, and memory systems have to run very fast.[15]

---

[12]The calculation involves remainders of polynomials in base 2.

[13]This is like the "casting out nines" that many of us learned about in elementary school, where we added up the digits of all the numbers, added up those digits until there was just one digit left, and did the additions, subtractions, and multiplications on those single-digit values. Essentially, the remaining digit is a checksum with a 1 in 9 chance of error. This is good enough for elementary school, but packet data need stronger protection.

[14]Some readers may remember struggling to make modem connections by setting the number of data bits to 7 or 8 and parity to odd, even, or zero because different early computer networks had incompatible standards.

[15]Ultimately, computer memory became so reliable that the parity-checking chip generated more errors than the memory chips did, and memory parity checking was eliminated. After that, memory became inexpensive enough that its capacity was increased to include some forward error correction (FEC) capability.

It is important to realize that these checks only work on the links where they are being used. IP systems often have host computers and routers at the ends of these TCP links, and there is no error recovery or error correction in those machines. In any data system, it is a good idea to have some kind of end-to-end error detection and correction.

17.4.2 Forward error correction (FEC)

FEC is generally used for real-time voice (and video) transmission. High-quality voice communication must occur in real time with the entire signal received in order, but the occasional dropping of small bits of sound is not terrible. FEC is better for these systems because the receiving system can assemble almost all the message accurately almost all the time.

This simple example illustrates FEC: We are going to send 8 bytes of data, 64 bits. Let us lay those bits out on an eight-by-eight checkerboard and for each row and for each column we will add a parity bit, a 0 if the sum of the row or column is even and a 1 if the sum is odd. If we have a single error in our original 64-bit square, then there will be one row and one column that do not add up, as shown in Fig. 17.1. A single error in the parity row will leave all the row sums intact, and the same is true for the parity column. If there are two errors in the 80 bits, then there is no way of knowing the original 64 bits.

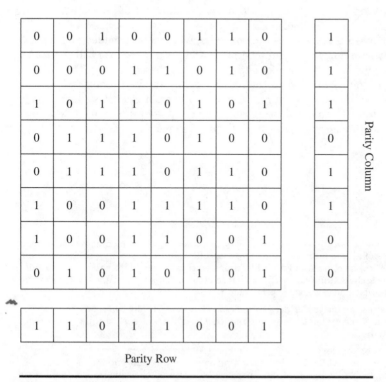

Parity Row

Figure 17.1 Simple forward error correction (FEC).

Thus our *error-resolution algorithm* at the receiving end is simple: We look at our 80 bits and check all 16 row and column sums.

- If they all match, then we are happy with the 64 bits.

- If just one row or one column is wrong, then we assume that it is a 1-bit error in a parity column or row, and we are still happy with the 64 bits.

- If one row and one column are each wrong, then we assume that the single bit at the intersection of that row and that column is wrong. We flip the bit, and we are happy with the corrected 64 bits.

- Finally, if more than one row or more than one column is wrong, then we conclude that there is more than one error, shrug our shoulders, and do whatever our system does with unresolvable errors. We may take an educated guess with some likelihood of being right, or we may report some kind of error condition depending on our data system's architecture and needs.

A system sending real-time voice or video has to send *something,* even if it is wrong. If two errors are rare enough in this eight-by-eight matrix, then the *wrong* sound is sent, but the problem is small enough to be disregarded. On the other hand, a computer file transfer system likely will be better off reporting irrecoverable errors and having the data resent. In this case, the computer system relying on this eight-by-eight matrix error-correcting system would have limited FEC plus additional error recovery. FEC would suffice to ensure accurate transmission in the case of a single error in one matrix. Error recovery with a request for retransmission would be used as the fallback in case there were two errors in a single matrix.

We can measure our bit error rate (BER), therefore, by counting the number of times our error correction has to resolve an error and the number of times it cannot resolve an error. Occasionally there will be a set of errors that cancel each other out, the system is fooled, and the wrong 64 bits are sent without our knowledge. However, this does not happen very often even with this primitive error-correction scheme. The mathematical structure of an error-correction scheme allows us to create an accurate estimate of the rate of undetected errors.

There are two measurements of BER, and it is important to specify which one is being offered. A radio channel with 3 percent BER (0.03 BER) before error correction may have a 10^{-4} BER after error correction. The customer probably is only interested in the final numbers, the data rates and BER after error correction, and does not care how many bits and what internal error rates were used to get that result. Engineers and designers, on the other hand, have to understand the channel and the error correction and therefore pay attention to the BER before error correction, the rates of corrected and uncorrectable errors. The bits in the message are the *information bits* or *encoded bits,* and the larger number of bits being sent are *coded bits.*

The eight-by-eight matrix, using 80 bits to carry 64 bits of data with FEC information, is useful as a simple example of how error correction works, but it is not very robust. The design of actual error-correction schemes depends on the use of the information, the differing needs for accuracy depending on the use of the information, and the expected quality of the transmission system.

For example, the audio compact disks (CDs) we listen to use a compression scheme of 28 to 16 bits that resolves multiple errors and, very important for an audio signal, is most likely to make its mistakes on the bits that affect the sound least. However, such a design would be very poor for computer data transmission, where no one bit can be predetermined to be less important than any other. CD audio and data formats are different.

Just as different channel architectures may use different modulation schemes for different radio conditions and different customer requirements, the developers of cellular systems use different error-correction schemes for different needs. One of the challenges facing cellular systems today is that the systems were designed and built for voice quality, which allows a certain degree of error. However, customers are wanting to use the cellular networks for data transmission more and more and, as a result, are often dissatisfied because optimization for voice is different from optimization for data.

The FEC scheme illustrated in Fig. 17.1 is a form of *block code* where each collection of k bits is mapped into n bits $(n > k)$, and the receiver turns n bits back into k bits even if one or two of them is wrong.[17] In this case, $k = 64$ bits are turned into $n = 80$ bits. This example was chosen because it is easy to see how it works; there are much better block codes for FEC, codes with better error-recovery rates.

The other kind of error correction is a *convolutional code,* where each k bits become n bits using a moving window of K bits. The ratio $r = k/n$ is the *code rate* and the ability of an FEC system, block or convolutional code, to correct errors increases as the code rate r increases. The Viterbi algorithm is a commonly used convolutional code for error correction.

There are longer and more efficient convolutional-code FEC algorithms called *turbo codes.* Because these codes require longer sequences of bits, turbo codes create long latency times for low bit rates. Therefore, they are used for high data rates or when delay times are not critical.

Service providers often offer data rates that are not exactly the same as the channel rates. This is often an opportunity for stronger FEC algorithms, but sometimes it is simpler just to modify the bit stream.

When the rate is a simple fraction, then the bits can be repeated, such as sending 4800 bits twice on a 9600-bit channel. This multiple sending of the same bits is essentially more spreading gain in the code division multiple access (CDMA) channel and usually results in a power reduction as a result.

When the offered rate is slightly lower, 8550 bits per second on a 9600-bit-per-second channel, some of the bits are repeated. Because most of the bits are sent just once, there is no economy of power available. We end up wasting the extra data capacity.

When the offered data rate is slightly *higher,* 980 kbps on a 921-kbps channel, then we can employ a trick known as *puncturing,* where we steal a fraction of the bits and use them for extra data. The 921-kbps stream is turned into a larger stream of FEC coded bits, and some of these bits are deliberately replaced with a coded version of the other 59 kbps, but with few enough changes so that the FEC algorithm is not broken. The receiver detects the extra bits and uses them, whereas its FEC is robust enough to recover the original bit stream being sent. This does require a lower BER and more CDMA channel power because the error correction is weakened by puncturing.

---

[17]The various flavors of block codes are beyond the scope of this book, but the reader may encounter, for example, linear block codes, binary cyclic codes [also called *cyclic redundancy check* (CRC)], and Bose-Chadhusi-Hocquenghem (BCH) codes.

17.5   Conclusion

Defining quality makes the essential link between good engineering and good business. Over the last 30 years, manufacturing companies have moved to systems that measure and ensure quality all the way from raw materials to satisfied customers. Defining the quality of services, rather than that of manufactured products, is more challenging, and the transition is recent. However, with convergence and an increasingly mature, competitive market, implementation of fully defined and measurable QoS is essential to business success.

18

Speech Coding

The *digitization* of voice is a two-step process. In the first step, a *microphone* converts sound waves into electrical waves following the same pattern. This is an analog signal, a signal where the electrical pattern is analogous to the original pattern. Prior to digital technology, in telephone communications the analog signal was transmitted and then retranslated back into sound at the far end. Vinyl phonograph records created an analog pattern of bumps in the vinyl that reproduced sound waves. Magnetic audiotape recorders created an analog pattern of magnetized iron on the tape that could generate an electrical signal that reproduced the audio tone.

If we take the electrical wave pattern and translate it into a digital signal, this is *encoding,* which is the second step of creating a digital signal. Digital telephony, including digital cellular telephony, creates a digital signal for transmission. Digital encoding also can be used for storage. Audio compact disks (CDs) use digital coding. Whether the system is used for transmission, for storage, or for both, the sound going in is the *audio input,* and the sound coming out is the *audio output.*

Turning an audio or video signal into a code is performed by a device called a *coder,* and the reverse process is performed by a *decoder.* The output of a decoder is an analog audio signal that can be sent to an amplifier or speaker. Typically, a single physical device called a *codec* performs both functions. Sometimes the term *codec* is used to refer to the coding scheme as well as to the physical device.

In this chapter we will explore the engineering issues of digital encoding. There are several fundamental concepts that will come into play as we look at the various coding schemes that have been used in telephony and other audio technologies:

- There is an inverse relationship between the quality of reproduction of audio sound and the bandwidth (bits per second) of the signal. In a given coding scheme, better sound takes more bits.

- Different coding schemes have different amounts of intelligence built into them. A scheme that takes into account how human beings speak or hear can be more efficient, getting the same quality at lower bandwidth than a less intelligent coding scheme.[1]

---

[1] This may seem like an odd use of the word *intelligence.* We would suggest that even a typewriter keyboard has intelligence built in. Knowledge of the frequency of letters in the English alphabet was used in choosing where to place the letters. That is intelligence. In fact, the standard keyboard was designed to be slow so that fast typists would not jam the keyboards of mechanical typewriters. Electronic keyboards and other higher-speed devices can use a different layout, such as the Dvorak keyboard, which is optimized for speed, allowing for faster typing. Similarly, audio coding schemes can be more intelligent if they use our knowledge of how people speak and hear in order to optimize quality or maintain acceptable quality with reduced bandwidth.

- Once a signal is encoded, it is a digital bit stream. It can have error correction added. It can be packaged, stored, and transported in any number of ways. Whatever is done to it, if the bits present at the communications receiver or output device are the same as the bits at the input device, then the signal is transmitted accurately.

- For audio transmission, the code present at the end of the process does not have to be a perfect match for the original code. Some loss or distortion is acceptable.

In this chapter we will look at several types of *pulse code modulation* (PCM), a common type of encoding for telephony that is also used in the production of digital music CDs. Then we will look at linear predictive coding, including *code-excited linear prediction* (CELP), the coding scheme used for code division multiple access (CDMA) cellular telephony. We will close the chapter with a discussion of how the quality performance of speech coders is evaluated.

18.1   Pulse Code Modulation (PCM)

The most basic way to send an audio signal as a sequence of bits is called *pulse code modulation* (PCM). Traditional circuit-switched telephone lines use PCM, as do audio compact disks. To perform PCM encoding, we start with an audio signal in electrical form, a voltage level as a function of time, as shown in Fig. 18.1.

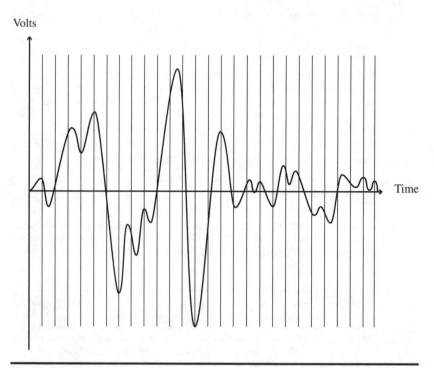

Figure 18.1   An audio signal, voltage as a function of time.

First we *sample* the timeline by dividing it into discrete intervals, which we call *samples*. The number of samples per unit time is called the *sampling rate*. For each sample, the audio input measures the signal voltage at that instant in time. The audio output starts with the sample measurements and reconstructs the audio waveform by filling in the gaps, as shown in Fig. 18.2.

The sample reconstruction loses wiggles in the audio signal that are smaller than the sampling interval. There is a fundamental limit called the *Nyquist frequency,* half the sampling rate. Since a CD samples its audio input 44,100 times per second, its Nyquist frequency is 22.05 kHz. Our telephone PCM systems sample their audio 8000 times per second, so their Nyquist frequency is 4000 Hz, more than adequate for the 300- to 3300-Hz frequency range of a telephone voice signal.

It is important that the audio input be filtered carefully to stay within the Nyquist range. At a sampling rate of s, the audio output has no way of telling frequency f from frequency $ks + f$ or $ks - f$, for $k = 1, 2, \ldots$. All these frequencies will have the exact same values in the stored samples. The term for this incorrect mapping of frequencies is *aliasing*. In our telephony example of 8000 samples per second, a 5000-Hz component of the audio signal samples the same as a 3000-Hz component and plays back as 3000 Hz. This can be a very annoying distortion, and we employ antialiasing filters to make sure that the audio input bandwidth stays within the Nyquist limits.

The Nyquist frequency is an unattainable upper limit for sampling theory the same way the speed of light is an unattainable upper limit for the theory of relativity. Getting to two-thirds of the Nyquist frequency is easy, but getting closer requires more

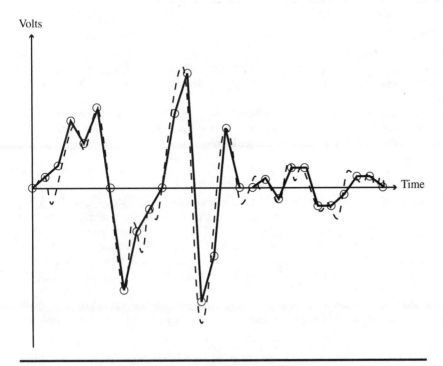

Figure 18.2 Reconstructing the audio wave from samples.

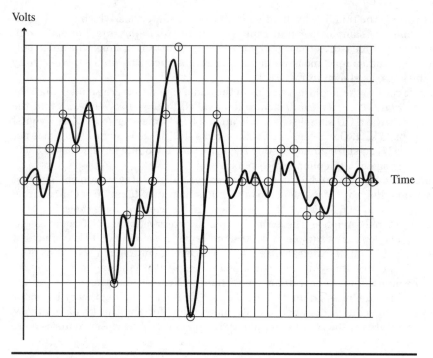

Volts

Time

Figure 18.3 Linear PCM sampling of an audio signal.

effort. Close to the Nyquist frequency, very small perturbations in the sample values make large errors in the audio output. Using 8000 samples for 3300 Hz gives the telephone system some extra room for sampling inaccuracies that using 7000 samples would not.[2]

Now that we have discussed sampling of time, we consider dividing the voltage range into discrete levels, first linear and then logarithmic. *Quantization* is the term for dividing continuous measurement into discrete levels.

18.1.1 Linear pulse code modulation

We can quantize the voltage levels by picking a discrete set of values and assigning each value a number. When these levels are uniformly spaced from a minimum to a maximum voltage, we call the process *linear PCM*. Figure 18.3 shows the sampling process for the audio signal that we saw in Fig. 18.1. Each of the circles represents the closest level of the voltage at the sampling time. The figure shows nine distinct levels, whereas a CD is linear PCM with 65,536 levels (16 bits). When telephone systems use linear PCM, they use 8192 levels (13 bits).

Each sample is converted to a binary number that goes to the audio output, which converts it back into analog electrical signals, as shown in Fig. 18.4. Linear PCM has

---

[2]This issue comes up in digital audio design regarding the CD, where designers try to achieve a top end of 20 kHz and make compromises in other parts of audio performance.

Volts

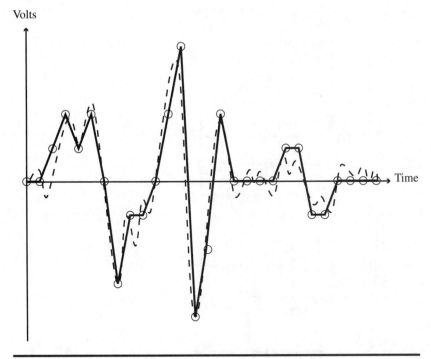

Time

Figure 18.4   Reconstructing the audio wave from linear PCM.

voltage quantization errors as well, in addition to the time-sampling errors we saw in Fig. 18.2.

The bit rate required for linear PCM is the sampling rate times the number of bits required for each sample. Thus a CD audio stream requires 44,100 samples times 16 bits per sample times 2 for stereo for a grand total of 1,411,200 bits per second.[3] Using 8000 samples times 13 bits per sample, we get 104,000 bits per second for a linear PCM telephone link.

The 104,000-bit-per-second linear PCM link described here is never used in telephone transport.[4] The linear PCM is converted to a logarithmic scheme (μ-law or A-law) or to adaptive differential PCM (ADPCM), as described in next two subsections.

18.1.2   μ-Law and A-law PCM

Looking at the reconstruction in Fig. 18.4, we notice that the small signal on the right is lost in the voltage quantization error. Both in music and in speech, normal audio has loud and soft periods. The audio waves shown in these figures have a loud period

---

[3]In fact, a CD has a robust error-correction scheme that uses 14 bits to get 8 correct bits, so the total raw bit count is more like 2.5 Mbps.

[4]We are loathe to say "never" here because somebody, somewhere, somehow probably is using 104,000-bit linear PCM on a telephone link. But we have not heard of anybody using it.

Volts

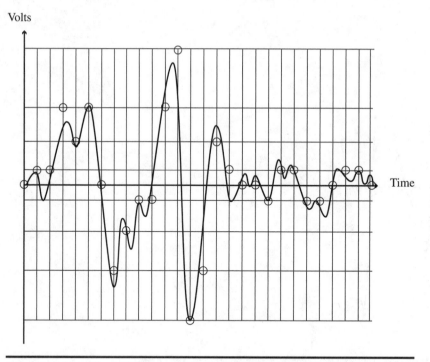

Time

Figure 18.5   Logarithmic PCM sampling of an audio signal.

followed by a soft period, with voltage about four times higher during the loud part. A factor of 4 in voltage is about 12 dB according to Eq. (1.4). We say that this audio signal has a *dynamic range* of 12 dB. Our ears care about errors that are *proportional* to the signal, so we mind the greater relative magnitude of the quantization error during the quiet period more than we mind the same error during the loud period.

We can improve the performance during the softer periods with only minor degradation of the loud periods by having more levels near 0 V, as shown in Fig. 18.5. There are two different logarithmic PCM standards, μ-law PCM used in DS-0 voice links and A-law PCM used in E-0 voice links.[5]

The μ-law formula calculates the signal function $s(t)$ from the input function $i(t)$ using Eq. (18.1):

$$s(t) = \text{sgn}[i(t)] \frac{\ln[1 + \mu\,|\,i(t)\,|\,]}{\ln(1 + \mu)} \tag{18.1}$$

The input function $i(t)$ goes from -1 to 1. The sgn(x) function is $+1$ for positive x and -1 for negative x, so the calculation is symmetric for positive and negative input values. The parameter μ is set to 255 in the United States.

---

[5]μ-Law is sometimes spelled out *mu-law* and is pronounced "myoo-law."

Volts

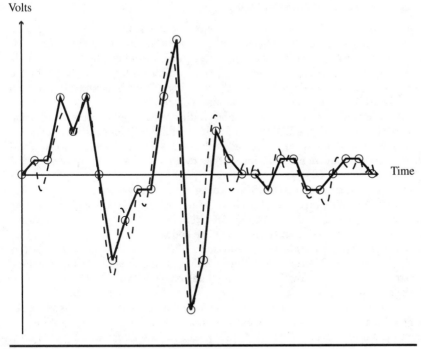

Time

Figure 18.6 Reconstructing the audio wave from logarithmic PCM.

The A-law formula calculates its signal function $s(t)$ using Eq. (18.2):

$$s(t) = \text{sgn}[i(t)] \frac{1 + \ln [A \; |i(t)|]}{1 + \ln A} \qquad \frac{1}{A} \le |i(t)| \le 1$$

$$s(t) = \text{sgn}[i(t)] \frac{A \; |i(t)|}{1 + \ln A}, \qquad 0 \le |i(t)| \le \frac{1}{A}$$

(18.2)

As in the earlier equation, the input function $i(t)$ goes from -1 to 1. The value $A = 87.6$ is typically used in Europe.

When we put the logarithmic PCM bits back into audio output, we see better resolution of the soft passage in Fig. 18.6 than we saw earlier in Fig. 18.4. The logarithmic PCM schemes, both μ-law and A-law, allow 8 bits per sample to work as well as 13-bit linear PCM. These 64,000-bit-per-second schemes are full-rate PCM telephone links.[6]

The A-law function $s(t)$ is logarithmic for $|i(t)| \le 1/A$ and linear for $|i(t)| > 1/A$. A-law provides better signal performance than μ-law at higher levels at the expense of some performance at lower levels.[7] Telephone communications between μ-law systems and A-law systems must have conversion routines in their transmission paths.

---

[6]The PSTN in the United States reserves 1 bit per sample, 8000 bits per second, for its own signaling, so DS-0 speech coding is usually 7 bits per sample, 56,000 bits per second.

[7]Is this because Europeans talk louder or more consistently loud than Americans?

18.1.3 Adaptive differential pulse code modulation (ADPCM)

While logarithmic quantization brings our requirement down from 13 to 8 bits per sample, we can do better by using differential coding. Instead of encoding the voltage for each sample, we can make a prediction of the next sample based on the recent sample history and encode the *difference* between the actual sample voltage and the predicted voltage. This technique is called *differential* PCM and can provide comparable speech quality with only 4 bits per sample, 32,000 bits per second.

The common differential speech coder, adaptive differential pulse code modulation (ADPCM), converts linear 13-bit PCM to 15 levels, 4 bits per symbol. If the source is analog voice, then it first must be converted to linear PCM. μ-Law and A-law PCM streams are converted from their logarithmic steps to linear PCM with equal steps between levels.

It will help us understand the ADPCM encoding process if we look at the decoder first. As each ADPCM 4-bit sample A comes into the decoder, it goes through an inverse adaptive 15-level quantizer, as shown in Fig. 18.7. The output of the inverse adaptive 15-level quantizer is a difference calculation D. The sequence of D samples is fed into an adaptive predictor whose output is a prediction P. When the prediction and difference terms are added, $P + D$ becomes the linear PCM output. This linear PCM stream can be converted to μ-law or A-law PCM for transmission, or it can be converted to analog audio for somebody to hear.

Now let us turn our attention to the ADPCM encoding process shown in Fig. 18.8. As each linear PCM sample L comes into the encoder, it is compared with the adaptive prediction P for that sample. The difference $L - P$ goes into the adaptive 15-level quantizer and becomes the ADPCM output sample A. This same output sample goes into an inverse adaptive 15-level quantizer and an adaptive predictor exactly like the ADPCM decoding process to form the prediction for the next sample. In this way, the prediction samples P used in the encoding difference calculation are the *exact* same sequence of digital bits as the prediction samples P used in the decoding process. This feedback process continuously steers the differential calculations so that the output of the ADPCM decoder $D + P$ is as close as possible to the linear PCM stream L that went into the encoder.

How is ADPCM able to deliver the same high-speech quality in 4 bits per symbol that takes 8 bits per symbol with μ-law or A-law PCM? The extra 4 bits worth of information is coming from the intelligence built into the adaptive predictor. If the predictor

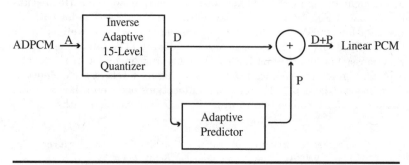

Figure 18.7 ADPCM decoder.

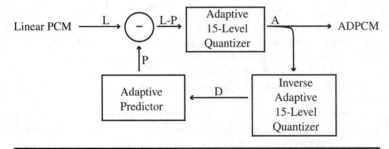

Figure 18.8 ADPCM encoder.

does a good job, then the $L - P$ terms will stay small, and 15 levels of difference will cover almost every sample. ADPCM relies on the adaptive predictor's ability to anticipate the next sample.

The intelligence used in ADPCM is the result of study of the typical variations over time of the frequency and volume of human speech. As a result, ADPCM can create sound quality that is very close to μ-law PCM using half the bandwidth. However, ADPCM would not be best for modem or fax transmissions because these signals are quite different from human speech patterns.[8]

18.2   Linear Predictive Coding

The PCM schemes operate in the time domain. The next steps in speech coding efficiency take us out of the time domain into the frequency domain. Speech waveforms are decomposed into their component frequencies and amplitudes, as we discussed in Sec. 1.2. By modeling speech as a frequency spectrum varying over time, we can use the redundancy of human speech audio signals to send speech with a bit rate that is lower than ADPCM.

18.2.1   Code-excited linear prediction (CELP)

The CELP speech coder used in one form or another in all our CDMA wireless telephone systems uses a table of speech sounds called a *codebook*. The decoder looks up the codebook index I and multiplies it by a loudness-level factor G to get a basic sound. This basic sound then goes through a synthesis filter with three parameters called *pitch lag L*, *pitch gain b*, and *pitch spectral lines i*. The output of this goes through a final sound cleanup, a spectral postfilter and gain control. Since the CDMA coders uses variable bit rates, there are different parameter coding schemes for the different speech data rates. The lowest rates use a simplified algorithm appropriate for the case where the subscriber is not speaking.

The CELP encoder is more complicated than PCM-based encoders because it must determine the values of I, G, L, b, and i to send over the CELP speech coding link. As

---

[8]Although one of the authors (Kemp) has met some computer technicians who are capable of reproducing modem signals with their mouths, just as some auto mechanics can illustrate the sounds of a car in distress.

in the case of ADPCM, the encoder has a complete decoding system within it, and it compares the output of the decoder with its own input to refine the signal being sent. The encoder computes line spectral pairs (LSPs) and pitch spectral lines i values. Then it uses an analysis-by-synthesis optimization employing a weighted error measure in a closed-loop feedback system to find the best values for pitch lag L and gain b. Finally, the selection of codebook index I and loudness G are made with a similar optimization based on i, L, and b values.

CELP is significantly more complex than ADPCM, and it uses a great deal more computer memory. Most of the computational complexity is in the encoding stage, where the optimization computations occur. Also, CELP is decidedly more tuned to the human voice than the PCM coding schemes by putting more information in its prearranged codebooks rather than in its bit stream. The CDMA development group (CDG), an industry association promoting CDMA and developing standards, has adopted CELP for CDMA. As a result, the 13-kbps rate set 2 (RS2) cdmaOne coding scheme is sometimes called *CDG encoding*.

18.2.2 Vocoders

The extreme low-bit-rate end of the speech coding spectrum is the *vocoder*. The forms of human vocal sounds are modeled and formed into an electronic vocal tract to produce human speech. The vowel and consonant sounds of speech are simulated by a set of source-filter parameters based on a speech-production model. Voiced speech is produced by generating a periodic pulse to simulate the opening and closing of vocal cords, whereas unvoiced speech is produced by exciting the same filter with random noise to simulate air rushing through the vocal tract. The receiver decodes the bit stream by using a speech synthesizer based on the same model.

The key to the vocoder is the perceptual model of speech, so vocoders do very poorly in sending nonspeech sound and particularly poorly for complex signals like analog modems and fax machines. Because it forms the sound in its own mathematical simulation of a human throat and mouth, the vocoder creates intelligible speech at very low bit rates. The vocoder speech, however, has a mechanical, robot-like quality. The speaker's soul seems filtered out. This is a speech coder that we can get away with for a few seconds without anybody noticing the difference, but extended use of a vocoder is not satisfactory for a wireless telephone system.

18.3 Speech Coding Performance

We evaluate speech coders by their mean opinion score (MOS). A large community of human listeners rates the voice quality in a scientific, controlled test process. While it

TABLE 18.1 Mean Opinion Score (MOS) Definitions

MOS	Rating
5	Excellent
4	Good
3	Fair
2	Poor
1	Bad

TABLE 18.2 Mean Opinion Scores of Speech Coding Systems

Coding scheme	Score
64-kbps μ-law	4.30
CDG 13 kbps for cdmaOne	4.15
32-kbps ADPCM	4.10
8-kbps EVRC	4.10
Stationary AMPS, 26 dB S/I	4.00
8-kbps IS-96	3.65
13-kbps GSM	3.65
2400-bit vocoder	2.30
50 km/h AMPS, 17 dB S/I	2.30

is expensive and time-consuming to have a large number of listeners evaluate a speech coding system, it does give statistically valid comparisons. The MOS is defined in International Telecommunications Union–Telecommunications Standardization Sector (ITU-T) Specification P.800. The quality of reproduced speech is ranked by the listeners on a scale from 1 to 5, as shown in Table 18.1.

Table 18.2 shows a range of speech coders and their MOS values ranging from the very high quality of full-rate DS-1 μ-law encoding at 4.3 down to a marginal Advanced Mobile Phone Service (AMPS) channel in a moving vehicle at 2.3. It is noteworthy that 8-kbps enhanced variable-rate codec (EVRC) has the same high MOS of 4.1 as 32-kbps ADPCM. Speech coding technology keeps getting better sounding with more complicated and sophisticated algorithms.

18.4 Conclusion

In this chapter we have explored the coding schemes used most commonly in landline and wireless telephony. The ability of the 13-kbps CDG CELP standard and, more recently, 8.55-kbps EVRC to provide sound quality nearly as good as standard μ-law PCM is significant to the cellular industry. Providing essentially equal quality in less than one-quarter the bit rate per call means that each call takes less than one-quarter the bandwidth of the carrier than would be needed if the standard PCM coding scheme were used. One-fourth the bandwidth per call translates directly into a factor of 4 increase in the call capacity of the cellular carrier.

In the next chapter we will turn our attention to the methods of transporting digital data over networks that were first developed within the information technology industry and are now used for both computer data transmission and telephony.

Hybrid Voice-Data Networks

The basic advantage of a single network carrying both voice and data is clear: If one network carries both, then we incur the expense of creating and maintaining only one network instead of two.

When low-speed data transfer was acceptable, hybrid local access and hybrid networks were easy and common. For local access, we used a modem over the telephone line instead of getting an expensive data line. Back in the early 1990s, Columbia University had a hybrid voice-data network for the entire campus. The university installed a telephone system that also supported digital data at 19.2 kbps. Over short distances, the wiring could even be used to support 4-Mbps token ring local area networks (LANs). We find similar examples in wireless with the use of low-speed modems over the cellular system.

In the twenty-first century, a hybrid voice-data network that meets the demands of high-speed data is a greater challenge. We have discussed the reasons for this in earlier chapters. And we have introduced the technologies that are overcoming these challenges. In this chapter we take a brief look at the status of actual implementations of hybrid voice-data networks, public and private, in the year 2002.

19.1 Quality of Service (QoS)

There are three basic approaches to creating a hybrid voice-data network: adding voice on an existing data network, carrying data on an existing voice network, or building the network from the ground up. In the last category, we include any thorough reengineering of existing facilities to upgrade the capability of the network.

All three approaches face the same challenge: providing voice quality for voice while providing data quality for data. The voice channel is time-sensitive, requiring low delay and low jitter, but it can tolerate a higher error rate. The data channel can tolerate more delay and jitter but requires a much lower error rate.

Due to the different methods of achieving these different classes of service, the network must function differently for voice and for data. For voice, the signal includes forward error correction (FEC). As a result, it is best to reassemble the entire data stream, even if it contains errors. It is better to have a corrupted bit than to discard an entire frame or packet. However, for data services, a bad packet should be discarded as soon as possible, allowing a request for retransmission. Very low-level protocols, at the Internet Protocol (IP) packet level and even at the data-link layer, need to be redesigned to accommodate this flexibility. Switches and routers operating at these levels must be able to identify the type of service needed for the data in the packet and act differently depending on that need, that class-of-service request.

As networks become more sophisticated, the class-of-service request has other functions as well. On networks with multiple paths, it is possible for routers and switches to pass information back and forth, which informs each piece of equipment of the state of the network. Routers can then identify paths with low latency and jitter that should be used for real-time service. Using these paths maintains a low-delay, low-jitter circuit, or virtual circuit, for telephone signals. Data packets are sent on slower routes to keep the high-speed route available for voice. We see an analogy for this in the operation of our highways at rush hour. Frequent radio reports of accidents and delays allow drivers to select alternate routes during rush hour, optimizing their travel time to their destinations. This system has been enhanced in the last few years as drivers with cell phones have been calling to inform radio stations of accidents and delays. The sooner intelligence is passed to the agents capable of selecting routes, the more flexibly the network can respond, reducing traffic jams.

Similar technology is being developed for networks. Several different designs are being tested. There are two basic approaches. One involves using some type of central control for the entire network, and the other involves building intelligent routing capability into each network device. In both approaches, it is crucial to have a method of signaling that informs devices of network conditions.

The methods that use intelligent routers and switches rely on providing each device with information about network timing and delays. The devices use this information to make decisions regarding where to send voice packets and where to send data packets. Some systems report status using signals among switches and routers, and others use timing codes included in the data packet headers. Timing methods are, at present, used on Asynchronous Transfer Mode (ATM) networks but not on IP networks. In ATM, ATM adaptive layer 1 (AAL1) cells can provide information to the network, but the functionality of that information is somewhat limited. The signaling approach uses network overhead, but it has an advantage over the timing-code approach. When the timing codes are read by a router or switch, the node can calculate what is happening on the part of the network from which the packet came. However, the information only describes what happened to the packet as it was heading toward the switch. It does not provide any information about the status of the path away from the switch. This limitation of the timing-code approach means that the device, in deciding which path to send packets on, must assume that information regarding the status of the incoming direction is useful in deciding what to do when packets need to go back in that direction on the opposite side of the full-duplex channel. This assumption is not always warranted. Networks can be clogged at a certain point, but only in one direction, just as a highway can have a traffic jam in only one direction.

19.2   Voice and Data Sharing Physical Transport

The simplest case of mixing voice and data would not be seen as a single network at all by some, but it still has some of the economic benefits. If a pipe is laid with multiple channels, some of those channels can be used for voice and others for data. In the public switched telephone network (PSTN), the larger trunks, DS-3 and higher, carry many DS-1 circuits via time division multiplexing (TDM). The channels are independent. As a result, it does not matter if the DS-1 link carries 24 voice calls or one 1.536-Mbps data signal. At each end, the voice calls are routed to a voice switch, and the data calls to TDM equipment designed for data.

We find a similar approach used with fiberoptic transport. Companies that own fiber lease capacity in two forms. *Dark fiber* is a set of glass tubes with a certain carrying capacity from one point to another. The owner of the fiber leases it, and the lessee adds equipment and uses it for any purpose. Or the owner can set up a service, such as ATM or Synchronous Optical Network (SONET), and lease active capacity on a fiber backbone. Sometimes three companies are involved. One company that owns the physical transport leases dark fiber to a local exchange carrier, long-distance carrier, or data network services company. This company lights up the fiber and leases services to a third company. Some large companies may lease dark fiber directly from the owner of the physical transport and maintain their own networks as well.

The same approach is used in wireless carriers. If a company has capacity on a microwave channel, it can use that capacity for a mix of voice telephone calls and data service. The company also can lease some of the line's capacity to other customers.

In the cellular network, the current air interface is used primarily for voice calls, but there is some use for low-speed data and fax services. With the advent of third-generation (3G) data services, the cellular network will become a hybrid channel with high-speed data capacity. When this is done using data-only (DO) carriers, the approach is analogous to sharing a single pipe, with some optical fibers used for data and some for voice. Let us take a look at the way in which hybrid voice-data networks are evolving over time. The earliest approach was discussed earlier: Using a single physical transport with different data-link-layer protocols has the advantages of making use of existing capacity and reducing the need to install new physical facilities. However, this approach also has two significant limitations:

- The terminating and routing equipment for voice and data is different and separate. This increases the equipment cost substantially.

- Once the carriers are set up, each carrier is locked in for use for only voice or only data. There is no possibility for dynamic allocation of bandwidth across types of service.[1]

This second point is worth elaborating on. A flood of telephone calls could create blocking, even while the data stream has open channels. Similarly, if the data capacity is used up, the data transport network cannot make use of excess capacity on the transport dedicated to voice service. Because the pipes using data equipment are not suitable for voice, the full capacity of the pipes cannot be flexibly allocated on demand.

There are two common ways of managing around this problem, but neither of them is ideal. The first method is to provide excess capacity for peak demand on the voice service and let the data network deal with delay due to low bandwidth. This may be acceptable because data systems have a higher tolerance for delay than do voice networks. This approach is commonly taken by telephony carriers who want to use some

---

[1]There is a partial exception to this rule. Some carriers can carry voice and also carry low-speed data, such as modem or low-speed cellular data transmission, and allocate dynamically on demand. However, these lines are not best for data service; they are merely sufficient.

of their excess capacity to serve data customers. However, it still leaves a large part of the capacity assigned to voice use unused except at peak hours.

The other approach is to have an alternate route for voice calls. Of the total transport capacity available, enough is given to data services to satisfy peak demand. Excess transport capacity is given to a limited number of voice channels. When these voice channels are used up, the switches at each end adopt an alternate route for any new telephone calls that come in. This solution is used commonly by private companies. The point-to-point links, which are leased for wide area network (WAN) data uses, have excess capacity beyond the current demand for data, usually because they were leased with future capacity in mind. The excess capacity is given to the internal private branch exchange (PBX) telephone switches. It is the preferred route because it offers long-distance service within the office without a per-minute charge. Whenever the internal network reaches capacity, the next request for an office-to-office call that comes to the PBX is sent to the chosen long-distance telephone service. Some companies have enhanced this functionality by having the PBX use the internal lines to serve certain outgoing calls, as well as intercompany calls.[2] This approach has the advantage of making very good use of available capacity, but it requires the existence of a reliable alternate route.

Sharing physical transport media and the physical layer of the Open Systems Interconnect (OSI) model does have some cost-saving advantages. The techniques described earlier make good use of capacity within the limits of the technology. However, they also illustrate the limitations of this approach.

New approaches allow dynamic sharing of a single pipe defined at the data link layer or network layer for a mix of voice and high-speed data. For reasons explained in Chap. 23, high-quality voice networks almost always have unused capacity. Technologies that dynamically allocate channels to voice or data as needed can make use of this capacity, reducing transport costs. However, this requires that voice and data both be carried over the same equipment, as well as being carried over the same physical transport medium. In the remaining sections of this chapter we will look at several solutions to this problem.

19.3   Voice and Data Sharing ATM

ATM, with its defined QoS levels, its ATM adaptive layers (AALs), and its high-speed cell-based transport service, provides a data-link layer that can carry a mixture of voice and data services. The various AAL protocols translate the service being offered. If the customer has asked for frame relay, the customer sees frame relay, and the same is true for customers who want voice channels or other services.

Major corporations use ATM for this function. One cellular carrier built an ATM network that provided data management services, allowing its network operations center (NOC) to monitor the entire computer network as well as the switch status at its mobile switching centers. The company also routed customer-service calls from cellular subscribers who dialed 611 over the same ATM network. As a result, the majority of

---

[2]For example, if a company has offices in Atlanta and San Antonio and an employee in Atlanta makes a call to a destination in the local San Antonio area, this call can be routed over the internal network from the Atlanta PBX to the San Antonio PBX and then routed from the San Antonio PBX to the local exchange of the PSTN, eliminating long distance charges for the call.

customer-service calls from its own subscribers nationwide reached the call center without incurring any long-distance charges. The ATM network handled these voice calls whenever it had the capacity to do so. If the ATM network was at capacity, the customer calls were routed to the PSTN. The result was a significant reduction in the corporation's monthly long-distance bill with no loss of call quality.

ATM would seem like an ideal solution to the problem of optimizing capacity, and it is, up to a point. The limitation arises when the network nears capacity. ATM, as currently implemented, has two shortfalls that can be improved on in future standards.

The first area for improvement is in the arena of managing QoS within the call path. Although the network knows the QoS class requested for each cell of data, it does not have advanced alternate routing capability within the network. Standards and systems are under development to provide better management in situations in which bottlenecks could occur. Some of these improve ATM, and others operate with new services that compete with ATM. ATM performance will improve as two specific issues are addressed. The first is an improved ability to create virtual circuits providing QoS specific to time-sensitive service, and the second is the ability to adjust dynamically to changing network conditions even during a call or session.

The second area for improvement is related to the encapsulation of higher-level-protocol data units (PDUs) into ATM cells. If a way can be found to create networks that send IP packets with low latency and jitter without encapsulating them, it would have two advantages. First, the processing time for encapsulation of the IP PDU would be eliminated. Second, effective capacity would increase because ATM has a relatively large header size for its payload.[3]

ATM was a well-suited technology at the time it was developed. But now it may be possible to implement technologies that use variable-sized packets and also are able to reduce latency and jitter to the point where the network can support real-time voice and video when operating at or near capacity.

So far we have talked about alternative physical media, physical-layer implementation, and data-link-layer implementation. However, the voice channels have still been DS-0 circuits, and the data channels have used IP packets. In the next section we will change this scenario by looking at voice services over the IP at the network layer of the OSI model.

19.4   Voice over IP (VoIP)

VoIP in packet pipes improves resource utilization because, for each call, only the capacity needed is actually used. In circuit-switched technology, a full-duplex channel is always provided for each call. Switching to VoIP increases the total capacity of the pipe significantly for several reasons:

- In most phone calls, most of the time, only one person is talking at a time. Voice activity detection (VAD) is a natural way for VoIP to send fewer bits and packets when the one party is not speaking.

- Depending on a variety of conditions, low-bit-rate codecs reduce the resources needed to carry voice-quality audio.

---

[3]The ATM header-to-payload ratio is 9 to 44. Additional bytes are lost due to padding when the last cell containing a PDU has fewer than 44 bytes.

- On cellular networks, such codecs are already in place for use on the air interface. Keeping the signal encoded all the way to the mobile switching center (MSC) makes sense.

VoIP can be used by large corporations on their own networks. The functionality is similar to the case of an internal ATM network carrying both data and voice: If the capacity is there for data, why not put voice on it when we can? As of 2001, one major company answered this question: Because we cannot ensure voice quality well enough with any product now on the market. But this is changing, if not this year, then very soon. Companies are beginning to adopt VoIP for internal long-distance services to reduce long-distance telephony charges. In addition, VoIP is available for general use on the Internet, and it will be coming into use on the PSTN.

19.4.1   VoIP on the Internet

VoIP on the Internet first arrived in the late 1990s as a form of free long-distance telephone service. Using a microphone and speaker attached to a personal computer, special software, and an Internet connection, people could talk to other people over the Web and not pay any long-distance charges. This gave it a small niche, especially for international telephone calls. However, early VoIP simply relied on the fact that the Internet had available bandwidth and would deliver acceptable call quality for the price.

Some efforts have been made since then to create a viable commercial product. These efforts must overcome several technical hurdles:

- Service must be available to telephones, not just computers. The callING party must be able to place a call from a telephone to a telephone switch that then moves the call to Internet transport.

- The service must reach all telephones. It must become possible for any callED party with a computer, landline phone, or cellular phone to be reached. This requires that the network establish a point of presence (POP) in every local access transport area (LATA).

- Some level of call quality must be ensured. If this level is lower than traditional telephony, there will be a market as long as price is also lower.

- An alternate route via a long-distance carrier should be available if the Internet cannot support the call. If the call will be billed at a different rate when it goes over a traditional long-distance carrier, the user will need to be prompted and given a choice as to whether to place the call.

Overcoming these hurdles requires a major business investment as well as significant technical expertise. On the business side, establishing a POP in every LATA being served and negotiating with local telephone companies (nationally or internationally) to become a long-distance service provider mean satisfying many regulatory requirements and negotiating major contracts. It also requires a significant investment in infrastructure.

On the technical side, even confirming rapid delivery of packets for time-sensitive services on the Internet is difficult. Guaranteeing rapid delivery, with current technology, is impossible. It may be possible to measure latency and jitter end to end during call setup and only connect the call if the service works at that moment. However, the Internet has no provisions for latency or jitter management during the call. Traffic bottlenecks, equipment failures, and denial-of-service attacks would all threaten active calls. If calls are dropped in the middle, people are unlikely to accept the service.[4]

In the meantime, the cost of long-distance and international telephone calls continues to drop. In North America, long-distance service is now readily available for 5 cents a minute or less. International rates are dropping as well. It is difficult to make a business case for developing VoIP on the Internet when competitors with infrastructure already in place have such low costs.

Another challenge to the business case for VoIP on the Internet is the historical unreliability of the Internet. The Internet is becoming more reliable, but this is more the result of overprovisioning capacity than it is the result of improvements that define QoS at the network layer. If vendors overprovision capacity and, based on this, guarantee QoS, this may help with the business case. However, this option to some extent means that VoIP is working but that it is not creating greater efficiency in the use of transport capacity. With all of this, two key facts remain. First, the Internet is used for many things, and demand for its capacity is shared among these functions. Without improved management of the network supporting defined QoS, it is difficult to ensure support for time-sensitive voice and video services. Second, the Internet is more vulnerable to attack than the PSTN. As a result, the system a VoIP on the Internet provider is relying on (and trying to sell to venture capitalists) is not under the control of the provider that must guarantee quality to its customers. Investing in VoIP on the Internet might seem like investing in a competitor for Greyhound Bus Lines that plans to save money by using bicycle repairmen instead of licensed mechanics.

The third challenge to the business case comes in the form of other, more robust implementations of VoIP. We will discuss some of these in the next subsection.

19.4.2   VoIP on the PSTN

The problem with VoIP on the Internet is not VoIP, it is the Internet. The most likely place for VoIP long-distance and even local services to appear as a viable business service is on the PSTN itself. As of summer 2002, Lucent Technologies, Cisco, Nortel, and Avaya are selling PSTN switches supporting VoIP. In the PSTN, the physical infrastructure is already optimized for voice-quality calls with low noise, low latency, and low jitter. Designing IP switches is a major job, and retrofitting the network by replacing traditional circuit switches with IP packet switches is an even bigger job. However, the job is internal to the local exchange company or long-distance provider who takes on the challenge. The other components of the network and their quality are already determined. It is much easier to ensure quality when switching out only one component of a network than when trying to make an inherently unreliable network reliable enough for real-time service.

In addition, the business case is much simpler. There is no need to attract new customers. In fact, the company might be better off if the customers do not even know that VoIP has arrived. The company, whether it offers local or long-distance service or both, simply can continue to provide service to the same customers at the same rate while saving money through reduced demand for transport capacity. Over time, the reduced

---

[4]We say this with some confidence when we consider the number of subscribers who change cellular providers if their calls are dropped by their current provider.

demand for capacity will generate enough savings to justify the investment in the purchase and installation of new switches. Added value can be realized in that these switches support IP, which means, in addition to supporting VoIP, that they are also a reliable environment for data services. If a customer is given a duplex IP packet-based virtual circuit over the PSTN, the customer can add Transmission Control Protocol (TCP) error correction at each end and receive reliable high-speed data service. This configuration is good for the provider because transport capacity can be allocated dynamically for voice or data. In current networks, transport must be transferred from voice circuit switches to data packet switches or ATM switches or routers to support frame relay and other customer data services.

19.4.3   VoIP on the PPDN

Major companies have been leasing dedicated transport from telephone companies and other service providers for both voice and data for decades. The term *public packet data network* (PPDN) is new, but the functionality is simply the expansion of leased-line service plus the infrastructure of the Internet. With the advent of VoIP, companies now have a new way of sharing capacity between voice and high-speed data on leased lines. These lines are available from long-distance carriers and also from local exchange carriers, where they serve metropolitan areas on *metropolitan area networks* (MANs). For example, a company could lease IP services on a MAN to connect two offices. The company could then use this virtual connection to support both corporate data and corporate voice calls, reducing its cost for local calls. The service could be allocated dynamically for voice and data, utilizing capacity effectively.[5] In addition, it would be possible to monitor usage levels at the packet level. In the future, MAN and long-distance packet service providers could offer per-packet rates based on actual usage rather than rates based on maximum line capacity.

19.4.4   VoIP on the cellular backhaul network

There is one other place where VoIP is coming into use: the cellular telephone networks. Over the air interface, low available bandwidth already requires that the voice signal be compressed, and variable compression is optimal. As a result, the duplex voice channel is already a variable-rate channel. Moving that signal on a fixed-bandwidth circuit hardly makes sense. Some cellular networks already use packet pipes for backhaul. Those that do not, by and large, are converting to IP pipes as they deploy equipment to support 3G services.

As 2.5G and 3G services are being deployed, the capacity of the air interface is increasing, and data in IP packets are being carried. In this environment, it is advantageous to turn the backhaul network into IP pipes rather than keep it as voice circuits. In some cases, the increased demand for capacity would require upgrading the capacity of the backhaul network. Prior to upgrade of the network to 3G, the backhaul was designed to meet the capacity requirements of the cdmaOne air interface. If the capacity of the air interface increases with the advent of 3G, then it may be necessary

---

[5]Some dynamic allocation is already available through the use of multiplexed channels and subrates. However, VoIP will provide greater flexibiltiy and efficiency.

to modify the backhaul network. Otherwise, the backhaul network could become the bottleneck. If the backhaul network is not already using IP, then converting it to IP pipes will increase its capacity.

The deployment of new radio air-interface technology at the base station requires visits to the base station now. It is most cost-effective to upgrade the base station's end of the backhaul network now. This is especially true in the case of cdma2000 1x, where the plan for later versions of cdma2000 is to install new cards that are software-upgradeable. If this plan succeeds, upgrade trips to base stations will be rarer in the future. It makes the most sense to get all the hardware changed out now. As a result, the cellular backhaul network is moving rapidly to IP pipes.

The base stations are not the only locations in need of equipment changes in this process. Changes at the MSC are fairly simple because all the cards supporting the backhaul network are in one building. In addition, cellular providers will need to request change of service at the local offices where backhaul connections are terminated. Our logical diagrams show links from base station to MSC, but the reality of the backhaul network is that it is largely composed of leased lines routed through the PSTN.

There are three other issues to address in developing IP backhaul network solutions. One is the issue of IP networks versus IP pipes. This is fairly easy to address. Even when the backhaul network is a network, it is a fan-out network, not an interconnected network with alternate routing. As a result, in terms of latency and jitter, it can be considered a single-route packet pipe.

The second issue is alternate transport, such as point-to-point microwave transmission. Here, due to the layering of OSI protocols, the problem is relatively simple. The physical layer of the microwave channel does not need to change. It is a matter of putting IP over microwave instead of putting voice circuits over microwave. The problem is analogous to the situation in the landline backhaul network, and IP over microwave solutions have been in use for a number of years.

The third issue is the issue of the capacity of the backhaul network. This will be addressed in Chaps. 35 and 44.

19.5   Conclusion

This chapter on hybrid voice-data networks completes the picture of voice and data convergence. Many books on convergence include video transmission and provide more information about use of the cable TV network as data transport.

A brief discussion of the role of video transmission in convergence is appropriate here. Including video data in the convergence picture does realize economy through the sharing of facilities, and video can be packetized for this purpose. For broadcast-quality video, the data rate is much higher than that for telephony, the sensitivity to latency is less (because the transmission is one way, and delay will not be noticed as easily), and the sensitivity to jitter is high. In many situations, such as the download of video clips and short movies, there are two techniques that reduce QoS requirements below what broadcast television requires while still being acceptable to customers. One is to provide images at lower resolution, which are either shown on a smaller screen, appear more grainy, or both. The other is to use video streaming, which uses buffering and allows the start of play of the video image prior to completion of file download. Streaming also can be used for compact disk–quality music transmission. For our purposes, this brief introduction to the role of video in convergence suffices. In cellular networks, video is not yet an important issue. Until the high-capacity standards of full 3G implementation prove their cost-effectiveness, video transmission over the cellular network is not a significant part of our capacity planning. When these services become available, data rates will be close to what is found in landline digital subscriber line (DSL), and it is likely that technologies such as video streaming, which are acceptable in that environment, will be acceptable for cellular subscribers as well.

With our background in data systems and convergence complete, we now turn to the transmission of data over cellular networks. There are two sets of standards that support user data on cellular networks: the short message service (SMS) and its descendants, which carry limited data over the signal channel without establishing a call, and the data-carrier services of 2.5G and 3G networks. We will look at each of these in turn.

20

Short Message Service (SMS)

As you may have noticed, pagers are almost a thing of the past. This is only partly due to the fact that people prefer cellular telephones to pagers. The other major reason is that cellular telephones have become pagers. Cellular telephones act as pagers due to the short message service (SMS), which carries short text messages over the signaling channel, delivering messages to users without setting up a cellular voice or data channel.

On the air interface, signaling is carried over the paging and access channel in the forward direction and over the access channel in the reverse direction. Within the code division multiple access (CDMA) network, SMS messages are carried on the Signaling System 7 (SS7) network as specified by the ANSI-41 protocol.

From the business perspective, SMS is a vertical service that provides paging-type message service to subscribers while using very little air-interface capacity and using network signaling capacity, which is readily available. There is often no charge for SMS services, but there is value. Offering SMS makes it easier to sell subscribers the switch from paging services to more expensive cellular service because subscribers do not have to give up anything when they upgrade.

The latest versions of SMS-type services add the ability to transmit small data attachments, including alternate ring tones and small images. As the cellular market becomes increasingly competitive, these services draw some consumers away from one brand of cellular phone to another. We say consumers rather than subscribers because so far these services are not offering much of value to business customers.

This chapter will provide the history, benefits, and specifications of SMS. We also will introduce the successors to SMS, enhanced message service (EMS) and multimedia message service (MMS).

20.1 History

In 1980, pagers were all the rage, and they had significant value to business customers. It is not surprising, therefore, that early second-generation (2G) cellular service providers came up with a system designed to compete with, and perhaps supplant, pagers.[1]

---

[1]For the paging services industry, the result is that pagers are now a niche market rather than a growth market. Signals to pagers do a better job of reaching their destination than signals on the cellular network, particularly inside buildings. Customers who want to be sure to receive their pages have kept pagers, even when they also have gotten cellular phones.

This was the initial purpose of the *short message service* (SMS). SMS was a detail specification of the Global System for Mobility (GSM) time division multiple access (TDMA) cellular telephone standard. As we explained in Chap. 7, the GSM standard defined more than the air interface and, as a result, contained many detail specifications. The goal of GSM was to standardize enough interfaces to allow multiple companies to compete in the development of equipment and services while all being compatible across Europe. SMS was included in the package, and the design of the standard took this approach. Interfaces essential to compatibility were specified in detail. Interfaces not essential to compatibility, such as the user interface, were left open, allowing for competitive design of different services. Component design was left open, as long as the components delivered the results for the interface standard so that they would communicate successfully on the network. With the exception of a few small glitches, this worked very well.

SMS has been adopted worldwide, operating on almost all 2G and 3G cellular networks. SMS operates at Open Systems Interconnect (OSI) layer 7 (the application layer), riding on top of GSM in the signaling channel. Due to layer independence, SMS has migrated easily to ANSI-41 cellular networks, making it operable in North America and compatible with 3G wireless systems.

20.2   Benefits for Subscribers and Vendors

In addition to making it easier for subscribers of paging services to switch to cellular telephones, SMS is offering two other benefits for cellular subscribers. The first benefit is that a variety of business-to-business and business-to-customer transaction services can be managed over SMS. The openness of the standard for entities that send short messages allows integration with a variety of computer systems, including point-of-sale terminals (cash registers), bank computers, and the Internet.

The second benefit is that SMS and its successors, EMS and MMS, have created fads that have made cellular telephones more popular, especially in Japan. While it is unlikely that anyone bought a cellular phone to send text messages, SMS-based services have driven sales in two ways. First of all, companies can gain market share against their competitors by offering more trendy services than their competitors. Second, any fad can bring peer pressure and increase sales of the equipment needed to participate. We will take a look at some cases where SMS and related services have driven sales and altered market share in Chap. 49.

What are the benefits of SMS to vendors?

- SMS messages can be used by the cellular provider to inform subscribers of waiting voice mail, roaming status, or other account or network conditions.

- Because SMS messages ride on the signaling channel, SMS delivers messages on forward paging channel capacity that would otherwise remain unused.

- Some vendors charge for the activation of SMS service or for its use.

- Even if cellular subscribers are not charged for consumer use of SMS, vendors can charge corporations that use SMS transactions for delivering their messages or for gaining access to their network.

- Wireless service providers can partner with information services to provide everything from stock quotes to soccer scores.

- When a subscriber receives an SMS page and wants to reply, the logical way to do it is to use the cellular telephone, and the subscriber pays for that.

- SMS messages can be used to reconfigure the user terminal over the air with over-the-air parameter administration (OTAPA), reducing support costs.[2]

- The buzz about SMS increases sales so much that some vendors have used special SMS, EMS, and MMS services to drive up market share.

20.3 Technical Specifications

SMS provides a store-and-forward message service for short text messages, up to approximately 160 characters. The store-and-forward functionality allows the system to continue to try to send SMS messages if the user terminal is not available (e.g., powered off or out of range) when the message is first sent. SMS also offers delivery confirmation by returning an SMS message to the sender that reports receipt of the original message.

SMS delivers these messages using an out-of-band packet delivery system, sending low-bandwidth transmissions over the signaling channel. As a result, any registered user terminal can receive or send an SMS message. There is no need to set up a voice or data channel, and in fact, an SMS message might get through in radio conditions that would not be sufficient for a telephone call. This out-of-band signaling system makes SMS competitive with paging systems for corporate dispatch services. SMS messages can be received by the user terminal even when the subscriber is on a voice call or using data services.

The process of sending an SMS message can be divided into five steps: origination, store and forward, transport, receipt, and confirmation. The key components of the SMS system are short message entities (SMEs) and short message service centers (SMSCs).[3] SMEs are devices that can send or receive short messages via SMS. Cellular user terminals with SMS capability are SMEs. SMEs that access the cellular network via modem, Internet, or other landline connection are called *external SMEs* (ESMEs). An ESME could be a corporate mainframe carrying a roster of on-duty technicians that pages them when they are needed, or it could be an Internet server that allows someone who is browsing the World Wide Web to send a quick note to a friend.

The SME components need to interact with key cellular network components. The interface between the SMSC and the cellular network is the *signal transfer point* (STP), the access point for Signaling System 7 (SS7) networks, described in Chap. 14. Through the STP, the SMSC interacts with the mobile switching center (MSC) via the ANSI-41 Q interface and with the home location register (HLR) via the N interface.[4] The message itself is carried across the cellular backhaul network to a base station, where it is transmitted across the air interface to a user terminal. The cellular user terminal is the SME receiving the message.

Let us take a closer look at SMS message format, SMS logical components, and the process of sending an SMS message. The advanced details of the SMS specification are

---

[2]OTAPA is described in Sec. 15.5

[3]That would be short message service centRE in Europe. So much for worldwide compatibility.

[4]The diagram specifying these interfaces may be found in Fig. 15.2. In the ANSI-41 specification, the SMSC is sometimes referred to as the *message center* (MC).

of interest only to SMS equipment designers and programmers creating custom SME functions. As such, they are beyond the scope of this book.

20.3.1 Message format

There are four major types of SMS messages: submit, deliver, command, and status report. The submit format goes from the originating SME to the SMSC. The SMSC changes the message to the delivery format and sends it to the receiving SME. The command format allows an originating SME to give instructions to the SMSC. The status report message is used to confirm receipt of a delivered SMS message.

Each message has a header. Some data elements within the header vary with the message type, and most are of interest only to programmers. We will highlight the more important elements here, focusing on the submit and deliver formats.

- Three addresses are included: originating, SMSC (for the center the message is sent to, which handles its transmission), and destination.

- The validity period is an interval during which the SMSC will try to deliver the message. After this time, the message is purged. The maximum is set by the SMSC and is usually 48 or 72 hours. A message sender can select a shorter interval so that the message delivery attempts can be canceled if the message does not get through within a few minutes.

- The data coding scheme describes the format of the message contents.

- The message reference is a counter that supports concatenation of messages.

After the header comes the message body, in one of three data formats. A simple text message in a western European language is encoded in 7-bit ASCII format, allowing about 160 characters per message. Worldwide character sets that support multiple alphabets and languages are encoded in a 16-bit text format, but not all user terminals support these character sets. The third option is binary data. The data coding scheme setting in the header identifies which of these formats is being used for the body of the message and also specifies whether the message is encrypted. The exact length of the text in a single SMS message varies depending on format and compression. For example, the message size limit is 140 bytes, which allows about 160 seven-bit ASCII characters without compression. If a simple compression scheme is used, this can be increased to about 200 characters.[5] However, a 16-bit character set is limited to only 70 characters unless compression is used.

SMS messages can be concatenated, allowing an SME to send a message that is longer than what can fit in a single message. The specification allows concatenation of up to 255 messages. However, it is strongly recommended that SMS users and applications concatenate no more than three or four SMS messages. The service rides along as an extra on the cellular network's signaling channel, and that channel is not intended to carry large quantities of data.

There are a variety of other elements to the SMS message format. Many of them allow interaction with personal digital assistants (PDAs), fax devices, specialized wireless devices, or computers. Software application developers can make use of such

---

[5]For the most part, complex compression schemes have not been implemented because they exceed the processing capacity of most user terminals.

features to develop custom applications that integrate SMS with other tools. However, technical limitations and some vagueness in the data transmission standards for SMS from cellular devices to other devices have made development of these services somewhat challenging.

20.3.2 Logical components and message delivery

The first logical component of the SMS system is the originating SME. This can be a user terminal or an ESME. In configuring an originating SME, a serving SMSC is defined. There are a wide range of possibilities for the birth of an SMS message. A cellular telephone user can type in a message to a friend and add cute "emoticons." Or at the other extreme, a corporate mainframe can generate thousands of messages and send one to every employee in the corporate database. In either case, the message is prepared in submit format and submitted to the serving SMSC. If the originating SME is on the cellular network, the short message travels across the cellular network, through an STP, and to the SMSC. If the originating SME is an ESME, then it reaches the SMSC through the Internet, via modem, or on some other landline data connection. There is a special category of ESMEs that are not wireless but are on the wireless network: ESMEs that belong to the wireless service provider. The voice message entity (VME) that holds voice messages for cellular subscribers is one example. The VME can send a subscriber an SMS message informing the subscriber that voice mail is waiting. SMSCs do not offer broadcast services, so if a user wants to send a message to multiple recipients, one message must be prepared and sent for each recipient.

The SMSC receives messages from the originating SME and ensures their delivery or proper disposition. If the destination SME is on the cellular network, it communicates through the STP to the destination home location register (HLR), asking for the location of the destination SME. If the destination SME is registered on the network, the HLR sends its location and status to the SMSC, and the message is sent. If the destination SME is not available, then the SMSC leaves a message at the HLR asking it to notify the SMSC when the destination SME registers.

If the destination SME is available, the SMSC sends the message to the serving MSC of the SME. If the destination SME is not available, the SMSC holds onto the message and waits to hear from the HLR. If the expiration time is reached before the recipient is available, the SMSC handles the failure to deliver according to a protocol programmed in by the originating SME. The message simply may be abandoned, the SMSC may inform the originating SME that the message did not go through, or the SMSC may send the message on a designated alternate service. The alternate service might be a paging service capable of reaching a subscriber who was out of range of the mobile network.

The SMSC has some other capabilities. For example, it is able to prioritize messages. If a high-priority message is received, it can be handled before low-priority messages received earlier. The SMSC is defined as a logical entity. Its physical implementation in hardware and/or software is left to the vendor.

The role of the HLR was discussed earlier. It assists the message delivery process by informing the SMSC regarding the location of the destination SME. It also performs the same function for the MSC because the MSC rechecks the location of the destination SME before sending the SMS message. When the MSC receives the short message, it checks the current location of the destination SME with the HLR and then sends the message to the serving base station, which delivers it to the receiving SME.

The receiving SME receives the short message. If it is a typical text message, it signals the subscriber with a light, vibration, or tone depending on the device and user settings. It displays the message on screen. Usually, it also displays the name of the sender and the time the message was received at the SMSC. These may appear above or below the message. Of course, other types of messages will be handled differently by the SME. For example, control codes from the cellular service provider sent by SMS could set or reset parameters of the cellular telephone via binary SMS messages for over-the-air parameter administration (OTAPA). In addition, the receiving SME can send a status-report short message indicating receipt of the original message. This is passed back through the MSC to the SMSC and then to the originating SME.

20.4   Advanced Uses of SMS

Beyond the obvious functions of paging service technicians and sending cute love notes, SMS has some interesting applications:

- Binary SMS messages can be used to configure or reconfigure cellular phones via OTAPA. If a subscriber calls in with a problem, the vendor's technician can reprogram and, in some cases, upgrade the user terminal over the air interface.

- Customer-service information, such as instructions or frequently asked question (FAQ) notes can be sent via SMS.

- There is an internetworking function between e-mail systems and SMS. This allows for SMS notification of new e-mail and even for forwarding of e-mails to the cellular telephone.

- SMS can interact with the Wireless Application Protocol (WAP), which provides a scaled-down World Wide Web interface for user terminals with limited screens, such as cellular telephones. WAP applications can give SMS users direct access to specially designed Web pages.

- Using WAP or other programming approaches, news items can be sent via SMS. SMS can transmit stock quotations. Recently, one wireless provider offered a promotion based on its ability to provide World Cup soccer scores in short messages.

SMS also supports interaction between cellular telephones and other user devices through interfaces other than the air interface. It is possible to download SMS messages from a cellular telephone into an appropriately designed personal computer or fax machine. However, this is one point where the otherwise excellent SMS specification was a little too vague. Some of these interfaces are implemented differently by different vendors, and others are proprietary. This makes developing applications for these interfaces difficult.

One interesting feature of SMS on GSM is that the SMS message can be stored either in the memory of the cellular telephone itself or in the memory of the personal identification card belonging to the subscriber. If the message is stored in the card, then it can be read on a different GSM cellular telephone when the card is inserted in that phone. This is an important feature because some GSM subscribers rely on cellular telephone rental to get coverage in different frequencies when traveling to remote countries. With on-card storage, they can take their messages home with them, even after they drop off their rented cell phones.

20.5   EMS and MMS: Newer Short Message Standards

EMS, the *enhanced message service,* is SMS with bells and whistles (literally). EMS allows ring tones, icons, and graphics to be sent to user terminals. EMS actually changes very little in the SMS system except for the receiving SME. The receiving SME must be equipped to receive, interpret, and display EMS messages. EMS has two functions, both intended for consumer use rather than for commercial business use. The first is that any subscriber who upgrades to an EMS-compatible cellular telephone can download custom ring tones, icons, and display images. The subscriber can customize the EMS cellular telephone to suit personal tastes. This is all done without any changes to the specifications for the SMSC or other SMS, GSM, or ANSI-41 network components. EMS simply uses concatenated, binary-encoded SMS messages to create the bells, whistles, icons, and images.

The other use of EMS is that a subscriber equipped with an EMS-capable terminal can send EMS messages to other subscribers. These messages can include text and also have graphics or sounds inserted at any point. If the receiving SME is EMS-capable, then the message will display properly. However, if the receiving SME is a standard cellular telephone with traditional SMS capability, the message is likely to be garbled or to fail to display. Older units that are not EMS-compliant are unable to interpret concatenated messages that are part text and part binary. It is also quite difficult to create EMS messages on small handsets. On the other hand, they are rather easy to create on the Internet, and some people download images or sounds and send them to their friends. EMS is a step on the path to MMS.

MMS, the *multimedia message service,* is under development as part of the 3G standard. It can be implemented on some 2.5G networks, and one existing service, Japan's i-mode service, offers functionality similar to MMS. MMS replaces the SMSC with a multimedia message switching center (MMSC), which is IP-based. It supports a wide variety of graphic and audio formats and can handle graphic resizing and audio compression to create compatibility across different user terminals. As the MMS standard develops, there is a very strong effort to make it compliant with a wide range of existing and forthcoming standards for Internet, multimedia, and home services. The result is that all kinds of still images and sounds and, when the higher-speed data rates are in place, perhaps even moving images will be available on cellular telephones, on the Internet, and on all kinds of other devices as well.

What is the concept behind such a major development effort? Proponents of MMS make the analogy to the shift from DOS-based personal computers to Microsoft Windows. They argue that the entire world of personal computing changed when the personal computer became able to handle graphics and that screen savers were popular and profitable software packages. MMS proponents think that MMS can do for cellular telephones what Windows and the After Dark screen saver did for the personal computer.

This may be true, but it may be the wrong analogy. The personal computer worked primarily with printed documents, where adding pictures and text formatting to plain text made a big difference. Screen savers sold well as a novelty item for a while, but eventually they became free giveaways included in Windows. The real value that made Windows sell more personal computers was in the business market, with features such as multitasking, WYSIWYG (what you see is what you get) on-screen images, better font handling, and desktop publishing. So far MMS proponents have not shown any value added to businesses or home businesses in the MMS service proposals.

Another analogy might be this: The PicturePhone was first demonstrated in the early 1960s, and it never caught on. A cellular telephone is, after all, a telephone. Its primary purpose is talking. Will people really care that much about the look and feel of their cell phones?

And we need to consider one other factor. When Microsoft Windows arrived, there were plenty of people, and even plenty of business offices, that did not have personal computers yet. The personal computer market was still expanding. In contrast, the European market for user terminals is reaching saturation, and North America is not far behind. Even if a feature that has absolutely no effect on the primary purpose of cellular telephony does draw subscribers, it will only draw them to buy new phones or change service providers. MMS is not likely to create a new market full of people wanting to buy their first cellular telephone and subscribe for service. If MMS is indeed "the key driver of the 3G business case," as the MobileMMS group would like us to think, it is hard to believe that 3G will provide revenues justifying the overhaul of every MSC and base station on the network because subscribers want pictures on their cellular telephones.

20.6   Conclusion

SMS is a well-designed, robust enhancement to 2G cellular networks. It helped the cellular industry grow more rapidly by speeding customers of paging services into the cellular telephone market. It offered cost-saving features, such as over-the-air technical support, which reduces operations, administration, and maintenance (OA&M) costs. Its reliability and consistency made it interoperable, and at the same time, the user interface and interactions with other systems were left open, allowing for innovation. Competition was fostered because interfaces were standardized, but devices were not specified.

EMS and MMS expand on SMS but only add features that may enhance the consumer's experience of the cellular telephone. They in no way enhance the experience of placing a telephone call. It is unclear if these services will increase revenue in the short term as a fad and then either fade away or be offered as standard features at low cost or no cost or if they will have lasting benefit as a source of revenue. Even if they do continue to bring in revenue, will that increase revenue for the entire cellular industry or just help each vendor in the battle for market share? And will that revenue justify the cost of designing new user terminals and installing MMS-capable networks, not to mention the cost of other 3G components?

One other set of questions needs to be answered: Is there a limit to the capacity available for SMS, EMS, and MMS? The original design assumed that these would be low-volume services, short messages squeezed in between essential control signals. As messages grow larger and more common and they are concatenated, what happens? Our research has not turned up any discussion of these issues. At what point would SMS, EMS, or MMS messages exceed either the capacity of the ANSI-41 signaling channel or of the paging and access channel on the air interface? If this capacity is exceeded, what happens? If a priority system is in place, then short messages simply will wait behind control signals. However, if it is not, then short messages might interfere with the signaling channel. It is unclear what effect this would have on a user terminal in the registered but inactive state, and perhaps the effect would not be significant. However, short messages also can be received while users are on calls. What effect will excessive short messages have on a user terminal in the middle of a call? Are there any possible glare conditions that would cause equipment failure or confuse the user? And what effect, if any, will SMS, EMS, and MMS messages have on total cell capacity?

Also, as more messages are sent, some will have errors, even if the error rate is very low. Should error detection and retransmission be enabled in MMS? If not, how will components and users detect and respond to failed messages?

Chapter 21

Wireless Data Services

The drive for enhanced second-generation (2G) and third-generation (3G) cellular services is due to the perception that there is demand for transmitting data over the cellular network. This was the only one of the two original goals of the 3G standards committee to survive the politics that arose as the standards were developed. The other goal, a single worldwide standard for cellular networks, will not happen any time soon as cdma2000 and wideband code division multiple access (W-CDMA) battle it out in the marketplace and in the media.

21.1 QoS over the Air Interface

As the detail standards that actually will implement cdma2000 and W-CDMA are developed, many issues will still need to be resolved. Standards are not intended to resolve all questions but to answer some and define others, which are then resolved in design specifications and real-world experience. As 3G services become a reality, we will learn more and improve standards, specifications, and equipment. At a minimum, vendors will need to allocate enough carriers to data-only (DO) or to voice-and-data use to support variable demand for voice and data.

21.1.1 Voice

The addition of a reverse pilot and of other aspects of the 3G standard is expected to increase voice capacity per cell by about half as cdma2000 replaces cdmaOne.[1] However, fully realizing this capacity gain relies on user terminals that use the reverse pilot, and it is reasonable to expect that the existing base of cdmaOne phones will not vanish quickly, especially in the current economy. If demand increases before this increased capacity is realized, bottlenecks could create poor service over the air interface.

Another uncertain element is the effect of cdma2000 3x services. These services will offer 5 MHz of bandwidth spread over three 1.25-MHz cdmaOne carriers (with guard bands). A cdma2000 3x needs to operate at its own frequency; it cannot be deployed on

---

[1]Some industry estimates say that capacity may double, but we believe that a 50 percent increase is more likely.

top of existing 1x carriers. However, there is an option called *cdma2000 multicarrier* (MC) that would coordinate three 1x carriers, supporting an increased data rate. In the multicarrier specification, orthogonal codes are maintained, so interference on each 1x band and on the multichannel band, which is effectively 3x, should be managed. However, the 3x and multicarrier standards need to be fully developed and then tested in real-world situations. W-CDMA services at a bandwidth of 4.8 MHz or higher will be able to offer flexible higher data rate services up to 1.92 Mbps in increments of only 100 bits per second.

21.1.2 Data

There are two challenges in developing data capacity. The first is to manage the capacity of individual data channels to meet the data-rate requirements of subscribers while maintaining an acceptable error rate. With the data services supported by the Transmission Control Protocol/Internet Protocol (TCP/IP), the bit error rate (BER) can be relatively high because retransmission is acceptable. However, retransmission adds delay, and total delay limits may preclude a lot of retransmission even when TCP/IP is being used. See Chap. 29 for more on this tradeoff. Even so, maximum data rates are achieved in current technology only by creating DO carriers. This solution is far from optimal because it means reserving fixed bandwidth for data and different carriers' bandwidth for voice (or voice plus low-speed data). This misses the whole point of convergence, which is to obtain optimal capacity through the ability to allocate resources dynamically to either voice or data as needed. High-speed data services push the envelope of the theoretical capacity of both CDMA and time division multiple access (TDMA). For this reason, the 3G standards allowed that the highest data rates would only be available within a limited radius of the base station and for stationary or slow-moving users, that is, users who are walking and not riding in a vehicle. Early 2G cellular systems offered data rates comparable with slow modems, under 9.6 kbps. 2.5G systems offered data rates approaching 64 kbps. The minimum speed that can be certified as 3G is significantly higher, at 144 kbps. As of summer 2002, the only detail standard that has attained this certification and is being deployed is cdma2000 1x EV-DO, which uses a DO channel. If the higher data speeds of 3G are achieved, then we will see data services in the range of 384 kbps in microcells where users are within 500 m of the base station, or perhaps farther, and 2048 kbps (2 Mbps) in picocells where users are stationary or walking slowly inside buildings or perhaps on small campuses.

If these data rates are achieved, another capacity issue still needs to be addressed: How many users will be able to achieve these rates in a single cell? If usage is bursty, and therefore intermittent, it might be possible to serve many subscribers. However, if each subscriber is tuning into a different streaming video program, there may be a capacity problem.

User demand will be for high-quality data as well as for speed. If subscribers are playing the stock market over the Web, the costs of long delays or failed connections could be very high. Given the intrinsic tradeoff of capacity and quality on CDMA networks, this could be a problem. Engineers may have to make decisions to limit the number of users on a cell to ensure sufficient quality of service (QoS) for current active subscribers on that cell. The interference among multiple cells also will become a more difficult issue to manage as cells become smaller.

21.2   Changes at the Base Station and MSC

cdma2000 migration begins with a significant overhaul of base station equipment. New channel cards are installed to support cdma2000 voice and data services. Other new equipment improves support for IP at the base station. In addition, IP support turns the backhaul link to the mobile switching center (MSC) from a traditional circuit into a packet pipe if the 2G network is not already using packet pipes.

The good news is that if things go according to plan, this may be the last overhaul of base station electronics for a while. The new cards being installed are software-upgradeable. If all goes well, the new standards will be implemented on the same physical cards, and future 3G service improvements can be realized by remote software configuration without requiring a visit to each base station.[2]

21.2.1   IP support

As base station equipment providers innovate in their development of IP support, there are several different approaches to IP support that might be taken. The simplest model is to receive the IP packets for data, to packetize the voice into IP, and to run a single IP pipe back to the MSC. An alternative approach might be to install the equivalent of a digital subscriber line access multiplexer (DSLAM) at the base station and have the IP packet data separated from the voice stream, just as a DSLAM at a local office puts data from digital subscriber lines (DSL) onto the public packet data network (PPDN), reducing the load on the public switched telephone network (PSTN). However, a DSLAM would only be useful in cases where soft handoff was not being used for data, such as the shared supplemental data packet channel. A third approach might be to install a TCP/IP router at each base station. Having a router at each base station would allow more precise management of packet transport over the air interface and the backhaul network, possibly reducing delay and jitter. Such a system would require defined QoS classes for voice and data over the air interface.

21.2.2   Smart antennas

Adaptive phased arrays (described in Sec. 4.1.1), known as *smart antennas,* are another rapidly improving technology. In summer 2002, the CDMA Development Group (CDG) gave only one award for the annual 3G CDMA Industry Achievement Award, and it went to Sprint PCS and Nortel Networks for the development and testing of an innovative smart antenna. This antenna doubles capacity for cdmaOne cells without requiring any other changes to the cellular network or to cellular telephones. The antenna uses adaptive antenna beam selection and is compatible with the cdmaOne, cdma2000 1x, cdma2000 1x EV-DO, and cdma2000 1x EV-DV air interfaces. Deploying smart antenna arrays reduces same-cell interference by making use of space division multiplexing (SDM). This can lead to significant cost savings or at least can delay a major expense because it delays the need to deploy new base stations, a major cost in cellular network deployment. Smart antennas also may help extend the range

---

[2]The authors shudder as we remember years of promises saying that we would be able to plug new CPU chips into our personal computers without buying a new motherboard.

of high-speed data services so that higher rates are available further from the base station.

21.3 Conclusion

With this chapter on the deployment of wireless data services, we have completed our exploration of the core principles of information technology systems and their application to wireless telephony. 3G services put data technology underneath the wireless network by providing packet pipes, smart antennas, and other high-tech components. At the same time, 3G services include data service as an integral part of the wireless telephone system so that subscribers can download custom ring tones and e-mail to their cellular telephones.

In Part 5 "Capacity and Quality Principles," we look at the core concepts we will apply to the remainder of this book, where we focus on the application of all these ideas to our goals: the planning and optimization of capacity and quality on CDMA networks.

Capacity and Quality Principles

In Part 5, "Capacity and Quality Principles," we bring expertise from mathematics, telephony, and other fields that we can apply to the design and modification of cellular systems. In Chap. 22, "Capacity and Quality Tradeoffs," we look at business and engineering tools that help us specify requirements, make decisions, and define problems. Then we look in greater depth at the process of creating mathematical models and performing quantifiable optimization of those models. The results of modeling and optimization can guide real-world design and implementation choices. Many of the models described in this chapter have been used in cellular design, in network design, and in the creation of modeling tools for cellular systems.

In Chap. 23, "Traffic Engineering for Voice and Data," we provide a historical overview of the traffic engineering principles and models behind the public switched telephone network (PSTN) and then look at what happens to capacity when data services are added. We cover both circuit-switched and packet-switched networks.

We then turn our attention to the mobile switching center (MSC). In Chap. 24, "Switching Capacity," we define the issues behind capacity calculations for the MSC switch main processor and other components. The next two chapters, "ANSI-41 Signaling Capacity" (Chap. 25), and "Capacity Calculations for Cellular Networks," (Chap. 26), provide the conceptual background for estimating capacity for the cellular signaling network, for every significant component of the cellular system, and for the network backhaul transport from base stations to MSC.

In Chap. 27, "Conventional Reuse Principles," we look at the methods for optimizing the air interface of Advanced Mobile Phone Service (AMPS) and Global System for Mobility (GSM) networks.

The last three chapters of Part 5 are dedicated specifically to code division multiple access (CDMA). In Chap. 28, "CDMA Principles for

Multicellular Systems," *we apply the CDMA equations to derive realistic capacity estimates for the CDMA air interface in both forward and reverse directions. We explore how a myriad of environmental factors and equipment choices affect CDMA capacity.*

In Chap. 29, "CDMA Data Capacity Principles," *we derive the data rate and bit error rate for second- and third-generation CDMA data services. Finally, in Chap. 30,* "Capacity Issues Specific to CDMA," *we explain how unique aspects of CDMA technology, such as soft and softer handoff and the speech coding algorithms specific to CDMA, affect overall CDMA capacity. We close Chap. 30* "Capacity Issues Specific to CDMA," *with a section on radio capacity estimation, where we derive reasonable capacity figures for each of the CDMA standards. This lays the groundwork for the practical application of these principles in Part 6,* "Planning for CDMA Capacity," *and Part 7,* "Increasing Capacity."

Capacity and Quality Tradeoffs

This chapter explores the fundamental principles involved in making decisions that will change the way a system operates. Such decisions are essential to the process of improving or optimizing code division multiple access (CDMA) cellular networks.

There are some general business and engineering guidelines for making decisions that are useful in cases where we cannot create a mathematical model of the system we want to improve. We can think of these as nonquantified decisions.

If we are able to build a mathematical model of the system, then we can optimize the model and use the results to guide our design and upgrade of real-world systems. The principles used in making general business and engineering decisions also apply to the design of mathematical models, but the optimization models also have more stringent requirements. Mathematicians also have a unique skill at picturing systems in ways that might appear counterintuitive but which make problems easier to solve.

In the first section of this chapter we will look at some business and engineering tools that help us define problems and improve systems as we plan them or modify them. The business models are important because, more and more, engineers are expected to be able to present business cases, express value in business terms, and commit to business requirements and goals.

The engineering tools are valuable because organizations such as the Institute for Operations Research and the Management Sciences (INFORMS) and the Institute of Electrical and Electronics Engineers (IEEE) have developed approaches that can save us time and improve the quality of our work. These conceptual tools allow us to achieve better results simply because we do not have to reinvent the wheel if we use them.

When it is possible to create models of systems and quantify the parameters, we can engage in optimization. We use the term *optimization* to refer to the mathematical process of finding the best solution in a model of a system that can then be used to guide real-world implementation. After our discussion of business and engineering tools, we will define quantitative optimization and then discuss several elements of optimization, including tradeoffs, network flows with pipes, bottlenecks, and alternate routes, constrained optimization, optimization variables, quality, capacity, and cost.

22.1   Business and Engineering Tools

The biggest complaint about bad business decisions is that some key element was left out of the decision process. Therefore, business decision tools often focus on the process of defining the question and ensuring that no key factors are left out. One of these tools is *total cost of ownership* (TCO), developed by the Gartner Group. In the TCO model, all costs of developing, maintaining, and decommissioning a system are included in the cost of the system. For example, the total cost of an employee includes not only salary but also hiring costs, the employee's share of all overhead expenses (including the cost of the rental of the floorspace of the employee's cubicle), and all costs related to the employee's retirement or departure.

In recent years, TCO has been applied to information technology (IT) and telephony projects. In evaluating the cost of a new product or service, executives are requiring that project managers estimate the project cost; the operations, administration and maintenance (OA&M) cost for the life of the product or service; and the cost of decommissioning. This is then annualized and offset against the value of the product or service.

The primary value of a product or service is its effect on net revenue. If a change to cellular service is expected to bring in more money or reduce operating costs, then these contribute to the hard dollar value of the project. *Hard dollar value* denotes value that can be estimated and then measured. Businesses also recognize value that cannot be quantified and call it *soft dollar value*. Once we have defined the value and the TCO, we can calculate the *return on investment* (ROI). ROI is measured in months or years, and it is the length of time in which the amount invested in the new product or service, the TCO, will be recognized as revenue through hard dollar value. A project is clearly worthwhile if the ROI is under 1 year and worth considering if the ROI is 1 to 3 years and strong soft-dollar value can be shown. Projects with an ROI of over 3 years require some other justification. For example, if a company will go out of business if the project is not done, then the investment in the project may be essential, even if the rate of return is low. For example, the cost of a North American time division multiple access (TDMA) cellular provider switching to third-generation (3G) CDMA may be very high, but it may be essential in order to compete and to stay in business.

For any new system design, as well as for redesign, repair, or upgrade, we may be asked to prepare a business case justifying our engineering choices by showing business benefits. There is a significant difference between a business case for initial design and a business case for redesign, repair, or upgrade. If a decision is made during design, it is a choice among options: Which way will we do this? As a result, the comparison may be among different outcomes for more than one equal-cost alternative or may be among alternatives with different costs. Any of these options can be compared with the option of doing nothing and not creating the new system at all.

However, if a choice is being made regarding changes to an existing system, then all options must still be compared with the option of doing nothing, but this option itself has a set of costs associated with it. For example, if the current system has an intermittent failure, then the costs of fixing the failure each time it occurs and of losing business during times of failure need to be factored into our thinking. If a permanent repair or upgrade will cost less than repeated failures, then it might be worth doing. Otherwise, it may be less expensive to live with the intermittent failure.

One of the most useful engineering tools for project planning and system modeling is the IEEE standard for a software quality metrics methodology (IEEE-1992). Although it was created for use by software developers, it is valuable in the design of any system where attributes required or desired by users must be turned into engineering specifications.

IEEE-1992 defines a set of attributes that might be important to users, including availability, efficiency, flexibility, integrity, interoperability, reliability, robustness, and usability. It is important to have the customer or customer representative define each of these terms as they apply to the product or service. The goal is to create a verifiable set of attributes that define quality to the customer. If possible, these attributes should be quantifiable; if not, then it is valuable at least to obtain the customer's priorities and preferences. In addition, IEEE-1992 defines a set of attributes that are important to developers or engineers. These include portability, reusability, and testability.

Once we have defined these attributes, we can create a requirements specification. A *requirement* is a document that defines a condition or capability needed by a user to solve a problem or achieve an objective or a condition or capability that must be achieved by a system or system component to satisfy a contract, standard, specification, or other formally imposed document.

It is valuable to prepare two versions of a requirements document, a requirements statement and a requirements specification. The *requirements statement* is more general, written so that the customer and nontechnical executives can follow it. The *requirements specification* is designed for engineers. The two should be coordinated carefully so that they are consistent, and engineers designing or implementing the system should be comfortable with both documents. It has been found that engineers who understand the system from the customer's perspective do a better job at technical design. A high-quality requirements statement will be complete, correct, feasible, necessary, prioritized, unambiguous, and verifiable. If a requirement is necessary and verifiable, then we can define which set of users or interested parties requires or will benefit from that attribute. The technical requirements specification should have these qualities and also should be consistent, modifiable, and traceable. Modifiability and traceability allow the document to be changed and the sources of those changes to be traced to resolve any future difficulties such as inconsistency or incompleteness.

The process of gathering requirements from users and technical sources and creating a requirements document is a huge topic in itself, beyond the scope of this book. The essential points are, first, only the customer knows what the customer wants, and second, a design cannot be improved if it is not clear and well defined in the first place. These points may seem obvious, but real-world experience of failed projects and extensive studies of why telecommunications and information technology (IT) projects fail indicate that we often do not take care of these basics.

Based on the requirements, we design a new system or modifications to an existing system. As we make decisions during the design process and later in development or deployment, we must make tradeoffs. Before we look at quantitative models and the mathematics of tradeoff, a few practical rules of thumb might be useful. There are many cases where we cannot model a problem in a quantitative way, and we still need to make good decisions in those situations.

Project managers have a saying: "You can have it quick, you can have it good, or you can have it cheap: Pick any two of the three." By this we mean that there is always a tradeoff among time, quality, and cost. Our options are limited by the constraints that reality imposes.

One approach to such decisions is to look at each attribute and decide if it is a driver, a constraint, or a factor with a degree of freedom. When we have several attributes and we have to make a tradeoff decision, it is easiest if we can define one driver. We then do the best we can on that one attribute given other constraints and working with degrees of freedom. For example, if we are asked to make something as good as it can be for under $500 and it does not matter if we have it ready tomorrow or in 6 months, then quality is the driver, cost is the constraint, and time has a large degree of freedom.

Of course, practical situations are rarely this clear-cut. A customer may give me the preceding requirements and then, when I call him or her back with a solution that costs $502, be perfectly happy. On the other hand, another customer may genuinely have an absolute limit, perhaps due to regulatory requirements on expenses. In communicating with our customers, it is very important to be clear about each requirement, its constraints, and its degree of freedom.

We are often given problems with multiple drivers. We might receive these instructions: Make it good, but keep costs down. In this case, it is best to propose scenarios with costs to the customer, trying to turn one of the drivers (cost) into a constraint.

22.2   Quantitative Optimization

Optimization is the science of making things as good as they can be. We can break the process up into three component parts:

- Figuring out what is good.

- Figuring out what can be.

- Figuring out the best solution.

Actually, there are many sciences and technologies that fit under the umbrella of optimization, and those who practice the art of optimization usually have a subspecialty.[1] What area of optimization is relevant for a particular problem depends on the particular circumstances.

What is good can be a single variable or a single expression, as most of us saw in our high school mathematics classes:

- Find the largest area . . .

- Maximize profit . . .

- What is the shortest distance . . .

- Minimize the cost . . .

- Calculate the shortest time . . .

- How fast can we . . .

Or what is good can be a combination of goal factors, such as capacity, performance, and cost in wireless telephone systems. Sometimes the factors are harder to quantify, such

---

[1]We know few scientists who consider themselves just "biologists." They are immunologists or hematologists or in some other subfield. One of us (Rosenberg) is in the field of operations research, a field of mathematics specializing in constrained optimization. We will have more to say about this in Sec. 22.5.

as goodwill, community support, customer satisfaction, and reliability. We have to remember that goals are important even when they are hard to quantify.

What can be, what is possible, also can be a single variable or an easily described mathematical set:

- Where x is a real number
- Where n is a positive integer
- Where j and k are integers from 1 to n

Often, figuring out the optimization choices is the hard part of the optimization, especially when thousands or even millions of variables are in play. Sometimes the interactions of the system parameters are the important part of the problem, as we often see in complex aerodynamic systems. More often in economic problems, the system constraints and their relationships are the interesting part of the optimization.

The calculation of an optimal solution depends on the formulation of the problem. Many interesting business problem formulations have optimization calculations beyond the realm of practical computing. More important, many interesting business problems have important factors not formulated at all in the mathematical models that are computationally manageable. While one can define an economic model of goodwill or customer satisfaction, getting a community of business professionals to agree on such a formulation is often a challenging problem in itself and not a mathematical one. Thus it is usually an incorrect use of jargon to refer to many of our solutions as *optimal*, and we often hear people refer to later refinement as a *more optimal* solution, a term that grates on the ears of a professional mathematician. In fact, we are looking for good answers, then better answers, but not necessarily some kind of perfect optimum.[2]

Consider a complex business system, such as a wireless telephone network, that has many component parts and many decisions to be made. In the case of a large wireless telephone network, the decisions include the placement of hundreds of base stations, the design of the backhaul network from those base stations to the mobile switching centers (MSCs), the services to be offered, and the choice of regions where advertising should be concentrated. The wireless service provider swims in a sea of decisions ranging from small day-to-day decisions to big long-term decisions. The ability to run the network at a profit depends on the quality of these decisions.

Rather than experiment on the physical system itself, we conduct thought experiments, make changes in a mental picture of the system, and try to visualize the results. This mental picture is often a computer simulation or a mathematical model; it can be a profit-and-loss business case or an engineer's visualization of the entire system. It is in the realm of the thought experiment that we optimize our decisions.

Some of the decisions are made among a limited set of choices. We call these choices *discrete variables*. They can be as simple as yes or no or as sophisticated as selecting the handoff neighbor lists of each cell sector. In any case, discrete variables have a finite number of choices. Other decisions are *continuous variables* and have an infinitely adjustable set of values. The height of a tower or the angle of an antenna are continuous

---

[2]Operations research is the art of getting bad answers to problems that otherwise would have had even worse answers.

variables. The combination of discrete and continuous variables makes a problem very difficult to solve in a mathematically closed form.[3] Often our choice of discrete or continuous variables depends more on the practical aspects of our model than on the physical reality. A radio power level might have 128 discrete decibel choices in the actual system, but we can model it as a continuous variable and round it off at the end of the optimization because it makes the calculations easier. Or we might take a set of base station antenna heights and model them as integer heights in half-meter increments so that we can use a search algorithm to find better solutions.

22.3   Tradeoff

So we have a system on the drawing board, in a computer model, or in physical reality, and we consider a change to that system and ask how the goal factors will change as a result. For any given change, some of the factors may improve, some may get worse, and others stay the same. If we are lucky, then the change will make some improvements and degrade nothing, a universally good change. Of course, we will execute any change that does nothing but good to our system.[4]

Once we make all the changes we can find that are unequivocal improvements, we come upon changes that are more ambiguous in their results. Most of the changes we think about for a system, wireless or otherwise, are good for some things and bad for other things. They may improve capacity but cost money, or they may improve quality of service at the expense of reliability. We can anticipate that the easy changes are already done and that any change we are contemplating is going to make something worse.

As we consider changes that do something good and something bad, we have to consider the *tradeoff* between the good and the bad. We can think of the options available to us as some kind of geometric space called the *feasible region*. Figure 22.1 shows a two-dimensional picture, whereas reality often has dozens, hundreds, thousands, or even millions of variables.[5] Let us assume that both x and y in Fig. 22.1 are good things to increase. If we are currently at point **P**, then there are still changes to make that increase both x and y, so these are unambiguously good changes to make, easy decisions. Once we have exhausted the easy choices, we are at some point on the

---

[3]In mathematical calculation, we use the term *closed form* to mean an exact equation, or a specific set of computations, used to obtain an answer.

[4]This statement may grate against the ears of two sets of readers: those who work in economics and are familiar with opportunity costs and fans of Robert Heinlein, the science fiction author. For the economists, please note that we are taking opportunity cost into account. If we are in the planning stages, we can change our plans and find a system that will be better than the one we had previously planned at no greater cost. Even if systems already exist, we may find changes that are better in all ways. For example, replacing a defective component that has high maintenance cost with a reliable component may have a net dollar cost over time of less than zero, actually reducing total cost as well as increasing reliability. For Heinlein fans, the great author was correct when he gave us TANSTAAFL (there ain't no such thing as a free lunch). But some lunches pay for themselves.

[5]As an example of a large problem one of us (Rosenberg) has solved, think of an airline with 1500 flights and 100,000 possible passenger routings. For each routing and for each fare class (first class, full fare coach, two levels of discount fares, and frequent flier), the airline has to decide how many passengers it is willing to book.

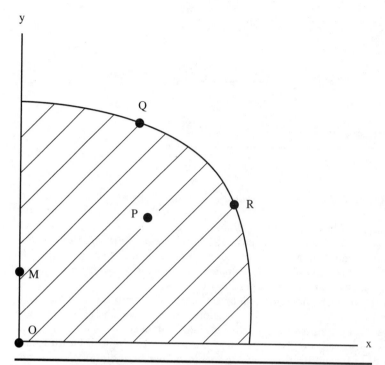

Figure 22.1 Two-dimensional feasible space.

boundary of the feasible region, such as points **Q** and **R**, where all choices have some negative components.

We use the term *frontier* to describe the set of points where all the decisions have some negative consequences. The frontier is not the entire boundary because such points as **M** and **O** are boundary points that are not frontier points as we are using the terms here. If we are trying to make x and y larger, then there are clearly all-good things we can do from points **M** and **O**. Points **Q** and **R**, on the other hand, have clear tradeoff consequences if we try to make changes from them.

This geometric picture represents the state of a system where we make one variable better only by allowing some other variable to get worse. The casual term we use is "push down, pop up" because we can push one problem down only to have something else pop up elsewhere. It is an important part of decision making and decision support to understand what pops up when we push down on specific problems.

22.4 Pipes, Bottlenecks, and Alternate Routes

Suppose that we have a flow network from a source **S** to a target **T**. Each of the dots in Fig. 22.2 is a junction point, and each of the arrows has some flow capacity from one dot to another. It could be a telephone network, it could be water flowing through pipes, it could be cars driving on highways, or it could be a workflow plan for a factory. It does not matter what the application is; the network structure is the same.

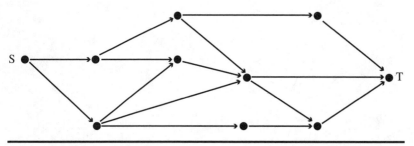

Figure 22.2   Flow network.

Mathematically speaking, this is a combinatorial graph defined and described in Sec. 26.1.2. Each dot is a *node* or *vertex,* and each arrow is an *arc* or *edge*. The arcs have maximum flow capacities, and the object of the network flow game is to get as much stuff to flow from source **S** to target **T** as possible. The network flow structure is particularly amenable to computational solution, so mathematical modelers work hard to put practical problems into the network flow model. As a result, network flow computations can get quite large, some with millions of arcs.

We start solving the network flow problem by finding routes from source **S** to target **T** that have flow capacity available. We add flow along those routes until they can handle no more, until some of their arcs are saturated. We think of these as bottlenecks in the solution process. Rather than give up at the first bottleneck, we look for alternate routes from source **S** to target **T**, other ways to increase flow. Sometimes we find that we have to decrease flow along one route so that we can use capacity more effectively on another route.

At some point in the solution process, it becomes impossible to add more flow from source **S** to target **T**. There is a set of arcs that are at maximum capacity, final bottlenecks that are called *critical resources* in the solution. The other resources are called *slack resources*. Expanding the critical resources can improve the flow in the network, whereas expanding the slack resources has no effect on the solution. It is the job of network designers to find ways to add capacity to the critical resources. Sometimes the internal flows have their own costs so that the solution is a tradeoff between the value of the flow and the cost of the resources.

The power of the network flow problem is that the flows can be metaphors for a wide variety of things. Some problems involve a serious stretch of the imagination to see what is flowing through what kind of system, but the mathematical solution is just as valid as the maximization of water flow through a network of pipes.

22.5   Constrained Optimization

We have seen optimization in mathematics class. Usually there is some variable, x, and some function, $f(x)$, and the object is to find the value of x that gives the maximum value of $f(x)$. The function we are optimizing is called the *objective function*. The usual tactic is to use calculus to find the derivative $f'(x)$ and to set that derivative equal to zero and to solve for x. The math teacher usually makes some passing remark about checking the end points if x has a limited range in case the maximum of $f(x)$ is at a boundary point instead of a nice hump on the (x, y) graph of $f(x)$.

In our business optimization problems, there are multiple objectives, so there are many variables to be traded off. We go back to the feasible region in Fig. 22.1 and realize that the character of the solution depends on the shape of the frontier. While both x and y are good things, the decision as to whether **Q** or **R** is a better solution depends entirely on the relative values of increasing x and y.

In our business optimization problems, the constraints are typically very important. The character of the feasible region determines what solutions are available to be traded off in the solution process. In the network flow example, the feasible region is all the possible flows in the network, all the possible allocations of resources in the system.

The role of the decision maker, or the decision support system supporting the decision maker, is somehow to select the frontier's best point, which will be the best feasible point. This is the best possible selection of all the system parameters that satisfies all the constraints. These decision support systems are typically large and complex computer programs embedded in even larger and more complex data systems.

There are several methods for finding the best solution to constrained optimization problems. Many of these are iterative schemes where the decision support system works its way from worse to better. Let us look at the framework of an iterative scheme for constrained optimization without going into the details of any one particular method. We create thought experiments for the system, hypothetical mathematical solutions that we try and keep trying until we find our optimal solution.

- *Step 1:* Start by finding some feasible solution. Sometimes there is a good starting point, which can be from previous experience with a similar problem, but often the starting point is some zero point known to be feasible.

- *Step 2:* Look around for something better, some direction we can move, some change we can make, so that the solution stays feasible and gets better. Usually such improvement is a mixed blessing, a tradeoff where something good is given up to get something more or something better.

- *Step 3:* If no possible change within the feasible region makes things better, then we conclude that we have found the optimal point, and we can stop.

- *Step 4:* We make the improvement from one feasible point to a better one.

- *Step 5:* Go back to step Two.

This iterative scheme is not the only solution technique for constrained optimization problems, but it is a common theme underlying many optimization algorithms. For continuous problems, the feasible point changes in step 4 go in some improvement direction as far as they can go while the solution improves or as far as they can go and stay feasible. For discrete problems, the feasible point changes are typically some rearrangement or permutation. The case of a linear objective function with linear constraints is called a *linear program.* The traditional solution technique for linear programs, called the *simplex method,* is a combination of the continuous and the discrete, a series of steps from one combination of constraints to another along a sequence of linear moves.[6] As the optimization involves more complex functions

---

[6]Our operations research friends will forgive us, we hope, for trying to sum up a vast area of research in a single sentence.

and constraints, the solution technologies get more complicated and use more computer time.

In any case of a constrained optimization, the optimal solution will divide the constraints into slack and critical resources. The slack resources have a value of zero so far as the optimal value is concerned. However, the critical resources have some positive value, some positive effect on the optimal solution, if we can increase them a little bit. The understanding of the incremental value of relaxing constraints is an important part of the solution of a constrained optimization. These incremental values, sometimes called *prices,* describe the impact and importance of the constraints in a constrained optimization problem. If all the constraint variables are slack, then the constraints are irrelevant, and the optimization is really an unconstrained optimization.

22.6   Optimization Variables

Deciding on the relative values of the various components of a business solution is a difficult and often political process. We study a problem, we look at as many aspects of it as we can, and we try to model as much of this as we can in a mathematical formulation. The list of objective function parameters can be long, and their relationships can be confusing. Getting consensus on the relative importance of various aspects of an optimization problem can be a difficult problem in human negotiation as well. Members of a team who are trying to find the best solution are likely to think that their own parts of the problem are the most important.

There is another side to the optimization formulation. Decision support problems often have different formulations for the same goals and constraints. One formulation of a problem might take a computer thousands of times longer to solve than another formulation of the same problem. The difference in computational effort can be particularly great when an integer solution is required, one consisting of whole numbers.[7]

Sometimes we can gain computational efficiency by interchanging objectives and constraints. We may be trying to serve a given customer community at the lowest cost, but the problem may be far easier to solve with an objective of maximizing users served at a fixed cost. Solving one problem is the practical equivalent of, or at least similar to, solving the other problem, but the mathematics and computation may be dramatically different.

22.6.1   Multiple dimensions of quality

It is important to remember that objectives can take many forms, particularly when we are trying to improve quality. Pick any industry, any business, any technical area, and there are almost certainly many aspects to consider in optimizing quality. Some problems get virtually the same answer as long as all the important aspects appear in the objective function whereas others have more specific tradeoffs to be optimized. These other problems are the ones where it is most important to define the objective function and to understand its relative emphasis on the different parts of quality.

---

[7]We can set a radio to transmit at half power, but we cannot put half a base station at a specific location.

In the case of a wireless telephone system, the are many objectives and tradeoffs vying for a decision maker's attention.

- Better service versus lower cost

- Volume of traffic versus sound quality

- Voice calls versus data services

- Faster data transmission versus higher availability

- Lower cost today versus higher revenue tomorrow

- Assured gains versus greater opportunity

The answers that a decision support system provides to wireless telephone system engineers depend greatly on the input parameters to the optimization tools being used.

22.6.2 Capacity

Even something as apparently obvious as capacity can have ambiguity of meaning. Coming from the telephone world, it is natural to define capacity as the number of voice calls a system can support over a specific period of time. In the next chapter, "Traffic Engineering for Voice and Data," we explore the relationship between the amount of subscriber demand and the system's ability to serve a high enough fraction of the call attempts. There are other issues of geographic distribution of subscriber demand and demand at different times of day.

Also, as we move from 2G to 3G, we have a growing market for data traffic. A wireless telephone system is unlikely to be optimized for voice and for data simultaneously. Making a better packet data environment almost certainly means making compromises in voice-call performance. Perhaps the system will allow fewer calls, or perhaps those calls will have poorer sound quality. We cannot evade these choices by looking away: If the engineers deploying a wireless telephone system do not address these issues, then they will be decided by the gods of chance.

In the design of a CDMA system, capacity is not an absolute feature. Rather, voice capacity will be determined by quality-of-service (QoS) performance criteria. These include the rate of ineffective attempts (call blocking), lost calls, and sound quality. We can serve more calls if we are willing to subject our subscribers to more static. Television commercials to the contrary, many subscribers would rather have some fuzz on the line than to be denied cellular service outright.

22.6.3 Cost

If capacity is one side of the optimization coin, then the other side is cost. The bottom line of the capacity versus cost debate, however, is really money versus money. After all, capacity means subscribers who pay for the service, so the tradeoff of capacity versus cost is really revenue in the future versus capital expenditure now. Wireless financial planners have to make the decision of how much they are willing to spend in capital equipment to justify an increase in revenue over time. Revenue versus cost is the unifying tradeoff in many wireless engineering decisions.

We can model this tradeoff as a maximization of capacity in the face of a fixed cost and fixed parameters of system operations. The QoS is determined upfront, requirements of ineffective attempts, lost calls, and voice quality. Once we decide how much money we can spend on switches, base stations, radio equipment, and telephone transport, the wireless engineer tries to maximize the number of calls per hour.

Or we can model the same tradeoff as a minimization of cost in the face of a fixed subscriber base. At the given QoS required, we decide how many calls per hour we have to serve, and the wireless engineer tries to minimize the equipment and transport cost.

Of course, the mix of subscriber services makes a tremendous difference in cost. We know that voice calls compete with data services for resources. However, we may not realize how alternative data services, perhaps for different customer communities, compete with each other for system resources. Having enough common channels for rapid-access, high-speed packet access may limit not only voice traffic but also e-mail and dedicated virtual-circuit data services.

22.7 Conclusion

Many of the practical rules of thumb we follow in making good business and engineering decisions actually have a source in quantitative models from economics, engineering, and other disciplines. When we have a problem that has a precise definition and we have enough time and money to spend on it, creating a model of our system and making decisions based on optimization of the model are extremely beneficial. However, in the far more common case of needing to make decisions quickly when a number of elements are not well defined, we need to be able to fall back on less precise tools such as common sense and clear thinking.

Traffic Engineering for Voice and Data

To engineer any telephone system, wireless or landline, we need some way of estimating telephony demand and criteria for satisfying that demand. *Traffic engineering* is the process of assigning facilities to satisfy a demand. Traffic engineering is particularly important inside the public switched telephone network (PSTN), the Internet, the cellular backhaul network, and other networks where demand from multiple users is carried over shared channels. Statistical methods can be applied to estimate demand and design facilities to meet peak demand. In this chapter we will explain those methods. First, we will look at the methods traditionally used in telephony, which assumed that all circuits were the same size and that the average length of telephone calls was unlikely to change very much. Then we will discuss the implications of data usage and data services and how they require new models. After that, we will discuss the traffic engineering of packet networks.

23.1 Voice Calls Assuming Fixed-Size Channels

The most fundamental traffic engineering job in telephony is the assignment of transport facilities to serve fixed-size voice channels. There is a demand of voice telephone calls to be served, some community of subscribers making calls. While telephone usage patterns are not random to the subscribers themselves, so far as the telephone network is concerned, the demand is a random statistical process, a distribution of calls over time.

When each call attempt is made, the telephone network serves the call if it has facilities to do so, or it blocks the call. Our measure of network performance in traffic engineering is the fraction of calls that are blocked. The fraction of calls that are blocked can be viewed as the probability that any one call will be blocked. If the blocking probability is low enough, then the network is adequately designed. Each voice channel uses a single DS-0 or E-0, and it is the traffic engineer's job to ensure that the network has enough telephone lines to keep the blocking rate low enough during the busy period.

23.1.1   Erlangs, the units of demand

The units of telephony demand are *erlangs,* named after Danish telephone pioneer Agner Krarup Erlang. Roughly speaking, the demand in erlangs is the average number of calls that would be in progress if there were no blocking. We say that a call is *blocked* if it is not served because there are not enough resources to serve it. The demand in erlangs is the sum over a large subscriber base of the small probabilities of making a call times the durations of those calls. If each of s subscribers $j = 1, 2, \ldots, s$ has a probability γ_j of making a call of average duration d_j during a time period t, then the number of erlangs is ϕ in Eq. (23.1):

$$\phi = \frac{1}{t} \sum_{j=1}^{s} \gamma_j d_j \tag{23.1}$$

For our discussion on telephone system capacity, we can ignore the fact that some subscribers are more inclined to use the telephone than others[1] and give every one of the s users the same probability γ of making a call of duration d during time period t. This gives us Eq. (23.2):

$$\phi = \frac{s\gamma d}{t} \tag{23.2}$$

The deep and magnificent mathematical insight is that once s, the number of subscribers, is large and γ, the probability of a subscriber making a call, is small, the behavior of the telephone system is completely determined by the erlang measure ϕ. The individual component values of s, γ, and t lose their importance once ϕ is known.

23.1.2   The Poisson distribution

Let us examine a state-space diagram of the telephone system. Let the state of the telephone system at time t be the number of calls j in the system at time t. From any given state, two things could happen:

1. Somebody starts a new call.

2. Somebody ends a call.[2]

Let λ be the rate of users placing calls, the average number of new calls per unit time. And let μ be the rate of calls ending once they are started.

The state of the telephone system is the number of calls j. We turn the system on with the state j set to 0 and wait for a call to start and to change the state j from 0 to 1. Then another call could come along and change the state j from 1 to 2 or the first call could end and set the state j back to 0.

---

[1]Never mind that the loquacious telephone subscribers are the loud talkers in the room next to you. In this discussion, we can ignore them too.

[2]This is not possible, of course, from the zero-calls state.

New call behavior is completely independent of the state the telephone system, so the rate of changing from state j to $j + 1$ is a constant value, λ, for every state j. The rate of changing from state j to $j - 1$, on the other hand, depends on the number of calls in the system. Each call in the system has a rate μ of ending. Since state j has j calls in progress, the total rate of a call ending in state j is j times μ, and therefore, the rate of the system going from state j to $j - 1$ is $j\mu$.

The state diagram is shown in Fig. 23.1. The arrows indicate the rates between states. If we think of each state as having some probability P_j, then we can picture the probabilities as functions of time $P_j(t)$ flowing along the arrows in the state diagram and generating a set of time derivatives in Eq. (23.3):

$$P_0'(t) = \mu P_1(t) - \lambda P_0(t)$$

$$P_j'(t) = (j + 1)\mu P_{j+1}(t) + \lambda P_{j-1}(t) - j\mu P_j(t) - \lambda P_j(t) \quad \text{for } j > 0$$

(23.3)

Let us look at these differential equations for a moment to understand what they are telling us. Suppose that we have some knowledge of the system's state at a specific time $t = 0$. For example, we may know with 100 percent certainty that there are no calls on the system when we first turn it on, so $P_0(0) = 1.0$ and $P_j(0) = 0.0$, for $j > 0$. Or our knowledge of the system may be statistical in nature, that each state j has some probability $P_j(0)$ of being occupied by the system at the specific time $t = 0$. Once we have a set of probabilities for each state, we can computationally integrate the differential equations to get these probabilities over a period of time. This is the same sort of calculation early astronomers used to integrate the differential equations of gravity and inertia to predict planetary motion.

The *equilibrium* case occurs when the probabilities are not changing as time passes, when all their time derivatives are zero. When $P_j'(t) = 0$, Eq. (23.3) reduces to the simpler Eq. (23.4). We can think of Eq. (23.4) as zero net probability flow between each pair of adjacent nodes.

$$\lambda P_{j-1}(t) = j\mu P_j(t)$$

(23.4)

This gives us the infinite set of equations (Eq. 23.5) for the fixed-probability sequence P_j:

$$P_j = \frac{\lambda}{j\mu} P_{j-1}$$

$$\sum_{j=0}^{\infty} P_j = 1$$

(23.5)

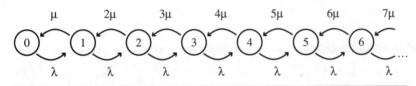

Figure 23.1   State diagram for counting calls.

This sequence of probabilities comes from the single value $\alpha = \lambda/\mu$, and it is called the *Poisson distribution*. The solution to Eq. (23.5) is Eq. (23.6)[3]:

$$P_j = e^{-\alpha} \frac{\alpha^j}{j!} \tag{23.6}$$

The Poisson distribution of number of channels used depends only on $\alpha = \lambda/\mu$, the total likelihood of the entire user population to be using the telephone. As long as the population is large and no individual makes up a significant fraction of the total telephone usage, the distribution on channels used will follow this Poisson distribution with mean α.

23.1.3 Blocking probabilities

Real telephone systems have a limited number of available channels k. When all k channels are in use, new call attempts are refused, and we say the telephone system is *blocked*. The upper bound of k channels can be modeled in two reasonable ways. The blocked callers can continue to try and retry, still in the system but not using any channels, or the blocked callers can go away and not retry their calls.

In the retry case, the distribution for $j > k$ depends on the behavior of blocked callers. If the callers keep trying and retrying and never give up, then our system becomes an $M/M/c$ queue, as described in Sec. 23.3.1. If they coincidentally just happen to give up at the same rate as they hang up when they are talking on the telephone, then we can represent the system as a full Poisson distribution with every state $j \geq k$ as a blocked state. In this case, Eq. (23.7) says that the probability of a call being blocked is the probability that the Poisson distribution has at least k calls in it:

$$P(\text{blocking}) = e^{-\alpha} \sum_{j=k}^{\infty} \frac{\alpha^j}{j!} \tag{23.7}$$

In the no-retry case shown in Fig. 23.2, there is no activity beyond state k, and the blocking probability is as in Eq. (23.8):

$$P(\text{blocking}) = \frac{\alpha^k/k!}{\sum_{j=0}^{k} \alpha^j/j!} \tag{23.8}$$

This is called the *Erlang-B model*. Any time the system is in a state $j < k$, there is room for one more call, but when the system is in state k, there is no more room for another

---

[3]The notation $j!$ is j *factorial,* the product of all the whole numbers from 1 up to j.

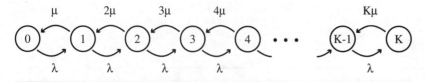

Figure 23.2 State diagram for k telephone channels.

call, and the system is in a blocked state. The probability of a new call being blocked is precisely the probability that the system is in state k.

The difference between having blocked callers try again and having them go away quietly can be important in congested systems or systems with small numbers of channels. For large values of k and for low blocking probabilities, it makes little difference whether or not blocked callers try again.

23.1.4 Blocking tables

Table 23.1 shows Erlang-B capacity as a function of the number of channels (trunks) and the blocking probability. For example, 25 channels can support 16.1 erlangs at 1 percent blocking and 20 erlangs at 5 percent blocking. Insisting on the higher standard of 1 percent blocking costs us 4 erlangs of capacity. Another way of looking at the same 20 erlangs is that 25 channels can serve that demand at 5 percent blocking, but it takes 30 channels to offer 1 percent blocking to the same user community.

Let us look closely at the first, second, and last rows of Table 23.1 to help us understand what this table really means. Consider a single server, one telephone line, one Xerox copier, or one bank teller, and consider customers who refuse to wait in line. How much business can this server support so that 90 percent of the time a new customer sees an available server? This means that the server is only busy 10 percent of the time. A demand of 10 percent of an erlang will keep one server busy 10 percent of the time.

TABLE 23.1 Erlang-B Capacity Table

Lines	Erlang-B capacity			
	$P(B) = 0.01$	$P(B) = 0.02$	$P(B) = 0.05$	$P(B) = 0.10$
1	0.01	0.02	0.05	0.11
2	0.15	0.22	0.38	0.60
3	0.46	0.60	0.90	1.27
4	0.87	1.09	1.52	2.05
5	1.36	1.66	2.22	2.88
6	1.91	2.28	2.96	3.76
8	3.13	3.63	4.54	5.60
10	4.46	5.08	6.22	7.51
12	5.88	6.61	7.95	9.47
14	7.35	8.20	9.73	11.47
16	8.87	9.83	11.54	13.50
18	10.44	11.49	13.39	15.55
20	12.03	13.18	15.25	17.61
25	16.12	17.50	19.99	22.83
30	20.34	21.93	24.80	28.11
35	24.64	26.43	29.68	33.43
40	29.01	31.00	34.60	38.79
45	33.43	35.61	39.55	44.17
50	37.90	40.26	44.53	49.56
60	46.95	49.64	54.57	60.40
70	56.11	59.13	64.67	71.29
80	65.36	68.69	74.82	82.20
90	74.68	78.31	85.01	93.15
100	84.06	87.97	95.24	104.11

Well, this is almost correct. Since the customers turned away are out of the system, we can raise the demand levels so that the remaining 90 percent keeps the server busy 10 percent of the time. Solving the algebra[4] tells us that a demand of *one-ninth* of an erlang for one server has nine-tenths of it keeping the server busy one-tenth of the time. And the other one-tenth is the traffic turned away by the busy server.

The same reasoning applies to 1, 2, and 5 percent blocking for a single server. The correction terms are smaller because fewer customers are being turned away, but the reasoning is the same. A single server at 5 percent blocking serves just a little over 0.05 erlang, not very efficient.

Going to two servers improves our efficiency enormously. Picture two servers serving 0.2 erlang each. Ignoring the turn-away correction terms, we estimate that each of these servers has 20 percent blocking. If that 20 percent can overflow to the other server, then the only time a customer is blocked is when *both* servers are busy. With a little more mathematical hand waving, we can figure that two independent servers are busy 20 percent of 20 percent of the time, a net blocking rate of 4 percent for 0.4 erlang of demand. As loose as this approximation is, it helps us believe that two servers can serve 0.38 erlang at 5 percent blocking.

The last line of Table 23.1 has the puzzling number of 100 channels serving 104.11 erlangs at 10 percent blocking. How can this be? The answer is that only 93.7 erlangs are actually *served,* and the other 10.4 erlangs are turned away, the 10 percent who are blocked.

Another view of system capacity is *occupancy,* the fraction of the time the channels are in use. Table 23.2 shows the same information in terms of occupancy rather than erlangs of demand. The relationship is

$$\text{Occupancy} = \frac{\text{erlangs}}{\text{channels}} (1.0 - \text{blocking})$$

Tables 23.3 and 23.4 show the Erlang-B capacity and occupancy data for 24-trunk DS-1 and 30-trunk E-1 links.

We also can do this calculation for very low blocking rates and larger numbers of channels in Tables 23.5 and 23.6. These are more typical values for the large trunk groups found inside the public switched telephone network (PSTN).

Knowing the demand is a critical part of wireless telephone system. Planning engineers rely on forecasts for initial planning, but wireless system growth can be based on *measured* demand.

Measuring demand is easy when blocking rates are low. If nobody is blocked, then usage is the same as demand. The demand in erlangs is the total of the call durations divided by the time interval chosen. Typical telephone company engineering is for the highest-demand time of day, the "busy hour."

Measuring blocking rate by itself is not useful because we do not know how often callers retry when their calls are blocked. Instead, we can measure the occupancy and use the formulas that generated Tables 23.1 through 23.4. We measure the call minutes used during the busy hour, divide by the available call minutes, and use that ratio as the occupancy in estimating demand.

---

[4]While this is high school algebra, the reader who does not remember it should not feel bad. Most of us in the wireless telephone business have been away from high school for a while. You can trust us—it all works out.

TABLE 23.2 Erlang-B Occupancy

Lines	Erlang-B occupancy			
	P(B) = 0.01	P(B) = 0.02	P(B) = 0.05	P(B) = 0.10
1	0.010	0.020	0.050	0.100
2	0.076	0.109	0.181	0.268
3	0.150	0.197	0.285	0.381
4	0.215	0.268	0.362	0.460
5	0.269	0.325	0.422	0.519
6	0.315	0.372	0.469	0.564
8	0.387	0.444	0.539	0.630
10	0.442	0.498	0.590	0.676
12	0.485	0.540	0.629	0.711
14	0.520	0.574	0.660	0.738
16	0.549	0.602	0.685	0.759
18	0.574	0.626	0.706	0.777
20	0.596	0.646	0.724	0.793
25	0.639	0.686	0.759	0.822
30	0.671	0.716	0.785	0.843
35	0.697	0.740	0.806	0.860
40	0.718	0.759	0.822	0.873
45	0.735	0.775	0.835	0.883
50	0.750	0.789	0.846	0.892
60	0.775	0.811	0.864	0.906
70	0.794	0.828	0.878	0.917
80	0.809	0.841	0.888	0.925
90	0.822	0.853	0.897	0.931
100	0.832	0.862	0.905	0.937

TABLE 23.3 Erlang-B Tables for DS-1

Lines	Erlang-B capacity			
	P(B) = 0.01	P(B) = 0.02	P(B) = 0.05	P(B) = 0.10
24	15.29	16.63	19.03	21.78
48	36.11	38.39	42.54	47.40
72	57.96	61.04	66.69	73.47
96	80.31	84.10	91.15	99.72
120	102.96	107.42	115.77	126.08

Lines	Erlang-B occupancy			
	P(B) = 0.01	P(B) = 0.02	P(B) = 0.05	P(B) = 0.10
24	0.631	0.679	0.753	0.817
48	0.745	0.784	0.842	0.889
72	0.797	0.831	0.880	0.918
96	0.828	0.859	0.902	0.935
120	0.849	0.877	0.917	0.946

TABLE 23.4   Erlang-B Tables for E-1

Lines	Erlang-B capacity			
	$P(B) = 0.01$	$P(B) = 0.02$	$P(B) = 0.05$	$P(B) = 0.10$
32	22.05	23.72	26.75	30.24
64	50.60	53.43	58.60	64.75
96	80.31	84.10	91.15	99.72
128	110.57	115.23	124.01	134.88
Lines	Erlang-B occupancy			
	$P(B) = 0.01$	$P(B) = 0.02$	$P(B) = 0.05$	$P(B) = 0.10$
32	0.682	0.727	0.794	0.850
64	0.783	0.818	0.870	0.911
96	0.828	0.859	0.902	0.935
128	0.855	0.882	0.920	0.948

TABLE 23.5   Erlang-B Capacity Tables for Low Blocking Rates

Lines	Erlang-B capacity			
	$P(B) = 0.0001$	$P(B) = 0.0002$	$P(B) = 0.0005$	$P(B) = 0.0010$
100	69.24	70.88	73.24	75.24
200	156.14	158.72	162.46	165.62
300	246.39	249.72	254.56	258.64
400	338.34	342.32	348.09	352.99
500	431.35	435.91	442.53	448.15
600	525.11	530.20	537.60	543.89
700	619.43	625.01	633.14	640.06
800	714.19	720.23	729.05	736.55
900	809.32	815.80	825.25	833.32
1000	904.73	911.63	921.71	930.31

TABLE 23.6   Erlang-B Occupancy Tables for Low Blocking Rates

Lines	Erlang-B occupancy			
	$P(B) = 0.0001$	$P(B) = 0.0002$	$P(B) = 0.0005$	$P(B) = 0.0010$
100	0.692	0.709	0.732	0.752
200	0.781	0.793	0.812	0.827
300	0.821	0.832	0.848	0.861
400	0.846	0.856	0.870	0.882
500	0.863	0.872	0.885	0.895
600	0.875	0.883	0.896	0.906
700	0.885	0.893	0.904	0.913
800	0.893	0.900	0.911	0.920
900	0.899	0.906	0.916	0.925
1000	0.905	0.911	0.921	0.929

23.2   Effects of Adding ISDN Service

Adding circuit-switched data service to the demand for voice calls changes the traffic engineering in three important ways:

- The larger size of each data channel reduces the efficiency of our resource utilization.

- Having a mix of large and small users makes the traffic engineering models more complicated.

- Balancing the blocking requirements of large and small users requires strategic blocking of small users.

Consider a resource that handles 40 voice calls in which each voice call uses 14,400 bits per second. This same resource also can handle four Integrated Services Digital Network (ISDN) circuit-switched data calls at 144 kbps. To the ISDN user who is 10 times larger, the resource looks 10 times smaller.

Suppose, however, that we ask how much traffic this resource can support at 2 percent blocking. Our tables tell us that a resource with 40 channels can serve 31.0 erlangs of traffic at an occupancy of 76 percent. These same tables tell us that a resource with 4 channels can serve 1.1 erlangs of traffic at an occupancy of 27 percent. This means that the same resource with traffic engineering figured into the equation looks 30 times smaller.

Under these conditions, the ISDN rate would have to be 30 times the voice channel rate for the ISDN subscriber to pay an equal share.[5] In the CDMA world, this means that ISDN requires a share of the channel greater than 10 times the voice call.

Serving both types of users, voice and ISDN, requires a more sophisticated analysis. To make visualization easier, let us consider small users who use one unit of resource, big users who use three units of resource, and a total resource of seven units. At any given moment a small call can start, a small call can end, a big call can start, or a big call can end. Each of these events changes the available resource by one or three units one way or the other. It is tempting, but incorrect, to consider eight states, zero through seven units of resource available, as shown in Fig. 23.3.

Why is this picture not complete? Because it blurs the distinction between units of resource used for small calls and units of resource used for big calls. Consider state 4 in Fig. 23.3. There are four ways to leave state 4, a small call can start, a small call can end, a big call can start, or a big call can end. The distribution of new calls is independent of the four units of resource, but the distribution of calls ending depends on whether the four resource units in state 4 are four small calls or one small call and one big call. This is the important distinction lost in this picture.

The reality is two-dimensional, and the full state space for two services is a two-dimensional picture, shown in Fig. 23.4. There are 15 states rather than 8 if we correctly count the different combinations of small and big calls using the resources. Each of the 15 states represents some combination of small calls and big calls that fit in the total resource of 7 units. Let us denote by $4s + 0b$ the state with four small calls and no big calls, and let $1s + 1b$ be the state with one small call and one big call. While both

---

[5]In fact, the ratio of usage may be greater than 10 because ISDN requires a lower bit error rate (BER) and, therefore, a higher E_b/N_0. This relationship is described in Chap. 29.

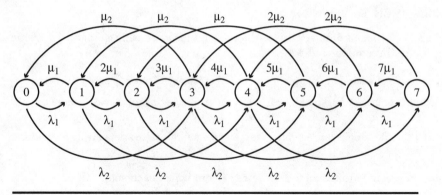

Figure 23.3 Incorrect state diagram for two types of users.

of these are using 4 units of resource, they are different states, and Fig. 23.4 shows them as different. While we can compute the equilibrium probabilities by solving 15 linear equations in 15 unknowns, there are some conclusions we can draw about the system without actually doing the calculation.

We have marked with black circles the three states that block both small and big calls, $7s + 0b$, $4s + 1b$, and $1s + 2b$. These are the states that use all 7 units of resource. The blocking rate for small calls is the sum of the equilibrium probabilities for the three black circles because these are the states that will block a small call. We have

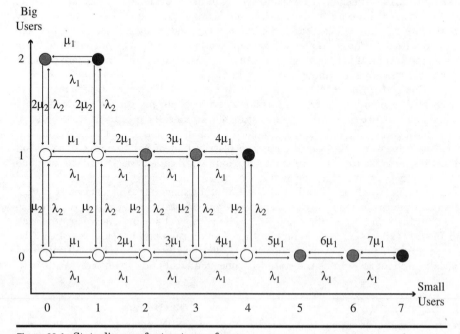

Figure 23.4 State diagram for two types of users.

marked with gray circles the five states that block big calls and not small calls, $6s + 0b$, $5s + 0b$, $3s + 1b$, $2s + 1b$, and $0s + 2b$. These are the states that use 5 or 6 units of resource, so there is room for a small call but not a big call. The blocking rate for big calls is the sum of the probabilities of the three black circles and the five gray circles because these are the states that will block a big call.

If we have been directed to offer both small calls and big calls the same blocking rate, then this is not an efficient scenario. If the target blocking rate is fairly low, then the gray circles will have equilibrium probabilities at least as high as the black circles. In the general case with big calls b times bigger than small calls, a low blocking rate, and a resource large enough for several big calls, the blocking probability for big calls will be at least b times higher than the blocking probability for small calls.

We can balance the service blocking rates between small calls and big calls by deciding to block *all* calls in any state where the physical system would block *any* calls. The dark circles in Fig. 23.5 are states where we are blocking both small calls and big calls even though the physical system has the resources to serve a small call in some of those states. This has the effect of raising the blocking rate for small calls, but it reduces the blocking for big calls without having to increase the resource. The same technique of balancing service blocking rates is useful when there are more than two services.

In fact, having a single condition for blocking calls is a natural thing to do in a CDMA system. The system can use some radio condition as a blocking criterion. For example, if the forward transmit amplifier power level exceeds some predetermined fraction of the total amplifier power, then the wireless system would be instructed to deny all new calls on that cell sector. Both low-rate voice calls and high-rate data calls would be denied service once the transmit power reaches this level.

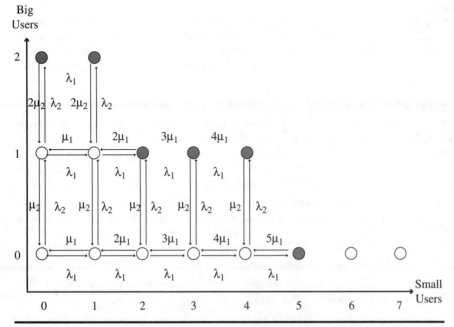

Figure 23.5 State diagram for balanced service for two types of users.

23.3   Traffic Engineering for Packet Data

When we move from circuit-switched to packet traffic engineering, we move our analysis from the status of a call to the status of a packet. A circuit-switched link sends all its content (analog or digital) on the same route to the same destination. Once the circuit-switched link is established, traffic engineering has no connection to the quality of the link. Packets are separate entities that can go separate ways to the same destination. This allows packet data transport to use resources far more efficiently at the expense of bringing more complex issues into the design stage of a packet pipe.

The quality-of-service (QoS) issues of a data link are latency, jitter, and loss, as described in Sec. 17.3. As a congested link increases packet delay, there is a direct connection between the size of a packet pipe and the QoS of each packet connection using that pipe. Packet switches use store-and-forward technology to manage data traffic, essentially lining the packets up in a queue for transport, so our packet traffic engineering model is a mathematical queuing model. We will examine the $M/M/1$ queue, a simple queuing model that fits pretty well to packet-switch behavior.

If a packet data service has any kind of delay requirement, then there is some delay beyond which a packet is too late to be useful. After we have described the $M/M/1$ queue, we will show its distribution of delay probabilities and how that can be used for traffic engineering a packet data link.

23.3.1   Queuing theory

The physical picture of a queue has some kind of service being provided to people who arrive to be served. If there are no servers available when a person shows up for service, then that person waits in some kind of queue, usually a line, until a server is available.

The mathematical notion of a queue, at least at the basic level, is a system where there is an incoming flow of work and a collection of servers. The incoming flow is usually a collection of discrete arrivals where each arrival has some workload associated with it. We denote the various mathematical queuing models with a three- to five-letter notation, $A/B/c/K/N$.

A. The first letter is the interarrival time distribution. The letter D denotes constant interarrival times, an arrival every 2 minutes without exception. The letter M denotes exponential interarrival times, a *memoryless* process where the time distribution until the next arrival does not depend on how much time has passed since the last arrival. The letter G denotes a general interarrival distribution. For queuing theory specialists, there are E_r, r-stage Erlangian, and H_R, R-stage hyperexponential, distributions as well.

B. The second letter the service time distribution. The letter D denotes constant service time; every customer requires exactly 5 minutes of work. The letter M denotes exponential service time. And the letter G denotes a general service time. There are E_r, r-stage Erlangian, and H_R, R-stage hyperexponential, distributions for service time as well.

c. The third symbol is the number of servers. This can be a constant such as 1, 2, or 3 or a variable c that appears in the solution equations.

K. The fourth symbol is the storage capacity of the queue, the maximum number of arrivals that can wait in line. Any arrivals beyond K are dropped from the system. If the last two symbols are left out (e.g., $M/M/1$), then there is no limit; the storage capacity is infinite, and $K = \infty$.

N. The fifth symbol is the customer population, the size of the community that can request service. If the last symbol or the last two symbols are left out (e.g., $M/M/1$), then there is an unlimited population, and $M = \infty$. The significance of a noninfinite value of M is that a bigger queue has fewer customers out there and has a lower arrival rate.

Let us concentrate on the $M/M/1$ queue whose full five-letter form is $M/M/1/\infty/\infty$. This is a queue with random exponential interarrival time and random exponential service time. We also assume this is a first-come, first-served (FCFS) queue, where the order of service is the order of arrival. There are prioritized queues, but $M/M/1$ is not one of them.

Do we really believe that packet data have perfectly memoryless interarrivals and an exactly exponential distribution of packet size? Of course not. However, it is a good starting point, and research has shown that $M/M/1$ is a reasonably good model for network traffic.

There are only two parameters for an $M/M/1$ queue. The mean interarrival time is λ, and the mean service rate is μ, so the mean service *time* is $1/\mu$. We use the service rate rather than time to follow the same pattern used in Fig. 23.1 for circuit-switched traffic engineering. The *utilization,* or *traffic intensity,* of an $M/M/1$ queue is the fraction of time the server is busy. The utilization ρ is the mean interarrival time λ divided by the mean service rate μ, as shown in Eq. (23.9):

$$\rho = \frac{\lambda}{\mu} \tag{23.9}$$

That ρ is the utilization makes sense because the server, on average, does $1/\mu$ work every λ time interval. Each arrival coming into the queue will receive immediate service when the server is not busy, $1 - \rho$ of the time, and will have to wait in the queue when the server is busy, the other ρ of the time. The $M/M/1$ queue does not make sense if $\rho \geq 1$ because the queue backs up endlessly.

The distribution of the length of the $M/M/1$ queue over time is an exponential distribution as shown in Eq. (23.10):

$$\text{Prob}(n \text{ in queue}) = (1 - \rho)\rho^n \tag{23.10}$$

$$\text{Average in queue} = \frac{\rho}{1 - \rho} \tag{23.11}$$

Notice that the average number in the queue in Eq. (23.11) rises sharply to infinity as ρ gets close to one.

23.3.2 Waiting probabilities

The most important law in queuing theory is Little's law. It says that the average waiting time in a queuing system is the average service time multiplied by the average number in the queue, as shown in Eq. (23.12):

$$W = \frac{L}{\lambda} \tag{23.12}$$

where W = average time in the system
L = average number in system
λ = average interarrival time

The queuing system can be very general; it does not have to be $M/M/1$ for Little's law to hold. While this sounds obvious, the average of a multiplication product is typically not the same as the product of their individual averages. Queues are a special case.

Equation (23.13) shows that the average number waiting in the queue L_q is the average number in the entire system L minus the average number being served ρ[6]:

$$L_q = \frac{\rho}{1 - \rho} - \rho = \frac{\rho^2}{1 - \rho} \tag{23.13}$$

Little's law gives us the average time in an $M/M/1$ queue in Eq. (23.14):

$$W_q = \frac{L_q}{\lambda} = \frac{\rho^2}{\lambda(1 - \rho)} = \frac{\rho}{\mu(1 - \rho)} \tag{23.14}$$

Notice that this average waiting time is proportional to the average service time $1/\mu$. In a packet system, this is the average packet size divided by the speed of the packet pipe. And ρ is the utilization, the packet traffic divided by the capacity of the pipe leaving the store-and-forward environment of a packet switch.

However, we want more than the average waiting time. We want to know what fraction of the packets have to wait more than a given amount of time. More specifically, we want to engineer the packet pipe to have enough capacity that the probability of a packet waiting too long is small enough to satisfy the QoS requirements.

Given that an arrival has to wait in an $M/M/1$ queue, the waiting time follows an exponential distribution, as shown in Eq. (23.15):

$$\text{Prob(wait} > t) = \text{Prob(wait)} e^{-t\mu(1-\rho)} \tag{23.15}$$

Since the probability that a customer has to wait Prob(wait) is just the server utilization ρ, we get Eq. (23.16) telling us the probability that a packet has to wait longer than a given time t:

$$\text{Prob(wait} > t) = \rho e^{-t\mu(1-\rho)} \tag{23.16}$$

Assuming a packet switch delay requirement t, and assuming that we know the utilization ρ and the average packet duration $1/\mu$, Eq. (23.16) gives us the fraction of packets that will not clear the packet switch within time t.

As we said earlier, this is not the entire story of packet traffic engineering. The distribution of packet sizes is not exponential, and the interarrival time between packets has a bursty characteristic that the $M/M/1$ model ignores. We can deal with some of this by treating the packet bursts as customers in the queue and choosing μ so that $1/\mu$ is the average size of a packet burst. Also, packet routing often involves multiple packet switches, and several of these packet switches can be congested at the same time. To make the statistics problem more difficult, when one packet switch is congested, the others are more likely to be congested as well.

---

[6]If the server is busy ρ of the time and the server has one customer when busy, then the average number being served is ρ.

However, while we are looking at reasons why the $M/M/1$ queue model is not perfect, we might be missing the point. As long as ρ stays low enough, the packets will seldom have to wait very long. Increasing the capacity and speed of the packet pipe reduces both the average burst duration $1/\mu$ and the packet pipe utilization ρ. Equation (23.16) gives us a good indication of just how much extra capacity the pipe needs so that few enough packets have excess delay. And this is what traffic engineering is all about.

23.4   Network Routing for Increased Capacity

Consider the three nodes in a 2 percent blocking network in Fig. 23.6 with 5.0 erlangs of demand between each pair of nodes. According to Table 23.1, 10 channels are needed to meet a standard of 2 percent blocking for 5.0 erlangs. This means 10 channels on each of the three pathways, a total of 30 channels.

Suppose, however, that there is an alternative route, some other path to try. If its direct, *primary* route is not available, then a call is sent on a less direct, *secondary* route on the other two sides of the triangle. The telephony buzzword for a multiple-link path is a *tandem route*.

Let us now engineer each link for a primary blocking rate of 10 percent and do a "back of the envelope" calculation. Ten percent of the traffic is now using two links instead of one, so each link must be engineered for an extra 10 percent, 5.5 erlangs instead of 5.0. Ten percent blocking for 5.5 erlangs requires eight channels on each link.

When a primary link is blocked, each secondary link is blocked 10 percent of the time, so the secondary route has 20 percent blocking. Thus the network blocking rate is 10 percent for the primary route and 20 percent for the secondary alternative, a 2

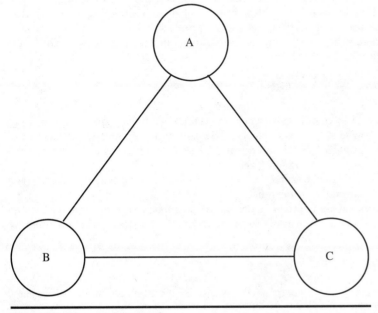

Figure 23.6   Three-node network.

percent blocking rate, $0.10 \times 0.20 = 0.02$. The overflow network achieves the same 2 percent blocking rate with 8 channels on each link instead of 10.

For the sake of this three-node example, we assumed that the distributions on the three paths are Poisson, that the three links have independent blocking behavior, and that we can add blocking probabilities for tandem paths. All three of these are reasonable enough approximations to make the point that overflowing calls to alternative routing can save facilities and reduce blocking.

- The Poisson distribution depends on time-independent call arrivals, that the likelihood of a call being placed is independent of other calls. This makes sense for millions of people independently using their telephones, but a link is going to be blocked for some period of time during which there will be a burst of overflow traffic.

- Since the extra 10 percent is bursty rather than smooth, the likelihood of multiple links being blocked is actually higher than if they block independently.

- The likelihood of a tandem link being blocked is *less* than the sum of their individual blocking probabilities because the sum double counts the case where both links are full. (At least this one works in our favor; the truth is better than our assumptions.)

With more sophisticated calculations taking these effects into account, we still realize an enormous benefit from alternative routing. As a telephone network gets bigger, there are more alternative routes, and the facilities can be used more effectively.[7]

In real telephone networks, the actual calculation of blocking rates with alternative pathways is a huge calculation. Most of the time networks are growing and changing fast enough that the engineer's job is to stay ahead of the new demand for service rather than trying to figure out exact blocking rates. It is still important to estimate the present and future point-to-point demand throughout the network, to enumerate primary and alternative routes, to develop internal routing tables for the network, and to provision enough facilities to meet the demand. However, with the rapid growth that telephone markets have seen for the last century, and with few signs of that growth ending anytime soon, overprovisioning is more often a matter of buying facilities too soon rather than buying too much transport.

23.5   IP Services (DSL)

With the rapid growth of digital subscriber line (DSL) technology, one would have to wonder about the traffic engineering issues. DSL generates a tremendous amount of packet traffic on the public packet data network (PPDN), and there is a significant traffic engineering job to be done managing the Internet.

However, DSL itself is an access technology. Most of us with DSL service were given some new terminal equipment, some junction boxes on the side of the house, and a DSL modem, but the service itself is provided on the existing loop to the telephone company switch. DSL installation is an access provisioning issue but not a traffic engineering issue.

---

[7]We say *effectively* here rather than *efficiently* because having 10 percent of the calls using 2 links may not be considered more *efficient,* but having 8 links instead of 10 for the same level of service is certainly more *effective.*

The advent of DSL did, however, solve a traffic engineering problem. In the late 1990s, the PSTN suddenly and unexpectedly encountered capacity problems. The problems were created not so much by having more calls but by the fact that with so many people browsing the World Wide Web via modem, calls were lasting much longer, on average, than ever before. Traffic engineering assumptions about the average length of calls, which had been accurate for decades, simply were no longer true. Parts of the PSTN, in California and elsewhere, were clogged with local calls. Capacity was being exceeded, blocking rates were too high, and because the calls generally were made on unlimited local service, no revenue was being realized.

Offering DSL mitigates the problem in two ways. The users who most want a lot of time on the Internet switch to DSL at about $50 per month. This increases the local access provider's revenue. The DSL data are routed outside the PSTN by a device at the local exchange office called a *digital subscriber line access multiplexer* (DSLAM), which moves the data off the PSTN onto the PPDN. As a result, PSTN capacity usage was reduced, and average call length dropped back toward its historical norm.

23.6   Reliability and Redundancy

Traffic engineering to meet demand at minimum cost for a growing system is only part of the problem. The other part is making sure the system meets its standard for reliability. Customers blocked because facilities are all in use can retry their calls a few minutes later, but customers can be blocked for hours by equipment failures.

The reliability requirement for a telephone system can be a sensitive issue because it depends on how people are using their telephones. As an example from another area, when toy pocket calculators evolved into electronic pocket planners, rigorous backup systems were added. Not having a pocket calculator handy to compute the tip on a restaurant check is in a different class from having a personal digital assistant (PDA) break down and lose all the information in its memory. Many cellular users rely on their mobile telephones, and fixed wireless local loop is typically a primary household telephone with the same reliability standards we came to expect of our landline telephones.

This becomes an important issue because reliability is expensive. If a base station stops working, then some of its users will be able to use other base stations, perhaps with a lower quality of service, and the others will lose their service entirely. Either alternative means unhappy customers. The equipment reliability modeling concepts for the base station in Sec. 4.8 apply to other equipment.

The network itself can be designed for reliability as well. The same traffic overflow that improves blocking probabilities also can improve reliability, but the facilities have to have redundant paths. Installing extra pathways in a telephone network means incurring extra cost, so the network designer has to decide how much it is *worth* to prevent a total failure for a community of subscribers. It is some consolation for the capacity engineer that there will be some favorable traffic engineering in a network with alternate pathways. Most of the time all the facilities will be working properly, and the system will enjoy favorable blocking probabilities. And when a facility fails, the extra expense incurred for reliability pays off.

Our focus has been on the big job of traffic engineering, calculating the facilities needed to support circuit-switched or packet-switched transport for voice calls on circuits or virtual circuits and data bandwidth on packet networks. However, the principles of traffic engineering also apply to calculating the facilities needed for signaling or other purposes. However, when the capacity calculation is done, the actual implementation of physical transport capacity for Signaling System 7 (SS7) networks is quite different from the implementation of transport capacity for the calls themselves. For transport capacity, we seek to provide capacity to produce an acceptably low blocking rate at low cost. For the SS7 network, we define our expected maximum capacity and then build a network that can handle that load when running at 40 percent capacity. To ensure reliability, SS7 is built with hot spares and redundant transport capacity. If any one link fails, its hot spare takes over and should be running at a maximum of 80 percent capacity even during the time of the infrastructure failure.

23.7 Conclusion

In this chapter we have explored the traffic engineering principles first developed for telephony and then expanded for use in planning hybrid voice-data circuit-switched networks as well as packet pipes and packet networks.

24

Switching Capacity

A mobile switching center (MSC) is a busy place, with processors, switches, speech handlers, and other equipment handling calls coming from dozens or hundreds of base stations.[1] Each component must have the capacity to handle the workload for all its various tasks. In this chapter we look at each logical component and enumerate its tasks, laying out the picture of what needs to be done to estimate capacity for each component of the MSC.

24.1 Main Processor Capacity

The MSC is a telephone switch that has extra responsibilities for the wireless part of the wireless telephone system. The main processor of the MSC is the computer processor that figures out what the switch should be doing. The processor itself has to have enough computing power to do everything the MSC has to do. Therefore, let us go over the functions that the MSC main processor has to perform, and let us keep our eyes open for how much workload each presents. These functions include call processing, providing short message service (SMS), and managing handoffs.

24.1.1 Call processing

The call-processing workload for the MSC processor is on a per-call basis. Once we estimate the number of calls the MSC has to handle, we have a handle on the amount of computation the processor has to do.

Each voice call originating from a cellular telephone served by the MSC uses processing resources. Each call involves analyzing dialed digits and deciding if the subscriber can make the call. For example, if a subscriber's telephone service has been canceled due to reported theft or loss of the cellular telephone or due to nonpayment of bills, the system will not allow most calls, but by law it must accept 911 emergency calls, even from these phones. The exchange of signaling messages with the home location register (HLR) needed for this type of validation uses processor resources. If the

---

[1]The term *MSC* is used ambiguously. Sometimes it refers to the central switch and its processor, and other times it refers to the entire building and all its contents.

MSC has a direct connection to an interexchange carrier for long-distance service, then the digits have to be analyzed for long-distance routing. Finally, once all the decisions are made, the processor has to assign resources to switch the call.

Each voice-call termination uses processor resources for call processing and for locating the user terminal. The MSC has to interpret the dialed digits, communicate with the HLR and base stations to pinpoint the MSC and base station serving the user terminal, and set up the call. Roamers use call-processing resources at their home and serving MSCs with some unavoidable duplication of effort.

Subscriber features use switch resources. Incoming call-waiting calls require the processor to notify the subscriber, to detect flashes, and to do the appropriate switching. Similarly, three-way calls keep the processor busy managing the extra connection, at least as much work as a new mobile origination.

Data calls are not terribly different from voice calls as far as the MSC processor is concerned. It has to establish an Internet Protocol (IP) link for the subscriber and direct the packet-switching components to send their packets into the landline data network. While the packets may need some processing of their own, this should be independent of the MSC main processor.

Voice-call releases use call-processing resources as well. Resources allocated have to be freed, and there is communication with the base station, the HLR, and perhaps the visitor location register (VLR). In capacity planning, we can consider the call release activity as part of the call origination and termination workload because every call has to end sometime.[2]

Accounting and statistics have to be maintained by the MSC for billing and for system analysis. While these functions are not part of call processing, we put them here because the workload they represent for the MSC processor is also proportional to the number of calls made.

24.1.2 Short message service (SMS)

SMS is another work item for the MSC main processor. Each SMS message may represent a workload comparable to a voice call.

An outgoing message comes from the base station and becomes a Signaling System 7 (SS7) signaling message to the SMS message center (SMSC). The MSC processor has to make sure that the message goes to the correct SMSC for the user terminal, so it has to process some subscriber information. And it has to maintain accounting statistics for billing and analysis. Incoming messages are more complicated. They involve locating the user terminal, which means that the MSC processor is managing communication with base stations.

As we will discuss in Sec. 25.4, SMS is a wildcard because short messages use about as much MSC resources as a voice telephone call but almost no radio resources, so the SMS traffic can grow considerably and tie up switches and signaling without exceeding the air-interface capacity.

---

[2]Airport planners can figure that the number of takeoffs and landings will be more or less equal over time. What goes up must come down.

24.1.3 Handoffs

Handoffs are a big item in MSC processing usage. The base station does most of the work, but the MSC still performs the switching and manages the resources for the handoffs.

A great deal of design effort went into keeping the handoff workload at the base station rather than at the MSC. Measurements for handoffs are done at the base stations, and the MSC is little more than a message switch to send handoff measurements from one base station to another. A softer handoff (see Sec. 8.7) is done completely within the base station, so the MSC has no role to play.

A hard handoff involves the MSC in communication with two base stations, a disconnect and reconnect, and any associated statistical bookkeeping. When the hard handoff is from one MSC to another, there is a lot more calculation to determine if it is a handoff forward, a handoff back, or a path minimization, as described in Sec. 15.1.

A soft handoff (see Sec. 8.7) keeps the MSC very busy. The switch designer has to decide where the workload goes, but some part of the MSC has to compare each frame from all the soft-handoff sources on the reverse link and select the best one. This frame-comparison activity is an ongoing effort unlike call processing, where the work ends when the call begins. cdmaOne calls can spend one-third of their time in soft handoff, so this is a major MSC resource issue.

The MSC processor is such an important part of the system that it should have comfortable performance margins. If a link runs out of capacity, then a few subscribers get lousy service for a few seconds. If a packet handler gets backed up, then a few subscribers get slow Internet response. However, if the MSC processor gets behind in its workload, then the entire system is in trouble.

24.2 The Flow of Bits Through the Switch

There is a limit to how many connections a switch can have. Wires in an analog switch have evolved into time division multiplexers in a digital switch, but the principle is the same. Any switch technology can handle some maximum bit throughput, some maximum number of simultaneous voice connections.

All the data pipes from all the base stations come into the MSC. Each base station needs one or more DS-1 or E-1 links, and the sum total of these is a major part of the total facilities load of the MSC.

The voice trunks to the public switched telephone network (PSTN) are also part of the switch-throughput picture. If subscriber demand is primarily voice calls, then these trunks may dominate the MSC facilities. The base station transport carries compressed voice and user data, whereas the PSTN voice trunks are not compressed. Compressed voice is 4000 to 13,000 bits per second, whereas full-rate pulse code modulated (PCM) voice is 64,000 bits per second.

We also have to consider the data going to and from the speech coders that convert the compressed voice to full-rate PCM. A system that is primarily voice traffic is going to have a lot of speech coding and a lot of data going into and out of the speech coding equipment.

We have to count the entire data service load as well. Wireless data service connects a subscriber's user terminal via IP to the landline data network. Subscriber data include packets in and packets out, and those packets use telephone facilities in the MSC.

We also have to count the signaling connections. The pipes from the MSC to the SS7 network should be a tiny fraction of the total data going through the MSC, but it has to be figured into the data flow equation.

There really is not any traffic engineering to be done here. Each component of the transport has to be designed to have enough capacity for adequate performance according to the service provider's quality requirements and the subscriber demand the provider wishes to serve. If each base station has to have 0.02 percent packet blocking, then this determines how many DS-1 or E-1 links are needed. If speech coders have to have 99.9 percent availability, then this determines how many packet and circuit links they need. If all of these data-throughput requirements add up to more than the capacity of the switch hardware, then we have to buy another switch or to lower our expectations.

24.3   Circuit Switching

Circuit-switching loads are pretty simple to estimate because they are directly related to voice-call activity. A call starts, a call stops, a call-waiting call comes in, a three-way call goes out, and a call ends. Each of these involves a fixed amount of switching activity.

Circuit-switched data channels such as Integrated Services Digital Network (ISDN) follow the same rules. The connection is made, the connection is broken, and not much more happens to an ISDN call.[3]

Direct data packet service is no work at all for the circuit-switching component of the MSC. Packets come in and go out through packet handlers, and no circuit switching is required.

The wireless environment adds one enormous monkey wrench to the circuit-switching load—handoffs. If the typical call averages 1.5 handoffs, then this multiplies the switching load by approximately 2.5 compared with a network with no handoffs.[4] Soft handoff may keep other MSC components busy, but it is no worse than hard handoff for switching. The estimate of handoff activity per voice call can be an important calculation for MSC capacity.

It is not clear, however, that handoffs must involve circuit switching at all. Some vendors maintain a pure packet link between the base stations and the MSC. In this case, the packet handler routes the compressed-speech voice packets based on their address. Even a handoff to another MSC may use a virtual circuit over a packet link and therefore may require no circuit switching.

24.4   Packet Switching

Going down the list of MSC functions and resources, there seems to be a pattern. We seem to be putting more and more work into the packet handlers. Therefore, let us examine what these packet handlers might be doing and how their capacity is engineered.

The packet component of the base-station-to-MSC link goes through the packet handler. If the speech is packetized, then the entire base-station-to-MSC link is a packet link and goes through the packet handler. The voice packets go to the speech coding

---

[3]Some telephone companies provide an *always-on* ISDN service that leaves a point-to-point connection permanently in place.

[4]In our old Advanced Mobile Phone Service (AMPS) days, we had about 1.5 handoffs per call, and the 1AESS analog switch that could handle 250,000 landline calls per hour only handled 100,000 cellular calls per hour.

components, whereas the subscriber data packets are routed to the landline data network, and SMS packets are routed to the SS7 network.

The packets still may require processing. The base-station-to-MSC link may be a packet link, but its packets may not be the same format as the speech coder and landline data network require. Packets are often encapsulated with an extra header to get them between the base station and the MSC.

The packet handlers are a queuing system, and we have to know their specific architecture to do the queuing analysis in Sec. 23.3. The model tells us how likely a packet is to wait longer than some specified time t based on the characteristics of the packets and the queue. The time t is how long we are willing to wait for a packet to get through the packet handler.

While this waiting time t may affect a data subscriber's Internet performance, the most important issue in determining t is how quickly packets have to be processed for the speech coding not to have gaps, assuming that the speech data come by packet. We can allow t to be longer only at the expense of lengthening the propagation delay of voice calls in the MSC. If we try to make the speech propagation too short, then we run the risk of losing too many packets to queuing delay.

24.5 Speech Coding

The speech coding process converts the low-bit-rate speech (QCELP for cdmaOne or cmda2000, AMR for W-CDMA) to standard telephone trunk PCM. Each wireless voice call requires one bidirectional speech handler. Having call waiting or a three-way call does not increase the amount of speech handling required.

Thus we have a simple Erlang-B traffic engineering problem: There is a stochastic demand for speech handling, some number of erlangs, and our job is to make sure that there are enough speech handlers to provide an adequately low blocking rate for the speech coding service.

The blocking percentages required, however, are typically much lower than the 1 to 10 percent we engineer on the air interface. We can take advantage of the larger numbers of voice calls for an entire MSC community and engineer the speech handling for very low blocking rates on the order of 0.1 or even 0.01 percent, as shown in Tables 23.5 and 23.6.

The blocking of speech handlers is virtually independent of air-interface blocking, so the two blocking rates are almost exactly added.[5] Because of the large numbers of speech handlers for the entire MSC voice-call community, we should be able to maintain busy-hour occupancy of around 90 percent even at low blocking levels of around 0.01 percent. Thus it seems that there is not much economy to be gained and significant performance to be lost by engineering fewer speech coders and raising their blocking rates.

24.6 Conclusion

Now that we have completed our general look at MSC switching capacity, we will turn our attention to the capacity required for mobile-specific activities, as defined by the ANSI-41 standard.

---

[5]So long as both air-interface and speech coding have low blocking rates, adding the two blocking rates is a reasonable approximation of the combined blocking rate.

25

ANSI-41 Signaling Capacity

This chapter provides the approach to estimating the capacity needed to support activity defined by the ANSI-41 standard for communications between mobile switching centers (MSCs) and between MSCs and other devices. We discuss capacity for inter-MSC handoffs and support for roaming, call processing, the short message service (SMS), and other switching activities. In some cases, we include activities that are outside the ANSI-41 specification. We note these occurrences. This chapter builds the basis in principles for the practical techniques we will offer in Parts 6 and 7.

25.1   Inter-MSC Handoffs

We will call sectors that can have handoffs to another MSC *border sectors*. The base stations serving these sectors can initiate inter-MSC handoff activity. Handoffs from nonborder cells may involve signaling and communication, but all of this is among base stations and their common MSC.

Each handoff will require some of these elements: handoff measurement, handoff forward, handoff back, path minimization, and call release. Let us take a look at how many signaling messages are involved in each case

Handoff measurement requests are messages sent from the serving MSC to candidate MSCs. Each candidate MSC measures the signal from the user terminal. While every candidate MSC is sent a handoff measurement request, only those which detect adequate signal from the user terminal to accept a handoff respond with a measurement message. Once the serving MSC has selected a target MSC for a handoff, it may decide to do a handoff forward, a handoff back, or a path minimization.

A handoff forward is three messages. The serving MSC sends a message to the target MSC, and the target acknowledges the message. Once the MSC-to-MSC channel is established and the handoff is complete, the target MSC sends a third message.

A handoff back is four messages. The serving MSC sends a message to the target MSC, and the message is acknowledged. Once the handoff is complete, the target sends a message to release the facility, and that message is acknowledged.

A path minimization is seven or more messages. The serving MSC sends a message to the anchor MSC that may go through one or more tandem MSCs. The serving MSC communicates directly with the target MSC using two messages to establish a connection for the handoff and then acknowledges the path-minimized handoff to the serving

MSC. Once the handoff is complete, the serving MSC and anchor MSC coordinate the release of their facility with two messages that may go through the same tandem MSCs.

When a handed-off call is released while the serving MSC is not the anchor MSC, the call release is two messages, one to release the facility and one to acknowledge. Each tandem MSC adds two messages to the tally.

25.1.1   Total handoff message count

What does this tell us about inter-MSC handoff signaling? It gives us an idea of how much inter-MSC signaling to expect from handoff traffic. If we can estimate the handoff activity, then we can add up the messages to estimate the signaling load.

The number of handoff measurements is related to the number of calls in border sectors. We can think of each call as having some likelihood of having a handoff measurement involving another MSC or more than one other MSC. By knowing the call volume and the rate of handoff measurements per call minute, we can estimate the number of handoff measurements. We also know that each estimate has one to two messages per neighboring MSC, so this gives us an MSC-to-MSC message load from handoff measurements.

The first inter-MSC handoff of any call is a handoff forward with three inter-MSC messages. Subsequent handoffs are not so predictable, but the difference in message count between a handoff forward and handoff back is only one message, so the message count from subsequent handoffs on a call mostly depends on whether there are many tandem-MSC handoffs and whether there are many path minimizations. And any call that ends with its serving MSC not the same as its anchor MSC generates two messages. The same analysis that estimates the ANSI-41 workload also can be used to estimate the demand on the MSC-to-MSC links dedicated to handoff traffic.

25.2   Roaming

We want to know how much signaling activity is associated with roaming subscribers. Handoffs only involve ANSI-41 signaling when a call moves from one MSC region to another, but roaming involves significant signaling for any subscriber activity in an MSC region other than the user's home MSC region.[1]

When a roaming user terminal is detected in a wireless telephone system, the serving MSC sends the home location register (HLR) a message through its visitor location register (VLR) requesting service qualification and expects a reply message. This message exchange takes place when the visiting user terminal is detected initially. If the original service qualification covers a particular coverage area and the user terminal is detected outside that area, then the serving MSC requests service qualification again. This does not have to be a new MSC region because service qualification may be limited to just a part of the entire MSC service area. Also, a service qualification can have a time limit, and a new service qualification is required when that time limit expires.

---

[1]In this chapter we are using the ANSI-41 definition of roaming, where a user terminal is roaming whenever it is being served by an MSC other than its home MSC. The subscriber may not be notified of or charged for roaming in this technical sense of the term.

The serving MSC sends a location cancellation message through its VLR to the HLR when the user terminal sends a power-down message or when the user terminal fails to register when it is required to do so. This message is acknowledged by the HLR.

When the user terminal registers in a new MSC region while still qualified for service in another MSC region, the HLR sends a registration cancellation message through its VLR to the HLR, which is acknowledged. In code division multiple access (CDMA) systems, there also can be an exchange of information between the old and new MSC regions directly, two more messages between their VLRs.

There are also messages and acknowledgments when the HLR revokes the service qualification it has sent or when the subscriber activates a feature. These are far rarer events, we hope, than the normal, successful registration of a user terminal.

Registration uses two messages from serving MSC to serving VLR to HLR and back to turn on the user terminal, four messages to move to a new MSC region, two more messages to move within the MSC region if subregions of service have been defined, and two more messages for a timeout. Finally, there are two more messages when the user terminal is turned off or otherwise loses connection with the serving MSC.

When two MSC regions have border cells as described in Sec. 15.2.3, user terminal registration can be detected by a base station served by an MSC other than the serving MSC. When the border MSC receives a registration message, it sends a message requesting service qualification to the HLR, so the HLR receives two messages. The HLR resolves the two MSCs based on their reported radio signal conditions and sends a message to the serving MSC, either an acknowledgment of its request or a cancellation. Once the serving MSC replies, the HLR sends the border MSC a message. Thus a border-cell detection generates two messages between the border MSC and the HLR and two messages between the serving MSC and the HLR.

Roamer call delivery is a very important and fairly common cellular event. And roamer call delivery can have many outcomes. The user terminal may be available to answer the call, the user terminal may be out of the service area, or the user terminal may be busy on another call. The subscriber may or may not answer the call. The subscriber may have the bad timing to initiate a call after the assignment of a temporary local directory number (TLDN) but before the call setup to the TLDN arrives through the telephone network.

In all these cases, however, there is a location request message from the originating MSC to the HLR and some kind of response back from the HLR. Also, unless the user terminal has powered down and sent a message to the base station saying so, there is a message from the HLR to the serving MSC requesting a TLDN and an acknowledgment, favorable or otherwise.

This means that we should estimate the volume of roamer terminated calls for each MSC. We should do this for each MSC in the role of home MSC and in the role of serving MSC. The traffic patterns in a major metropolitan area may have significant asymmetries between the two because commuters buy service in their suburban neighborhoods and use their cell phones in their city-center workplaces.

25.3   Call Processing

Virtually every time a subscriber does something with his or her user terminal, the serving MSC contacts the HLR, and the HLR sends some response. Whenever a call is made or received, the user terminal registers, and the HLR verifies its validity. When the subscriber picks up voice messages, the serving MSC signals the HLR, which, in turn, signals the voice-message system. When the subscriber makes a three-way call, the HLR confirms that the three-way calling feature is available to that user terminal.

If there is a normal case for terminating calls, then it is home-system call delivery, where the originating MSC is the serving MSC and there is a single exchange of verification between the MSC and the HLR. If the termination is a roamer call delivery and the user terminal has an active location, then there is another pair of messages between the HLR and the serving MSC through the serving VLR, as we discussed in the preceding section. If the call delivery involves a border cell, then there is an extra set of five or six messages between the border MSC and the serving MSC to coordinate the page response, to establish the call setup, and to alert the subscriber of an incoming call.

Call forwarding can be unconditional (CFU) or conditional. Conditional call forwarding can be busy (CFB), no answer (CFNA), or default (CFD). CFD forwards calls when either the line is busy or there is no answer. Calls forwarded to the subscriber come through the public switched telephone network (PSTN) and involve no extra signaling. Calls to the user terminal being forwarded someplace else, per subscriber instruction, are another story because they can generate ANSI-41 signaling messages if the user terminal has an active location. When the originating MSC gets the message from the serving MSC that the forwarding condition is met, the originating MSC requests and receives the call-forwarding directory number.

Call waiting only generates ANSI-41 signal messages when the call has been handed off to another MSC. The anchor MSC sends the serving MSC a call-waiting message, which is acknowledged. When the subscriber presses the SEND key, the serving MSC detects it, signals the flash to the anchor MSC, and awaits acknowledgment of the flash.

There are various kinds of subscriber-screened calls. These include calling number identification presentation (CNIP), calling name presentation (CNAP), selective call acceptance (SCA), and password call acceptance (PCA). The HLR sends a message to the Signaling System 7 (SS7) service control point (SCP), where it retrieves data on the subscriber's call screening. Then the HLR contacts the serving MSC with the screening information so that the user terminal can be paged and the subscriber can be queried. Do not disturb (DND) is just what it sounds like; terminating calls are denied as if the user terminal were turned off.

Mobile access hunting (MAH) goes through a group of mobile telephone numbers just like a hunt group of landline telephones, where the local office rings the first nonbusy line. In the wireless case, each user terminal in the sequence is checked for an active location and paged if it has one. Flexible alerting (FA) allows several wireless telephones *all* to alert when a call is placed to a designated telephone number. These two features place more calls to user terminals and, therefore, may increase the ANSI-41 message load. We suspect that use of these features will be low enough not to be a major issue.

What is the bottom line for call processing? Calls generate work for the wireless telephone system, obviously, and they also generate ANSI-41 message for the SS7 signaling network.

25.4   Short Message Service (SMS)

SMS is the signaling wildcard. SMS makes demands of signaling capacity without increasing the demand for voice channels on the network and while demanding very little capacity on the air interface. Generally, calls start and stop in predictable fashion, so we can predict the number of ANSI-41 messages that are related to calls per minute of air-interface use. SMS adds to the demand for ANSI-41 at a much higher ratio of signals per capacity demanded than voice calls do.

We expect that virtually every user terminal is a short message entity (SME) and has the capability of sending and receiving short messages. When a user terminal registers, the response message from the HLR to the serving MSC contains SMS service qualification information, but these are not extra messages because of SMS. The message loads are between a user terminal's serving MSC and the subscriber's SMS message center (SMSC).

When mobile subscriber A sends a short message to mobile subscriber B, that message is sent from serving MSC A to the SMSC associated with subscriber A. The SMSC then contacts MSC B and the HLR for B to route the call to the serving MSC for subscriber B. All these information requests and the transmission of the message itself have acknowledgments, and all use ANSI-41 capacity. Thus a short message from A to B generates two messages between serving MSC A and the SMSC and two messages between the SMSC and serving MSC B.

When subscriber A sends a message to an external SME (ESME), that messages is sent from serving MSC A through the SMSC to the outside world. There are two messages between MSC A and the SMSC and two messages between the SMSC and the public Transmission Control Protocol/Internet Protocol (TCP/IP) network.

When an ESME sends a short message to subscriber B, that message is sent to the SMSC for the ESME, which queries HLR B so that it can send the message to serving MSC B. There are two messages between the external SMSC and HLR B and two messages between the external SMSC and serving MSC B.[2]

The reason we call SMS a signaling wildcard is that it uses a high proportion of signaling to message size and also uses no voice or data channel resources on the cellular network. SMS messages are handled entirely on the signaling channel. If voice calls send 10 signals per minute of talk, then we might have 1000 bytes of ANSI-41 data for a million bytes of speech. High-speed data users are going to tip the proportion even further by sending tens of megabytes for each 1000 ANSI-41 bytes. However, SMS uses at least four ANSI-41 messages over the SS7 network for each short message of 160 bytes or less, and the short message itself is on separate switches and data links, not on voice channels. This means that a surge in SMS use will make major demands on the SS7 network without using any voice circuits (real or virtual) or data circuits. Also, SMS messages will increase the demand on switching capacity while requiring very little air-interface capacity, unlike voice calls and data services, which require a much higher proportion of air-interface capacity than signaling capacity.

25.5   Other Signaling Activity

There are many other parts to ANSI-41. There is user terminal authentication to prevent wireless telephone fraud. There is over-the-air service provisioning (OTASP) so that subscribers can sign up for service and change their feature lists over the radio link. System technicians use ANSI-41 messages to change the status of facilities for operations, administration, and maintenance (OA&M).

---

[2]One SMS message-routing model we analyzed indicated that, for some calls, there might be two SMSCs involved. If these methods are used, then two or four more routing messages would be added to the overhead for each SMS message.

These are all important ANSI-41 functions, but these are not heavy hitters in terms of message quantity. The authentication process is part of user terminal registration, we suspect that few subscribers make frequent service changes, and facilities should not need to go out of service for testing very often.

25.6   Adding It All Up

Once we have estimated all the significant contributions to the ANSI-41 signaling load, we have to add them up. Each handoff, roaming, call-processing, and SMS link has an estimated busy-hour message prediction based on our estimates and forecasts of activity.

Consider the small example in Fig. 25.1. The network has 13 logical functions for three MSC regions in 4 physical locations:

- The upper-left location has MSC *A*, VLR *A*, and all three HLRs.

- The upper-right location has MSC *B*, VLR *B*, and all three MCs.

- The lower-left location has MSC *C* and VLR *C*.

- The lower-right location has the SCP database.

Thus there are 6 possible paths connecting pairs of physical locations.

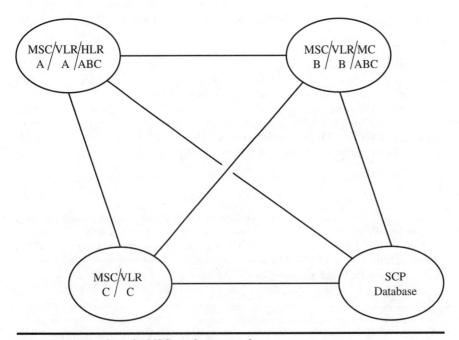

Figure 25.1   A simple multi-MSC wireless network.

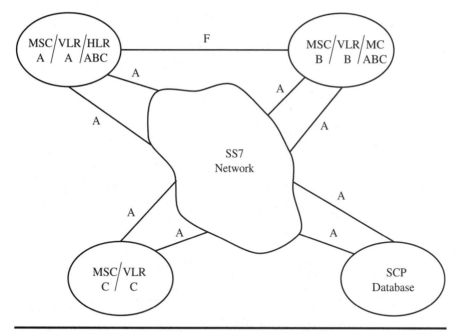

Figure 25.2   Starting to engineer a simple multi-MSC network.

Suppose that we have done the forecasting and calculating so that we have specific ANSI-41 load projections for handoff, roaming, call processing, and SMS. For each pair of logical functions, we have a load estimate, so we can add up the total expected ANSI-41 signaling traffic for each of the six point-to-point paths. From there, we can engineer a network.

In Fig. 25.2 we have the first steps toward capacity engineering for the ANSI-41 network. We have dual A-links to an SS7 network cloud from all four physical locations plus a dedicated fully associated (F) link connecting the location with all three HLRs to the location with all three MCs.

The next step is to analyze the SS7 network to make sure that it can handle the traffic this network is going to put on it. An operating requirement of SS7 is to have no link more than 40 percent occupied. In this way, the packet network has a comfortable margin even if one of the dual transport links fails.

25.7   Conclusion

Having laid out the principles of calculating capacity for ANSI-41 signaling, let us now turn our attention to calculating capacity for the cellular backhaul network and other components.

Capacity Calculations for Cellular Networks

In Chap. 25 we discussed signaling capacity for the cellular network. In this chapter we discuss the remaining capacity issues: transport capacity and capacity for various equipment components of the cellular network. Some of these design issues involve traffic engineering, whereas others require different tools for capacity estimation or calculation.

26.1 Backhaul Design Principles

The transport that connects the base stations to the mobile switching center (MSC) is called *backhaul*. In the analog days, the backhaul network consisted of dedicated analog trunks, one for each voice radio. The base stations were already traffic engineered with enough radios for the quantity and quality of service required, and each radio required a trunk, so there was no traffic engineering issue for backhaul. The backhaul in code division multiple access (CDMA) networks is not only digital but also digital packet data pipes. Voice links become virtual circuits between the base station and the MSC. This is a more efficient transport technology, and it gives us the opportunity to do some traffic engineering in backhaul.

Backhaul from hundreds of base stations to a single MSC is a complex transport network. We could satisfy the transport demand with a separate pipe from each base station to the MSC and have no interaction among these separate pipes. However, the cost saving of designing an efficient backhaul network can be quite large. There are economies of scale in transport: Doubling the capacity of a pipe between two points often costs far less than twice as much. A well-designed packet network for backhaul can take advantage of these economies.

In this section we explain the mathematical model for network optimization. We expect the minimum-cost backhaul network to be a *fan-out network* where each base station has a single path to the MSC and these paths join like tributaries in a river to the mouth at the MSC. This kind of network is shown in Fig. 26.1. We will explain the modeling assumptions and the mathematical reasons why we expect a fan-out network to provide backhaul at the lowest cost.

26.1.1 Backhaul traffic engineering

For each carrier on each cell sector, there is a two-way stream of data between the base station and the MSC. These data represent the voice calls, dedicated data sessions, shared packet data channels, short message service (SMS) messages, and all the paging and access overhead. The base station collects all these for all its antenna faces and CDMA carriers and forms a single stream of packet data. Each manufacturer determines what protocols are used and how efficiently these data are packetized, but the result is a two-way packet stream between each base station and its serving MSC.

It is important that this backhaul link not be a bottleneck. Delays in signaling information can degrade system performance even though all the radio and switch equipment is up to the job. It is also important that circuit-switched voice links have minimum delay to provide adequate sound quality. For time-sensitive service such as real-time voice or video, delayed packets are no better than lost packets.

The requirement of adequate backhaul capacity still leaves room for some traffic engineering on the packet links. Unlike conventional reuse systems (FDMA or TDMA), a

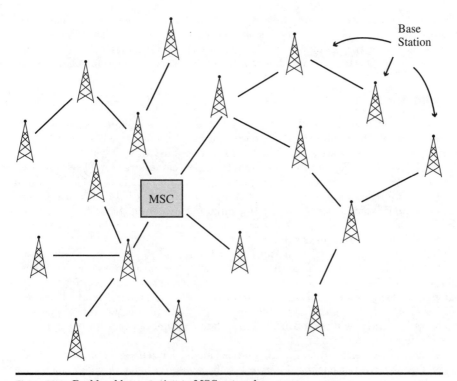

Figure 26.1 Backhaul base station-to-MSC network.

CDMA cell cannot use all the capacity of all the sectors.[1] Even at maximum CDMA power levels, full capacity, there is a statistical distribution of backhaul data load, and we can do some packet data traffic engineering to manage the cost of backhaul transport. We want to make sure that the backhaul link is engineered for very low packet delay. For real-time voice-link packets, we want to see almost real-time service because packets delayed more than a few milliseconds are treated as blocked. This blocking rate should be kept down in the range of 0.02 percent, far lower than the blocking levels in deploying base stations and radio links. Once the packet pipe sizes are determined for each base station, we have the more complex problem of building a transport network to get all these packets to the common MSC.

26.1.2 Graph theory

We model a transport network as a set of *nodes* or *vertices*[2] connected by *arcs* or *edges*. Such a collection of nodes and arcs is called a *graph*.[3] The study of these graphs is called *graph theory,* and this is part of a bigger field called *combinatorial mathematics,* or *combinatorics* for short.

Graph theory is a field of mathematics rich in useful optimization insights. Modeling a practical problem as a mathematical graph allows us to use these theoretical insights to find practical solutions. The variables in these problems can number in the millions, and combinatorial optimization is a powerful tool in computing good solutions far beyond a paper-and-pencil analysis. Often the graph representation of a problem is direct (for example, warehouses are nodes and trucking routes are arcs), and other times the relationship can be more subtle.[4] In our case, the base stations and the MSC are nodes, and the transport links are the arcs.

The easiest solution to visualize is a *star graph,* where all the arcs share a common node, as shown in Fig. 26.2. This is the solution where the wireless service provider calls the telephone company and leases separate transport links from the MSC to each base station. Even if we never plan to build a star-graph solution for a backhaul network, the star-graph is a useful starting point for a computational algorithm to work its way to a less expensive network solution.

The graph that contains every possible arc between every pair of nodes is the *complete graph* shown in Fig. 26.3. If there are n nodes, then there are $\frac{n(n+1)}{2}$ arcs in a complete, fully-connected, graph. We would not deploy a complete transport network, but we would use the complete graph to enumerate all the route possibilities. If nodes are

---

[1]In conventional reuse, a cell can have every radio channel of every sector in use because sectors do not share the same frequencies. In CDMA, however, the sectors of a cell share a carrier frequency and interfere with each other. A well-designed CDMA cell will have enough capacity on each antenna face so that its sector has low blocking, and the total capacity of all the antenna faces should be more than the cell as a whole can handle.

[2]The plural of *vertex* is *vertices* just as the plurals of *index* and *simplex* are *indices* and *simplices.* The grammatical waters are muddied, however, because the plural of the noun *complex* is *complexes.*

[3]This is not to be confused with the (x, y) plot of points, lines, and curves also called a *graph.*

[4]We recall one optimization model where the airline flights were nodes and the connections at the airports were arcs. This is contrary to the usual pictures of the nodes as "dots" and the arcs as "lines" between them, but this representation allowed us to use graph-theory methods to reduce maintenance expenses.

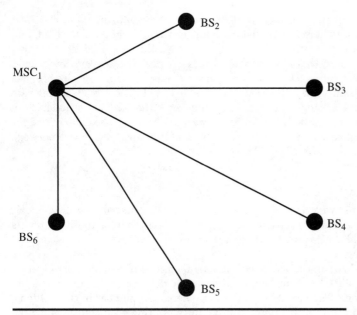

Figure 26.2 Star graph showing separate backhaul links.

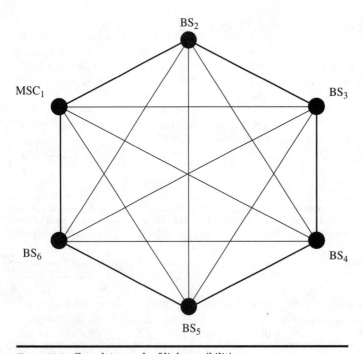

Figure 26.3 Complete graph of link possibilities.

airports and arcs are airline flight routes, then the complete graph represents a full schedule, and the star graph represents the hub-and-spoke pattern.

The third type of graph that interests us is a *tree graph,* a graph that has exactly one way to get from any node to any other node along the arcs. In a tree graph, we can pick any node as a root, and the arcs form a fan-out from that root. If the backhaul network is a tree graph, then we can make the MSC node the root of the tree so that the arcs of the graph are the backhaul links in use. Another feature of a tree graph is that it has no *cycles,* no sequences of distinct arcs that end up where they start.

26.1.3 Network flow optimization

There is a class of graph-theory optimization problems called *network flow problems.* These problems have a specific structure that makes them particularly attractive for computer-solving algorithms. Some network flow formulations are apparent, such as maximization the flow of something through some sort of network. Other model formulations are more subtle, where the nodes, arcs, and flows are mathematically symbolic representations of physical entities that look nothing like dots, lines, and stuff flowing along those lines.

The reason decision analysts make a strong effort to put business problems into the form of a network flow problem is that these problems are computationally easier to solve than similar-sized problems without the network flow structure. Also, the solutions that come from network flow optimization have one extremely nice property: If the constraints and objective are linear functions, which we call a *linear network*

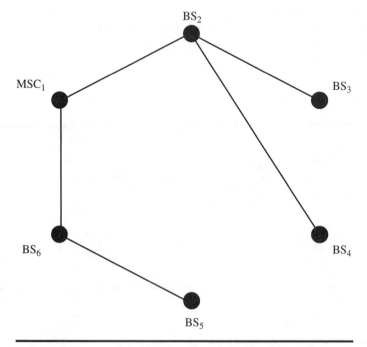

Figure 26.4 Tree graph represents a fan-out backhaul network.

flow problem, and the constraints have integer (whole number) coefficients, then the optimum calculated by a method known as the *network simplex method* will have integer values. There may be equally valid optimal solutions, not integer solutions, that other optimization algorithms will find as their solutions. However, the network simplex method runs fast and finds whole-number optimal solutions to whole-number problems.[5]

The network flow formulation is a graph with arcs that can carry flow from one node to another. The arc from node i to node j carries flow x_{ij}. Each arc can have a minimum flow a_{ij}, zero or some specified lower bound, and may have a maximum flow b_{ij} or may be unbounded. Each arc also has a cost $c_{ij}(x_{ij})$ that is a function of the flow x_{ij}. The object of the network flow problem is to get a specific flow from a source node S to a target node T for the minimum cost. In the case of backhaul, the lower bounds are all zero because there are no required links that we have to use. There are no upper bounds because we can use as much transport as we are willing to pay for.

When we solve the linear network flow problem, the arcs that have flow x_{ij} not at their minimum a_{ij} or maximum b_{ij} values will form a tree graph or a collection of disconnected tree graphs. This is another desirable feature of network flow formulations, since a tree graph has some nice mathematical structure.

26.1.4 Fan-out backhaul solutions

We can shape the backhaul problem into a network flow optimization by adding some phantom arcs to the graph. Figure 26.5 illustrates the network flow graph that we can use to show that backhaul is a network flow optimization. Since the backhaul problem

---

[5]As in Chap. 22, we beg the forgiveness of operations research professionals for such a simplistic presentation of a vast area of study.

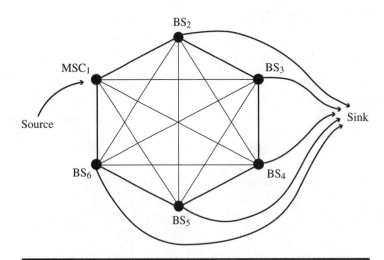

Figure 26.5   Network flow formulation of backhaul optimization.

has no upper bounds ($b_{ij} = \infty$ for all i and j) and the minimum values are all zero ($a_{ij} = 0$ for all i and j), the backhaul problem looks like Eq. (26.1):

$$\text{Minimize} \quad \sum_{i=1}^{n} \sum_{j=1}^{n} c_{ij}(x_{ij})$$

$$\text{subject to} \quad \sum_{i=1}^{n} x_{ij} - \sum_{k=1}^{n} x_{jk} = r_j \quad \text{for all } j \tag{26.1}$$

$$x_{ij} \geq 0 \quad \text{for all } i \text{ and } j$$

The value of r_j is the required transport at node j. The constraints are linear equations, so the entire system would be linear if the cost functions were linear, if $c_{ij}(x_{ij}) = c_{ij}x_{ij}$. The linear constraints and linear cost functions would be enough to ensure that there is an optimal solution that is a fan-out network like the solution shown in Fig. 26.4. The tree graph determines the flow pattern, and the link configuration determines how much capacity each link requires, the flow in each arc of the graph. Equations (26.2) show the flows for the example in Fig. 26.4.

$$x_{12} = r_2 + r_3 + r_4$$

$$x_{23} = r_3$$

$$x_{24} = r_4 \tag{26.2}$$

$$x_{16} = r_5 + r_6$$

$$x_{56} = r_5$$

Base stations at nodes 2 and 6 serve as cross-connects for the others.

There are two assumptions we can make about the backhaul costs that are enough to ensure that a fan-out network is still optimal. First, the link costs are separable, and the total network cost is the sum of the individual costs of all the links. When the costs of some links depend on the flow of other links, the solutions can become more complicated. Each facility, each link, each flow determines its own cost independent of the other links. If a transport vendor decided to offer a package deal, a single price for a bundle of transport, then the costs would not be separable. However, let us assume that even a package deal still has separate costs for transport within the package.

Second, the link costs have economy of scale, decreasing unit costs with increasing flow. We call a function $f(x)$ *concave* when it increases less as x increases. The technical definition of concave is that a function $f(x)$ is concave if and only if $f[\lambda x + (1 - \lambda)y] \geq \lambda f(x) + (1 - \lambda)f(y)$, where λ is between 0 and 1.[6] The effect of concave cost functions is to concentrate usage on a smaller number of links.[7] The concave assumption is not

---

[6]A linear function is considered concave for this discussion.

[7]When the cost functions are *convex*, the solutions tend to "spread the pain." We might expect such solutions in environmental impact problems where all the pollution in one place is more than twice as bad as half the pollution in each of two places.

quite true in telephone transport because 1.5 DS-1 links costs more than 1.5 times the cost of a single DS-1. Probably the cheapest way to get 1.5 DS-1 links is to pay for 2. However, for larger quantities of transport, we think the concave assumption is reasonable.

The optimal solution to a network flow formulation with linear constraints and a separable concave cost function is a fan-out network.

26.1.5 Other backhaul issues

The expenses of transport in the telephone network are more complex than a simple cost function. Some links offer the option of a microwave link in which the wireless service provider has the option of paying once for equipment rather than leasing facilities with a monthly tariff. The economics of owning versus leasing are beyond the scope of the network flow model itself; rather, these tradeoffs are part of the inputs to the optimization and must be weighed carefully in using backhaul optimization models.

Nothing in this discussion presumes that the costs depend on distance. The telephone network has high-capacity, low-cost transport backbones, so the cost of transport is almost entirely dependent on the distance from the base station to the backbone network. This means that the minimum-cost backhaul solution may look very strange on a map. The important thing is that the backhaul solution takes maximum advantage of the available transport options.

The traffic engineering of the backhaul links is typically based on load during the busy period. If the busy period is the same for all the base stations, then the network flow optimization gives a good solution for minimum-cost backhaul. If there are significant time-of-day variations, there may be further economies to be realized by traffic engineering the combined links in the fan-out based on their traffic loads throughout the day.

Our final backhaul issue has little to do with cost. If some of the transport links are not sufficiently reliable, then we may want to engineer the backhaul network to have some alternate routing. A backhaul solution with higher reliability would no longer be a tree graph, no longer be a fan-out, no longer be absolutely minimum cost. However, it might be the optimal solution for providing the best level of service to wireless subscribers.

26.2 Cellular System Components

In addition to transport and switches, our cellular network has a number of other components. In planning and increasing network capacity, we must specify capacity requirements for these items as well, including the home location register (HLR), the visitor location register (VLR), short message service (SMS) equipment, and voice-messaging equipment.

26.2.1 HLR and VLR capacity

The HLR has a record in its database for every subscriber based in its service area. The HLR database has to have enough storage capacity for all its customers. This is not necessarily a function of wireless system traffic load. If the service provider signs up a lot of people who got cellular service mostly for emergency use, then there will be a lot of subscribers who use their cell phones very little. The HLR must have records in its database for all these subscribers. The service provider has ample warning when this database is reaching its capacity as new subscribers are signed up.

The VLR has database records for every subscriber roaming in its territory. This is a dynamic allocation that depends on the momentary concentration of nonresident cell phones in the area. Because of the dynamic nature of VLR traffic, this database can overload without warning.

The registration workload for the HLR and VLR depends on the number of subscribers turning on their cell phones or moving into a new area. The HLR has somewhat more work to do when its subscribers are roaming than when they are home. Every call start and stop generates HLR work, as well as VLR work for roamers. When the call starts, the HLR has to record that the user terminal is in use so that subsequent calls can be handled correctly, and its available status has to be restored at the end of the call. The registration and call-processing loads are reflected by comparable workloads in the MSC and the base station, so these calculations already should be reflected in the allocation of computational resources in the MSC.

The HLR has a significant load in service validation as well. Every time a subscriber sends or receives a short message via SMS or uses some vertical service, the HLR is consulted to see if the subscriber can use the service. This process has to be fast so that the subscriber does not have to wait too long for service.

When an external event occurs, a traffic accident, a ball game, or even the landing of a Boeing 747 with 400 passengers, a single VLR sees the full workload of registering all the subscribers and setting up all the calls. The total HLR workload may be just as great, but it is distributed over many HLRs in many places. Thus the HLR traffic is comparatively smooth compared with the bursty workload at the VLR where an event occurs.

26.2.2 SMS message center (SMSC) capacity

In addition to the signaling load, the SMS message center (SMSC) has to process the messages and store them. The workload of the SMSC depends on the activity of SMS subscribers, leaving messages and picking them up. The rate of message arrivals depends on subscribers using SMS, and SMS usage is highly dependent on events. People leave messages when they feel like it, but they also leave messages when something happens. An unexpected local event, a traffic jam or a late flight, stimulates a wave of SMS messages arriving at the SMSC all begging to be processed. The SMSC workload picking messages up is far smoother because most subscribers pick up their messages when they turn on their cell phones.

The incoming message workload is a queue. Messages come into the SMSC, and they are processed and sometimes they can back up. The SMSC must have enough queue storage for the workload it cannot process immediately. The storage issue is important because SMS subscribers do not like to lose their messages. The time issue is less important because SMS messages usually are not time-critical, at least not for delays of only a few minutes. SMS does have a function for prioritization of message handling. A message marked high priority is moved to the top of the queue when it arrives at the SMSC and is handled before low-priority messages, even those which arrived sooner. This reduces latency for messages, such as dispatch messages for utility repairmen, which may be relatively urgent.

The messages themselves have to be stored until they are picked up. These messages are small, no more than 160 bytes each, but the SMSC has to keep not only the message but the status of the message. The SMSC has to store messages from the time they are left until the time they are picked up.

There are significant daily and weekly cycles in SMS usage, as well as bursty usage for unexpected local events. The SMSC queue for incoming messages and the storage for messages waiting for pickup have to leave enough room in the SMSC for the work. The rate of lost messages in the SMSC is the combined loss of messages lost in the incoming message queue and messages lost because there is not enough storage for them.

26.2.3   Voice-mail capacity

A voice-mail system is an entity separate from the wireless telephone system, not a base station, not an MSC. The voice messages have to get from the system to the voice-mail system when they are being left and have to get to their destinations when they are fetched. There must be enough full-duplex voice channels for both functions: leaving voice-mail messages and picking up voice-mail messages. This is a traffic engineering problem, an exercise in making sure that there are enough channels between the wireless telephone system and the voice-mail system that the blocking rate for voice mail stays low.

Voice mail is not a queuing system. The voice-mail system has to have the capability to support all the channels between it and the wireless system. This includes the processing capacity to support the channels and several internal capacity and speed design issues as well. The system must have sufficient storage space for all messages. The amount of storage needed may be hard to calculate as any given message may remain on the system anywhere from just a few minutes up to the maximum allowed, often 15 days. Each voice-mail message is stored in the voice-mail system as it is being left, and it remains in the system until it is deleted by the subscriber or exceeds the allowed maximum storage time. The system also must have a sufficiently rapid retrieval mechanism for messages when users call to pick them up. We view these as equipment design issues.

Voice mail is an attribute of an incoming call, so local events such as traffic jams and late flights do not generate the same local workload as they do for SMS. This is so because the recipients of all the calls being made are typically spread over many voice-mail service areas. However, there are still day-of-week and time-of-day cycles in voice-mail activity. One important issue is that business messages left on Friday afternoon typically remain in the system until Monday morning.

Voice-mail system will have a burst of usage when other components fail in the wireless telephone system. A base station outage or a facility failure will generate a surge of voice-mail activity for those subscribers.

26.3   Cellular Network Issues

Dedicated transport between MSCs can save significant cost over using the public switched telephone network (PSTN) for those calls. Two areas where dedicated facilities can save money are roamer-terminated calls and mobile-to-mobile calls from one MSC to another. We can calculate the traffic loads for these functions and do some traffic engineering to determine how much dedicated transport is required. However, the PSTN is always there as an overflow medium, so there is no degradation of service to the subscriber when the dedicated facilities are blocked. These dedicated facilities have virtually unlimited overflow capability in the PSTN. Therefore, the decision is a cost issue rather than a capacity issue.

Let us look at the distribution of usage for the dedicated connection we are considering. For any level of transport capacity, we are interested in the fraction of time, minutes per day, where the demand exceeds that capacity. Figure 26.6 shows a graph of the function we are describing.[8] There is some traffic nearly 18 hours per day (1050 minutes) and enough to fill six DS-1 links 6 hours per day (360 minutes). This means that one-quarter of the time the traffic is six DS-1 links *or more*. If the daily cost of a dedicated DS-1 link is the same as 360 minutes of PSTN usage, then the breakeven point for this connection is 360 minutes per day. Once we know the breakeven usage level, we can look at the usage graph and determine the number of DS-1 or E-1 links that minimize our transport costs for this connection, six DS-1 links in this example.

Estimating the usage function is the key to making a good decision about dedicated facilities. The simplest model is a normal bell-shaped statistical distribution based on total daily usage. We could take the total minutes per day as the mean μ and estimate the standard deviation σ based on the customer mix. Then we can compute the probability of exceeding capacity x from Eq. (26.3):

$$P(t > x) = \frac{1}{2}\,\text{erfc}\!\left(\frac{x - \mu}{\sigma}\right) \tag{26.3}$$

---

[8]Those who have worked in the field of revenue management (which used to be called *yield management*) will recognize this graph. It is the same as the expected marginal seat revenue (EMSR) graph the airlines use to determine how many seats they should sell at a discount and still have enough for the expected last-minute, high-price passengers.

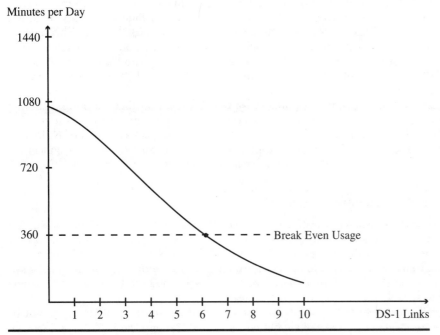

Figure 26.6 Facilities capacity usage.

The erfc function is the statistical cumulative distribution of the positive half of the normal distribution, the integral in Eq. (26.4):

$$\text{erfc}(x) = \frac{2}{\sqrt{\pi}} \int_{t=x}^{\infty} e^{-t^2/2} \, dt \tag{26.4}$$

This simplification ignores usage patterns by day of week and time of day. The average demand for the connection may be two DS-1 links, but most of those calls are happening during the 8-hour business day. It is the time-of-day concentration of traffic that makes the sixth dedicated DS-1 economical.

26.3.1   Capacity for roamer termination

Termination of roamer calls is a busy time for a wireless telephone system. The ANSI-41 signaling system sends messages, the HLR and VLR update their data records, the base station sends paging messages, and the user terminal responds so that the call can start. The signaling system, HLR, VLR, and MSC have all been designed to handle the workload of a roamer termination. The final stage is establishing a connection between the two MSCs for the call.

It is the MSC-to-MSC connection where we may derive some economy from dedicated transport links. The wireless system engineer should identify MSC-to-MSC links with enough traffic to justify dedicated transport. While there may be large wireless subscriber communities in Bogota and Dhaka, few Bogota subscribers are going to be in Dhaka when somebody calls them. It does not make sense to maintain a dedicated transport link from Columbia to Bangladesh for the tiny traffic it would support. However, a large wireless system in Chicago may have three or four MSCs serving it, and subscribers do not have to travel far to be outside their own MSC regions. This may not be roamer service on the subscribers' cellular telephone bills, but it is roamer service so far as the system is concerned; the call comes into one MSC and is served by another. These MSC-to-MSC connections may have dozens of dedicated DS-1 links to avoid paying PSTN connection rates.

We have to remember, in designing these connections, that high blocking rates may be perfectly acceptable. The issue is pure economics rather than customer satisfaction, and the minimum-cost solution may have 10 percent of the calls on the PSTN. Designing an MSC-to-MSC link with 10 percent blocking may make a telephone engineer uncomfortable, but it will be the right decision in many of these connections.

26.3.2   Adjusting capacity for mobile-to-mobile calls

The case of calls from one wireless user terminal to another is not, by itself, a wireless system function. A subscriber calls another subscriber, and the system has no obligation to handle the call in a special way. The call can be sent to the PSTN for normal handling. This is in contrast to roamer terminations, where the wireless telephone system has to handle the call in a special way. Mobile-to-mobile calls represent an opportunity to save a lot of money in PSTN charges.

If a call is between two subscribers at the same MSC, an intra-MSC call, then it really seems obvious that the call should not be routed through the PSTN. The MSC has loopback trunks so that the calls can go out and come back in without going anywhere and without incurring PSTN charges.[9] The MSC could be designed with special software for intra-MSC calls, but it is easier to use loopback trunks and have intra-MSC calls routed preferentially to those trunks. Since loopback trunks are very cheap, the economic decision is to install enough of them to handle the highest intra-MSC call rates. Estimating the growth in demand for intra-MSC mobile-to-mobile capacity requires evaluation of subscribers' calling patterns. Special service offerings, such as mobile-to-mobile minutes, also may affect demand.

Calls between subscribers served by different MSCs, inter-MSC calls, are a different story. If there is enough traffic to justify dedicated trunks between the two MSCs, then the MSC uses the PSTN to route calls only when the dedicated links are blocked.

As wireless systems grow, we expect that an increasing fraction of the calls would be inter-MSC calls. A system with one MSC grows to the point where it needs two MSCs, and calls from one side of town to the other that used to be intra-MSC calls are now inter-MSC calls. MSC regions multiply in number and shrink in area, so a greater fraction of same-system calls cross one of the lines.

26.4 Conclusion

This chapter provided an explanation of why fan-out networks are the optimal backhaul solution. We also explored the traffic engineering and capacity issues for all the major components of the cellular network. The approach presented here lays the groundwork for the practical calculation of network and network component capacity in Parts 6 and 7.

---

[9]This is not the same loopback trunk concept that we saw in Sec. 15.4. Those loopbacks were for testing rather than call handling.

Conventional Reuse Principles

To distinguish code division multiple access (CDMA) from other modulation schemes, we call frequency division multiple access (FDMA) and time division multiple access (TDMA) schemes *conventional reuse*.[1] In these conventional reuse wireless systems, the modulation schemes are designed to be highly efficient of radio spectrum, designed to squeeze the most calls into a given frequency bandwidth. Once we have our efficient radio carriers, we make our conventional reuse *systems* efficient by managing interference as well as we can.[2] The effect of interference on cellular call quality is determined by the modulation scheme and by the radio environment, including radio path propagation and Rayleigh fading.

27.1 The AMPS Channel (FDMA)

As we described in Chap. 6, the Advanced Mobile Phone System (AMPS) channel is a frequency division duplex (FDD) frequency modulation (FM) analog 30-kHz channel. It carries an audio signal from 300 to 3300 Hz.

The AMPS system maintains the link with a supervisory audio tone (SAT) that can have one of three frequencies, 5970, 6000, or 6030 Hz, all out of the audio bandwidth. The AMPS receiver mutes the audio when it receives the wrong SAT or when no SAT is detected. When an interfering channel dominates the desired channel and that interfering channel has the same SAT frequency, the subscriber hears another call. We call this highly undesirable interference *crosstalk*.

---

[1]We admit that this is our own term rather than an industry standard. We use it because it emphasizes the unique nature of spread spectrum.

[2]The reader may wonder why a book on CDMA is spending so much time on conventional reuse. There are three good reasons for this. First, CDMA systems are a technology evolution from FDMA and TDMA, and understanding the historical insights helps us understand the technology. Second, CDMA is strange and often counterintuitive, and we feel that it is often easier to understand what it is by understanding what it is *not*. And third, there are 700 million conventional reuse wireless telephones in use today, and a CDMA engineer should be well versed in their technology as well.

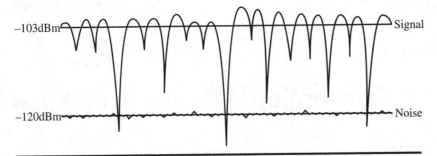

Figure 27.1 A 17-dB S/N ratio with Rayleigh fading.

At any given moment, the AMPS channel provides excellent audio quality as long as the signal exceeds the noise plus interference by the capture ratio described in Sec. 1.10.1. Since the capture ratio is only about 1 dB, the determining factor in voice quality is the effect of Rayleigh fading.

Consider a noise-limited setting, a 30-kHz narrowband channel with a moving transmitter or receiver. It does not matter which end is moving because the changing radio path generates a moving Rayleigh distribution in either case. According to the statistics of a Rayleigh distribution, the fraction of the time the signal is below the noise level is the average ratio of noise to signal. When the average signal-to-noise (S/N) ratio is 17 dB, the picture looks like Fig. 27.1. The Rayleigh fading signal dips below the noise level about 2 percent of the time. Listening tests at Bell Telephone Laboratories found that three-quarters of the listeners were pretty well satisfied with such a call, as we discussed in Sec. 18.3.

In an interference-limited setting, the same 17-dB ratio of signal to interference (S/I) may be somewhat worse because the interference is also varying in its own independent Rayleigh fading pattern, as shown in Fig. 27.2. This adds more variation to the

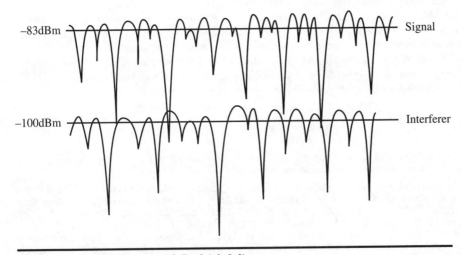

Figure 27.2 A 17-dB S/I ratio with Rayleigh fading.

instant-to-instant S/I ratio, but it still sounds good enough in listening tests to be an acceptable call standard for AMPS.

There is also interference from adjacent channels in AMPS. The 30-kHz frequency spacing is large enough that the adjacent-channel interference effect is 25 dB less than the same interference would be for cochannel. Thus an adjacent-channel interferer 8 dB stronger than the signal would have the same destructive effect as a cochannel interferer 17 dB below the signal. While cochannel interference causes dropouts from momentary loss of SAT, adjacent-channel interference adds noise to the actual voice, sometimes a whistling sound. While it is hard to make an exactly equivalent comparison, 25 dB seems to be an accepted value. This means that we can use adjacent channels on adjacent cells without too much worry, but the signal range of same-cell calls is too great to allow adjacent-channel use on the same cell, even on different sectors of the same cell.

The 25-MHz of AMPS spectrum allows 12.5 MHz forward and 12.5 MHz reverse for a total of 416 channels of 30 kHz each. An AMPS system reserves 21 channels for paging and access, leaving 395 two-way voice channels.

27.2   The GSM Channel (TDMA)

As we described in Sec. 7.3, the Global System for Mobility (GSM) channel is one of eight time slots in a 200-kHz carrier. Without being too cavalier, we can think of the GSM channel as 25 kHz of digital radio carrying 13 kbps of regular pulse excitation–long-term prediction (RPE-LTP) speech coding along with digital supervision and control.

This channel is tougher than AMPS for two reasons. First, its broader carrier bandwidth of 200 kHz makes it more resistant to Rayleigh fading. Second, it is sending only 13 kbps in 25 kHz of bandwidth after error correction, a fairly light data load. In comparison, a quadraphase phase shift keying (QPSK) modulated channel sends 2 bits per second per hertz of bandwidth, so GSM's speech channel of half a bit per hertz leaves a lot of room to manage interference. The result is a channel that can withstand as S/I ratio of 12 dB with the same audio quality as AMPS gets with an S/I ratio of 17 dB.

The 25-MHz of GSM spectrum allows 12.5 MHz forward and 12.5 MHz reverse for a total of 62 carriers of 200 kHz each, 480 channels. Each cell sector in GSM system reserves its own paging and access channels, so the voice channel count is about 450 channels.

27.3   Signal-to-Interference Performance

While Rayleigh fading is changing the signal and interference instant to instant, we can think of the S/I ratio as a signal value over some area.[3] As long as this area has a favorable S/I ratio, we expect a wireless telephone call to sound good.

The radio path between an antenna face at a base station and a wireless user terminal depends on many factors. The distance between the base station and the user terminal is the most obvious determinant of radio path gain, and we expect it to follow the fourth-power-of-distance rule from Sec. 1.7. Other factors affecting the radio path are terrain, buildings, and vegetation. The result of all these factors is a radio

---

[3]At frequencies of 900 MHz and higher, wavelengths of 33 cm (13 in) or less, an area can be very small, not even a meter across, and still have global statistical properties.

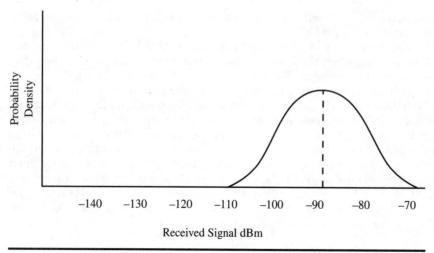

Figure 27.3 Signal distribution for a cell sector.

path *distribution* between a base station's antenna face and a community of wireless user terminals. This distribution is typically log-normal, as described in Sec. 1.6, and it looks like a normal bell-shaped distribution on the decibel scale.

Figure 27.3 shows a typical received signal distribution for AMPS or GSM calls. This is the signal S part of the S/I ratio, the *received* radio power, the product of the transmit power and the radio path gain, or their sum in decibels. In this case, the average received signal power is −90 dBm, 1 pW, and the standard deviation is about 10 dB.

Figure 27.4 shows the same signal distribution with an interferer I superimposed on the picture. In this case, the average received interferer power is −120 dBm, 1 fW, and the standard deviation is about 10 dB.

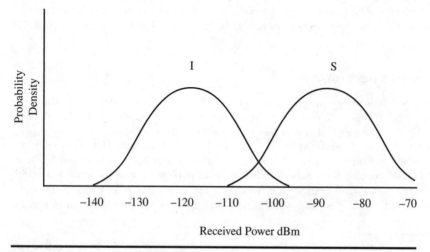

Figure 27.4 Signal and interference for a cell sector.

While it is tempting to suggest that the overlap of the two curves represents the lost call rate, the actual mathematical relationship is more complicated. If $s(x)$ is the signal density function and $i(x)$ is the interference density function, then the distribution of S/I ratio $q(x)$ is the *convolution* of $s(x)$ and $i(x)$, as shown in Eq. (27.1)[4]:

$$q(x) = \int_{t=-\infty}^{\infty} s(t)i(x + t)\, dt \qquad (27.1)$$

Even with the complexity of mathematical convolution, the bottom line is that the closer these curves are, the more signal and interference overlap, the more likely an AMPS or GSM system is to have poor sound quality and to lose calls.

The statistical convolution of two normal distributions is another normal distribution whose standard deviation σ_q is the root-mean-square of the first two standard deviations σ_s and σ_i, as shown in Eq. (27.2):

$$\sigma_q = \sqrt{\sigma_s^2 + \sigma_i^2} \qquad (27.2)$$

Thus we have two log-normal distributions with 10-dB standard deviations whose means differ by 30 dB. The S/I ratio, the difference in decibels, has a standard deviation of 14 dB ($\sqrt{10^2 + 10^2}$).

In evaluating wireless telephone service, we tend to use the *tenth percentile* of the S/I distribution, the level where 90 percent of the calls are better and 10 percent of the calls are worse. In the case of a normal distribution, the tenth percentile is 1.3 standard deviations worse than the mean, so this particular example has a tenth percentile S/I ratio of 12 dB, quite acceptable performance for GSM.[5]

It is an important observation that the signal and interfering radio paths are all between base stations and wireless telephones. The radio paths between base stations and the radio paths between user terminals are not a factor in the AMPS or GSM FDD radio environments.[6]

The signal distribution depends on the cell radius but generally is independent of the antenna radiation pattern or how the channel is reused. The interference distributions, on the other hand, are highly dependent on which cells reuse the channels and the antenna radiation patterns used at the base stations. In the next section we will examine how channel reuse is related to S/I distribution.

---

[4]There are two noteworthy comments here for those interested in the convolution integral. First, the units of $s(x)$, $i(x)$, and $q(x)$ are all logarithmic decibels. A calculus student could perform a change of variables from decibels to power using Eq. (1.3), but there is no reason to do so. Second, this is a convolution of the *difference* rather than the sum, so we have $i(x + t)$ instead of $i(x - t)$ as we would have in a normal convolution.

[5]This is a mean S/I ratio of 30 dB minus 1.3 times a standard deviation of 14 dB.

[6]Time division duplex (TDD) systems can have significant other interference components because both forward and reverse directions share the same frequencies at different times. If one cell is using its forward link at the same time and on the same frequency as another cell is using its reverse link, then there will be an interfering radio path from one base station to another.

27.4   Regular Channel Reuse

Our goal in conventional reuse cellular system design is to have a high S/I ratio at the tenth percentile. Having less variation in the signal distribution is an important part of this, so we want our cells to be as round as possible so that the maximum distance from the center is not too much more than the average distance from the center. Three regular shapes can cover (or *tessellate*) a geographic area, triangles, squares, and hexagons, and the shape that gives the least distance variation is the hexagon (Fig. 27.5). The hexagonal shapes are similar to the cells of a honeycomb.[7]

We divide the entire set of available channels into N *channel sets*. An AMPS $N = 7$ system has seven channel sets of 56 channels each for a total of 395 voice channels.[8] A GSM $N = 4$ system has four channel sets of 112 channels each distributed over 14 carriers of 8 time slots each.

The geometry of hexagonal tessellation allows N channel sets in regular reuse as long as $N = i^2 + j^2 + ij$ for some integers i and j. This means that we can have $N = 1$, $N = 3$, $N = 4$, $N = 7$, and $N = 9$, but not $N = 2$, $N = 5$, $N = 6$, or $N = 8$.[9] Figure 27.6

---

[7]Those of us who have worked in cellular for a while are really good at drawing hexagonal lattice grids. If the wireless telephone business goes away, then we will have a generation of engineers who can tile public washroom floors.

[8]Since $7 \times 56 = 392$, we can afford to have four channel sets of 56 channels and three channel sets of 57 channels.

[9]$N = 1$ means using the same radio channels at every base station, as CDMA does.

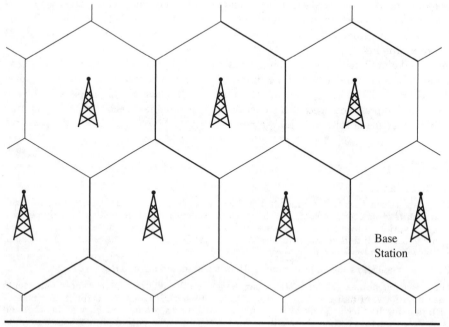

Figure 27.5   Cellular system layout.

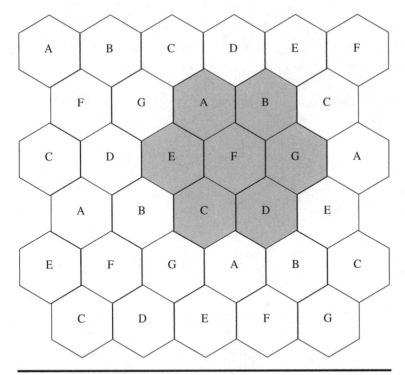

Figure 27.6 $N = 7$ reuse pattern.

shows the geometric layout of seven channel sets in an $N = 7$ system typical for AMPS. Notice that the trip from one cell to a cochannel cell is two steps out and one step to the right. This corresponds to $i = 2$ and $j = 1$ in the $N = i^2 + j^2 + ij$ formula. We have shaded a cluster of seven cells that represents all seven channel sets. Figure 27.7 shows the layout of four channel sets in an $N = 4$ system typical for GSM, where $i = 2$ and $j = 0$. We have similarly shaded a cluster of four cells representing the four channel sets.

In this geometric format we can describe the channel reuse by comparing the distance between cochannel cells and the cell radius. The distance-to-radius (D/R) ratio, shown in Fig. 27.8, works out to $\sqrt{3N}$, about 4.6 for $N = 7$ and about 3.5 for $N = 4$.

The radio performance effect of changing the reuse ratio N is to change the interference radio path and, therefore, to change the average S/I ratio, as shown in Fig. 27.9. Changing N has little effect on the *shape* of the interference distribution. Of course, as increasing N improves S/I performance, it decreases capacity by allowing fewer channels per cell. The improvement from AMPS $N = 7$ to GSM $N = 4$ is a factor of 1.75 (7 to 4) in channel capacity.[10]

---

[10]There may be more benefit when we take into account that the larger number of channels in the $N = 4$ configuration allows a higher occupancy at the same blocking rate, as discussed in Sec. 23.1.

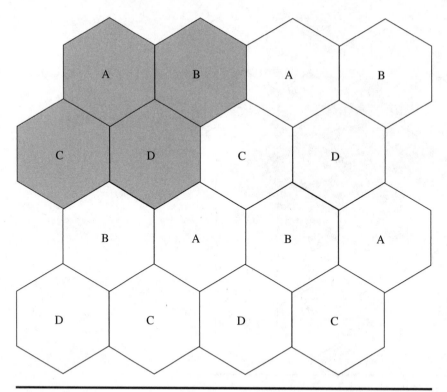

Figure 27.7 $N = 4$ reuse pattern.

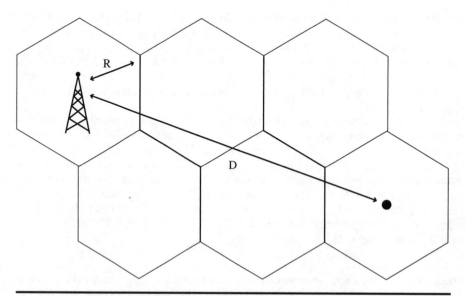

Figure 27.8 Distance-to-radius (D/R) ratio.

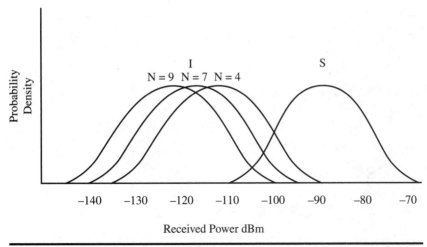

Figure 27.9 Signal and interference for different values of N.

27.5 Sectors

Using the fourth-power-of-distance rule from Sec. 1.7, we get an S/I ratio of $9N^2$. For $N = 4$, this works out to 21.5 dB.[11] While this sounds adequate for GSM, we would point out that this is just one interferer, not all six pictured in Fig. 27.7 and that terrain, buildings, and vegetation add more variability to the simple distance calculation. When we take all these factors into account, we get a tenth percentile of about 7 dB for $N = 4$, not at all adequate for GSM.

While we could get an extra 5 dB of S/I by going from $N = 4$ to $N = 7$, we could get the same 5 dB by dividing the cell coverage area into three sectors and using antennas with a 120-degree radiation pattern. Figure 27.10 shows how each sector only covers two of the six first-ring interferers. The factor of 3 works both ways. In the reverse direction, the receive antenna only picks up interfering user terminals in two of the six interfering cells. In the forward direction, the transmitting antenna faces point away from the user terminal community for four of the six interferers.

The 15 carriers and 120 channels are divided into three sectors of 5 carriers and 40 channels each served on a separate antenna face. We can perform the same three-way division on an $N = 7$ AMPS channel set to break the 57 channels into three sectors of 19 channels each. Figure 27.11 shows how the three-sector antenna plan has the same benefit for $N = 7$ as it does for $N = 4$.

Breaking cells into sectors is not a new cellular concept. At Bell Telephone Laboratories, the original High Capacity Mobile Telephone Service (HCMTS) papers explained three-sector directive cells. They actually defined the cell as the region surrounded by three base stations; these were called *corner-excited cells*. As HCMTS evolved into AMPS, engineers realized that *center-excited cells* were an easier way to view the cell sector plan.

---

[11]Since $9 \times 4^2 = 144$, this is 21.5 dB.

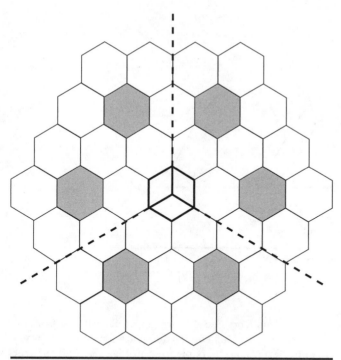

Figure 27.10 The three-sector plan removes two-thirds of interferers.

27.6 Power Control

Suppose that we could narrow the received signal distribution, the S curve in Fig. 27.3. Even if the average signal remains unchanged, making the distribution less variable ought to keep more calls above the interference level. We can narrow the signal distribution by having the base station measure its received signal to control the transmit power of both the base station and the mobile telephone. The transmit power is boosted when the radio path gain is low and cut back down again when the radio path gain is high. This has the effect of reducing the variability of the received signal.

The coverage of a power-controlled system is as good as that of a system running at full power. When the user terminal is in a fringe area with poor radio coverage, the base station detects a weak signal, boosts its own transmit power, and directs the user terminal to do the same. Thus power control gives the benefit of S/I management without sacrificing service in poor radio environments with low path gain.

If all calls are power-controlled, then the average S/I ratio should stay the same. However, it is not the average that we care about but the tenth percentile, the calls at the bottom of the performance range. The reduced variability of the received signal should keep more calls out of the low S/I ratio zone. And it does, to a point.

The power control that is reducing the signal variability accomplishes this by changing the transmit power. This change reduces variability because it is deliberately correlated, negatively, with the radio path gain between the base station and the user

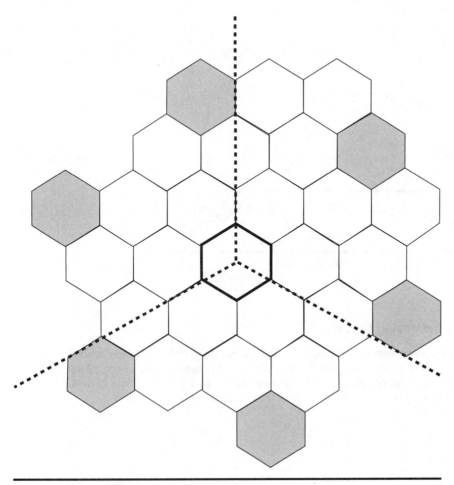

Figure 27.11 The three-sector plan for $N = 7$.

terminal. However, the interfering radio path typically is not correlated with the signal radio path, so power control *increases* the variability of the interference and changes the distributions from Fig. 27.4 to Fig. 27.12.

The initial effect of power control is positive. The reduction in signal variation outweighs the increase in interference variation and the tenth percentile S/I ratio decreases. As we increase the transmit power changes to narrow the received signal distribution further, the increase in interference variability does more harm than good, as shown in Fig. 27.13. Under ideal measurement conditions, the optimal power control removes half the decibel variation, so we should change both forward and reverse transmit power by 4 dB to compensate for an 8-dB change of radio signal path. Since the base station measurement equipment is not perfect, the ratio should be reduced. In a simulation study for AMPS, one of us (Rosenberg) found that the best solution was to change transmit power by 4 dB for a 12-dB change in the radio signal path. We expect the same optimal performance for GSM.

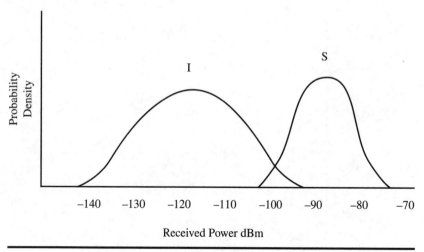

Figure 27.12 Signal and interference with some power control.

27.7   Growing for Increasing Demand

The cellular system is designed to grow. A successful wireless telephone service attracts more subscriber demand, and the tools for serving that increased demand are built into the conventional reuse cellular technology. Serving more subscribers almost always involves adding new base stations, and that can be costly.

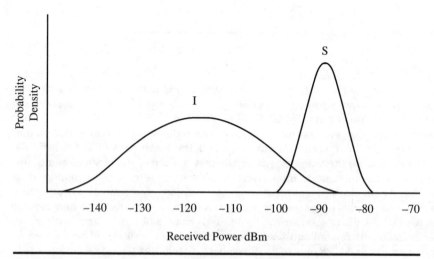

Figure 27.13 Signal and interference with more power control.

27.7.1 Smaller cell grid

If we have an $N = 7$ system serving its customers and we wake up one morning to find four users for every one we had before, then we can make a grid four time denser with cells half the size and one-quarter the area. Figure 27.14 shows the four-to-one cell splitting that has been used in many AMPS systems. The larger $N = 7$ grid on the left continues onto the smaller right-side grid. The shading of the cells on the right shows the original, larger $N = 7$ grid. Please note that the smaller grid is a full $N = 7$ grid with no obvious indication that it was split from a larger grid.

Similarly, should we wake up to find that our demand has tripled instead of quadrupled, we can make a grid three times denser with cells $\sqrt{3}/3$ the size and one-third the area. Figure 27.15 shows the three-to-one cell splitting that also has been used in many AMPS and GSM systems. Again, the larger $N = 7$ grid on the left continues onto the smaller right-side grid with the shading to show the original, larger grid. Again, the smaller grid is a full $N = 7$ grid on its own. While an $N = 7$ layout can split

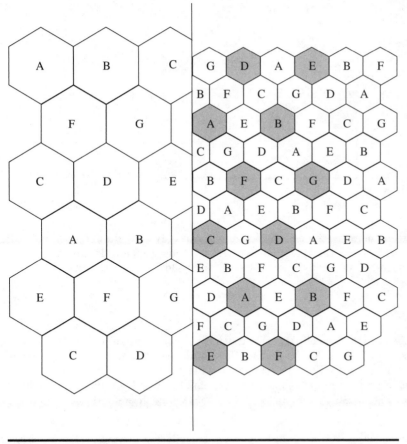

Figure 27.14 Four-to-one cell splitting.

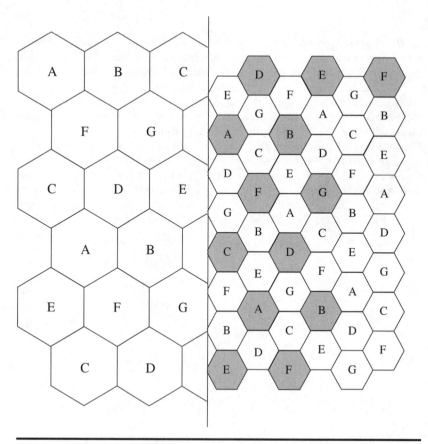

Figure 27.15 Three-to-one cell splitting.

either four-to-one or three-to-one, an $N = 4$ grid can only split three-to-one and still have the old cells retain their channel sets. At the time we were designing the first AMPS systems, retuning a radio transmitter was a big deal, and retaining channel sets was a serious issue.

27.7.2 Cell splitting

In real life, the growth in a cellular system usually is concentrated in some area or in a small collection of areas. As the demand for cellular service grows in the center of town near the river,[12] we split a few cells to meet the demand, and the new cells follow the new, smaller grid. More growth begets more small-grid cells to relieve the cellular congestion. In Fig. 27.16 we see an $N = 7$ system using three-for-one cell splitting

---

[12]There is almost always a river in the busy part of town. Denver, Colorado, is the only exception that comes to mind.

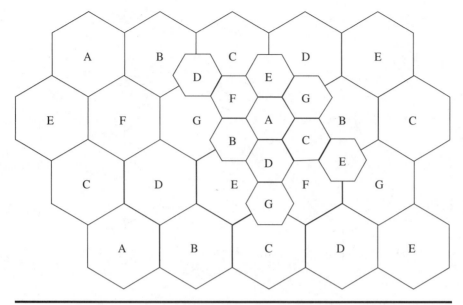

Figure 27.16   A growing $N = 7$ system with three-for-one cell splitting.

to serve its growing demand. The smaller grid is starting to become a cellular grid in itself with a full seven-cell cluster and three channel sets (D, E, and G) already showing reuse. The A cell in the center of the small-cell cluster is a large cell that became a small cell as it was relieved of excess traffic six times with six new small-cell neighbors. In its metamorphosis from large to small, this cell retained its A channel set.

Why did we add these new cells? The simplest reason is a measured increase in cellular congestion resulting in too many ineffective call attempts on that cell. A slightly more foresightful reason would be recognition that the traffic is increasing and that there *will* be congestion if we do not add relief to the existing large cells. A third reason might be to relieve other cells in a congested channel set. For example, consider the small cell D in the upper left of the small cell grid. It *may* be that its large-cell neighbors (B, C, and G) are getting too busy, but the addition of cells may be to free up some of the C channels used by the large C cell so that the small C cell nearby can have more C channels.

While it may be that channel set C cells can only use channel set C channels, a growing system does not guarantee that *any* C channel can be used in any C cell. Failure to meet the D/R requirement will cause unacceptable interference within the same channel when cell sizes mix. Thus each channel set in a growing cellular system becomes a mathematical exercise to figure out how to assign its individual channels.

As the demand continues and as the center of town (near the river) continues to be growth area, we continue to add cells. Figure 27.17 shows a system with three cell sizes because the small cells required even smaller cells to relieve their traffic loads. Since the small cells have been referred to as *secondary cells,* the even-smaller cells are sometimes called *tertiary cells.*

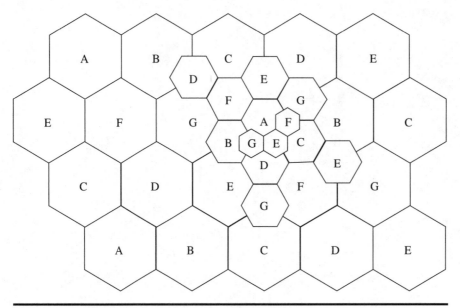

Figure 27.17   A growing $N = 7$ system with three cell sizes.

27.7.3 Overlaid cells

One technique used to increase the efficiency of cell splitting is to use *overlaid cells,* cells that share the same base station at the center but cover a smaller area. We do this by adding server groups and logical faces (see Sec. 5.2.3) to existing base stations. The large- and small-radius channels use the same antenna face at the same base station. The extra server group is a software change only because the extra cell radius is created by changing access and handoff parameters for some of the radio channels.

In Fig. 27.18 we have the same growth cells, but now we have overlaid all the cells that are too close by D/R rules. The large C cell on the right side of Fig. 27.18 cannot share channels with the small C cell a few steps to the left side of it. We have added an overlaid small C cell that *can* share those channels so that the capacity of the large C cell is no longer compromised by adding the small C cell.

Another way to use overlaid cells is *reuse partitioning.* Suppose that a reuse ratio of $N = 7$ is required to serve an entire wireless subscriber community. Then the user terminals nearer the base stations have radio path gain high enough that $N = 4$ reuse will serve them adequately. Because they are closer to the cell center, they can enjoy $N = 7$ D/R with $N = 4$ reuse.[13] In Fig. 27.19 we overlay an $N = 4$ system on top of an $N = 7$ system.

Reuse partitioning has two advantages. First, it increases capacity so that it can delay initial cell splitting of an existing large-cell grid and make a small-cell grid more efficient. Second, it gives the cellular carrier flexibility in tuning the reuse ratio of the

---

[13]While we refer to *close* and *far* as geometric terms and draw the figures that way, the actual discrimination in a reuse-partitioned system, or any overlaid cell with more than one server group, is based on radio signal path.

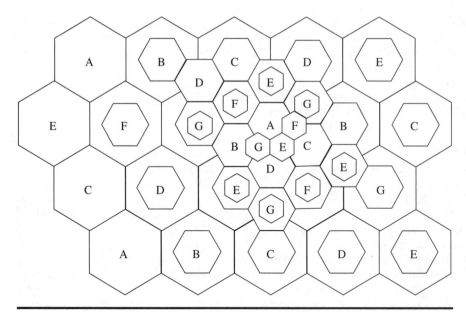

Figure 27.18 A growing $N = 7$ system with overlaid cells.

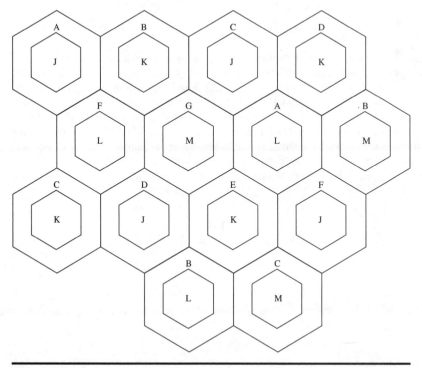

Figure 27.19 Reuse-partitioning $N = 7$ and $N = 4$.

system. There is no $N = 5$ regular reuse ratio, but a reuse-partitioned system with $N = 7$ and $N = 4$ can have more or less of the traffic on each value of N for an equivalent of $N = 5$ or $N = 6$ or even $N = 5.5$. The outer thresholds for the $N = 4$ part of the system determine the equivalent value of N.

The bad side of reuse partitioning is that it is a major engineering challenge to keep track of all the channel sets. The system shown in the figure, for example, has $N = 7$ and $N = 4$ for a total of 11 channel sets.

27.7.4 Economics of growth

Mixing cell sizes extracts a price in capacity. Maintaining D/R ratios with different D and R values partitions channel sets into large-cell channels and small-cell channels. Adding a third tier of even smaller cells only aggravates the channel-assignment problem. An aggressive AMPS or GSM cellular engineer can maintain about two-thirds of the channel set at each cell in a growing system. Thus our $N = 4$ system averages 80 channels per cell instead of 125. Using traffic engineering analysis in Chap. 23, such cells might average 70 percent occupancy, 40 erlangs during the busy hour.

The cellular provider gasps at this point, spending as much as a million U.S. dollars per cell site for just 40 average simultaneous busy-hour calls. The economics of a growing system only get worse as the system keeps growing, and no providers are anxious to tell their stockholders that they have decided to stop growing their cellular systems just so the company can make some money. The early days of cellular ran rivers of red ink as companies found themselves investing and investing in a future that stayed years away.

27.8 Power Balancing

As a conventional reuse wireless telephone system grows by splitting cells, the channel sets become complicated. Often a single radio channel is used for cells of different sizes. Consider a channel used in both cells in Fig. 27.20. The large cell on the left has an S/I distribution based on D/R_1, whereas the small cell on the right has an more favorable S/I distribution based on D/R_2. We can make the two distributions more equitable by reducing the power level for that channel on the smaller cell. In doing this, we spend some of the surplus performance on the small cell to improve the performance of the larger cell.

The technique of setting average power levels for cochannel cells is called *power balancing*, and it is useful even when cells are the same size. When one cell is atop a hill

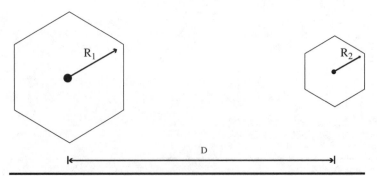

Figure 27.20 Different cell sizes using the same channel.

and another is in a valley, we can have an asymmetric interference environment that power balancing can compensate for.

When dozens of cells are using the same channel, the power-balancing problem can be a sophisticated exercise in mathematical optimization. Each channel should be power balanced in a wireless telephone system as different channels are used in different cells. If one channel is used in 20 cells and another is used in the same 20 and one more cell, then their power-balancing calculations are different. Usually the usage patterns of several channels are the same, so there may be 10 or 20 separate power-balancing calculations to be done for one system.

This power balancing is not the same as the power control described in Sec. 27.6. Power balancing is a one-time setting of system parameters, whereas power control is dynamic, moment-by-moment adjustment of power levels depending on radio conditions at the time. Power balancing sets the *average* power levels to improve the *average* S/I ratios among cochannel cells, whereas the power-control algorithm improves the low-end, tenth percentile performance once that average level is set.

27.9   Traffic Engineering

Wireless telephony operates at the extreme end of traffic engineering in two important ways. First, our trunk groups are small, and they stay small even when our systems grow large. Instead of having small trunk groups growing large, we have more small trunk groups to more base stations. Second, we have a lot of options for traffic overflow. A call blocked on its primary route often has a secondary option, sometimes one less efficient of system resources.

An $N = 4$, three-sector GSM system with uniform coverage of uniform-density traffic has 40 channels serving each sector. As the system grows and cells split, the channel sets are shared by large and small cells. Thus the sectors in a growing system are served by 16, 24, or 32 channels. As a third cell size is introduced, more of the cell sectors have smaller number of channels.

The overflow possibilities, however, are encouraging. As shown in Fig. 27.21, we can send blocked traffic to other servers in three ways.

- We can overflow from one antenna face to another at the same base station, sector-to-sector overflow.

- We can overflow from a small server group to a larger server group, overlaid-cell overflow.

- We can overflow from one cell to a neighbor cell.

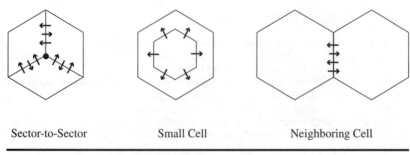

Sector-to-Sector   Small Cell   Neighboring Cell

Figure 27.21   Three kinds of traffic overflow.

These overflow techniques can be done actively at call setup by sending the user terminal to another server. Or they can be done passively in the handoff process by not handing off a call to a blocked server.

As an example of the kind of traffic engineering we do, let us examine a system where half the user terminal traffic can be served on two cells. The user terminal tries to access one cell, which is blocked, and is directed to retry another cell. Suppose that we are engineering this system for 2 percent blocking.

If we engineer all the cells for 4 percent blocking, then half the traffic will be blocked 4 percent of the time, whereas the other half will have two chances and will see a blocking rate of 0.16 percent, 4 percent of 4 percent. If half the traffic sees 4 percent blocking and the other half sees 0.16 percent, then the average blocking for the subscriber community is 2 percent.[14] Thus our example system with 50 percent overflow possibility can be engineered for roughly double the blocking rate.

We are relying on the fact that this is a *mobile* telephone system, that the subscribers are in varying places. A landline system with 4 percent blocking in half its service community would not be able to claim 2 percent blocking, but our subscribers move around and are just as likely, we expect, to be in the good areas as the in the bad areas, so they see the average blocking rate.

The actual traffic engineering for a system with split cells, overlaid cells, and neighbor cells is much more complicated than our example. Often there are many adequate solutions for traffic engineering a system, and the best choice depends on where facilities are available or where they can be obtained cheaply.

27.10   Cookie-Cutter Hexagons

It is fascinating to explore all these geometric algorithms. During the development of cellular telephone technology in general and of AMPS in particular, many sophisticated studies were done using these cookie-cutter hexagons to represent the cells of a cellular telephone system. These models allowed cellular pioneers to formulate the traffic and reuse models still in use throughout the world today.

The reality is quite different. It is important to keep in mind that all these cells are not the lovely, symmetric, equal-sided hexagons we draw in our geometric diagrams. Many analytic and simulation studies also were done using radio propagation models described in Chap. 47.[15] Figure 27.22 shows the contrast between the geometric hexagons we draw and the cell shapes that we actually see in the field. The boundaries between cells are fuzzy, actually, because of radio measurement error in call setup and handoff.

We have discussed the variation in radio propagation that would deform the even boundaries between cells, but there are other effects as well. The information an AMPS

---

[14]We are taking two liberties here, rounding 2.08 percent to 2 and assuming that blocking on the two attempts is independent in a statistical sense. While neither of these assumptions is absolutely true, they are close enough for our example to make sense.

[15]At this point in the discussion we will point out that one of us (Rosenberg) has been on both sides of the cookie-cutter hexagon fence. He worked with some amazing coworkers to develop software growth models that took regular geometric reuse as far as anybody has ever taken it (so far as we know). And he worked with many of the same people to develop simulation models and proscriptive tools that used radio propagation and measurement models to evaluate the true impact of cellular design proposals.

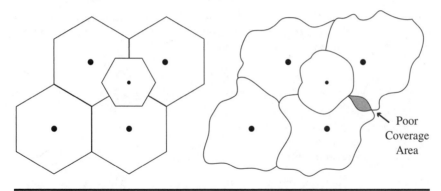

Figure 27.22 Cookie-cutter hexagons and more realistic cell shapes.

or GSM cellular system has to locate a mobile telephone is the radio signal measurements it makes. Initial call setup and cell-to-cell handoffs are all made based on radio power measurements and internally set threshold levels. Creating the threshold levels and handoff neighbor lists to operate an AMPS or GSM system with hundreds of cells is a daunting job.

27.11 Capacity Rules of Thumb

There are some simple formulas for capacity of conventional reuse cellular systems. Like the rules-of-thumb formulas for how to invest in stocks or how many calories you use playing tennis, these are not rigorously derived results. However, rules of thumb help us understand the *relationships* among various components of cellular design.

27.11.1 Channel bandwidth

The simplest rule is that narrow channels are more efficient. All other things being equal, a 10-kHz channel uses one-third the radio spectrum resource that a 30-kHz channel uses. For this calculation, we have to use the per-channel bandwidth rather than the per-carrier bandwidth, so the GSM channel counts as 25 kHz.

This means that GSM is 30/25 as efficient as AMPS at the same S/I distribution. That the GSM channel is one-eighth of a 200-kHz carrier does not affect this estimate.

27.11.2 6 dB of S/N is a factor of 2

The big background rule of conventional reuse is that 6 dB of radio performance is worth a factor of 2 in reuse.[16] Why is this? Because a factor of 2 in reuse means using each channel twice as often in the same number of cells. Since cells are spread out over a two-dimensional service area, doubling the reuse of each channel means reducing the

---

[16]This is a logarithmic relationship, so 10 dB is a factor of 3 (close enough to 3.162), 12 dB is a factor of 4, and 20 dB is a factor of 10.

cochannel reuse distance by the square root of two ($\sqrt{2}$). Using the fourth-power-of-distance rule from Sec. 1.7, reducing the distance by $\sqrt{2}$ creates four times the interference, 6 dB more. We can express the 6-dB rule is Eq. (27.3):

$$d_1 - d_2 = 20 \log\left(\frac{c_2}{c_1}\right) \tag{27.3}$$

where c_1 = first channels per base station
c_2 = second channels per base station
d_1 = first S/N ratio in decibels
d_2 = second S/N ratio in decibels

As an example, let us compare the AMPS and GSM channels. The GSM channel is 5 dB tougher than AMPS because it can withstand 12 dB of S/I at the same audio quality as AMPS at 17 dB of S/I. We can use this 5 dB to go from $N = 7$ to $N = 4$, an improvement of 1.75 in capacity. The 6-dB rule would suggest that a 5-dB performance improvement would get us a factor of 1.78, very close to what we actually get. If we factor in the 30/25 efficiency factor for GSM's smaller bandwidth, we get a combined factor of 2.1 for GSM over AMPS, so calling the GSM channel twice as efficient is a perfectly reasonable claim.

For another example, suppose that we have a 12.5-kHz channel proposal and we want to know what S/I distribution it would have to have to be equivalent in efficiency to the 25-kHz GSM channel with its 12-dB requirement. Well, the spectrum usage is two times better (20/10), so we can allow 6 dB more S/I, or 18 dB, for equivalent cellular efficiency. By the same relationship, if the channel were a narrowband 5-kHz channel, five times the bandwidth reduction, then we could tolerate 14 dB more S/I, or 26 dB, for cellular efficiency equivalent to GSM.

27.11.3 Sectors

Since three-sector cells remove about two-thirds of the interference, they are worth about a factor of 3 in radio performance, or 5 dB. By the previous rule of thumb, we would expect an $N = 7$ system with omnidirectional cells to be equivalent to an $N = 4$ system with three-sector cells.[17]

Six-sector cells are worth another 3 dB because they remove one of the remaining two first-ring interferers. Some AMPS systems use six-sector cells, but GSM does not, mostly because the very small channel groups required for six-sector cells would be difficult in the eight-channel GSM carrier environment.[18]

---

[17]The earliest AMPS deployment plans started off with $N = 12$ omnidirectional systems (for coverage) that would migrate into $N = 7$ systems with three-sector cells. This did not last long because the migration is an administrative nightmare and because the early growth rates were so rapid that it did not make sense to have a separate early coverage phase before the long-term growth phase.

[18]Again, in the early days of AMPS, there was contention between $N = 7$ with three-sector cells and $N = 4$ with six-sector cells. Going to six sectors gains 3 dB, whereas going to $N = 4$ loses 5 dB for a net performance loss of 2 dB. Cellular service providers had to decide whether the higher radio performance of $N = 7$ and three sectors was worth the higher cost.

27.11.4 Overlaid cells

Overlaid cells are worth another 30 percent, 2 dB. This is the incremental capacity of being able to use a second cell radius at one base station. This advantage comes from being able to reuse channels from large to small cells at closer range, and it seems to be a consistent advantage over a wide range of traffic demand density distributions and values of N. The same advantage of overlaid cells can be realized by using reuse partitioning in the early stages of cellular system growth.

Allowing more than two cell sizes, multiplying overlaid cells, is worth another 15 percent, 1 dB. A growing reuse-partitioned cellular system would have as many as six cell sizes, six server groups, served by the same base station. The engineering would be quite complicated, but the result would be that the cells could handle more traffic.

27.11.5 Nonregular channel reuse

Finally, we can leave geometric reuse altogether. Once the traffic and terrain are nonregular, the geometric regularity that inspired cellular thinking in 1966 can give way to a nonregular assignment that takes full advantage of every reuse opportunity. Analysis and simulation studies have showed this was worth another factor of 2 in capacity, 6 dB. Getting the benefit of this requires solving some terribly difficult mathematical optimizations, however.

27.11.6 Adding it all up

The interesting part of all these rules of thumb for conventional reuse is that we can multiply them together or add up the decibel count to get a reasonable estimate of capacity changes. Going from six-sector AMPS cells to three-sector GSM cells, for example, is worth 1 dB for the channel reduction from 30 to 25 kHz, 5 dB greater resistance to interference, and -3 dB for the sector change, a total of 3 dB. This suggests that this change will increase the capacity of the same cell layout by $\sqrt{2}$, or 40 percent.

27.12 Conclusion

This exposition of the principles used for managing capacity on conventional reuse systems gives a picture of the complex issues facing cellular engineers managing interference on rapidly growing systems. Of course, these issues are managed quite differently in the CDMA network, which uses an $N = 1$ reuse pattern where all carrier frequencies are used by every cell. We turn our attention to these issues in the next chapter.

Chapter

28

CDMA Principles for Multicellular Systems

In this chapter we present the code division multiple access (CDMA) equations for one server and for multiple servers and then discuss the assumptions held within the equations. We then apply the equations in general (to both directions) and specifically to the forward and reverse directions. This addresses many of the issues involved in moving the theoretical concept of CDMA into the practical world of cellular telephony. We close the chapter with a discussion of how the equations can be applied statistically to systems operating over time. Viewing a distribution over time allows us to address traffic engineering for the air interface.

28.1 The CDMA Equations

The CDMA equations allow us to calculate the theoretical capacity of one server or of a system that offers a multiserver environment. As you will see, one clear result of these equations is that the limit of CDMA capacity is quite pronounced. The number of users on a cell sector can increase steadily to a point without requiring a great deal more power from the base station and the user terminal. At some point, the power necessary to achieve a good call goes up very rapidly. And beyond a certain point, no amount of power in the world will serve all the users with high call quality. This result derived from the CDMA equations has a practical consequence: We can design CDMA systems that set a capacity limit by refusing all calls that require a power level higher than a predetermined setting.

28.1.1 One server

Let α be the maximum interference to signal (I/S) ratio our CDMA channel can allow. The value of α is the spreading gain divided by the required E_b/N_0 for the CDMA channel. It is the job of the CDMA system and user terminal working together to set power levels just high enough for sufficient E_b/N_0 so that the received power r_u for the user terminal u is determined by the received power r_v for all the other users $v \neq u$ and the noise level n. Equation (28.1) says that the total interference plus noise is α times the received signal for user u:

$$\alpha r_u = \sum_{v \neq u} r_v + n \qquad (28.1)$$

For a single antenna face serving ϕ users, this symmetric equation has a symmetric solution obtained by setting all their received power levels to the same value r. Equation (28.2) says that a system with ϕ users has $\phi - 1$ interferers for each user:

$$\alpha r = (\phi - 1)r + n \qquad (28.2)$$

which we can rephrase in Eq. (28.3):

$$mr = n \qquad (28.3)$$

where $m = \alpha + 1 - \phi$. We can solve for r in Eq. (28.4):

$$r = \frac{n}{m} = \frac{n}{\alpha + 1 - \phi} \qquad (28.4)$$

Naturally, this equation is meaningful only when the number of users ϕ is less than $\alpha + 1$. As ϕ approaches this limit, the system increases every user's transmit power to ensure sufficient E_b/N_0. Looking at the graph in Fig. 28.1, we can make an important observation. Until ϕ gets *very* close to $\alpha + 1$, the receiver power levels stay *very* low.

These equations (28.1 through 28.4) tell most of the story for the reverse direction. Each user terminal has to transmit with enough power to achieve r received power at the antenna face. Thus, if G_{ub} is the radio path gain from user u to antenna face b, then

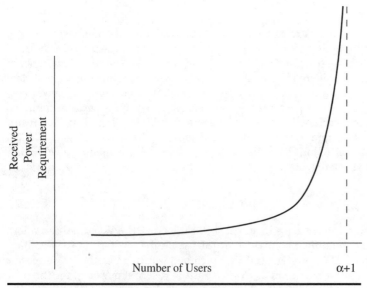

Figure 28.1 CDMA reverse power requirement as a function of number of users.

the received power is the transmit power multiplied by the path gain. Equation (28.5) shows the relationship in absolute terms rather than decibels:

$$r = t_u G_{ub} \tag{28.5}$$

Thus we can express the required transmit power for each user u in Eq. (28.6):

$$t_u = \frac{r}{G_{ub}} \tag{28.6}$$

The forward direction is a little more complicated. We start with Eq. (28.1) and use the radio path gain relationship $r_u = t_u G_{ub}$ to solve for the base station transmit power t_u for user u in Eqs. (28.7) and (28.8):

$$\alpha t_u G_{ub} = \sum_{v \neq u} t_v G_{ub} + n_u \tag{28.7}$$

$$\alpha t_u = \sum_{v \neq u} t_v + \frac{n_u}{G_{ub}} \tag{28.8}$$

Since our forward power limitation is the total transmitted power T for all the users served by the antenna face, we add up all the users' transmit power in Eq. (28.9) and realize that the double summation contains every user $\phi - 1$ times to get Eq. (28.10):

$$\alpha \sum_u t_u = \sum_u \sum_{v \neq u} t_v + \sum_u \frac{n_u}{G_{ub}} \tag{28.9}$$

$$\alpha \sum_u t_u = (\phi - 1) \sum_v t_v + N \tag{28.10}$$

where

$$N = \sum_u \frac{n_u}{G_{ub}}$$

The noise component N has a physical meaning: It is the total noise of all the user receivers as if all that noise were broadcast from the antenna face.

The total antenna face transmit power T for all the users is expressed in Eqs. (28.11) through (28.14):

$$\alpha T = (\phi - 1)T + N \tag{28.11}$$

$$(\alpha + 1 - \phi)T = N \tag{28.12}$$

$$Tm = N \tag{28.13}$$

$$T = \frac{N}{m} = \frac{N}{\alpha + 1 - \phi} \tag{28.14}$$

As in the reverse direction, the forward direction has a severe limit at $\alpha + 1$ users beyond which no amount of power is sufficient. The graph of total forward power T as a function of users shown in Fig. 28.2 has a different shape[1] with the same basic idea:

---

[1]The reverse power curve is a hyperbola with asymptotes at $r = 0$ and $\phi = \alpha + 1$, whereas the forward direction has an extra linear component. The forward noise value N increases linearly as ϕ increases, whereas the reverse noise n does not.

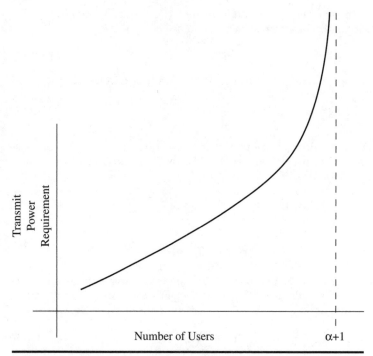

Figure 28.2 CDMA forward power requirement as a function of number of users.

Forward power requirements remain *very* low until ϕ gets *very* close to α. The reader should notice the pleasing symmetry of the same m value, $mr = n$ and $Tm = N$, in Eqs. (28.3) and (28.13).

28.1.2 Multiple servers

For multiple antenna faces, the algebra gets more complicated, but the basic idea stays the same. We need to maintain the distinction between the users in the cell sector being served and the base station antenna face serving them. We refer to the antenna faces as b and c and the sets of their users as B and C. Thus the notation $u \in B$ means all the users in the cell sector served by antenna face b.

It makes no difference for this analysis whether the two serving antenna faces b and c serve sectors in the same cell or sectors in two different cells. If a full-blown CDMA system has 500 cells and each cell has three sectors, then the total number of antenna faces and sectors, k is 1500. When we write about the interaction between antenna faces b and c, the reader should remember that sometimes they will be alternative sectors on the same cell and sometimes they will be on completely different cells.

In the reverse direction we get the matrix Eq. (28.15):

$$M \vec{r} = \vec{n} \qquad (28.15)$$

This becomes Eq. (28.16) when written out the long way:

$$m_{11}r_1 + m_{12}r_2 + \cdots + m_{1k}r_k = n_1$$

$$m_{21}r_1 + m_{22}r_2 + \cdots + m_{2k}r_k = n_2$$

. . . .   (28.16)

$$m_{k1}r_1 + m_{k2}r_2 + \cdots + m_{kk}r_k = n_k$$

r_b is the received power for b users (that is, subscribers served by antenna face b), and n_b is the receiver noise level at antenna face b.

The components of the matrix M in Eqs. (28.17) and (28.18) are the interactions of antenna faces and users of the same and other antenna faces:

$$m_{bb} = \alpha + 1 - \phi_b \qquad (28.17)$$

$$m_{bc} = -\sum_{v \in C} \frac{G_{vb}}{G_{vc}} \qquad \text{for } b \neq c \qquad (28.18)$$

The matrix diagonal terms m_{bb} are the same as the m variable we had in the single-server case where ϕ_b is the number of b users. The matrix off-diagonal terms m_{bc} for two different antenna faces b and c are the interference antenna face b causes for users served by antenna face c. m_{bc} reflects the interfering signal path from antenna face b to users served by c, as shown in Fig. 28.3.

Another way to look at M is that each antenna face b is represented by a row in the matrix, $m_{b1}, m_{b2}, \ldots$, and each set C of users served by an antenna face is represented by a column in the matrix, $m_{1c}, m_{2c}, \ldots$.

Let us look a little more closely at the summation in Eq. (28.18). m_{bc} is the (negative) sum over all C users of the path gain ratio of b to c. This makes sense, really. If the path from C users to face c is twice as good as the path from the same C users to another face b, then their effect on face b should be half as much as their effect on face c. Ten

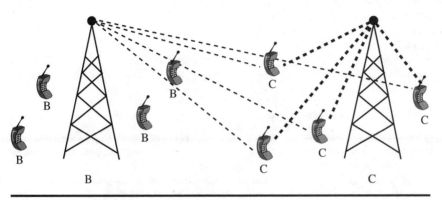

Figure 28.3 Interference path from base b to c users.

users with half the path gain (or twice the path loss) to another serving antenna look like five users over there.[2]

The forward direction for multiple servers is surprisingly similar:

$$\vec{T}M = \vec{N} \tag{28.19}$$

Equation (28.19) is the matrix equation, and Eq. (28.20) is the same system of equations written out the long way:

$$T_1 m_{11} + T_2 m_{21} + \cdots + T_k m_{k1} = N_1$$

$$T_1 m_{12} + T_2 m_{22} + \cdots + T_k m_{k2} = N_2$$

$$\cdots\cdots\cdots\cdots\cdots\cdots\cdots\cdots\cdots\cdots \tag{28.20}$$

$$T_1 m_{1k} + T_2 m_{2k} + \cdots + T_k m_{kk} = N_k$$

T_b is the total transmit power requirement for antenna face b, and N_b in Eq. (28.21) is the same noise component we saw in Eq. (28.10):

$$N_b = \sum_{u \in B} \frac{n_u}{G_{ub}} \tag{28.21}$$

That the M matrix for reverse and forward directions is the exact same matrix is a beautiful and wonderful result.[3] Mathematician types would call it elegant. The single-server analysis tells us that there is a single component m, or m_{bb} in the matrix case, for both reverse and forward same-cell interference. The multiple-server analysis tells us that there is a single term m_{bc} for the interference between antenna face b and users served by antenna face c. The analysis goes on to tell us that the interference from one cell sector to another is completely described by the difference in the radio signal paths, the ability of the wireless telephone system to separate the radio environments of its cell sectors.

28.1.3   Wireless local loop (WLL) capacity

These equations tell us that wireless local loop can have a fantastic capacity advantage over wireless mobile telephones. An antenna mounted on the side of a building can be a narrow-beam, directional, high-gain antenna pointed directly at the serving base station. A directional antenna at the user terminal reduces the interference from that user terminal to other cells in the wireless telephone system.

This appears as a reduction in the cell-to-cell interference terms, m_{bc}, in M. There is one strange effect in the capacity calculations for a system of WLL subscribers: The pri-

---

[2]The sanity test of this is to consider the case where b and c are collocated or even the exact same antenna face. The capacity equation should give the same answer if we divide the users of one antenna face into two groups and consider them as separate cell sectors b and c with the same radio paths to both sets of users B and C. Then $m_{bb} + m_{bc}$ is $\alpha + 1 - (\phi_b + \phi_c)$, the same term we would have if we had considered b and c together.

[3]Reverse and forward are actually in a *transpose* relationship because we represented $\vec{r}$ as a column vector on the right side of M, and $\vec{T}$ is a row vector on the left side of M.

mary interfering path from one cell to another is the signal path from user terminals to base stations on the other side of their serving cells because those base stations are in the same direction as the serving radio path. As we move further away from the serving base station, the radio paths to the two base stations become closer in signal path gain and, therefore, contribute more to the interference coefficient m_{bc}.

While WLL reduces cell-to-cell interference, face-to-face interference in the same cell is unaffected by the user antenna. The antenna faces of one base station are all in the same place and all enjoy the same radio path advantage of the WLL directive antenna beam.

28.1.4   CDMA overflow of blocked traffic

The equations tell us, also, that having an overloaded cell overflow its excess demand onto other cells with poorer radio paths is going to hurt more than it helps. If the users have better path gain to the busy cell than to the neighbor cells, then the m_{bc} components of interference actually will be *greater* than the same-cell m_{bb} contribution.

As an example, consider a system with two omnidirectional cells 1 and 2 with the matrix in Eq. (28.22):

$$M = \begin{pmatrix} \alpha + 1 - \phi_1 & -\beta\phi_2 \\ -\beta\phi_1 & \alpha + 1 - \phi_2 \end{pmatrix} \tag{28.22}$$

Cell 1 serves ϕ_1 users, cell 2 serves ϕ_2 users, and we are assuming for this example that the radio path gain for all users to their nonserving cells is β times lower than to their serving cells.

We are going to add k users in the serving area of cell 1. Thus their radio signal path gains to cell 2 are β times lower than to cell 1. Adding k users to cell 1 increases the first-column components of $M_1(k)$ as shown in Eq. (28.23):

$$M_1(k) = \begin{pmatrix} \alpha + 1 - (\phi_1 + k) & -\beta\phi_2 \\ -\beta\phi_1 - \beta k & \alpha + 1 - \phi_2 \end{pmatrix} \tag{28.23}$$

The upper-left coefficient of $M_1(k)$ reflects the full value of k extra users being served by base station 1, whereas the lower-left coefficient reflects the extra β contribution of their interfering path to base station 2.

However, for some reason, we may decide to serve the same k users with antenna face 2 instead. This increases the second-column components of $M_2(k)$ as shown in Eq. (28.24):

$$M_2(k) = \begin{pmatrix} \alpha + 1 - \phi_1 & -\beta\phi_2 - k/\beta \\ -\beta\phi_1 & \alpha + 1 - (\phi_2 + k) \end{pmatrix} \tag{28.24}$$

The lower-right coefficient of $M_2(k)$ reflects the extra user being served by base station 2, but in this example the upper-right coefficient increases by k/β, a number greater than k, because the interfering path has greater path gain than the serving signal path.

We can solve the CDMA equations using matrices $M_1(k)$ and $M_2(k)$ and notice that matrix $M_2(k)$ gives higher power levels or, even worse, negative power levels that indicate system overload. A simpler approach to see the same effect is to realize that the system becomes overloaded when the *determinant* of matrix M is zero. The determinant for a two-by-two matrix is given in Eq. (28.25):[4]

$$\det\begin{pmatrix} a & b \\ c & d \end{pmatrix} = ad - bc \qquad (28.25)$$

If we start with a well-behaved CDMA system, a matrix M with positive determinant D, then we can ask how many users k the two matrices $M_1(k)$ and $M_2(k)$ can take before their systems reach overload status.

Let k_1 be the overload point where $\det[M_1(k_1)] = 0$, and let k_2 be the overload point where $\det[M_2(k_2)] = 0$. The combination of Eqs. (28.23) and (28.25) and some algebra gives us Eq. (28.26):

$$\det[M_1(k)] = D - k[\alpha + 1 - (1 - \beta^2)\phi_2]$$

$$(28.26)$$

$$\det[M_2(k)] = D - k(\alpha + 1)$$

It is convenient that the factors of β and $1/\beta$ cancel out to unity in the equation for $\det[M_2(k)]$. We can solve for k_1 and k_2 to get Eq. (28.27)[5]:

$$k_1 = \frac{D}{\alpha + 1 - (1 - \beta^2)\phi_2} \qquad (28.27)$$

$$k_2 = \frac{D}{\alpha + 1}$$

So long as $\beta < 1$ and $\phi_2 > 0$, we have $k_1 > k_2$. This means that this simple CDMA system will serve more users on the right cell than on the wrong cell before it becomes overloaded and unstable.[6] This example gives us insight that there are unlikely to be any reasonable CDMA systems where overflowing traffic to other cell sectors is good for system capacity.

28.1.5 Assumptions

The above equations make several assumptions as we present them in this section. While these assumptions are not exactly true, these equations still give us insight into the behavior of channels within a CDMA radio carrier. In the rest of this chapter we explore the differences between these assumptions and real CDMA systems, how we can modify the equations to take these differences into account, and how these differences affect our calculations.

---

[4]If this brings back too many linear algebra classes from too long ago, then the reader can trust our analysis.

[5]At first glance, the reader may wonder how the formula $k_2 = D/(\alpha + 1)$ could have reference to user population ϕ_1 or ϕ_2 or radio conditions β. However, the original determinant D depends on ϕ_1, ϕ_2, and β.

[6]The condition $\beta = 1$ means that the two cells are exactly the same for both user communities, and the condition $\phi_2 = 0$ means that there are no users in the service area of cell 2. Thus we are comfortable that $\beta < 1$ and $\phi_2 > 0$ are reasonable assumptions for this example.

The channels are all the same size. Each CDMA channel in these equations is presumed to have the same spreading gain and E_b/N_0 requirement, the same value of α. This ignores the varying data rates used in voice activity detection (VAD), where the CDMA channel reduces its bit rate and power requirements when the subscriber's speech pattern permits it. It ignores adaptive multirate (AMR) speech coding, where the CDMA channel adapts its voice bit rate to the prevailing radio conditions. And it ignores a CDMA carrier that is a mix of channels using different data rates.

The noise values are all the same. The receiver noise levels in these equations are all set the same for all the user terminals and all the base stations. We do allow the forward and reverse directions to have different noise levels. In the reverse direction, there is a clear economy in designing a base station receiver with a lower noise floor to allow user terminals to transmit at lower power. In the forward direction, however, it is more economical to build a more powerful transmitter than to engineer thousands of user terminals with lower receiver noise levels.

There are no other channels. We have ignored the overhead of pilots, power-control channels, and signaling.

Each user has one channel. Each call is modeled here as a single two-way CDMA radio link. As described here, there are no soft or softer handoffs.

All interfering radio power is the same. This is an important modeling point in our concept of CDMA, perhaps the most important point: Interference from other CDMA calls is the same as additive white gaussian noise (AWGN) at the same power level. In fact, orthogonal codes and joint detection (JD) are ways for interferers to be far less intrusive than AWGN with the same radio power. And multipath in the radio environment can create intersymbol interference that can make it worse.

We have perfect power control. These equations presume that the CDMA system has set its power levels to their exact optimal levels. Of course, the power-control process is based on radio measurements, which can have measurement errors, and power control is a dynamic process that adapts to changing radio signal paths. As subscribers start calls, end calls, and move in and out of radio shadows, the power levels adjust to the changing radio environment.

28.2   Applying the Equations to Both Directions

We can address some of these assumptions, as well as some other issues, and adjust our equations and our model to understand CDMA capacity. In this section we will look at changes to our model that are relevant to both directions. In the next two sections we will look at issues where changes unique to each direction need to be made.

28.2.1   Variation in data rates

The most noticeable difference between the equations as presented earlier and real CDMA is the distribution of channel data rates. Voice channels are adjusted continuously by VAD as the subscriber's voice allows, wideband CDMA (W-CDMA) uses AMR to select a voice data rate depending on radio conditions, and there are circuit-switched and packet data services on the CDMA carrier.

VAD reduces the average voice call utilization of the CDMA resource. If speakers are silent half the time, then the primary effect of VAD is to double the spreading gain of voice calls. This has the effect of doubling the maximum I/S ratio the voice CDMA channel can allow, the value of α. We have to be a little bit careful because the equations were based on a constant channel, and VAD creates a varying channel, so there is a statistical distribution of interference, which is a variation factor that needs to taken into account. As long as each voice call is a reasonably small fraction of the CDMA carrier, we are probably safe ignoring that extra variation.

AMR is not the same as VAD because it depends on the radio environment rather than on subscriber's voice. This means that we can figure the system will be at its lowest AMR bit rate when the radio conditions are most hostile. In short, we can base our choice of α on the lowest bit rate available in AMR.

The presence of other CDMA subscriber traffic, circuit-switched and packet data, can be thought of as adding users to the equations with a multiplication factor. If a subscriber is using circuit-switched Integrated Services Digital Network (ISDN) data service with 20 times the average voice data rate and double the E_b/N_0 requirement, then that channel is the equivalent of 40 voice users. As long as each packet data user is using a reasonably small fraction of the total CDMA resource, less than one-third, then we should be able to incorporate their average usage into the equations. However, we do have to keep an eye open for the much more variable usage pattern of a packet data user.

28.2.2 Power control

The power-control process has two effects not directly addressed in the CDMA equations as presented earlier. First, there are power-control errors, inaccuracies in the open-loop and closed-loop power-control processes. Second, there is the overhead factor of the power-control data stream.

The best way to model the power-control errors is simply to accept that the system has a loss in efficiency, a reduction factor in α. If the power control is a statistically lognormal distribution with a standard deviation of 0.5 dB, then it reduces the CDMA efficiency of each channel about 0.5 dB, about 10 percent. We can model this effect by multiplying α by 0.9.

The power-control data stream can be punctured bits, a separate channel, or specifically allocated bits in the data stream. In any case, it is taking up CDMA channel space that is not subscriber voice or data. We can model this effect, too, as a multiplication factor in α. If a 7200-bit-per-second channel has 800 bits of power control, then we can multiply α by 0.9 to compensate.

The tone of the last two paragraphs makes it sound like power control is bad news. This is not accurate. The bad news is the near-far problem and changing radio conditions. Power control makes it possible to run a CDMA system in a hostile radio environment, and these multiplication factors reflect the cost of the problem and its solution.

28.2.3 Smart antennas

The effect of adaptive phased arrays, smart antennas, is to focus the two-way radio link on the user terminal being served. This is clearly a good thing that is not modeled directly in the multicellular CDMA equations. The problem in the analysis is that there is a notion in the equations of a fixed collection of antenna faces, and we refer to the radio signal path gains G_{vb} from user terminals v to antenna faces b.

In a system with smart antennas, each user terminal has its own separate antenna radiation pattern used for both forward and reverse directions. If we wanted to be perfectly complete, then we would have to generate a set of equations with a separate virtual antenna face for each user terminal. Clearly, this is an overwhelmingly complex mathematical computation, so instead we do some approximation approaching this ideal solution.

The best smart antennas present radiation patterns of a few tens of degrees. There are large and complex adaptive phased arrays that have a 40-degree arc of high signal path gain, but time division synchronized CDMA (TD-SCDMA) antennas are claiming path gain improvements of 8 dB, consistent with 60-degree arcs. If we divide the cell into 12 virtual sectors of 30 degrees each, then we have a collection of user communities with reasonably common radio characteristics. The set of equations has 4 times as many variables as the three-sector system, with 16 times as many terms in the matrix M, but it is still manageable to solve these equations for a large system on a personal computer.

28.2.4 Dynamic channel allocation

TD-SCDMA offers dynamic channel allocation to optimize CDMA performance. The system selects a carrier and time slot so that the TDMA/CDMA channel has minimum interference with other channels. Other CDMA systems also may select the carrier with the best radio conditions as each call is established. This certainly improves system performance when the wireless telephone system is less than full, but does it increase the maximum capacity of a system?

The answer is yes in the multicellular arena, but not for a single-cell sector. If all the carriers, time slots, and pseudonoise (PN) codes of a cell are as busy, as they can be, then no dynamic channel assignment is going to squeeze out one more channel. In other words, the value of α is not affected by a better choice of carriers, time slots, or PN codes.

It is the reduction of intercell interference that makes carrier choice important for each call. The ability of a system to keep new calls away from interference generated by existing calls should reduce (and therefore improve) the m_{bc} components of interference from one cell sector to another. As the system gets busy, this effect may not be large, but even a few percent improvement in capacity should be modeled.

28.2.5 User terminal movement (Doppler)

The motion of the user terminal creates Doppler distortion, as described in Sec. 2.1.5. The Doppler distortion contaminates the signal and adds its own noise factor. Unlike multipath, however, there is no rake filter to screen out the distortion. Furthermore, the motion of a user terminal changes the multipath delay spread, so the rake filter is constantly catching up. The effect of a moving user terminal is to reduce the value of α in both forward and reverse directions. The faster the motion, the greater is the effect on CDMA capacity. The third-generation (3G) CDMA carriers with their higher chip rates are more susceptible to this distortion from pedestrian or vehicular movement.

28.2.6 Time division duplex (TDD)

The equations presented earlier are separate for forward and reverse links. The assumption throughout is that forward radio paths interfere only with other forward paths and that reverse paths interfere only with other reverse paths. In time division duplex (TDD), base stations and user terminals are broadcasting radio in the same spectrum. This is not a problem in one cell because base stations and user terminals are time-synchronized.

If neighboring cells do not follow the exact same pattern of forward and reverse timing, then radiation from one base station during its forward phase will interfere with the reception at another base station during its reverse phase. Similarly, user terminals transmitting reverse signals will interfere with other user terminals receiving forward signals. To avoid this cross-interference, all the cells have to be time-synchronized, and all have to use the same ratio of forward and reverse time slots. Even if the cells are time-synchronized, radio signals travel at the speed of light, and this may be slow enough to interfere with other cells. These extra cross-interference terms are not modeled at all in the CDMA equations presented earlier.

The correction is not simple. We model radio propagation in these equations and in planning tools described in Part 8 by calculating an array of radio signal path gains from every base station location B to a set of (x, y) coordinates. Adding a set of radio paths from every b to every other B is not a significant addition to the list, but adding a set of radio paths from every (x, y) to every other (x, y) is a huge increase in the problem size. If there is a significant amount of TDD cross-interference as we described here, then these equations do not do an adequate job of helping us understand TDD CDMA.

28.3   Applying the Equations to the Forward Direction

As we move into the specific details of forward and reverse CDMA radio paths, some of the mathematical elegance and symmetry are lost. Once we take pilots, soft handoffs, orthogonal codes, and multipath into account, the matrix M is no longer exactly the same in forward and reverse directions. However, we stick to our original claim that the mathematical structure is the same and that the fundamental structure of CDMA matrix M is the same.

Because the forward link in cdmaOne uses coherent modulation and the reverse link uses noncoherent modulation (discussed in Sec. 1.10.2), a busy cdmaOne system is limited by the reverse direction. The 3G systems with their more balanced modulation schemes are limited in the forward direction. Soft handoffs and multipath are greater burdens in the forward direction, and these burdens outweigh the relative advantages of a common pilot and orthogonal codes.

28.3.1   Forward pilots and signaling

The forward pilot is a single source of radio power shared among all the user terminals. While it is an essential part of the receiving process, it is extra radio power, a noise component, as far as these CDMA equations are concerned. We need to add this noise component to the equations to model the effects of forward pilots and signaling channels.

Each base station puts out some amount of overhead radio power in its forward pilot and signaling channels. For each user terminal, we can regard the total received power of all the overhead channels of all the base stations as a component of its receiver noise, part of the n_u term in Eq. (28.10).

This makes sense from a physical perspective. Suppose that the forward pilot for antenna face b has transmit power t and reaches user terminal u with received power $r = tG_{ub}$. Then the component of noise N representing that pilot is $P = r/G_{ub}$, which is just t. What we are saying here is that the model for the forward direction in the CDMA equations treats the pilot power as another interferer when we include it in the noise coefficient N.

This model includes the adverse effect of having a region with several pilots with nearly equal received power competing to serve it. This effect is called *pilot pollution,* and it is discussed in Sec. 32.1.

28.3.2 Soft and softer handoffs

A call in soft or softer handoff is using more than one forward channel. The CDMA equations present earlier do not reflect having multiple CDMA channels for one call.

The areas with nearly equal radio paths are the areas in soft or softer handoff. When a call in the calculation has an alternative cell sector close to the highest radio signal path gain, the wireless telephone system almost certainly will put that call into a soft or softer handoff state and will replicate its forward signal on two antenna faces, c and d.

This means that the forward matrix M has the same call on two separate columns, c and d; the coverage areas C and D have some overlap.[7] If a single call served nearly equally by two antenna faces uses capacity resources equivalent to two calls, then that same call in soft handoff is using the resources of *four* calls. And there is no reason the number of simultaneous signals is limited to two. Some wireless telephone systems allow up to six simultaneous forward transmissions of the same call, so one of those calls would use the system resources of 36 calls.

The reality of multiple-cell soft handoffs may not be this bad. The user terminal combines the soft handoff channel replications with its rake filter to get a clearer signal. The total resource consumption of a k-cell soft-handoff call is probably less than k^2, but it still uses a lot of CDMA resources. Part of good CDMA design is avoiding large areas with multiple nearly equal serving cell sectors.

28.3.3 Orthogonal codes

Orthogonal codes also can be put into the CDMA system of equations by increasing the m_{bb} terms. In Eq. (28.17) we saw that $m_{bb} = \alpha + 1 - \phi_b$, where α is the spreading gain divided by E_b/N_0 and ϕ is the number of calls served by antenna face b. If the orthogonality is perfect so that the base station and user terminal do a perfect job of filtering out the other forward channels, then $m_{bb} = \alpha$ for all serving antenna faces b, no matter how many calls are being served up to the maximum number of orthogonal channels. In cdmaOne, the maximum number of orthogonal calls is about 60, depending on how many codes are used for paging and access. This number is about 250 for cdma2000 or W-CDMA.[8]

---

[7]In the CDMA equations without soft handoff, all the coverage areas are disjoint; the intersection of C and D, $C \cap D$, is the empty set $\varnothing$. Forward-direction soft handoff creates overlap, so the intersection is not empty, $C \cap D \neq \varnothing$.

[8]The reason the number of calls is *about* 60 or 250 is that some of the codes are reserved for signaling, and the number reserved for signaling can vary.

28.3.4 Multipath

Multipath has an amplifying effect on interference from the same radio source. The rake filter allows the receiver to see all the power of the transmitted signal even when the radio path spreads that signal over several chips. The price is time sidelobes, as described in Sec. 8.4, an extra interfering component generating in the receiver. We will let γ be the multipath factor, the time sidelobe power divided by the desired signal power in the rake filter.

Multipath, with the rake filter compensating for it, does in a bad way exactly what orthogonal codes do in a good way. The multiple paths in the forward direction increase the effect of same-sector interferers, whereas orthogonal codes reduce their effect.

The effect of multipath in the forward direction is to add γ interference for each radio signal in the same cell sector, so $m_{bb} = \alpha + 1 - (1 + \gamma)\phi_b$. In the orthogonal case, the good signals no longer interfere with each other, and we are left with only the multipath interference, so $m_{bb} = \alpha - \gamma\phi_b$.

28.4 Applying the Equations to the Reverse Direction

In cdmaOne, the reverse direction does not have orthogonal codes. In 3G systems that have reverse pilots, orthogonal codes are used, but the system has to support separate pilots for each user terminal. These are disadvantages for the reverse channel as compared with the forward channel. However, the reverse channel has advantages as well: soft handoff and the fact that multipath interference is coming from different locations rather than from a source co-located with the desired signal.

28.4.1 Reverse pilots and signaling

The reverse pilot is extra radio power for each user terminal using the wireless telephone system. This means extra bits, extra information, and extra power for each call. If a subscriber is combining voice and data in one session, then there is only one pilot for both. The overhead in the reverse direction is the pilot plus any signaling activity.[9]

The effect of the reverse overhead is to reduce the value of α. This reduction is not a fixed amount because the pilot is a bigger share of a voice channel than of a high-capacity data channel. In fact, the effect of overhead is a fixed increase in $1/\alpha$. This is so because $1/\alpha$ is the share of the total CDMA resource that a channel uses, and adding a fixed overhead will increase that share the same amount for different channels.

28.4.2 Soft and softer handoffs

The reverse soft handoff has no effect on the CDMA equations. It ensures that each user terminal is operating at the lowest power for several radio paths. We can think of reverse soft handoff as the process ensuring that the equations hold true. If the user terminal has to make a sequence of hard handoffs to maintain the highest radio signal path gain, then delays and inaccuracies in the handoff process will degrade the capacity from the equation solution.

---

[9]We have already accounted for power control, both forward and reverse, in Sec. 28.2.2.

Softer handoff, on the other hand, actually may improve the capacity beyond the equation values because the combination of two sector signals may allow lower power levels. We do not know if this combining effect of softer handoff is significant, but it is safe to say that soft and softer handoffs enforce the selection of the best cell sector instant by instant.

28.4.3 Joint detection (JD)

JD as described in Sec. 10.4.7 is a mathematically sophisticated approach that uses knowledge of all the CDMA signals in demodulating the reverse channel. It is useful with small numbers of signals, as seen in TD-SCDMA. The effect of JD is similar to orthogonal codes, but the effect is partial rather than total. We will let v be the same-sector interference factor.

In the case of normal CDMA demodulation, where we treat the interferers as AWGN and multiply by the channel PN code, we have $v = 1$. In the case of orthogonal codes, which we use in the forward direction, we have $v = 0$. The effect of JD is somewhere in between the two. The effect of JD is to attenuate the same-sector interference, so $m_{bb} = \alpha - v(\phi_b - 1)$.

28.4.4 Multipath

Multipath in the reverse direction has a small effect on CDMA capacity because each reverse CDMA channel is using a different radio path to the base station. As we said in Sec. 28.3.4, the rake filter generates time sidelobes, an extra component of interference γ for each CDMA channel sharing the radio path.

The effect of reverse multipath is to add γ interference, but only once for the signal channel time sidelobes. Thus $m_{bb} = \alpha + 1 - \phi_b - \gamma$. The combination of JD and multipath gives us $m_{bb} = \alpha - v(\phi_b - 1) - \gamma$. The effect of multipath in the reverse direction is to add γ rather than to multiply by γ as we did in the forward direction.

28.5 Applying the Equations Statistically

The CDMA equations represent a snapshot, a specific collection of users at a specific moment in time with specific radio paths to all the base stations. These equations give us significant insights into the capacity effects of various components of CDMA engineering.

By themselves, the CDMA equations do not tell us how to do traffic engineering in a wireless telephone system. We need to take them one step further and try to use them on a distribution of CDMA traffic.[10] We can extend the CDMA equations to a *distribution* of demand by replacing the sums with integrals in Eqs. (28.28) and (28.29):

$$m_{bb} = \alpha + 1 - \iint_{(x,y) \in B} \phi(x, y)\, dx\, dy \qquad (28.28)$$

$$m_{bc} = - \iint_{(x,y) \in C} \frac{G(x, y, b)}{G(x, y, c)}\, \phi(x, y)\, dx\, dy \qquad \text{for } b \neq c \qquad (28.29)$$

---

[10]One of us (Rosenberg) built a sophisticated mathematical model for CDMA WLL using these equations in a statistical form. The view presented here is simpler but may help the reader understand how these equations can be extended to CDMA traffic engineering.

where $\phi(x, y)$ = demand density at (x, y)

$G(x, y, b)$ = path gain from (x, y) to antenna face b

These equations give us power levels for a specified demand density $\phi(x, y)$. If $\phi(x, y)$ represents the average density of demand, then the solutions of $M\vec{r} = \vec{n}$ and $\vec{T}M = N$ still represent average power levels for reverse and forward directions. If we want a system to have 2 percent blocking, then we need the power level that satisfies the equations 98 percent of the time.

Solving the statistical distribution of a matrix equation is a difficult task.[11] Let us apply some loose, hand-waving mathematics to gain some insight into using these equations in traffic engineering a CDMA system.

Let's make life simple and consider whether each cell sector is able to stay out of an overload condition. If noise levels are low and radio signal path gains are high, then a base station power amplifier will get close to full power only when it is nearly overloaded. In the single-server case, that cell sector will overload when its traffic level ϕ_b reaches the CDMA limit, $\alpha + 1$. Let's make life simpler still and figure that the system overloads when a row sum or column sum of the matrix M adds up to zero or less.

Let's define the sector-to-sector radio path protection in Eq. (28.30):

$$\beta_{bb} = 1$$

$$\beta_{bc} = \frac{m_{bc}}{\phi_c} = \frac{\iint_{(x,\,y)\in C} \frac{G(x,\,y,\,b)}{G(x,\,y,\,c)}\phi(x,\,y)\,dx\,dy}{\iint_{(x,\,y)\in C} \phi(x,\,y)\,dx\,dy} \qquad \text{for } b \neq c \tag{28.30}$$

The column or row sum is in Eq. (28.31):

$$\sum_{b \text{ or } c} m_{bc} = \alpha - \sum_{b \text{ or } c} \beta_{bc}\phi_c \tag{28.31}$$

and the system is not overloaded as long as Eq. (28.32) is satisfied.

$$S_c = \sum_{b \text{ or } c} \beta_{bc}\phi_c \leq \alpha \tag{28.32}$$

Since ϕ_b is a Poisson-distributed distribution and S_c is a weighted sum of Poisson distributions, we can estimate the mean μ and variance σ^2 in Eqs. (28.33) and (28.34):

$$\mu = \sum_{b \text{ or } c} \beta_{bc}\overline{\phi_c} \tag{28.33}$$

$$\sigma^2 = \sum_{b \text{ or } c} \beta_{bc}^2 \overline{\phi_c} \tag{28.34}$$

If we consider the resulting distribution of weighted traffic and CDMA resource used to be a normal bell-shaped distribution, then the condition for 2 percent blocking is $\mu + 2\sigma \leq \alpha$.[12]

---

[11]It is hard enough when the determinant of the matrix M remains far from zero. Unfortunately for us, the CDMA equations are interesting *precisely* when the determinant of the matrix M is close to zero because that is when the CDMA carrier is close to its saturation level.

[12]If we are using more exact numbers from a statistical table, then the condition for 1 percent blocking is $\mu + 2.33\sigma \leq \alpha$ and the condition for 2 percent blocking is $\mu + 2.06\sigma \leq \alpha$.

This approach may have a lot of approximations, but it gives a good sense of CDMA capacity for small-cell, interference-limited environments. The β_{bc} values are the interference contributions from one cell sector to another. Taking this cross-interference into account, we find the average traffic μ and the standard deviation σ and make sure that the $\mu + 2\sigma < \alpha$ for 2 percent blocking.

When multiple channel sizes are involved, as with a mix of voice and data service, there is some scaling and adding involved, but the statistical model is basically the same. Let us define mean μ and standard deviation σ for the combined system in Eqs. (28.35) and (28.36):

$$\mu = \frac{\mu_{voice}}{\alpha_{voice}} + \frac{\mu_{data}}{\alpha_{data}} \tag{28.35}$$

$$\sigma^2 = \frac{\sigma^2_{voice}}{\alpha^2_{voice}} + \frac{\sigma^2_{data}}{\alpha^2_{data}} \tag{28.36}$$

We estimate that we have 2 percent blocking when $\mu + 2\sigma = 1$.

28.6 Conclusion

In this chapter we have introduced the application of the CDMA equations to estimation of cell capacity and to traffic engineering for the air interface. We have attempted to apply this to an interference-limited environment, that is, to a multiple-cell system serving many customers and, at times, operating near or at capacity. Understanding what happens to the CDMA system as its limits are approached is crucial to the practical design of these systems. In the next chapter we will look at the capacity issues specifically related to transmission of subscriber data.

29

CDMA Data Capacity Principles

Point-to-point data transmission service is defined by three things: rate, errors, and delay. The bit rate of a data service is typically well defined by a standard. The standard reflects the best data rate that can be provided reliably over a given interface. cdmaOne, cdma2000, wideband CDMA (W-CDMA), and time division synchronized CDMA (TD-SCDMA) each have data services defined in the standard. In code division multiple access (CDMA), data capacity, that is, the number of data links that can be supported, is determined by the data rate of the service and the bit-error-rate (BER) requirement. That is, the maximum capacity of a server, cell, or system can be determined by the data rate given each user and the BER requirement. The actual capacity will vary depending on the radio link conditions and the resulting E_b/N_0 requirement (the ratio between energy per bit and interference).[1]

The delay is mostly determined by the packet pipes between the base station and the public packet data network (PPDN). Jitter within the transmission channel is not significant to the user. The higher-level data services can ensure that the packets are processed in order even if they arrive out of order, and the slight addition to delay created by this management of jitter is not significant in data transmission.

29.1 Bit Error Rate (BER) and E_b/N_0

Let us start with a binary phase shift keying (BPSK) modulated channel with additive white gaussian noise (AWGN) as we did in Sec. 8.2. The BPSK transmitter sends a positive value for $+1$ and a negative value for -1, and the AWGN adds a normally distributed random component. The BPSK receiver sees a $+1$ if the sum is positive and a -1 if the sum is negative. The bit will be in error when the AWGN component adds enough error to the signal to push it onto the wrong side of the zero-voltage line, as shown in Figs. 8.1 and 8.2.

What is the probability that this will happen? The amplitude of the desired signal is the square root of the signal power, and the standard deviation of the noise will be the

---

[1]The E_b/N_0 requirement is a physical result of the BER requirement set in the quality of service (QoS) defined in the data transmission standard.

square root of the noise power. This means that the bit values will be $+\sqrt{E_b/N_0}$ and $-\sqrt{E_b/N_0}$ times the noise standard deviation. The bit error rate is the tail of the normal curve that falls on the wrong side of the zero-amplitude line. It is given by the formula in Eq. (29.1):

$$\text{BER} = \frac{1}{2}\text{erfc}\left(\frac{E_b}{N_0}\right) \qquad (29.1)$$

[The function erfc(x) is defined in Eq. (26.4) in Sec. 26.3. erfc(x) is the probability that a normally distributed random variable will be x or more standard deviations from its mean.]

In the case of quadraphase phase shift keying (QPSK) with AWGN interference, the receiver takes the transmitted two-dimensional, complex signal-plus-noise value and finds the closest point on the QPSK constellation, as shown in Fig. 29.1. The in-phase and quadrature dimensions can be treated completely independently in computing the BER of QPSK. The energy per bit E_b is divided equally between the two dimensions, and the noise power is also divided equally, so the formula for BER is exactly the same for QPSK as it is for BPSK. The graph of BER as a function of E_b/N_0 in the BPSK and QPSK cases is shown in Fig. 29.2.

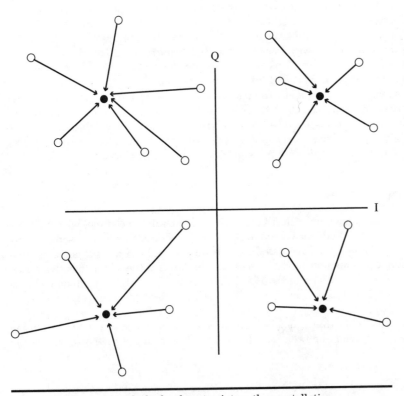

Figure 29.1   The receiver finds the closest point on the constellation.

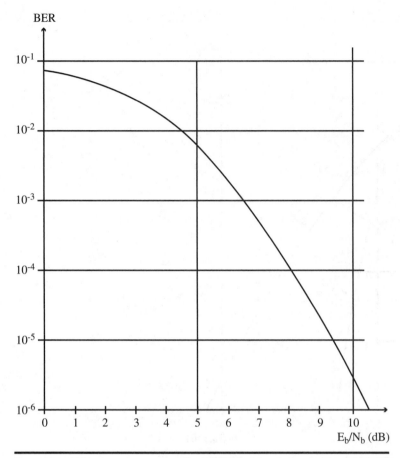

BER

Figure 29.2 Bit error rate as a function of E_b/N_0.

For more complex constellations such as 8-ary or 16QAM, the calculations are also more complex. Since the points on the constellation are closer together, each constellation symbol requires more energy than QPSK relative to the noise level. However, there are more bits in each symbol, so the energy per bit E_b relative to the noise level remains about the same as QPSK. The principle of code division multiple access (CDMA) processing gain is that interference from other CDMA channels can be treated the same as AWGN.

The performance of a data service is not the raw BER but the BER after forward error correction (FEC). This depends on the method of FEC, as shown in Fig. 29.3. The reduction in BER for typical convolutional coding ($K = 7$) is shown for two-to-one coding (rate ½) and for three-to-one coding (rate ⅓). The rate ⅓ coding requires 3 coded bits for every encoded bit, whereas the rate ½ coding only requires 2. The rate ⅓K = 41 line represents a very complex coder, computationally intense but theoretically possible.

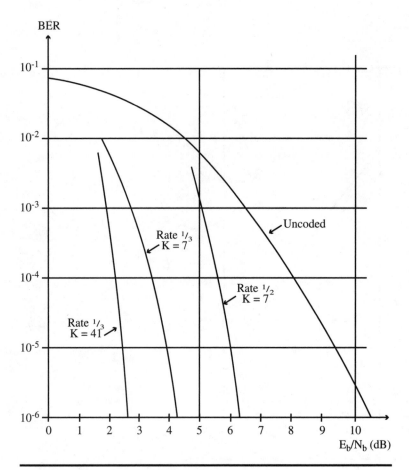

Figure 29.3 Bit error rate with forward error correction.

29.2 E_b/N_0, Rate, and Capacity

The number of channels a CDMA carrier can carry is the spreading gain divided by E_b/N_0. The spreading gain is simply the chip rate divided by the data rate. We can summarize this in Eq. (29.2):

$$\alpha = \frac{W/R}{E_b/N_0} \qquad (29.2)$$

where α = maximum number of interferers
W = chip rate
R = data rate

The capacity of the CDMA carrier for this data service is $\alpha + 1$.

If we want to find α, then the CDMA carrier determines the chip rate, and the data service defines the bit rate. The power-control process of the data service channel will adjust the radio power until the BER requirement is satisfied.

The bit rate is easy; the CDMA carrier has so many bits per second available under given radio conditions and E_b/N_0 requirements. Everything else being equal, a CDMA carrier can serve twice as many users at 32 kbps as it can at 64 kbps. We want to find the value of E_b/N_0 that corresponds to the data service BER so that we can use Eq. (29.2) to find the CDMA carrier capacity for a data service.

For mixed services, we can think in terms of each channel's share, $1/\alpha$, of the CDMA carrier pie in Eq. (29.3):

$$\text{Share} = \frac{1}{\alpha} = \frac{R}{W} E_b/N_0 \qquad (29.3)$$

This allows us to mix and match services, even services at the same bit rate that require different E_b/N_0, such as RS1 and RS2 voice calls. And it allows us to mix and match voice calls and a variety of data services at different rates and different BER and E_b/N_0 requirements.[2]

The connection between BER and E_b/N_0 is clear when the signal waveform is clean, where the connection is a single, stable radio path. As described in Sec. 28.3.4, the presence of multipath degrades data service capacity just as it degrades voice call capacity. And as described in Sec. 28.2.5, data terminal movement affects the data capacity as well. Since the waveform is affected by delay spread and Doppler distortion, the data capacity goes down.

There is another capacity tradeoff available to the CDMA data service provider. It is reasonable for the end-to-end data service to retransmit bad packets using automatic repeat request (ARQ). In this case, a higher frame error rate (FER) simply means more channel time sending packets again rather than a higher error rate to the data service subscriber. Thus the capacity α, with retries taken into account, is given in Eq. (29.4):

$$\alpha = \frac{W/R}{E_b/N_0} (1 - \text{FER}) \qquad (29.4)$$

Since a frame is a fixed number of bits, the FER depends on the BER.

For a given data rate, there is an optimal point where the tradeoff between minimum power (low E_b/N_0) and high FER reaches an overall least energy per bit. Lowering E_b/N_0 will create many more errors without saving much power, and raising E_b/N_0 will use more power without fixing many frame errors. Managing this tradeoff is an important way for the wireless data service provider to maximize data capacity. The optimal data performance depends on the speed of motion of the user terminal.

Another option to increase capacity is to use a longer interleaving length so that bursty channel errors are still corrected. The price of longer interleaving is a longer delay in packet transmission. Also, a longer interleaving length means that each service has larger chunks of data. If the packet is smaller than the interleaving length, then the system has to pad the rest of the interval.[3]

---

[2]The $1/\alpha$ estimate is not quite exact because it ignores the fact that a CDMA channel does not interfere with itself. While this can be significant for large-capacity users, $1/\alpha$ is a good estimate of the share of the CDMA carrier a data service uses.

[3]Personal computer owners may be aware of the issue of hard-disk cluster size, where each file uses a whole number of clusters. The smallest quantum of disk space is typically something like 16,384 bytes, so a 100-byte file wastes 16,284 bytes, the rest of the disk cluster.

29.3   cdmaOne Data Capacity (2G)

While cdmaOne is almost entirely a voice call system, it supports several distinct data modes, including short message service (SMS), asynchronous data (modems), facsimile (fax), and packet data. These are handled by the same CDMA channels as voice traffic with different handling at the user terminal and system ends. This makes sense for the services and data rates of second-generation (2G) systems.

SMS is handled by the paging channel, so user terminals can handle short messages even while engaged in a voice call. Since cdmaOne short messages are limited to 160 bytes and each character is a press of a button, we doubt that SMS will be a cdmaOne capacity issue as far as the air interface is concerned.

29.3.1   Data over voice-coded links

Asynchronous data and fax services are sent over cdmaOne channels set up as circuit-switched service with a dedicated path using the Radio Link Protocol (RLP) described in Sec. 10.1.1. The RLP manages the data communication with automatic repeat request (ARQ), (FEC), and flow control, so the user mobile data terminal and the equipment at the other end of the call see a normal asynchronous data link or fax signal.

While the data or fax service uses a normal 19.2-kilobit data stream, the cdmaOne system uses its own service option 4 for asynchronous data service and service option 5 for group 3 fax. Since there is no speech, there are no voice pauses, so the partial rates may not be used. Fax service is a continuous stream of bits, so it runs at full rate for the entire call in the direction the fax is being sent.

Using the IS-95B protocols, as many as eight voice channels can be bundled together to offer subscribers packet-switched data service with rates as high as 64 kbps using RS1 and 115 kbps using RS2.

29.3.2   Data over data links

Packet data service on cdmaOne is offered at 9.6 kbps using service option 4103 and at 14.4 kbps using service option 15. The packet links can be active or dormant, so bandwidth is not wasted when the data connection is inactive. The packet services of cdmaOne are managed by RLP, so they can use the air interface at cellular-type BER (similar to voice calls) and they can deliver data-type BER (around 10^{-6}) to their subscribers.

29.4   3G Data Capacity

While third-generation (3G) data capacity measurements will have to wait for 3G systems to go into service, we can estimate 3G data capacity using the same CDMA capacity principles we have been using. Data service takes a bigger bite of the CDMA apple because data subscribers typically want more bits per second than voice calls and because data services demand higher accuracy in those bits. A voice call with one wrong bit per thousand sounds fine, but one wrong pixel in a large graphic file can elicit loud complaints from the person viewing the image.

29.4.1   Normal data speeds

There are two kinds of data channels in 3G CDMA, supplemental channels dedicated to particular calls and shared packet channels. Both are administered over a fundamental channel for the data call. For a capacity example, let us use a 32-kbps data link. Let us consider cdma2000 3x, at 3.6844 megachips per second. Wideband CDMA (W-CDMA) at 3.84 megachips per second gives very similar numbers. With a rate ⅓ FEC scheme and QPSK modulation, the W/R value is 38.4. We could consider an 11-kbps data service for a similar 38.8:1 ratio for time division synchronized CDMA (TD-SCMDA), 1.28 megachips per second.

With a stationary radio path and no multipath, a BER of 10^{-5} can be maintained with E_b/N_0 equal to 4.1 dB, a factor of 2.6. Under these conditions, the value of α is 14.8, so a single call at this rate can withstand 14 interferers using this data rate in a single-cell-sector environment. Fifteen subscribers using this service will fill a 3G carrier with a little left over.

In the forward direction, we can use orthogonal codes. If we take the bit rate times the FEC rate, then we get 96 kbps, enough for a spreading gain of 32 and too many for a spreading gain greater than 32 on a 3.6844-megachip-per-second carrier.[4] Not only does the forward direction handle twice as many data channels on the 3G carrier, but those data channels also see no interference from the other channels on the same cell sector, so those 32 channels can withstand significant interference from other sectors.

If we put the user terminal in motion at 20 km/h, then the E_b/N_0 requirement goes up 2 dB, so the capacity goes down to 10 reverse and 26 forward. The reason for the reduction in forward capacity is multipath and Doppler interference not eliminated by the orthogonal codes. Increase the speed to 120 km/h, and the E_b/N_0 requirement goes up another 1 dB, the capacity going down to 8 reverse and 16 forward.

If we decide to use a higher channel FER and to manage the errors using ARQ, then we would have more delay but have a more efficient channel. The effective E_b/N_0 requirement goes down to 3.2 dB, a factor of 2.1, even counting the extra packets. The reverse capacity goes up to 19, whereas forward capacity remains 32 with orthogonal codes. At 20 km/h, the E_b/N_0 requirement is 3.8 dB, and reverse capacity goes down to 16, and at 120 km/h, the E_b/N_0 requirement is 4.8 dB, so reverse capacity goes down to 13. The forward capacity remains 32 even at the higher speeds, although the resistance to other interference goes down.

Lengthening the interleaving from 10 to 80 ms reduces the E_b/N_0 about 1 dB, so the reverse capacity goes up to 29 channels when stationary and up to 22 channels when moving 20 km/h.

29.4.2 High data rates

The high data rates around 2 Mbps of cdma2000 EV-DO and W-CDMA depend on close-range service from microcells or picocells. These services use high-order constellations with several bits per symbol, so having interference from other cell sectors would spoil the party.

We expect these high-rate wireless data services to be offered in office buildings or other specific targeted areas. Large-scale service at these levels would require an enormous number of serving base stations. These services do not support handoffs or even rapidly moving user terminals, so there is no obvious need for seamless service over a large area.

---

[4]The same arithmetic applies in our TD-SCDMA example. Thus 11 kbps times 3 is 33 kbps on a 1.28-megachip-per-second carrier.

The capacity of a 2-Mbps 3G CDMA carrier is one forward channel. This does not restrict the service to just one user, however, because that channel can be shared among several user terminals. A community of subscribers can all share high-speed, graphic Web access from the same carrier, although if multiple downloads are requested at the same time, service may be slow or somebody will have to wait.

29.5 Conclusion

In this chapter we derived the rates of data services offered to CDMA subscribers and showed the relationship between E_b/N_0 and BER. We also discussed the capacity of a server providing a particular data rate. In the next chapter we look at a number of issues affecting CDMA capacity and then derive equations that approximate real-world capacity values for each direction for each of the CDMA standards.

Capacity Issues Specific to CDMA

Code division multiple access (CDMA), with its $N = 1$ reuse pattern, has fundamentally different design and capacity issues than conventional reuse systems have. While both types of systems have sectors, handoff, and problems with multipath, the reasons for these features and the nature of the problems are very different on CDMA systems. In this chapter we explore the functions of various CDMA components and their contribution to capacity management. We close the chapter with a section in which we present capacity estimates based on the CDMA equations and resolve them for each of the four standards (cdmaOne, cdma2000, W-CDMA, and TD-SCDMA) in both directions. These applications of the equations are approximations based on the best available figures and do not provide guaranteed results. Our purpose in presenting them is not to guarantee a certain capacity level but to present the methods available for applying the CDMA equations to the standards. This last section, "CDMA Radio Capacity Estimation," provides a jumping-off point for the practical guides to capacity planning and increasing capacity in Parts 6 and 7.

30.1 Sectorized Cells

Sectors in CDMA are used for increased radio capacity. If we had a perfect base station antenna with 120 degrees of full gain and total attenuation in its backlobe, then a three-sector CDMA cell would have triple the capacity of an omnidirectional antenna. If our perfect antenna had 60 degrees and a perfect backlobe, then a six-sector CDMA cell would have six times the capacity.

More realistic antenna patterns have significant sector-to-sector overflow, as shown in Fig. 1.7 in Sec. 1.8. This means that there is a significant amount of user terminal traffic with nearly equal signal paths to two sectors. The CDMA equations in Sec. 28.1 tell us that there is a significant capacity cost to having equal-signal-path traffic. Even so, adding sectors adds some capacity to a CDMA system.

A CDMA rule of thumb is that three-sector cells carry 2.5 times as much traffic as omnidirectional cells and that six-sector cells carry four times as much traffic. The reduced rate of additional benefit is a result of the increase in sector-to-sector interference when there are more sectors. This is only a rough approximation based on some typical cellular sector antennas. Better antennas and carefully thought out sector orientation might do a little better than this.

There is nothing magic about three or six sectors in CDMA. Cellular engineers tend to think in terms of three or six sectors because of our conventional-reuse tradition. In conventional reuse, we use sectors to manage interference so that we can reuse channels more frequently. Since the conventional-reuse interferers are from other cells in the hexagonal pattern, we use three or six sectors. A CDMA cell could have 5 sectors, 9 sectors, or even 25 sectors if we wanted to put that much equipment at one base station.

So why don't we just keep adding sectors? Because each sector we add to a cell adds less capacity than the one before. Using our rule of thumb, doubling three sectors to six only adds 50 percent to the call capacity. This means that doubling the radio equipment cost does not double the subscriber revenue. At some point, adding sectors is less cost-effective than adding new base stations. The tradeoff between more sectors and more cells depends on the cost of building construction and of putting up a tower. Cost structures vary dramatically in different parts of the world. We have not seen a cost-benefit analysis to determine the optimal number of sectors for CDMA in any particular environment.

Adaptive phased arrays, or smart antennas, at the base stations may allow a CDMA system to gain the advantage of narrow-beam radio links without the serving provider having to pay for a lot of sectors. It depends on the relative cost and complexity of the smart antenna technology and having more antenna faces at the base station. Similar capacity advantages may be available using narrow-beam antennas at the user terminal, which is practical for wireless local loop (WLL), where the user terminal is in a fixed location.

30.2   Soft and Softer Handoff

The rigorous power-control requirements of CDMA led to the development of soft handoff to maintain the performance of a CDMA call able to be served by two antenna faces. This feature allows the system to maintain minimum power for these boundary calls.

The cost associated with soft handoff is significant. Typically, about one-half of all calls are in the soft-handoff state. This means that one-half of the calls are using two or more forward channels from two or more sectors. If exactly one-half of the calls used exactly two channels each, then the forward-capacity cost of soft-handoff traffic would be one-third of the total.[1] Since some of these calls are using three or more forward channels, we end up losing almost half the total capacity.

This is not a cost of the soft-handoff feature, however. It is a cost of having subscriber traffic nearly equal in more than one sector with rapidly changing radio propagation. The soft handoff is the technical solution that gets the best quality and most capacity from this adverse phenomenon of nature. Handling this with a rapid series of hard handoffs would be less satisfactory because it would have the wireless telephone system chasing the optimal radio path after the fact. At the cost of some capacity up front,

---

[1]Consider the simplified case of two cells, each of which, independently, has a capacity of 30 calls. If 20 calls are in soft handoff, each using two channels, then those 20 calls are using 40 channels. Each cell has an additional 10 channels for calls not in soft handoff. As a result, the two cells with a capacity of 30 calls each have a total shared capacity of only 40 calls. The capacity loss due to the need for soft handoff when a subscriber is near the border of a cell is one-third the total capacity of the two cells.

soft handoff maintains a continuous state of near-optimal communication between user terminals and antenna faces.

There are thresholds for starting and stopping the soft-handoff process in a CDMA system. Adjusting these thresholds tunes the tradeoff between suboptimal radio links and the cell-capacity reduction created by the up-front cost of soft handoff.

30.3 Multipath

As we discussed in Sec. 8.4, the multipath radio environment is another challenge in CDMA design. The broad bandwidth of CDMA and the yet broader bandwidth of 3G CDMA free us from the severe Rayleigh fading of narrowband radio links, but the higher chip rates increase the amount of intersymbol interference, which is dealt with by the rake filter. As a result, multipath is more of a capacity issue in CDMA and more of a call-quality issue in conventional reuse.

The capacity cost of multipath is highly dependent on the radio environment, which determines the amount of intersymbol interference. It is an unfortunate fact of nature that the regions with high demand densities are the same urban environments that have a lot of big buildings that generate multiple radio paths.

The rake filter does a better job of filtering out same-sector interferers on the reverse link where the radio paths are different for each user terminal. In the forward direction, the same antenna face that transmits the desired signal is transmitting most of the interference as well, so the rake filter does not discriminate between signal and same-sector interferers.

Unlike soft handoff, there is no way to tweak parameters to adjust the multipath loss. The only way to minimize the multipath problem is to avoid multiple radio paths. If we know where our demand is located, then we may be able to select base station locations and sector plans to minimize alternative radio routing. In-building microcells and picocells, for example, could mitigate the multipath problem.

30.4 Power Control and Pilots

The CDMA equations assume perfect power control. Even the aggressive combination of open-loop and closed-loop systems used in CDMA systems is not going to do a perfect job of managing the near-far problem. As we said in Sec. 8.5, this is seldom an issue of having enough power. Rather, the issue usually is managing the relative power levels of a user terminal community. Thus, adding a 1.5-dB margin to the power-control process does not add 1.5 dB of safety; rather, it reduces the capacity of the carrier by 1.5 dB, which is 40 percent.

This is primarily a problem in the reverse direction. When a user terminal experiences a sudden change in its radio signal path, when going into a building, for example, its signal to the base station is suddenly far worse in comparison with its interferers. The more rapidly radio signal paths change, the greater is the impact on the reverse link.

While the forward signal path to the user terminal is also reduced, the interfering signals from the same antenna face are similarly reduced, so the effect is less pronounced. There is still a need to increase forward power to maintain adequate signal quality, but the urgency is not so great. The cdmaOne system has a faster power-control loop for the reverse link than for the forward link for this reason, whereas cdma2000 and wideband CDMA (W-CDMA) have rapid feedback for power control in both directions.

The design tradeoff in CDMA power control is that we can have more precise power control only by spending more power on power-control messages. When the required rate of data transfer is relatively low, as is the case with cdmaOne and cdma2000 voice calls, the capacity devoted to extra power control costs more than the advantage that it provides. However, with the higher data rates offered on 3G systems, cdma2000 and W-CDMA are designed to have high-speed power control in both forward and reverse directions. The more data required on the channel, the more the tradeoff favors having more signal-control information. Signal-control information includes both forward power control and reverse pilot going from the user terminal to the base station. The smaller voice channel does not justify the extra overhead.

Let us work it out by an example. Suppose that we are considering whether to add an 800-bit-per-second power-control stream to the 9600-bit-per-second cdmaOne speech channel. To have this power-control stream be as accurate as the speech channel means sending 10,400 bits per second instead of 9600 bits per second with the same accuracy, a cost of 8.3 percent in power, 0.35 dB. If the benefit of that 800-bit-per-second power-control channel is a 0.30-dB reduction in power variation, then it is not worth the extra overhead. Everything becomes capacity in CDMA: The increase in power-control power costs 0.35 dB of capacity, and the decrease in power variation gains 0.30 dB in capacity. A losing proposition, but not by much.

The same decision for a 64,000-bit-per-second data channel comes out the other way, however. The same 800 bits per second gains the same 0.30 dB, but now our total data requirement goes from 64,000 to 64,800 bits per second. This is a 1.25 percent increase, which is a 0.05-dB change in total power. Another way to think of this is that spending 800 bits per second for signal improvement becomes a better deal when the signal being improved is bigger. Both cmda2000 and W-CDMA offer high-speed forward power control because they support higher data rates than cdmaOne.

This is similar to the decision for a reverse pilot. The extra overhead of a pilot is traded off against the signal improvement of coherent demodulation. In cdmaOne, the small reverse signal does not justify a reverse pilot, and the cdmaOne base station receiver uses a noncoherent demodulation scheme. Both cdma2000 and W-CDMA maintain a pilot signal in the reverse direction, in different ways, allowing the system to use a coherent demodulation scheme on the higher data signal with its higher rate.

30.5 Speech Coding

CDMA, like the Global System for Mobility (GSM), supports a low-bit-rate speech coding system. The air interface is a naturally scarce resource, and its bit-rate capacity is defined by bandwidth and signal-to-noise (S/N) ratio. Therefore, the designers of voice channels for wireless telephone systems want to make every bit count. We start with a low-bit-rate speech coder and reduce the rate further when we can. The cdmaOne and cdma2000 systems use Qualcomm code-excited linear prediction (QCELP), whereas W-CDMA uses adaptive multirate (AMR) speech coding.

cdmaOne and cdma2000 have two different speech systems, a mandatory 8000-bit-per-second speech coder and an optional 13,000-bit-per-second coder that offers fewer calls with higher speech quality.

The bit rate is reduced when the speech does not require the full rate. While GSM turns off the speech data stream when the speaker is idle, the CDMA air interface varies the rate in steps and makes full use of this rate reduction by broadcasting less power. This gives the carrier the immediate opportunity to handle more traffic on the same radio carrier.

Conventional-reuse systems also benefit from suppressing transmission when the speaker is quiet, but the benefits are a statistical reduction in interference in another cell, an engineering opportunity in overall system design. The difference is that CDMA allows us to take advantage of speaker pauses moment by moment and to realize immediate capacity improvement.

30.6   Soft Blocking

When a sector is unable to serve more traffic, we would like to overflow its traffic onto neighboring sectors. In Sec. 23.4 we saw how much more efficient a telephone network becomes with traffic overflow instead of blocking. In the small channel groups of wireless telephone systems this traffic overflow is a very important positive capacity factor.

The CDMA channel is not so cooperative. Having a call served on a nonoptimal sector is actually *worse* for the optimal sector. The interference is actually more damaging to the nonserving sector than overloading it with the extra call.[2] Thus, once a sector's CDMA channel is full, its traffic has to be blocked rather than overflowed.

There is some good news in the CDMA blocking picture. The capacity of a CDMA sector depends on its neighbors, so there is a notion of the capacity of a community of CDMA sectors and cells. This notion of the CDMA carrier having a pliable upper limit is called *soft blocking*. This gives the CDMA channel some more latitude than the hard upper limit of the Erlang-B values in Table 23.1.

There is another soft side to CDMA blocking. When a conventional-reuse sector is full, it is full. It is not like socks in the laundry bag, where we almost always can squeeze in just one more. However, a CDMA carrier will let us put one more call on it if we do not mind some deterioration of call quality. Thus a wireless carrier may decide to allow lower radio conditions once in a while because the condition is temporary. If the average call lasts 2 minutes and the carrier supports 20 calls, then the expected time until somebody hangs up is 6 seconds, not terribly long, probably not long enough for subscribers to notice. Even if the service provider does not deliberately overload the CDMA carrier, radio conditions can change. A subscriber can walk into a building or a truck can drive between a user terminal and its antenna face. The CDMA channel is flexible enough that it probably will be able to maintain all the calls, perhaps in a slightly degraded state, until somebody hangs up and relieves the overloaded condition.

30.7   CDMA Radio Capacity Estimation

In this section we look at eight CDMA carrier environments, forward and reverse for cdmaOne, cdma2000, W-CDMA, and time division synchronized CDMA (TD-SCDMA). The capacity values we get will have fractional values such as 20.7 voice users for cdmaOne. The fractional call, the extra 0.7, would not be important if this value were

---

[2]While this is not good news about CDMA, we should not forget how much more efficient CDMA is than conventional reuse.

derived for a single serving sector in isolation. After all, it is of no use to the twenty-first caller that there is room for seven-tenths of a voice call. However, these values represent CDMA carrier capacity taking into account interference from other cells and other sectors on the same cell, so a carrier with 20.7 call capacity is 0.7 users *better* than one with only 20.0.

Let us look at several variables that determine CDMA carrier capacity. The values we used in our capacity analysis are in Table 30.1.

- W is the chip rate.

- R is the information data rate.

- E_b/N_0 is the energy per bit divided by the noise level, the CDMA equivalent of S/N ratio.

- d is the effect of Doppler distortion and any other distortion due to motion, so a stationary call has $d = 0$ dB and a call from a moving car has $d = 2$ dB.

- f is the interference contribution from other cells relative to the carrier on the serving cell sector.

- g is the interference contribution from other sectors of the same cell relative to the carrier on the serving sector.

- h is the overhead for pilot and power control relative to the carrier on the serving sector.

- p is the effect of power-control delays and errors, so perfect power control is $p = 0$ dB ($p = 1$), and values less than 1 reflect lower performance.

- v is the voice activity factor, so $v = 1$ represents channels that are always on and $v = \frac{2}{3}$ represents channels that are powered off one-third of the time.

- s is the effect of soft and softer handoff in the forward direction, counting each extra serving antenna, so a system with one-half the traffic served by two cells has $s = \frac{1}{2}$ and a system with half that traffic served by three cells has $s = \frac{3}{4}$.

- j is the orthogonality or joint-detection (JD) effect, so an uncorrelated CDMA carrier has $j = 0$, a fully orthogonal carrier has $j = 1$, and a system using joint detection has $j = 0.8$ ($j = -1$ dB).

TABLE 30.1   CDMA Capacity Parameters

	Direction	W, Mchps	R, kbits	E_b/N_0, dB	d, dB	f	g	h	p	v	s	j	m
cdmaOne	Fwd	1.2288	9.6	5.0	1.0	0.5	0.3	0.1	0.6	0.4	0.8	0.0	0.3
cdmaOne	Rev	1.2288	9.6	7.5	1.0	0.5	0.3	0.0	0.8	0.4		1.0	0.3
cdma2000 voice	Fwd	1.2288	9.6	4.5	1.0	0.5	0.3	0.1	0.9	0.4	0.8	0.0	0.3
cdma2000 voice	Rev	1.2288	9.6	4.5	1.0	0.5	0.3	0.2	0.8	0.4		1.0	0.3
cdma2000 data	Fwd	1.2288	160.0	9.0	0.5	0.5	0.3	0.0	0.9	0.2	0.8	0.0	0.3
cdma2000 data	Rev	1.2288	160.0	9.0	0.5	0.5	0.3	0.0	0.8	0.2		1.0	0.3
W-CDMA voice	Fwd	3.84	4.8	4.5	1.0	0.5	0.3	0.3	0.9	0.6	0.8	0.0	0.4
W-CDMA voice	Rev	3.84	4.8	4.5	1.0	0.5	0.3	0.2	0.9	0.6		1.0	0.4
W-CDMA data	Fwd	3.84	160.0	9.0	0.5	0.5	0.3	0.0	0.9	0.2	0.8	0.0	0.4
W-CDMA data	Rev	3.84	160.0	9.0	0.5	0.5	0.3	0.0	0.9	0.2		1.0	0.4
TD-SCDMA voice	Fwd	1.28	8.0	4.5	1.0	2.0	0.3	0.1	0.7	0.6	0.0	0.0	0.3
TD-SCDMA voice	Rev	1.28	8.0	4.5	1.0	1.0	0.3	0.1	0.5	0.6		0.8	0.3
TD-SCDMA data	Fwd	1.28	160.0	9.0	0.5	2.0	0.3	0.0	0.7	0.2	0.0	0.0	0.3
TD-SCDMA data	Rev	1.28	160.0	9.0	0.5	1.0	0.3	0.0	0.5	0.2		0.8	0.3

- m is the contribution of multipath time sidelobes (after the rake filter described in Sec. 8.4) relative to the carrier on the serving cell sector, so $m = 0$ is no multipath interference and $m = 0.4$ is significant time sidelobe interference.

The general capacity formulas are Eqs. (30.1) and (30.2):

$$\text{Capacity}_{\text{forward}} = \frac{W/R}{E_b/N_0} \frac{p}{dv(1+s)} \frac{1}{(j+f+g)(1+h)+m} + j \qquad (30.1)$$

$$\text{Capacity}_{\text{reverse}} = \frac{W/R}{E_b/N_0} \frac{p}{dv} \frac{1}{(j+f+g)(1+h)} + j(1+h) - m \qquad (30.2)$$

We also should take into account a forward carrier in an environment so clean that it uses up all its orthogonal codes.[3]

The bit rate R and the E_b/N_0 values we are using are information rates, before forward error correction (FEC) is added. Let us use the cdmaOne voice channel with an information bit rate of 9.6 kbps as an example. Rate ½ FEC brings that to 19.2 kbps, and that FEC-coded bit stream requires an E_b/N_0 of 2 dB to sustain a sufficient bit error rate (BER). This is equivalent to a bit rate of 9.6 kbps with an E_b/N_0 of 5 dB. The extra 3 dB is the factor of 2 in the rate ½ FEC. We can look at R and E_b/N_0 from either the information viewpoint ($R = 9.6$, $E_b/N_0 = 5$ dB) or the FEC-coded viewpoint ($R = 19.2$, $E_b/N_0 = 2$ dB). These are two ways of looking at the same channel; the CDMA resource used is the same either way. We are using the information viewpoint in these calculations.

The traffic engineering we do is based on a soft-blocking model where the interference from other cells is from three equal sources. If $f = 0.6$, then we model that contribution as three sectors with the same demand as the serving sector with a radio path gain five times less (0.2) than the serving path. We also divide the other-sector contribution g over two neighboring sectors on the same cell. As a reasonable point of reference, we are doing the traffic engineering for 20 MHz of radio bandwidth in each direction.

The measure of radio spectrum efficiency is how much call activity each call can provide per unit bandwidth. We measure this efficiency in how many erlangs each cell can serve for each megahertz of bandwidth, so the unit of radio spectrum efficiency is erlangs per megahertz per cell. Advanced Mobile Phone Service (AMPS) $N = 7$ manages about 70 erlangs in each three-sector cell using 20 MHz of radio spectrum, so AMPS gets about 3.5 erlangs/MHz/cell.[4] At the same level of signal quality, the Global System for Mobility (GSM) gets 8.2 erlangs/MHz/cell.[5] CDMA, with rates varying from 49.5 to 79.6 erlangs/MHz/cell, is far more efficient at voice calls, as we can see in Table 30.2.

Since we are looking at the radio spectrum efficiency for an arbitrary amount of bandwidth available, we are ignoring guard bands between a CDMA system and whatever other technology may be on either side of the radio spectrum. For example, we are using 16 cdmaOne or cdma2000 carriers in 20 MHz even though a service provider with

---

[3]The orthogonal codes in the reverse direction of 3G CDMA are only orthogonal for individual channels from the same user terminal for one call. Calls are not orthogonal to each other.

[4]This value looks larger than typical AMPS systems because those AMPS systems use 10 or 12.5 MHz in each direction, and we are using 20 MHz for comparison.

[5]The GSM carrier itself is twice as efficient as AMPS, and traffic engineering adds another 17 percent.

TABLE 30.2   CDMA Capacity Estimates for Voice and Data

	Forward capacity per sector	Reverse capacity per sector	Limiting direction	Erlangs for one carrier	One-way erlangs per sector 20 MHz	Erlangs per MHz per cell
$N = 7$ AMPS	30	30	Neither		23.3	3.5
$N = 4$ GSM	64	64	Neither		54.8	8.2
cdmaOne	22.7	20.7	Reverse	16.2	317.0	47.5
cdma2000 voice	38.2	34.3	Reverse	28.3	530.9	79.6
cdma2000 data	2.0	2.6	Forward	0.9	25.6	3.8
W-CDMA voice	130.5	157.4	Forward	119.3	504.2	75.6
W-CDMA data	5.6	7.3	Forward	3.4	17.6	2.6
TD-SCDMA voice	18.6	16.9	Reverse	12.9	198.4	59.5
TD-SCDMA data	1.2	1.6	Forward	0.4	10.7	3.2

exactly 40 MHz of frequency allocation might only be able to use 15 carriers to avoid stepping on the radio neighbors' toes.

Because we are confining our estimate to the saturation capacity of CDMA under small-cell, high-density conditions, we have made three other simplifications. First, we are ignoring the limitations of transmit power in both forward and reverse directions. Since a CDMA carrier runs out of capacity by running out of transmit power, this would appear to be a major oversight. However, in the saturation mode of small cells and high density, we are climbing far up the vertical asymptotes of the power curves in Figs. 28.1 and 28.2, where even large increases in available power make little difference in capacity.

Second, we are ignoring any coverage advantage that multipath gives a CDMA channel. Having more than one path can increase the range by increasing the power at the CDMA receiver, but it degrades the top capacity by introducing time sidelobes in the output of the rake filter. The time sidelobes affect the maximum CDMA capacity when multipath is present.

Third, we have not included an independent variable for the quality of user terminals. The variability of the quality of user terminals can be accounted for by adjusting E_b/N_0 or p, the variable for power-control inefficiency.

Our data service example is a subscriber who is active one-fifth of the time, $v = 0.2$. The service is a data rate of 160 kbps with a low BER of 10^{-5}. With rate $\frac{1}{3}$ FEC, that BER requires an E_b/N_0 of about 4.2 dB. Counting the factor of 3 (4.8 dB) for the rate $\frac{1}{3}$ FEC, this gives us an equivalent E_b/N_0 of 9 dB.

In the sections that follow we discuss the assumptions that generate these capacity estimates. It is important to be clear that these are estimates rather than experimental results. For cdmaOne, definite figures derived from experimental results are beyond the scope of this book because results will vary with different proprietary equipment. For 3G systems, experimental results do not yet exist because the systems have not yet been deployed.

30.7.1   cdmaOne

The chip rate W is 1.2288 megachips per second, and the voice bit rate we are using for this calculation is the information rate R set to 9.6 kbps with E_b/N_0 set to 5.0 dB in the

TABLE 30.3   cdmaOne Summary

		Parameter values in CDMA capacity estimation												
	Direction	$W,$ Mchps	$R,$ kbits	$E_b/N_0,$ dB	$d,$ dB	f	g	h	p	v	s	j	m
cdmaOne	Fwd	1.2288	9.6	5.0	1.0	0.5	0.3	0.1	0.6	0.4	0.8	0.0	0.3
cdmaOne	Rev	1.2288	9.6	7.5	1.0	0.5	0.3	0.0	0.8	0.4		1.0	0.3

	CDMA capacity estimates					
	Forward capacity per sector	Reverse capacity per sector	Limiting direction	Erlangs for one carrier	One-way erlangs per sector 20 MHz	Erlangs per MHz per cell
cdmaOne	22.7	20.7	Reverse	16.2	317.0	47.5

forward direction and 7.5 dB in the reverse direction. Rate set 1 (RS1) uses a rate ½ FEC, so the equivalent coded numbers would be R = 19.2 kbps and E_b/N_0 set to 2.0 dB. We believe that most users will be using the lower-bit-rate RS1 enhanced variable-rate codec (EVRC) speech coder rather than the higher-bit-rate RS2 coder, which does not sound much better.

Stationary subscribers have d = 0 dB, and moving subscribers have d = 2 dB, an extra 2 dB of E_b/N_0 requirement. We decided to split the difference with d = 1 dB with some subscribers stationary and others moving. We are estimating other-cell interference at one-half the power of the serving CDMA carrier, f = 0.5, and other-sector interference at 30 percent of that same power, g = 0.3. We estimate the forward-direction overhead of global pilot, power control, and signaling to be one-tenth the total voice channel power, so h = 0.1 in the forward direction. The reverse direction has no pilot and has negligible power-control information, so h = 0.0. Power control is slow in the forward direction, so we use p = -2 dB to reflect the inefficiency of power often being out of balance. The reverse direction uses rapid power control, so we set p = -1 dB. The four-rate speech coder in cdmaOne and cdma2000 allows the channel to be aggressive in conserving voice call activity, so we set v to 0.4. If half the system is in a state of soft handoff or softer handoff and half of those calls involve three or four sectors, then the soft-handoff factor used in the forward direction should be about s = 0.8. Orthogonal codes give us j = 0 in the forward direction and j = 1 in the reverse direction because cdmaOne has no JD in the base station receiver. Finally, the multipath factor depends heavily on the radio environment, and we are figuring that the densest and busiest areas are fairly hostile, so m = 0.3 is our estimate for this chip rate.

In computing the values in Table 30.3 (the same as those in Tables 30.1 and 30.2), we put these parameters into Eqs. (30.1) and (30.2). We used three-sector cells and 16 carriers in 20 MHz for these calculations. These values are consistent with the reported experience of engineers using cdmaOne in practice.

30.7.2   cdma2000 1x

The chip rate W is 1.2288 megachips per second, and the voice bit rate we are using for this calculation is the information rate of R set to 9.6 kbps, the same EVRC speech coder we used in cdmaOne. The EVRC speech coder is a standard part of cdma2000. In the forward direction we are using an E_b/N_0 value of 4.5 dB, half a decibel better than cdmaOne, to reflect the more efficient FEC algorithms.[6] In the reverse direction, we gain the same improvement due to FEC and another 2.5 dB for the coherent demodulation for an E_b/N_0 value of 4.5 dB.

We continue to split the Doppler difference with $d = 1$ dB, with some subscribers stationary, $d = 0$ dB, and others moving, $d = 2$ dB. Our data subscribers are likely to be less mobile, so we set their Doppler factor to $d = 0.5$ dB. We are estimating other-cell interference at one-half the power of the serving CDMA carrier, $f = 0.5$, and other-sector interference at 30 percent of that same power, $g = 0.3$, the same as cdmaOne. The forward overhead for global channels is about the same as cdmaOne, $h = 0.1$. The reverse direction, however, has significant overhead in cdma2000 for pilot and power control, $h = 0.2$, that was not present in cdmaOne. (We pay 1 dB in overhead for a 4-dB benefit from coherent demodulation and fast power control.) We consider the overhead insignificant compared with the high-speed data channel, so $h = 0$ for data. cdma2000 has fast power control in both directions, so we set $p = -0.5$ dB in the forward direction and $p = -1$ dB in the reverse, where relative radio signal path changes are greater. The four-rate speech coder in cdmaOne and cdma2000 allows the channel to be aggressive in conserving voice call activity, so we set v to 0.4. Our data service example has $v = 0.2$. If half the system is in a state of soft or softer handoff and half of those calls involve three or four sectors, then the soft-handoff factor used in the forward direction should be about $s = 0.8$. Orthogonal codes give us $j = 0$ in the forward direction and $j = 1$ in the reverse direction because cdma2000 has no JD in the base station receiver. Finally, the multipath factor depends heavily on the radio environment, and we are figuring that the densest and busiest areas are fairly hostile, so $m = 0.3$ is our estimate for this chip rate.

In computing the values in Table 30.4 (the same as those in Tables 30.1 and 30.2), we put these parameters into Eqs. (30.1) and (30.2). We used three-sector cells and 16 carriers in 20 MHz for these calculations. We did not perform a separate set of calculations for cdma2000 3x because, although it supports higher data rates, it does so by operating at greater bandwidth. As a result, the capacity for cdma2000 3x data (whether multicarrier or single carrier) is the same as the capacity for cdma2000 1x data when measured in erlangs per megahertz per cell.

30.7.3   W-CDMA

The chip rate W is 3.84 megachips per second, and the voice bit rate we are using for this calculation is the information rate R set to 4.8 kbps, with E_b/N_0 set to 4.5 dB, the lowest AMR bit rate, with rate ½ FEC. The speech coder will use its lowest data rate when the system is reaching its capacity limit. The data example has R set to 160 kbps, with E_b/N_0 set to 9 dB for rate ⅓ FEC and 10^{-5} BER.

We continue to split the Doppler difference with $d = 1$ dB, with some subscribers stationary, $d = 0$ dB, and others moving, $d = 2$ dB. Our data subscribers are likely to be less mobile, so we set their Doppler factor to $d = 0.5$ dB. We are estimating other-cell interference at one-half the power of the serving CDMA carrier, $f = 0.5$, and other-

---

[6]This also includes some improvement due to cdma2000's better processing of soft-handoff signals.

TABLE 30.4   cdma2000 Summary

		W, Mchps	R, kbits	E_b/N_0, dB	d, dB	f	g	h	p	v	s	j	m
	Direction												
cdma2000 voice	Fwd	1.2288	9.6	4.5	1.0	0.5	0.3	0.1	0.9	0.4	0.8	0.0	0.3
cdma2000 voice	Rev	1.2288	9.6	4.5	1.0	0.5	0.3	0.2	0.8	0.4		1.0	0.3
cdma2000 data	Fwd	1.2288	160.0	9.0	0.5	0.5	0.3	0.0	0.9	0.2	0.8	0.0	0.3
cdma2000 data	Rev	1.2288	160.0	9.0	0.5	0.5	0.3	0.0	0.8	0.2		1.0	0.3

Parameter values in CDMA capacity estimation

CDMA capacity estimates

	Forward capacity per sector	Reverse capacity per sector	Limiting direction	Erlangs for one carrier	One-way erlangs per sector 20 MHz	Erlangs per MHz per cell
cdma2000 voice	38.2	34.3	Reverse	28.3	530.9	79.6
cdma2000 data	2.0	2.6	Forward	0.9	25.6	3.8

sector interference at 30 percent of that same power, $g = 0.3$, the same as cdmaOne and cdma2000. The forward overhead for global channels is about the same as cdmaOne and cdma2000, but there is a substantial overhead within each channel for pilot and power control. W-CDMA maintains global forward channels that are not orthogonal to the subscriber channels, and it maintains both a global pilot and a pilot within each channel. There is also high-speed power control in both directions, so we set the voice overhead to $h = 0.3$ in the forward direction and to $h = 0.2$ in the reverse direction. We consider the overhead insignificant compared with the high-speed data channel, so $h = 0$ for data. W-CDMA has very fast power control in both directions, so we set $p = -0.5$ dB in both directions. The AMR speech coder uses a lower bit rate than cdmaOne or cdma2000 when the radio channel is reaching saturation, but it does not have four-rate speech coding for voice activity detection, so we set v to 0.6 for voice, 1.5 times higher than the cdma2000 value. Our data service example has $v = 0.2$. If half the system is in a state of soft or softer handoff and half those calls involve three or four sectors, then the soft-handoff factor used in the forward direction should be about $s = 0.8$. Orthogonal codes give us $j = 0$ in the forward direction and $j = 1$ in the reverse direction because W-CDMA has no JD in the base station receiver. Finally, the multipath factor depends heavily on the radio environment, and we are figuring that the densest and busiest areas are fairly hostile, so $m = 0.4$ is our estimate for this higher chip rate.

In computing the values in Table 30.5 (the same as those in Tables 30.1 and 30.2), we put these parameters into Eqs. (30.1) and (30.2). We used three-sector cells and four carriers in 20 MHz for these calculations.

30.7.4   TD-SCDMA

The chip rate W is 1.28 megachips per second, and the voice bit rate we are using for this calculation is the information rate R set to 8 kbps with E_b/N_0 set to 4.5 dB. The data example has R set to 160 kbps, with E_b/N_0 set to 9 dB for rate ⅓ FEC and 10^{-5} BER.

A major difference between TD-SCDMA and the other CDMA carriers described here is the amount of interference from other cells. Because TD-SCDMA does not have soft

TABLE 30.5 W-CDMA Summary

		W,	R,	E_b/N_0,	d,								
	Direction	Mchps	kbits	dB	dB	f	g	h	p	v	s	j	m
W-CDMA voice	Fwd	3.84	4.8	4.5	1.0	0.5	0.3	0.3	0.9	0.6	0.8	0.0	0.4
W-CDMA voice	Rev	3.84	4.8	4.5	1.0	0.5	0.3	0.2	0.9	0.6		1.0	0.4
W-CDMA data	Fwd	3.84	160.0	9.0	0.5	0.5	0.3	0.0	0.9	0.2	0.8	0.0	0.4
W-CDMA data	Rev	3.84	160.0	9.0	0.5	0.5	0.3	0.0	0.9	0.2		1.0	0.4

Parameter values in CDMA capacity estimation

CDMA capacity estimates

	Forward capacity per sector	Reverse capacity per sector	Limiting direction	Erlangs for one carrier	One-way erlangs per sector 20 MHz	Erlangs per MHz per cell
W-CDMA voice	130.5	157.4	Forward	119.3	504.2	75.6
W-CDMA data	5.6	7.3	Forward	3.4	17.6	2.6

handoffs, we expect calls to reach higher levels of intercell interference before they are handed off the hard way. Thus we estimate a higher value for intercell interference, $f = 1.0$, in the reverse direction.

In addition, the forward-direction time division duplex (TDD) allows user terminals to interfere with each other on the TD-SCDMA carrier. This means that there will be interference not just from other base stations at the approximate centers of other cells but also from user terminals that can be much closer. In the absence of real-world experience, we are estimating a forward-direction intercell value of $f = 2$, but this could be dramatically lower or higher in a TD-SCDMA system.

We are using six sectors as an approximation for the adaptive phased arrays (smart antennas) used in TD-SCDMA. We estimate other-sector interference at 30 percent of total signal power, $g = 0.3$, the same as cdmaOne, cdma2000, and W-CDMA.

Orthogonal codes give us $j = 0$ in the forward direction, but TD-SCDMA uses joint detection in the reverse direction, which saves about 1 dB, so we use $j = -1$ dB.

We continue to split the Doppler difference with $d = 1$ dB, with some subscribers stationary, $d = 0$ dB, and others moving, $d = 2$ dB. Our data subscribers are likely to be less mobile, so we set their Doppler factor to $d = 0.5$ dB. Because of the low-rate power control, TD-SCDMA has less channel overhead, $h = 0.1$ for voice channels. We consider the overhead insignificant compared with the high-speed data channel, so $h = 0$ for data. The lower rate of power control means more variation in power levels compared with the optimal values, so we set $p = -1.5$ dB in the forward direction and $p = -3$ dB in the reverse direction, where relative radio signal path changes are greater. Like W-CDMA, the TD-SCDMA speech coder does not have four-rate speech coding for voice activity detection, so we set v to 0.6 for voice. Our data service example has $v = 0.2$. Without soft handoff, we can use $s = 0$. Finally, the multipath factor depends heavily on the radio environment, and we are figuring that the densest and busiest areas are fairly hostile, so $m = 0.3$ is our estimate for this chip rate.

In computing the values in Table 30.6 (the same as those in Tables 30.1 and 30.2), we put these parameters into Eqs. (30.1) and (30.2). We used six-sector cells as an approximation for smart antennas used in TD-SCDMA. And we used 12.5 carriers in 20 MHz for these calculations.

TABLE 30.6 TD-SCDMA Summary

		Parameter values in CDMA capacity estimation												
	Direction	W, Mchps	R, kbits	E_b/N_0, dB	d, dB	f	g	h	p	v	s	j	m
TD-SCDMA voice	Fwd	1.28	8.0	4.5	1.0	2.0	0.3	0.1	0.7	0.6	0.0	0.0	0.3
TD-SCDMA voice	Rev	1.28	8.0	4.5	1.0	1.0	0.3	0.1	0.5	0.6		0.8	0.3
TD-SCDMA data	Fwd	1.28	160.0	9.0	0.5	2.0	0.3	0.0	0.7	0.2	0.0	0.0	0.3
TD-SCDMA data	Rev	1.28	160.0	9.0	0.5	1.0	0.3	0.0	0.5	0.2		0.8	0.3

	CDMA capacity estimates					
	Forward capacity per sector	Reverse capacity per sector	Limiting direction	Erlangs for one carrier	One-way erlangs per sector 20 MHz	Erlangs per MHz per cell
TD-SCDMA voice	18.6	16.9	Reverse	12.9	198.4	59.5
TD-SCDMA data	1.2	1.6	Forward	0.4	10.7	3.2

By doing these calculations in just one direction, we have ignored the flexibility of the TDD carrier. If a TD-SCDMA has a total allocation of 20 MHz for both directions, then it can use six-sevenths of that, 17.1 MHz, for just one direction if the subscriber demand in the other direction is light. The frequency division duplex (FDD) carriers do not permit this variability on a minute-by-minute basis.

30.8 Conclusion

We have now explored and explained the underlying principles that determine capacity and quality on CDMA systems. This final chapter of Part 5 gives the equations and estimated values that allow us to move from principles to the application of those principles. In the next two parts, "Planning for CDMA Capacity" and "Increasing Capacity," we apply these principles to the practical day-to-day work of cellular system planning and growth.

In an ideal world, we would be able to model our entire network, optimize the model, and then tweak our systems based on accurate models. Unfortunately, this is not always possible. Sometimes, using algorithms and computer programs, we can work from models that apply these principles. All too often, however, factors that are difficult to quantify, plus time and cost pressures, require us to follow rules of thumb and make our best engineering judgments without the advantage of precise modeling. In such cases, our understanding of the principles we have discussed in this chapter guides our thinking and improves our ability to make good decisions, improving the capacity, reliability, and profitability of the systems we design and improve.

6

Planning for CDMA Capacity

Planning a major code division multiple access (CDMA) network is a huge job. A network is a set of interconnected parts, and the capacity and reliability of the network depend on the capacity and reliability of each component. There may be more and fewer important parts of a network, but there are no unimportant parts. In Part 6, "Planning for CDMA Capacity," we apply the principles and key concepts of radio, telephony, and data engineering to the job of planning all parts of a CDMA nework.

In addition to ensuring sufficient capacity and reliability, we must work to keep costs low. It is very easy for a technical or management error in one part of network planning to create a situation where costs for that one component, or one interface, spiral out of control. Cellular networks operate on a relatively low profit margin, so each failure to manage costs has a significant effect on the profitability of the entire network. Evidence gathered throughout all kinds of projects—from manufacturing to construction to software development—indicates that errors in planning, if not caught during planning, will cost approximately 10 times as much to fix during implementation. And if the errors are not caught during implementation, they increase operations, administration, and maintenance (OA&M) costs by a factor of 100. This rule, called the 1/10/100 rule, makes it clear that it is hard to overestimate the value of applying the best principles we can to network planning and double- and triple-checking our work before we go from planning to implementation.

In Chap. 31, "Estimating Wireless Telephone Demand," we estimate voice call demand and demand for data services. From these estimates, we derive a proposed value for the number of cells in the region. Once we know how many cells we are creating, we need to decide where to put them. In Chap. 32, "Planning Locations for Base Stations," we look at issues of the air interface, as well as cost issues for cell sites, and discuss how to make the best possible choice for placing each base station, balancing these two issues. Chapter 32 also introduces tools such as cell splitting, microcells, picocells, and repeaters for cost-effective ways to provide good coverage, as well as tips for reducing the cost of base stations and backhaul transport.

In Chap. 33, "Base Station Planning," we look into the details of planning each base station. We discuss coverage, downtilting antennas to reduce interference for nearby cells, sector plans, and carrier assignment. We also discuss how to handle peak traffic demand and how to migrate from frequency division multiple access (FDMA) or time division multiple access (TDMA) to CDMA and from cdmaOne to cdma2000. We end the chapter with discussions of quality, cost, and reliability issues in base station planning.

In Chap. 34, "Mobile Switching Center (MSC) Planning," we turn our attention to planning the mobile switching center (MSC). MSC switches are expensive, so we show how to estimate MSC count to determine the minimum number of switches necessary to support a region while allowing some room for growth. We discuss location planning, equipment provisioning, assigning base stations to MSCs, and planning inter-MSC borders. We also help you plan the auxiliary components that are typically housed at the MSC site but are not part of the MSC switch.

In Chap. 35, "Backhaul Planning," we turn our attention to transport. We discuss fan-out networks, backbone backhaul networks, transport for packet data that can bypass the MSC, reliability engineering, planning for growth, and alternative backhaul technologies.

In Chap. 36, "Signaling Capacity Planning," we turn to the process of planning our equipment and transport to ensure the network's ability to meet the demand for signaling, including handoffs, call processing, roamers, and short message service (SMS). In Chap. 37, "MSC Transport Planning," we discuss inter-MSC transport, as well as transport between the MSC and the public switched telephone network (PSTN) and the public packet data network (PPDN). We wrap up Part 6 by addressing special situations such as asymmetric demand and market-entry planning issues in Chap. 38, "Special Situations."

31

Estimating Wireless Telephone Demand

The demand for wireless telephone services from a given service provider is the primary determinant in the size and design of a wireless network. The number of cells, the capacity of each cell, and the approximate location of each base station should be planned with this one variable as the key factor.

Since we need to plan and deploy networks ahead of demand, we must have a way of estimating the demand for voice and data services. Historical data, demographics, and market analysis are all important elements in calculating demand.

31.1   Estimating Voice Call Demand

We start our estimate of wireless telephone demand with the demand for wireless voice telephone service. And we start our estimate of voice telephone demand by defining our market. The business plan for a code division multiple access (CDMA) wireless telephone system is based on some picture of who is going to use wireless service rather than landline service to make telephone calls and then going to pay their telephone bills so that we can make money. There are four sources of voice subscribers.

The first source of voice subscriber business is a base of new subscribers. Wireless telephone demand is still growing as we write this in 2002, and we expect that growth to continue. The *rate* of growth is declining as the market for wireless telephony is saturating, but there are still more cell phones each month. The CDMA air interface offers an efficient medium for offering lower rates to attract new subscribers. It is important to realize that in Europe and America, these are not actually new subscribers to telephone service but rather people who are adding wireless service to landline service or replacing landline service altogether with wireless service. The use of cellular telephones, in part, will be calls that would have been landline and are now wireless and also will be, in part, calls the subscriber would not have made before but will make now with a cellular telephone.

The second source is churn from other services from the same wireless vendor. *Churn* is the tendency of customers to move from one service to another. Since the CDMA air interface is more efficient than frequency division multiple access (FDMA) and time division multiple access (TDMA) technologies, when existing subscribers decide to re-place their old cell phones or when they want more features, we have a chance to switch them to CDMA service that can offer higher quality, lower rates, or both. As the market for new subscribers dries up, technology migration will become the dominant source of new CDMA voice subscribers.

The third source is competition from other vendors of wireless telephone service in the same area. Lower rates can attract price shoppers, and higher quality of service can tip the balance for subscribers who are less price-sensitive. The shift from second-generation (2G) to third-generation (3G) CMDA technology and the use of more efficient CDMA engineering techniques will allow vendors to compete for existing CDMA customers by offering the more efficient 3G air interface at lower cost and higher quality. In a few years, the wireless market will be almost entirely 3G CDMA, and the source of subscribers for one company will be competing for the subscriber base of another company.

The fourth source of wireless subscribers is our own service's current subscriber base. In making a market estimation, this is our starting point, but we cannot assume that it is static. We may be losing or gaining customers due to churn. Estimating future churn is part of our estimation planning for the demand for voice service.

This tells us that the long-term estimate of wireless voice call demand for any given provider depends primarily on the estimate of wireless voice market share. As the market size stabilizes, a service provider's share of the market becomes the primary deter-minant of the subscriber base. It is important for a CDMA vendor to form a clear idea of its position in the CDMA voice market in each region it serves.

While studies of market share are a critical function, it is important not to lose sight of the ultimate objective of voice market estimation—figuring out the total demand for telephone service and how much of that demand will be carried on wireless networks. The investment in cells, transport, switches, and other equipment is an amount of money that depends on the absolute number of subscribers.

Another important variable in the initial demand estimate is the planning horizon for the initial system. Planners have to decide how far into the future growth of the system they are planning for in the initial system design. By building in enough capacity to handle a period of initial growth after startup, we give ourselves time to get things running smoothly before we have to deal with growth issues. Getting a system up and running may keep first-time planners and engineers busy for 6 months, so they do not want to be combating growth issues during that time. More experienced CDMA vendors may be comfortable with a 2-month initial planning horizon because they can be ready to deal with growth issues sooner.

Taking into consideration the four sources of subscribers and the other issues just mentioned, the planners for the service provider should form a clear estimate of the number of voice subscribers they expect to be using the system at the end of their initial planning horizon. This is important not only for system planning but also for estimating how many cell phones are needed.

The next step in estimating demand is to form a time-of-day picture of telephone usage. There will be several distinct groups of voice subscribers, commuters, shoppers, mobile businesspeople, and so on. Each of these groups has its own time-of-day behavior and its own busy-hour usage pattern. CDMA planners should form a clear forecast of the different groups using cellular telephones and their usage patterns. One factor in estimating time-of-day usage is the variety of calling plans and packages offered by the service provider. Demand for voice capacity is different depending on whether subscribers are using minutes they view as a limited or an unlimited resource or if they are paying for calls by the minute.

31.1.1 Estimating voice call demand in other world markets

In estimating demand outside Europe and North America, the same four elements apply, but the market situation is very different. In the former Soviet Union, called the *second world,* cellular telephony is a growth market. The People's Republic of China is the world's largest potential market for cellular telephony. Rapid growth is beginning, and at least two companies are competing with significant market share. The largest company is based in Hong Kong, which is now part of the People's Republic, and much of the equipment production is occurring in factories on the mainland. In the *third world,* those nations that are largely agrarian but are developing technologically and economically, wireless telephony is still a growth market, often in its early stages. In the *fourth world,* those nations and regions that are largely agrarian or are the homes of indigenous tribal populations, cellular telephony is a future market, yet to be developed.

Cultural customs vary greatly, and this needs to be taken into account. There is a thriving wireless service in Pakistan based on the idea that those who can afford wireless service themselves will make money by allowing others to use the phone on a per-call basis. The billing model is set up to support multiple users per user terminal. In rural India, where television has been accepted as a communal service for decades, there are now plans to deploy satellite or wireless telephones, one per village.

In these parts of the world, there is often a very great difference between urban and rural lifestyles. In many nations, cellular service has been focused only in urban or metropolitan areas. This may begin to change, and coverage for large regions of some countries may become important.

In addition, we must be able to estimate the demand for wireless local loop (WLL). WLL will be used as an alternative to traditional landline access methods, but the user will view the service as landline service. Thus we can use methods for estimating landline demand when estimating WLL demand. However, we may allow a higher blocking rate for WLL than was allowed traditionally for landline service in North America and Europe for reasons described in Sec. 3.2.

31.1.2 Calculating demand

Once we know the number of subscribers and their peak-hour usage patterns, we have the basis for a total voice demand in erlangs. This demand in erlangs is the primary determinant of how much CDMA capacity is required to serve our voice telephone calls.

Another important component of voice traffic in CMDA is subscriber motion. The CDMA channel is 2 dB less efficient for a moving user terminal than for a stationary one. Drivers and rail commuters are going to use 1.6 times as much CDMA resource (2 dB more) than callers sitting or standing still.

Once we have the busy-hour voice traffic loads in erlangs, with subscriber motion taken into account, the next step is to convert that demand into CDMA resource consumption. In Sec. 30.7 we described the estimation of CDMA resource used. We can quantify the CDMA resource as a *gross chip rate y* in Eq. (31.1):

$$y = eR \frac{E_b}{N_0} v \qquad (31.1)$$

where y = chip-per-second estimate
$\quad\ e$ = demand in erlangs
$\quad R$ = data rate
$\quad\ v$ = activity factor

We can say the gross chip rate of a service is the value of y for one erlang, $e = 1$. Later on (in Eq. 31.2) we will turn the gross chip estimate y into a cell count estimate.[1]

The final part of the estimation process is to derive a geographic picture of the voice call demand. Knowing the total demand is important for estimating equipment requirements, including cell counts. However, the CDMA planning process is going to need a lot more than this. Later stages of system planning depend on having as accurate a forecast picture of the regional distribution of demand as possible.

31.2   Estimating Data Service Demand

The market for data service is harder to predict than the market for voice. We have historical data that tell us what kinds of people got cell phones and how much those people used their cell phones. As we move into the 3G world of data services, we move into uncharted territory.

We must not let our lack of market experience keep us from formulating forecasts and deriving decisions from those forecasts. The decision to proceed with a 3G CDMA system should be based on firm data service forecasts, and we should use these forecasts in forming our estimates of data service requirements.

As with voice, we start by estimating the subscriber count for wireless data services. There are several candidate data services with different bit rates and different quality-of-service (QoS) requirements. For each of these, we need some estimate of the subscriber count, and we need a more detailed forecast than the voice telephony estimate. In data services forecasting, we need to form a more detailed picture with less historical experience.

The interactive data services take priority in their use of air-interface resources. Subscribers browsing the Internet do not like to wait too long for their Web pages and images to download. The demand for these low-delay services at peak times determines the required capacity for wireless data service.

Other data services are less time-critical. E-mail and nightly database downloads can accept delays that interactive subscribers would not tolerate. Even though these data can tolerate longer delays, they still have to get to their destination, and the CDMA air interface has to have enough capacity for the total data service demand. Thus we should form an estimate of these higher-delay service subscribers as well,

---

[1]There are a variety of measures we can use for CDMA capacity. The gross chip rate used here is our own way of calibrating all these CDMA 3G services with a common measure of capacity utilization. These are not actual chip rates but an equivalent chip rate.

mostly to make sure that they are not the limiting factor in our system design. However, it is highly unlikely that the high-delay-tolerant data services are going to be a limiting factor in capacity planning.

With an enumeration of data services, subscriber count estimates for those services, and usage pattern estimates, we can form estimates of the peak-time demand in erlangs e for each low-delay wireless data service. We need to focus on the busy hour having adequate data capacity for those services. For the other services, those that can tolerate higher delay, we can ensure enough capacity for them by counting up the total requirement spread out over peak and nonpeak times.

Motion increases the required E_b/N_0 for the same bit error rate (BER), typically by about 2 dB, for data, just as it does for voice. Thus it is important to estimate the fraction of data subscribers who are moving while they are using wireless data service.

We can use Eq. (31.1) to get the CDMA capacity needed to serve each data service. The value of y is a proxy value. While y does not represent the true number of chips, it *does* represent a measure of air-interface capacity required to deliver each type of data service, as we will see in Eq. (31.2). The E_b/N_0 for each service should be adjusted to take into account the fraction of users who are stationary and the fraction who are moving.

For each data service, we estimate the number of subscribers and the peak-hour usage per subscriber. The product of the subscriber estimate and the peak-hour usage per subscriber gives us a total usage for each data service. Once we have that total for each data service, we use Eq. (31.1) to add up all these estimates for a gross number of chips per second y that we can use to estimate cell count later on in Sec. 31.4.

We complete the estimation process for data service by creating a geographic picture of data service demand. The total demand gives us an estimate of air-interface capacity requirement, but the geographic information will be needed further down the road in the CDMA planning process.

31.3 Estimating Demand for Other Services

There are other 3G CDMA services. We do not yet know which of these will be popular and which will fall by the wayside. But any of these could be popular features in the brave new world of 3G.

One service that may become popular is sending images using multimedia service (MMS), as described in Sec. 20.5. This is a visual extension of short message service (SMS). The subscriber might be able to point a cellular phone, press a button, take a digital photo, and send the picture to another subscriber.

As we said in the same Sec. 20.5, videotelephony was tried 40 years ago in the form of PicturePhone. It was terribly expensive, and it was tied to a home telephone. Perhaps in today's more mobile and more visual environment, real-time videotelephony will catch on. Bit rates tend to be high for video, and delay requirements are very low, so a video call demands a lot of CDMA capacity.

In estimating our total demand, we must develop a picture of the future that includes these services that go beyond voice telephony and data services. We need a picture of the subscriber base that will be using video phones and sending MMS images.

There are other services not on the beaten path that may become popular in the 3G environment. The 3G CDMA planners should do the best job they can of predicting what these services are.

For each of these services, we need a prediction of how many subscribers will be using the service and its busy-hour erlang count e. As hard as it is to predict data service subscriber counts, these other services will be harder to forecast. After all, we have some idea of what a realistic data service is and who might be using it, but we have little idea what services people will fancy when all the 3G choices become available. In making these predictions, there are two important issues we can take into account. First, it is important to distinguish business services from consumer services. The methods for predicting demand and for pricing for each of these are very different. Second, it is important to realize that demand is a function of price. As North American 3G data services are rolling out, vendors are taking very different approaches on this issue. Some, concerned with rapidly recouping the high cost of deployment, are setting a high price for data services. Others are charging a low rate, hoping to capture market share and more willing to risk being swamped with demand. In addition, some vendors are choosing to offer very limited usage in the basic package and a high charge for additional use. Others are offering a flat rate for unlimited or high-limit use, which allows them to estimate demand at a fixed price.

As before, in voice and data, we should use Eq. (31.1) and add up the gross estimated usage of these other services. Also, we should create a geographic distribution estimate for use in later stages of planning our system layout.

31.4 Estimating Cell Count

For voice telephony, we compute the total peak-time demand in erlangs e. Then we multiply that by the voice data rate R, E_b/N_0, and activity factor v in Eq. (31.1) for a gross usage in chips per second. The values of R and E_b/N_0 are determined by the air interface described in Sec. 30.7. This gives us a gross chip rate y for voice.

For data services, we have to add up multiple services. We concentrate on services that have low delay requirements. For each low-delay data service, we use our forecast demand in erlangs e, the data rate R, its E_b/N_0 requirement, and its activity factor v. The voice telephony y value is added to the sum of the gross chip rate y values for all the data services.

Other services also have demands in erlangs e, data rates R, E_b/N_0 requirements, and activity factors v. These video, messaging, and other services add their gross chip rates y to the pot.

Thus we have a total consumption of CDMA resource for voice, data, and other services in the form of our gross chip rate y. This value y is a measure of total CDMA resource required to serve the demand for the full suite of 3G services. From the mathematical models in Part 5, summarized in Sec. 30.7, we can estimate the number of cell sectors required.

From Eqs. (30.1) and (30.2) we can derive Eq. (31.2) to estimate the number of cells we need:

$$n = \frac{1}{k} \frac{y}{cW} \frac{(1+s)}{p} (j + f + g)(1 + h) \frac{1}{tu} \tag{31.2}$$

where n = estimated cell count
  k = number of sectors per cell
  y = gross chip rate
  c = number of CDMA carriers

W = CDMA chip rate
s = soft- and softer-handoff factor
p = power control effect
j = orthogonality or joint detection
f = other cell interference
g = other sector interference
h = overhead factor
t = occupancy
u = utilization

The same equation works in both forward and reverse directions, although the terms may have slightly different meanings. For example, j represents orthogonal codes in the forward direction and joint detection in the reverse direction. The soft- and softer-handoff factor s should be set to zero in the reverse direction, where there is no soft handoff.

Four new terms appear in Eq. (31.2), the number of CDMA carriers c, the number of sectors per cell k, occupancy t, and utilization u. Most CDMA systems have more than one carrier, so we added the term c to Eq. (31.2) to take this into account.

The number of cells is $1/k$ times the number of sectors, according to the sector plan of k sectors per cell. The value k is the average number of sectors per cell, so if half the cells have three sectors and half the cells have six sectors, then we use $k = 4.5$. For time division synchronized CDMA (TD-SCDMA) smart antennas, we recommend $k = 6$.

The occupancy term t takes traffic engineering into account. As discussed in Chap. 23, a system designed to serve traffic at a low blocking rate maintains some unused capacity most of the time. As long as the services have gross chip rates significantly less than cW, the occupancy should be close to 1, say, $t = 0.9$.

Utilization u is a different story. It reflects the fact that a growing system or a new system designed to grow is not going to make full use of all its capacity. When a cell reaches full capacity, we add a new cell, and we have two cells where we had one before. While the actual process of splitting cells is a bit more complicated than the mitosis of one cell splitting to two, it is hard to keep the utilization of a growing system close to 1. A utilization value of $u = \frac{2}{3}$ may be a more realistic value for planning purposes.

Another issue increasing the initial cell count is the planning horizon for the initial system setup. As we said in Sec. 31.1, we would like to have some time for our new system to settle in and for us to get used to it before we have to respond to its growth needs. Thus we should add a few months of anticipated growth to the cell count estimates.

The combination of all these factors, estimates of demands of all the services, performance factors for CDMA, traffic engineering, and utilization due to growth, gives us an estimate of how many cells are required to serve the demand we expect for our CDMA system.

Coverage is another factor increasing CDMA cell count. When traffic densities are low in the outer areas of a CDMA system, the larger cell radius reduces the radio signal path gain and increases the relative importance of receiver noise. Link budgets (described in Sec. 1.9) for specific technology and equipment tell us the cell size and capacity tradeoff, so there is no general value we can supply here. The more spread out the system, the greater is the cell count increase for coverage. Unlike conventional reuse systems, capacity and coverage play against each other in a CDMA system: Bigger cells have less capacity.

The result of the calculations in this chapter are geographic maps of time-of-day demand for voice, data, and other services. From these, we derive an initial cell count.

31.5 Conclusion

In this chapter we have illustrated the process of moving from a business and marketing plan to the beginning of an engineering plan for a CDMA cellular system. With geographic maps that plot time-of-day demand and an initial cell count, we are ready to take the next step: planning locations for base stations.

Planning Locations for Base Stations

Once we estimate the demand for wireless services and derive the appropriate number of cells for a given region, we then need to decide where to put the base station at the center of each of those cells. Base station location is a compromise between the ideal placement for managing interference and practical concerns of site cost, installation time, and ease of maintenance.

32.1 Pilot Pollution

The code division multiple access (CDMA) air interface is a complicated beast, and the capacity of a CDMA carrier depends on many factors. Even in simplified form, the estimates of CDMA capacity depend on many factors, as we saw in Sec. 30.7. Besides the obvious factors of bit rate and bit error rate (BER), the capacity depends on overhead, power control, orthogonality, joint detection, multipath, and Doppler distortion due to motion. Some of these are within the control of CDMA wireless system engineers.

The CDMA carrier capacity also depends on interference from other cells and interference from other sectors in the same cell. These two factors are linked directly to base station placement and planning for coverage and managed interference, the topics of this chapter. The other factor in Sec. 30.7, soft and softer handoff, is also directly linked to cell placement and planning of managed interference. Thus we will concentrate on these contributors to CDMA capacity estimation. In Chap. 33 we will address methods of creating sectors within each cell.

Any air interface, frequency division multiple access (FDMA), time division multiple access (TDMA), or CDMA, works at its best when cells are isolated from each other. The management of intercell interference is the challenge of cellular radio engineering. In CDMA, the primary interference comes from other users on the same carrier from the same cell sector. The next most important interference comes from adjacent sectors and nearby cells. We want to place cells to manage the regions that have multiple potential serving cell sectors.

The CDMA equations in Sec. 28.1 tell us that the carrier loses capacity in proportion to the number of nearly equal servers. A user terminal midway between two cells is going to create interference on both cells, so it effectively taxes the system capacity as much as two calls near the base station. Soft handoff aggravates this inefficiency in the forward direction by having two cells transmitting CDMA channels for one call. Under ideal conditions, the combination of the soft-handoff signals in the user terminal receiver would allow each transmitter to send half the power, so the total effect would be the same factor of 2 as we have in the reverse direction. If the two soft-handoff signals are no better than two separate calls, then the total effect of having one call on two cells is four times worse than a single call near the base station. The true effect of soft handoff is somewhere between the ideal and the worst case. When a call is equidistant from three cells, the reverse-direction cost of the call is three times that of a call close to the base station, and the forward cost with soft handoff is somewhere between three and nine times. Still, soft handoff is more efficient than trying to maintain the correct CDMA carrier through a sequence of hard handoffs.

Every cell layout has boundaries between cells, so avoiding two-cell coverage zones is impossible. And every cell layout has to have some place where those cell boundaries come together in triple-points, so avoiding three-cell coverage zones is impossible. These two- and three-cell zones are shown in Fig. 32.1.

It is the zones with four or more nearly equal cells that we are trying to avoid. In such zones, each CDMA channel is interfering with too many carriers, and user terminals are offered too many choices. In the planning stage, we look at the *fourth-pilot delta*, $p_1 - p_4$, the term used for the difference between the strongest pilot p_1 and the fourth-strongest pilot p_4. When $p_1 - p_4$ is low, we have an area where four or more cells are competing for each CDMA channel. The term used for this is *pilot pollution*, although the pollution goes beyond the pilot to the CDMA call channels.

A *valley* is an area covered evenly by many cells, an area with pilot-pollution problems. The ring of six cells in Fig. 32.2 creates a severe valley in the middle, a six-cell coverage zone. Not only is the fourth-pilot delta too small in this area, the *sixth*-pilot

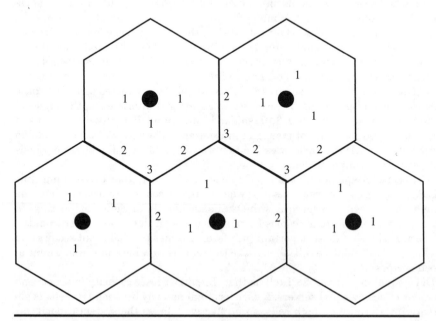

Figure 32.1 Regular cell layout with two- and three-cell regions.

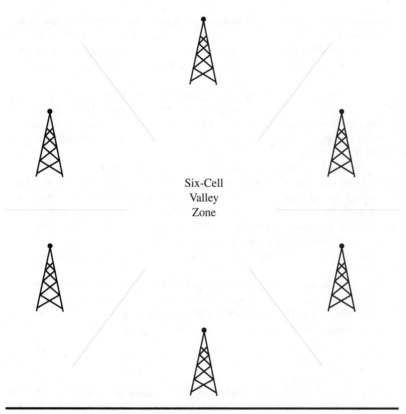

Figure 32.2   A cell layout with a six-cell valley.

delta is small as well. The pilot-pollution problem is not that the pilots are equal; it is that several radio signal paths are nearly equal; that is, they are within a few decibels of being equal. Thus the valley is a zone of difficult CDMA capacity performance, a problem area even if there is no single point where all the competing base stations are exactly equal in radio path gain.

The worst valley is the most likely to occur, a valley with low radio path gain. This means that four or more cells are competing for service not only with each other but also with the thermal noise floor of the receiver. A call from such an area can use the resources of 20 or more calls from near the base station. And the low radio path does not have to be visually obvious. We can picture a subscriber equidistant from five base stations making calls in a wooded hollow (a geologic rather than an electromagnetic valley). However, the same radio phenomenon can happen deep in the twentieth floor of a centrally located office building.

While it is important to avoid valleys in the cell layout, it is more important to avoid having subscriber demand in the valleys. If six cells circle a lake, then there is a valley on the map, but only a few boating enthusiasts are likely to be making cellular phone calls from there. It is the combination of radio conditions and user demand that creates CDMA capacity problems. It can be just as fruitful to adjust the cell layout for sparsely populated valleys as it is to eliminate valleys altogether.

We do not recommend solving the pilot-pollution problem by reducing pilot power levels. It is the comparative radio path gains that are causing the capacity loss, not the pilot power levels being equal. If it is deemed necessary to control the size of a cell, then we can reduce its size by downtilting and lowering its antennas, as described later in Sec. 33.2.

A cell layout passes the pilot-pollution test if there are very few areas with low fourth-pilot delta values. When cell coverage maps are printed using the radio propagation planning models described in Chap. 47, there should be few zones with four or more cells serving them. And the four-cell zones that are on the map should be areas of relatively low subscriber density.

It is important to understand that the pictures we draw, and even the propagation maps, are visual aids only. As we discussed in Sec. 27.10, the lines we draw showing geometric coverage shapes are representations of complex radio propagation paths. We use these simplifications in our book because we believe that they represent radio reality well enough to help planners make their CDMA systems work better. However, CDMA planners should remember the complexity of signal path interaction and try to gain an intuitive feel for the radio issues in the particular wireless telephone system they are planning.

Sector boundaries are also a contributor to pilot pollution. Once the cells are laid out to minimize four-cell valleys, the sectors should be oriented so that their boundaries do not cross the triple-points and valleys. Otherwise, the pilot pollution and the combination of soft and softer handoffs may overwhelm the CDMA carriers in the area.

32.2   Keeping the Grid

To keep pilot pollution to a minimum, the cell layout plan should avoid valleys and keep fourth-pilot delta values large. In geometric terms, the layout should avoid having areas where four or more cells meet at one point. Putting cells on the map without a consistent plan is likely to create the four-cell areas shown in Fig. 32.3. As we said earlier, these four-cell areas need not be exact junctions of cell boundaries, but they are areas where four cells are nearly equal in radio signal path.

One way to maintain large fourth-delta values, to avoid valleys, is to maintain the hexagonal layout that we used in the conventional-reuse systems (FDMA and TDMA). We can picture a four-cell zone as two triple-points that have been allowed to drift together by deforming the cell grid. Keeping the grid regular keeps the triple-points from drifting together into four- and five-cell zones, as we saw in Fig. 32.1.

A regular hexagonal grid pattern of cell placement not only keeps valleys away, but it also makes handoff lists easier to manage. We have a clear visual picture of the neighbors, and the visual neighbors are likely to be the best handoff neighbors as well. In addition to easing the administrative task of assigning neighbors, the regular cell grid also keeps the neighbor lists shorter.

In laying out cells for CDMA, we want to maintain the local character of the hexagonal cell grid. It is not so important that distant cells follow the local pattern.[1] Thus

---

[1] In conventional reuse, the primary interferer is several cells away, and we rely on the grid spacing to maintain D/R, as shown in Fig. 27.7. We do not have that kind of distant reuse issue in CDMA.

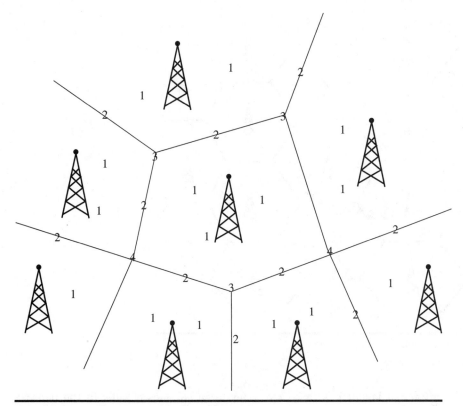

Figure 32.3 Random cell layout has four-cell areas.

the cell grid can be modified to follow the contours of geography or subscriber demand as long as small, nearby clusters of cells follow a local grid pattern. An example of a warped grid is shown in Fig. 32.4. This warped grid is regular enough to avoid pilot pollution.

A cell layout is a good grid when it looks like a good grid. The dots on the map should follow regular hexagon spacing, at least locally, and the radio propagation map should show a clear grid pattern. This helps the initial system avoid valleys, and it helps the growing system avoid valleys, as we will see in the next section.

32.3 Cell Splitting for High Traffic Density

A growing system needs more cells to serve more traffic, so areas with higher subscriber density will have a higher density of base stations as well. In the earliest days of cellular, we laid out a large-cell system and waited for subscribers to sign up. As demand increased, we would increase the cell count by adding growth cells to the infant system. Today's wireless service providers do not have the luxury of deploying a skeleton system and waiting for demand to fill it. The initial subscriber base of a new system could be hundreds of thousands of subscribers, all expecting excellent service,

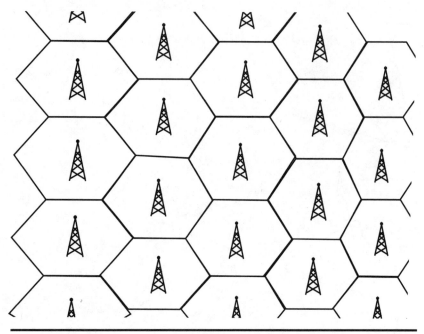

Figure 32.4 A cell layout that follows a warped grid.

so the initial cell layout has to be ready for large volumes of calls in the densest areas.[2]

The goal of efficient cell placement is to put cells where the demand is going to be. We have forecasts and maps, and we want to add enough cells to meet the demand. We also want to have each cell pulling its own weight, each cell carrying its share of the revenue-producing subscriber demand. The result of good planning is that the number of cells in a neighborhood follows the subscriber demand for service. This may sound obvious, but cellular engineers experienced in FDMA and TDMA may remember when cell splitting was dictated by channel sets as much as by local demand. The CDMA air interface allows us to follow subscriber demand much more closely in deploying base stations.

Pilot pollution is a concern for cell splitting. The same issues that apply when laying out coverage cells also apply when adding cells for expected demand. The growing grid maintains low pilot pollution and simpler handoff lists. It is not as nice as the perfect honeycomb pattern without any growth cells, but a growing grid is going to be easier and better than adding cells without the grid structure.

---

[2]In North America, the Federal Communications Commission (FCC) required that winners of grants or auctions for broadcast licenses in the cellular band guarantee service to a certain percentage of the population in the area covered by the license. To meet this requirement, first- and second-generation cellular providers built very large coverage cells in areas of low demand. If such a regulatory requirement is not present in current plans, cellular engineers are free to develop plans based primarily or exclusively on marketing and business plans.

In Fig. 27.15 we showed a cellular system adding cells on a smaller grid at the triple-points of its existing large-cell grid. Since the small-cell grid has three times as many cells as the large-cell grid, we call this *three-to-one cell splitting*. In the grid representation we show the new, small cell as a smaller hexagon, as shown in Fig. 32.5. The radius of this hexagon is $\sqrt{3}$ times smaller than the large hexagons surrounding it, and it has one-third the area.

The reality of a three-to-one cell split is not a small hexagon, however. It is the triangle shown in Fig. 32.6. The good news about the triangle cell shape is that it is bigger than the hexagon, so instead of serving one-third of a large-cell area, it serves one-half a large-cell area. Three large cells are reduced by one-sixth by the new cell, so they cover 83 percent of their former area.[3] The bad news about the triangle cell shape is that it creates three four-cell zones at its corners. These particular areas will go away as the small-cell grid expands, but the interface between the small-cell grid and the large-cell grid will continue to have four-cell zones.

The alternative grid-based growth scheme is four-to-one cell splitting, as we saw in Fig. 27.14. Instead of adding cells at triple-points, we add small cells midway between existing large cells. These cells form a grid half the radius and one-quarter the area of the large-cell grid, so we call this *four-to-one cell splitting*. In our grid mindset, we represent the new cell with a hexagon four times smaller, as shown in Fig. 32.7.

---

[3] In some ideal vision of cellular growth, we would like to add a new cell to relieve n busy cells so that $n + 1$ cells each serve an equal share. This would be a very efficient growth scheme.

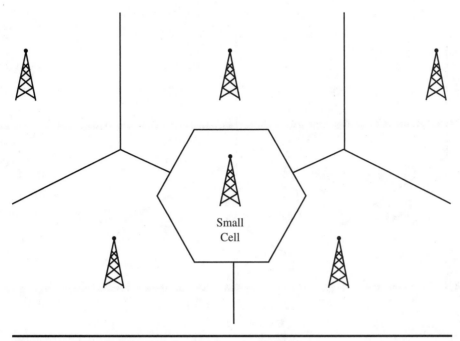

Figure 32.5 Grid representation of a three-to-one cell split.

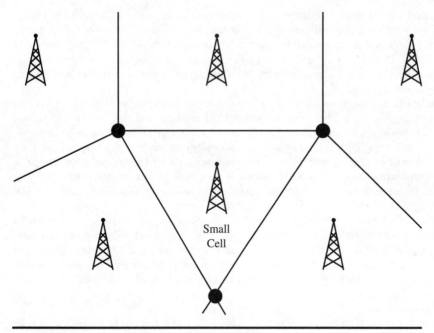

Figure 32.6   Three-to-one cell split creates a triangular cell.

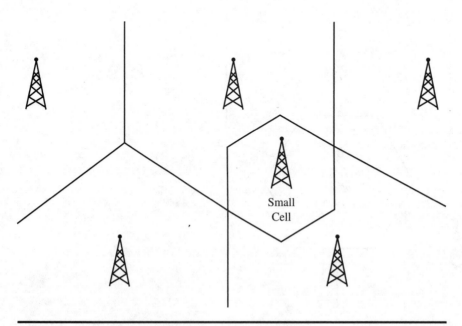

Figure 32.7   Grid representation of a four-to-one cell split.

Just as three-to-one splitting produced triangular rather than hexagonal cells, four-to-one cell splitting produces oblong, rectangular cells, as shown in Fig. 32.8. These rectangles eventually are whittled down to hexagons as their neighboring cells are added to the small-cell grid, but the initial growth cell is a long rectangle. The good news about these rectangles is that they are large, half the area of a large cell. The two cells being relieved are reduced to 79 percent of their former area, so the growth cell provides substantial capacity relief. The other good news is that the rectangular growth cell usually does not produce any four-cell zones.

When using four-to-one cell splitting, we can create four-cell zones. This happens when the small-cell grid has holes in it, as shown in Fig. 32.9. As long as the small-cell grid regions are *convex areas,* regions without any boundary U shapes or internal holes, then four-to-one cell splitting should grow a CDMA system with very few four-cell zones of high pilot pollution.[4] Its ability to grow a CDMA system with less pilot pollution than three-to-one splitting causes us to favor four-to-one cell splitting for CDMA growth.

The cell layout with coverage cells and growth cells should avoid valleys, that is, four-way zones of CDMA coverage. The lack of valleys will be visually apparent in a good layout and can be confirmed with planning tools that look for small fourth-pilot delta values. This procedure gives us a good layout for macrocells, the regular big-tower

---

[4]The term *convex area* is mathematically defined as an area where line segments between any two points in the area are entirely contained in the area. This describes a region with no holes and no U shapes in its boundary.

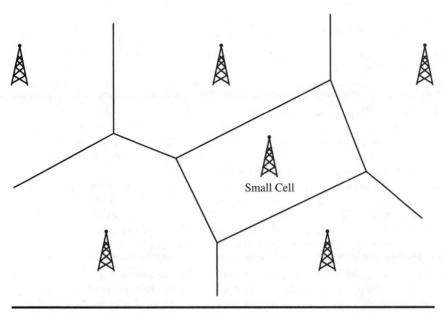

Small Cell

Figure 32.8 Four-to-one split creates a rectangular cell.

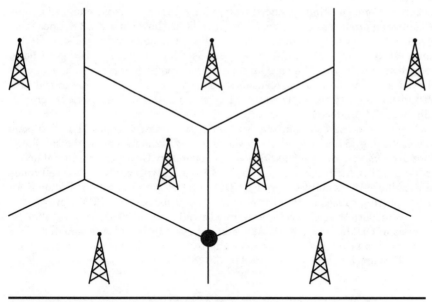

Figure 32.9   Four-to-one cell splitting with a four-cell zone.

cells that cover large geographic areas. We now turn our attention to spot engineering with microcells, picocells, and repeaters.

32.4   Microcells for Highly Concentrated Demand

Once we have covered the geographic service area of a wireless telephone system, we look at local hot spots. We can use the high-reuse capability of CDMA to treat local high-density areas with local equipment, including microcells, picocells, and repeaters.

The estimation process that gave us the geographic distribution of wireless demand also should identify local areas of high density. The high density could be a large community of voice calls, or it could be a community of very active data subscribers. It takes only a few high-rate data users to use a lot of CDMA capacity, especially if they require low delay and high accuracy. Once we identify these subscriber communities, we can use local equipment to meet their needs.

If a high-demand area is a few hundred meters in radius, then we can use a microcell to serve it. Microcells typically are mounted low to limit their radio range and to restrict their coverage. These microcell areas could be dense residential areas or heavily traveled commuter routes. In either case, we want to position a microcell to cover the area rather than try to find a macrocell solution that puts those subscribers in competition with others for CDMA resources.

When the high-demand area is concentrated in a particular building or part of a building, we can use picocells. These create radio service areas confined to a radius of a few tens of meters. Large office buildings are obvious picocell targets, but train stations and airport waiting areas may generate enough demand for picocells as well.

When there is a community of demand in an area without telephone transport, we can think about using repeaters. These are two-way reradiators with a line-of-sight path to the base station and high radio path gain to the subscriber community. We think of repeaters as a solution to be used in remote areas in countries where telephone transport is not universal as it is in the United States and Europe. The repeater might sit on a hilltop to serve people living between two mountains. Of course, repeaters require electrical power, so we cannot use them in truly remote areas.[5]

Anyplace where there is a large and concentrated subscriber community is a good candidate for local equipment to serve local CDMA demand. We look at neighborhoods, office buildings, train stations, airports, or any place where we expect a lot of wireless telephone demand in one place. These are the areas where microcells and picocells can provide cost-effective service. Engineers should identify these areas in the planning process and evaluate the impact of serving the demand locally.

Another opportunity for microcells is in patching pilot-pollution valleys. While we just got through all these discussions of how to avoid pilot pollution, it is still going to happen. The principles of grid-based cellular growth depend on uniform radio propagation and availability of good sites for base stations. When these do not occur, the CDMA system will have four- and five-cell zones that may be patched with microcells, particularly when those zones fall in high-density demand areas.

The third candidate for a microcell, or perhaps a repeater, is a tunnel, a very deep canyon, or a similar place with a lot of commuters and poor radio access. The spot coverage of a microcell allows us to provide seamless service in the macrocell gaps.

The local equipment planning job is done when all the opportunities for microcells, picocells, and repeaters have been evaluated. Microcells and picocells are usually small enough to save the cost of separate buildings and towers, so they cost about half as much as full-blown macrocell base stations. They also allow the wireless service provider to promote service in high-density areas that might not be well served with just a macrocell grid.

32.5 Good Cell Locations

Now that we have discussed how to make an efficient CDMA cellular grid, it is time to turn our attention to what are good base station locations. Good radio properties for a base station involve clear radio paths to the locations being served and weak radio paths outside the cell. Nature may not be kind enough to offer locations that have perfect radio performance right to the cell boundary and perfect radio silence outside, but we can look for base station locations that reach their audience well while creating relatively little interference for surrounding cells.

The result of the analysis of the radio properties of a given area can be plotted on a map. One way to create this plot is to put a mark at the ideal location for each base station and then to place a small circle around it indicating that any location within the circle is a good choice. This is then surrounded by a second, larger circle. Any location within the larger circle but outside the inner circle is an acceptable choice but not a good one. Planners can then seek cost-effective sites first within the good area and then within the acceptable area. However, if a particular site does not fall within its good

---

[5]Perhaps there are wind-driven or solar-battery systems that make a stand-alone wireless repeater practical.

area in terms of radio properties, it may be necessary to adjust the locations of other cells or to use microcells, picocells, or repeaters to make up the difference.

A vital part of site selection is finding cost-effective locations. A wireless service provider's ability to compete in the marketplace depends on its ability to offer its service for less. Base stations are a major cost component of a wireless telephone system, and it is in our interest not only to provide good radio performance but also to provide that service at low cost.

The third part of site selection is being able to install sites quickly. Years of courtroom zoning battles are a poor substitute for building base stations and realizing revenue from subscribers. We should identify sites where base station construction will be particularly easy and fast. Decent service soon can be a better business decision than great service that takes years longer to establish.

When all the base station locations are considered, the final decision is usually a tradeoff between really good radio performance and cheap, fast base station installation. The final judgment in this decision should involve the radio performance map described in Sec. 48.5 with its delineation of fourth-pilot delta values, four-cell zones, and radio signal valleys.

In making these decisions, it is also important to consider the future consequences of cell placement decisions. A few off-grid cells early in the service-providing game may not be a problem for the first year, but then it may become more difficult to grow the system to meet demand. The system plan should consider the short-term installation issues and the longer-term issues of growth. A system plan is also a financial plan and should take into account financial details such as revenue streams and cost tradeoff decisions.

32.5.1 Sharing existing towers

The best starting place for cheap base stations is towers that are already there. Building towers is expensive and takes a long time, and getting zoning permission for towers can be even more expensive and more time-consuming. It is much quicker, easier, and cheaper to add our equipment to an existing tower (or a building that already serves as a base for radio services) than it is to get zoning permission and then build a new tower.

If the tower has indoor room for base station equipment and electrical power readily available, this substantially increases the cost savings over building a new facility. Good tower locations with adequate power should be on the base station list. About half the cost of a base station is in towers and buildings, so being able to use existing physical structures may allow us to as much as double the number of cells the initial system capital can pay for.

Another consideration for using existing buildings and towers is finding those locations with access to low-cost telephone transport, perhaps on a high-capacity transport backbone. We should make sure that there is capacity available for backhaul on the backbone network of the public switched telephone network (PSTN). Backhaul transport is an ongoing expense for a wireless service provider, and having base stations where cheap transport is available is an important opportunity.

Another consideration for base station location is access to maintenance. If there is already some kind of electronics maintenance at the proposed base station site, then it is going to be easier and cheaper to get service for our own base station.

The best candidates for new wireless system base stations are the current base stations from an existing wireless telephone system. In Sec. 33.6 we discuss the migration of radio bandwidth from FDMA or TDMA to CDMA. We also should consider the economic benefit of reusing base stations as well. It may be worth adding a new wireless telephone system in new radio spectrum to an existing set of base stations without replacing the old system.

There are also innovative solutions for base station installation that reduce costs or allow costs to be shared. Readers may have noticed that many malls and industrial parks across the United States have gotten large American flags on tall poles in recent years. Part of the cause may be patriotic fervor, but many of these flagpoles have cellular antennas inside. In addition to avoiding zoning issues altogether, the cellular company gets a lower lease cost by adding a desirable feature to the site.

The end result of an existing-site survey is a list of base station possibilities and the relative advantage and disadvantages of each. Towers, buildings, transport, maintenance, and existing wireless telephony are all good reasons to consider a location for a base station.

The combination of the site survey and the plot of good and acceptable cell locations allows planners to decide which sites to select for the placement of base stations. Of course, zoning problems, political conflicts, and other difficulties may intervene. When they do, this may mean moving the base station to an alternate location. In such a case, it may be necessary to change equipment at that base station and also perhaps to change the equipment or the location of surrounding base stations. These risks can be minimized if the site-survey team maintains good relationships with supportive business, community, and political groups.

32.5.2 Using existing transport routes

The other criterion for locating base stations is the availability of backhaul transport between the base station and the mobile switching center (MSC). If the local telephone network has a transport backbone, then it certainly is worth our while to look for sites that easily and cheaply connect to it.

If the base station site is not already on the PSTN backbone, then how close it is to the backbone does matter. The cost of leased transport lines is often based on the distance from the facility to the backbone. Since the MSC almost certainly will be housed on the backbone, the distance from the base station to the backbone is a major factor in the cost of transport. Local exchange offices (LEOs) are often the connection point to the PSTN backbone. Before siting a base station near an LEO, it is also important to determine whether or not the LEO has the available capacity to support the additional lines you want to lease.

Backhaul can be a very expensive part of operating a wireless telephone system, so it is important to find opportunities to get transport without paying a lot for it. Getting a few DS-1 or E-1 links from a base station to the backbone can be terribly expensive, so we should favor sites that already have low-cost backbone transport available to the MSC. Another factor is that it can take a long time to get off-backbone transport installed. Having base stations located where transport is already available means not having to work within the schedule constraints of installing new transport, perhaps to remote locations.

An excellent candidate transport network is a cable television system. It already has high-capacity routes covering a large community of residential users. We could use some bandwidth that the cable company is not already using, perhaps in the lower frequencies below the very high frequency (VHF) television band. Or we could use the physical wiring locations to run fiber optic cable for our own transport. In either case, a cable television network is a good place to consider for a collection of base stations, or perhaps microcells, to serve a residential community.

Business locations also have telephone transport. Microcells and picocells located in business districts likely will have transport already available for backhaul. This applies not only to office complexes but also to shopping malls and other commercial centers. Even housing developments and other busy residential areas usually have available telephone transport that can be used for wireless telephone system backhaul from base stations. It is not enough that telephone transport exists; it also must be available for our use.

The difficult areas for backhaul are the low-occupancy areas between high-occupancy areas, the commuting and travel zones. These zones could be highways or railroads. They are potentially lucrative wireless markets, for both voice and data, but do not normally have transport available. Serving these areas means finding base station sites that can be connected to the MSC in areas that are not normally connected to the telephone network.

When we are done evaluating good backhaul opportunities, we will have a geographic layout of areas where telephone transport is cheap and available at the times when we will need it. These are the preferred places to put base stations based on transport opportunities.

32.6   Conclusion

The selection of base station locations should be based primarily on these four inputs:

- The geographic plot of projected peak demand described in Chap. 31.

- The map identifying good and acceptable base station locations derived from the demand map and the principles of site location for managed interference.

- The site survey identifying low-cost base station sites such as existing towers or buildings with radio facilities. Projected maintenance as well as construction costs should be included in the site survey.

- The availability of low-cost transport.

Considering all these elements, we can create a plan for the location of all the base stations serving a particular region. The plan should be validated to ensure that it meets the peak capacity with acceptable coverage and acceptable blocking rates in all areas. The plan also can be used to generate a detailed project plan and budget for deployment and a projected cost budget for operations, administration, and maintenance (OA&M).

Now that we have decided where we will put each base station, we turn our attention to the design of the equipment at each base station.

33

Base Station Planning

Now that we have decided where each base station is going to be, we need to decide what equipment goes into each base station. Our goals are to provide good or excellent coverage, reduce interference to other cells, maximize capacity, handle peak-traffic demand, reduce cost, and maintain reliability. We need to do this for new deployments, as well as for code division multiple access (CDMA) systems we create by migrating existing systems from frequency division multiple access (FDMA) or time division multiple access (TDMA) to CDMA. In this chapter we will explore the tradeoffs among these goals and identify techniques we can use to achieve them.

33.1 Coverage

The first mission of a wireless telephone system is to reach its subscribers, providing radio coverage. Subscribers in areas not covered well will find wireless telephone service from someone else, someone who *does* cover those areas well. Even subscribers who spend most of their time in good coverage areas will find occasional poor coverage a good reason to switch from one wireless service provider to another.

In a CDMA environment, poor coverage means lower signal levels competing with thermal noise in the receiver. As long as the CDMA signal is strong, we can ignore thermal noise as a factor in capacity planning because CDMA power control can turn down the power levels. When the CDMA signal is weak due to poor radio coverage, the power levels are turned up as far as they can go, but fewer channels can be served when the transmitters run out of power.

In the older technologies, FDMA and TDMA, coverage and capacity are distinct issues. Once the radio signal gets from transmitter to receiver, better radio conditions do not increase system capacity. FDMA and TDMA wireless planning methods have distinct phases for coverage and capacity. The capacity of CDMA is based on its ability to operate with less signal than noise, so a relative weakening of its signal is going to reduce its capacity. The result is a continuous tradeoff between coverage and capacity. As a result, a larger cell can support fewer CDMA channels.

For a particular environment, when we know the transmitting and receiving equipment, antenna gains, and radio signal paths, we can do a link-budget analysis that shows the relationship between cell size and CDMA capacity. The maximum cell size for coverage is highly dependent on these factors, so there is no global rule of thumb that says half the capacity is gone at such and such a cell size. There is some distance, some radio signal path, where a single voice call will use the entire capacity of a cell sector because it takes the full power of the transmitter, even with antenna gain and processing gain, to produce a signal that surpasses the thermal noise of the receiver. Clearly, we do not want to operate our CDMA system in this mode often.

Radio signal path gains vary considerably, so it is not enough that most subscribers are in good coverage areas. If a wireless telephone system is going to compete in the marketplace on the basis of superior service, then it must have very few areas of poor radio performance. Those weak regions are the areas that will lose calls first when the system gets busy. The subscriber will be making a perfectly ordinary cellular phone call with perfectly good signal, but when somebody else starts a data session, the first subscriber's call gets noisy for a few seconds and then is dropped. Or a few calls from a weak zone use up the entire CDMA capacity, and a large community of subscribers near the base station is blocked. We recommend making sure that the coverage picture is good enough that events such as these seldom happen.

The bad news is that poor radio signal paths erode CDMA capacity. The worse news is that these weak radio areas are typically three- and four-cell zones with pilot pollution adding to the capacity problem. The radio coverage maps described in Sec. 47.9 and the pilot-pollution maps described in Sec. 48.5 are invaluable aids in engineering a system to avoid these circumstances that consume capacity.

The bottom line in CDMA cell coverage is that high-demand areas should have high radio signal path gain. It is not so important to make good radio paths better, but it is terribly important to make bad radio paths good. Once we have good enough signal to climb onto the steep part of the curves in Figs. 28.1 and 28.2, we are operating in a high-efficiency CDMA mode. Making a good radio path better climbs higher on the curve and gains little on the horizontal axis, which is CDMA capacity.

When a cell layout has areas of poor radio coverage, we can use more powerful base station amplifiers and higher-gain antennas to reach those areas. It may make sense to have a thin-wedge sector, using a narrow-beam antenna with high gain. We lean toward antenna gain rather than amplifier power as a solution because the user terminal is limited in its transmit power. While the busy areas in CDMA are forward-direction-limited, the reverse direction is often the radio range limiter of CDMA. Higher-gain antennas extend the range in both forward and reverse directions, but higher transmitter power helps only the forward direction.

When more power and more antenna gain are not sufficient, we can consider adding microcells for coverage. The obvious cases are commuter tunnels and geographic valleys, where the serving radio signal has a hard time reaching. A more subtle use for a microcell is a busy four-way zone with four weak radio paths that is hard to avoid in macrocell planning. A microcell placed in such an area can ensure coverage, perhaps for significantly lower cost than a full-fledged macrocell.

When the weak radio area is far from telephone transport, we can use a repeater and connect it to an existing macrocell. But we tend to prefer the microcell solution to repeaters because microcells serve their own subscribers and do not use capacity from nearby macrocells. However, sometimes the lack of available transport facilities makes the repeater a more attractive solution.

Finally, we can add cells in the coverage rough spots. We recommend treating these as growth cells and placing them on the cell grid as growth cells using the four-to-one cell-splitting scheme described in Sec. 32.3. The most tempting place for a new cell on the coverage map is the triple-point of three cells, so keeping the grid means resisting temptation. We believe that seeing the longer view will pay off when demand for wireless service starts to grow in these areas, and the reduction in pilot pollution means being able to serve more demand with the same number of cells.

The coverage job is complete when the service area is covered by enough cells to serve the forecast demand with good radio signals and enough CDMA capacity to ensure low blocking.

33.2   Downtilting and Lowering Antennas

In the cellular world, good radio performance is not always a good thing. In fact, the entire cellular principle is one of managed interference, having a strong signal from the serving base station and comparatively weak signals from others. We want to reduce the range of a base station as its cell acquires new, nearby neighbors.

Radio propagation can do strange things depending on terrain and ground surface features. Radio valleys can appear when a far-away base station has a clear line of sight to a hilltop or building being served by nearer, local base stations. We need to relieve this pilot pollution by reducing the range of the far-away base station.

Turning down pilot power sounds reasonable, but it seldom works on CDMA systems. When an antenna face is attracting too much traffic, lowering its pilot power will send that traffic to other sectors served by other antenna faces. Most of the time this is a bad thing to do because the interfering component in CDMA is not the pilot signal but the channels being used by other subscribers. In Sec. 28.1.4 we explained that serving these signals on an alternative antenna face actually uses *more* of the CDMA radio resource than serving them on their highest-path-gain face. The only way to reduce the interference is to reduce the radio signal path gain of interfering cell sectors.

We can shrink a cell radius by reducing the path gain of the antennas. As long as we do not make similar changes at neighboring base stations, lowering the antenna path gain will move the equal-radio-path border zones closer to the modified base station and will reduce its service area. Reducing antenna path gain, however, also reduces the ability of the base station to serve hard-to-reach areas. This will aggravate any coverage problems we have and may create new ones. Keep in mind that coverage problems are not all distance-related. Underground malls and subway trains may have low radio signal path gains even to nearby base stations.

Tilting the serving antennas downward can offer the best of both worlds, as shown in Fig. 33.1. Downtilting reduces the radio signal path gain further from the base station by moving the high-gain, narrow-beam vertical lobe of the antenna away from those areas.[1]

Downtilting also increases the path gain nearer the base station because that high-gain vertical lobe is now serving user terminals closer in. We think of downtilting as a part of CDMA system evolution. In its early days, a base station is designed to reach as much traffic as possible with the highest path gain possible for maximum large-cell capacity.[2] As neighbors appear, the same base station has a new mission, to serve the

---

[1]Recall that high-gain antennas get their high gain by concentrating their transmission and reception in a beam with a narrow vertical angle.

[2]We cannot use downtilting on an omnidirectional antenna, but we are figuring that most CDMA systems are serving enough subscriber demand that they are using sectorized cells with multiple antenna faces, as described in the next section.

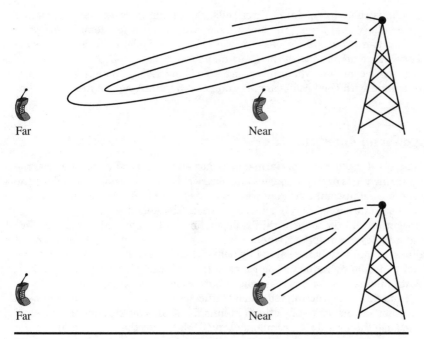

Figure 33.1 Downtilting antennas concentrates signal close to the base station.

same amount of traffic in a smaller area and to reduce interference with its neighbors. With this change in function, the high-gain antennas that served well for wider coverage use their narrow-vertical-beam property to concentrate on their smaller service area and to reject signals from further away.

Lowering antennas on their towers also reduces the radio range and shrinks cell size. We lean more toward downtilting because downtilting improves local radio performance, whereas lowering antennas does not. Lowering antennas reduces everybody's radio path gain but affects shorter paths less than longer ones. Also, lowering antennas is more expensive because it requires moving the antennas and possibly installing a new platform, whereas downtilting requires only adjusting the antennas on their current platform.

Aside from lowering existing antennas, we do sometimes choose to place antennas at a lower height than our planning would indicate is best. The primary reason for this is political: Neighborhood residents often do not like high towers nearby. They like cellular phones and they want good service, but they can fight hard to keep the radio towers out of their neighborhoods. Using a shorter tower is a good way to make neighbors feel better by reducing the visual intrusion of cellular in the communities. In the initial system deployment phase, a shorter tower may make the difference between a successful and an unsuccessful zoning application.

When downtilting or lowering antennas, each sector can be treated separately. It is possible to downtilt the antennas on just one face, affecting only one sector. This could be done to resolve a pilot-pollution problem at the edge of that sector without making changes to other sectors on the same cell. A program of downtilting and lowering antennas is complete when the cell layout covers the service area with high radio signal path gain and controls intercell interference.

33.3 Sector Plans

Each CDMA sector is a separate server with its own CDMA carriers serving their own wireless subscribers. More sectors means more subscribers and more revenue for the service provider.[3]

The capacity increase of adding sectors is not linear because sectors on the same cell interfere with each other. The amount of sector-to-sector interference depends on the radiation patterns of the antenna faces. All the antenna faces of a cell are at the same place, give or take 2 m, so techniques such as downtilting that protect one cell from another are not going to protect one sector from another on the same cell. A three-sector cell can get about 2.5 times the traffic of a single omnidirectional antenna. A six-sector cell is limited to about 4 times the omnidirectional traffic.

The second and third sector we add to an omnidirectional cell add nearly their full capacity.[4] The loss due to intersector interference is about 15 percent of the total capacity. Given the high fixed cost of installing a base station, having at least three sectors on a cell seems like an obviously good decision.

Adding more sectors still adds capacity, but with decreasing economy of scale. The fourth, fifth, and sixth sectors on a cell add about half as much capacity as the first three sectors, from 2.5 to 4 times a single omnidirectional cell. This is so because more sectors generate more interference and antenna radiation patterns do not have sharp edges. A 60-degree sector has proportionally more of its area in sector border zones.

We are under no obligation to stick to three or six sectors as we did in FDMA and TDMA systems. Nor are we required to have equal-angle wedges for all the sectors of a cell. The sector plan for each cell can be a separate design problem depending on the individual requirements of that cell and its subscriber community. There are administrative advantages in a system with hundreds of base stations to having the same number of sectors in all the cells.

The sector plan has to work for the cell layout. Sector boundaries are pilot-pollution zones, just as cell boundaries are. Each sector is its own CDMA server with its own carriers and pilots. Thus we should orient the sector antennas so that the equal-path-gain lines between sectors do not cross cell triple-points and four-cell zones. The good news for cellular planners is that the sector boundaries are clean, straight lines. Neighboring sectors are in the same place, served by the same base station, so the radio path gains should be exactly the same. The only difference is the comparative antenna gains, and this difference should depend only on the serving signal direction. Thus, while cell boundaries are made fuzzy by terrain, obstructions, and vegetation, realistic sector boundaries should be drawn easily on a map.

---

[3]As we discussed in Sec. 30.1, this is in direct contrast to conventional reuse (FDMA and TDMA), where sectorization is used to manage interference.

[4]Note that the omnidirectional cell in this discussion is hypothetical. We do not expect to find many omnidirectional CDMA cells actually out there serving calls, but we start with the omnicell as a point of comparison for our capacity calculations.

When we need a high-gain antenna for a tough area with low radio path gain, we can create a narrow sector for coverage. Or when there is a small area of high demand, we can create a narrow sector for capacity. The narrow sector goes *between* two existing sectors as part of the sector plan, not as some kind of overlay sector in the middle of another sector.

The next technology step in cell sectorization is introducing adaptive phased arrays, or smart antennas. Time division synchronized CDMA (TD-SCDMA) has made smart antennas an integral part of its specification, but other CDMA air interfaces might benefit from them as well. We estimate that they are able to produce a 60-degree beam, the equivalent of six-sector cells, but they could be more beneficial than that. The technology may develop to form narrower beams that would increase the capacity through reduced interference. Another possible benefit is that each forward path has a slightly different source, and this may reduce some of the same-path multipath degradation discussed in Sec. 30.3. The downside of smart antennas is that they add complexity and cost to base station design. However, radio equipment manufacturers have shown an amazing ability to come up with simple and inexpensive systems to tackle hard problems in the past.

A good CDMA sector plan is one that reaches all the areas of demand with enough radio path gain and CDMA capacity to serve the traffic. A good sector plan does all this with efficient use of radio equipment. And a good sector plan orients and shapes its sectors to keep pilot pollution to a minimum.

33.4 Carrier Assignment

In the dense-demand area of a CDMA system, there is no issue of radio spectrum management. We want to use every bit of radio spectrum in every sector of every cell to squeeze the maximum revenue out of our CDMA equipment. We can think of no reason a service provider would want to back off from the best utilization of the resources of equipment and bandwidth.

As cells spread into the lower traffic densities, we find that coverage can dictate a maximum cell size. When cells get larger than the maximum coverage size, the CDMA capacity declines very rapidly. In low-density areas, we have a cell grid determined by local radio conditions and equipment design. And many of these cells are not serving enough demand to warrant using all the carriers on all the faces. Since we want to ensure radio coverage and we want to enable the system to grow, we want to maintain the grid in the low-density areas because they may grow into higher-density areas over time. Putting radio equipment in a base station for all the carriers costs a lot more money than just installing the carriers needed to serve the traffic. In any case, there is typically a community of base stations that are not using all the CDMA carriers.

We could economize on equipment by using fewer antenna faces, perhaps even using omnidirectional cells. We recommend against this for two reasons. First, when the wireless demand does grow, adding sectors means reconfiguring many antennas on top of a high tower. Second, cells that are coverage-limited need higher antenna gain to reach their tough areas. Sector antennas have higher gain because they cover a narrow arc. Thus, when we do not need sectors for their narrow-beam capacity, we need them for their high-gain coverage.

The fringe of the wireless telephone system will have cells with just a single carrier. These cells with just one carrier frequency are called F1 cells. As we progress toward the busy region, we run into F2 cells with two carrier frequencies. These increase as we work our way to the denser demand areas. Even if F1 cells cost a lot of money and do not earn enough revenue to pay their way in the early system stages, we expect our wireless system to grow, and we expect these outer cells to become F2 and F3 cells over time. Even if they never grow, they can add value by providing seamless service so that high-density-area subscribers see no gaps in their service when they travel further out or because they help the cellular carrier remain in compliance with government requirements to provide coverage.

We call the carrier used in F1 cells the *first carrier* and the carriers added to cells after the first *higher carriers*. The first carrier's service area is the entire system. Each higher carrier should have a convex service area, as we described in Sec. 32.3 for growth cells. This means that there may be a few transition cells using a carrier to make its coverage region rounder and to reduce pilot pollution for that carrier. Each carrier must be engineered to reduce pilot pollution. The result is a system with nested carrier coverage, as shown in Fig. 33.2. The fringe areas are served by one-carrier F1 cells, and we move through F2, F3, and F4 cells. When we get to the densest area

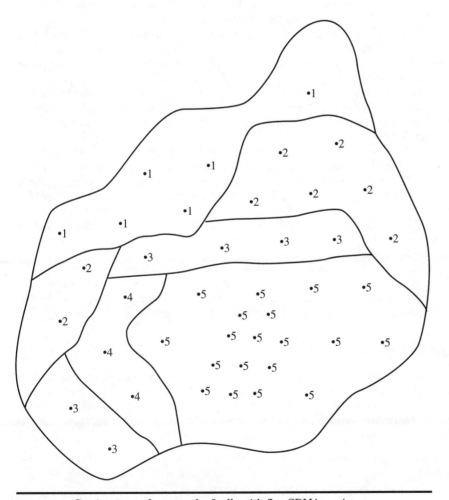

Figure 33.2 Carrier count of a network of cells with five CDMA carriers.

served by F5 cells, we start to see cell splitting to serve greater demand. Each carrier service area is convex, no holes and no dimpled U shapes on the boundary. The small-grid area also is convex. This grid as shown has very little pilot pollution.[5]

The two carriers used in F2 cells and the three carriers used in F3 cells are adjacent in frequency. While the concern of adjacent carrier interference is real, filters are easier to design and manage for adjacent carriers than for carriers with spectrum space in between.

We stack the carriers so that all the F1 cells use the same carrier. It might seem reasonable to stagger the carriers, to use different carriers in different cells, so that each cell has fewer cocarrier cells, fewer cells interfering, and less cell-to-cell interference system wide. Staggering carriers is almost certainly a bad thing to do. The effect of staggering carriers is to make the individual cells for each carrier larger and to create coverage problems in addition to neighbor-cell interference problems. Also, staggered carriers would only support hard handoff, not soft or softer handoff, between cells with different carrier frequencies. CDMA works best on a small grid where short radio links have high path gain. Thus the first carrier is used systemwide, the second carrier is used for all F2 and higher cells, the third carrier is used for all F3 and higher cells, and so on.

Reducing pilot pollution is an engineering issue for each carrier in a CDMA system. If demand patterns dictate an F3 cell layout with a hole in it, an F2 valley in the F3 cells, then we should plug that hole with an F3 cell as a transition cell for the third carrier. Otherwise, we have the capacity erosion due to the pilot pollution that we have been trying to engineer our system away from. The overall system should be convex, each carrier's coverage should be convex, and the growth-cell grid should be convex. Not only does this keep system capacity high, but it also makes handoff neighbor lists easier to manage.

When a call moves from one cell to a cell with fewer carriers, it may need to change carrier with a hard handoff. We say that the call is *handed down* to the lower carrier.[6] Different CDMA equipment vendors hand down calls differently, so it is important to understand how the equipment actually works. We may have to rely on hard handoff between carriers because soft handoff between carriers is not supported in cdmaOne and may not be supported in the third-generation (3G) systems. If a carrier boundary falls on a busy traffic route, then it may be wise to add a carrier so that the hand-down boundary is not in such a busy place.

The carrier assignment is done well when it meets the traffic, convexity, and hand-down requirements. Each cell should have enough carriers to meet its demand, each carrier should have its coverage area designed to minimize pilot pollution, and the carrier boundaries should avoid the highest rates of hard handoffs.

33.5 Handling Peak Traffic Demand

In designing a wireless system, CDMA or otherwise, we calculate a peak traffic load for each cell sector and make sure that the serving antenna face has enough capacity to serve the traffic at the chosen blocking rate. The suite of 3G services makes this job more challenging. We have to add up the variety of voice and data services and any

---

[5]Real systems may have more pilot pollution simply because we cannot get perfect grid locations for base stations.

[6]The carrier may not be lower in frequency, just lower in our F1, F2, F3 list.

other services we think subscribers will want and make sure that each cell sector can handle the load.

We therefore compute the peak traffic for each cell sector based on the forecast time-of-day distributions of each service. We map out the hours of the day and total the demand for each hour. Different-sized services can be compared using the gross chip rate we used in Chap. 31. Voice service may be busy at commuting times, whereas data service may reach its peak at lunchtime, so it is important to develop and forecast a clear picture of each service for each hour of the day. We inspect all hours of the day to be sure that we know which time of day is the peak-demand period. When we have the peak hour demand, we take a close look to make sure that it is our best estimate.

Neighboring cells and sectors have their own CDMA activity in each carrier frequency. These are the f and g terms we used in Eqs. (30.1) and (30.2). We use f to represent the total interfering power received from other cells, and we use g to represent the total interfering power received from other sectors in the same cell, both compared to the serving sector. In Sec. 30.7 we assumed values of $f = 0.5$ and $g = 0.3$ except for the forward link of TD-SCDMA, where we assumed a higher value of $f = 2.0$. We expect our total capacity to be reduced by a factor of $1 + f + g$ by this interference.

There is a statistical effect called *soft blocking* (described in Sec. 30.6) that mitigates the capacity reduction due to f and g. If we are engineering several sectors for a low blocking rate, then most of them will not be at their maximum capacity most of the time. In Sec. 28.5 we saw how to use that statistical smoothing of soft blocking to engineer more aggressively when there is a significant amount of intercell and intersector interference. More aggressive engineering means serving more demand with the same CDMA carriers. The benefit of soft blocking does not mean that the interference is good for capacity, only that the damage can be reduced by relying on the statistical behavior of a larger community of interfering user terminals.

If the adjacent cells and sectors have different time-of-day distributions, then we can be even more aggressive in our traffic engineering. For each sector we are planning, we should use f and g values that represent the other-cell and other-sector demand at the peak time of the serving sector. The shopping mall is busy at lunch, the office building is busy during the afternoon, and the commuter route is busy at night. If these three cells are next to each other, then we may be able to plan on serving more demand with each of the three cells because the neighbors are not busy at the same time.

Deciding when to block calls is a fundamental engineering decision. A busy 3G CDMA system usually will run out of forward capacity first, so we can set the system to block calls when the forward transmit power reaches a certain level. Since we do not want to block handoffs of any sort, soft, softer, semisoft, or hard, we should leave some extra power in reserve for handoffs. Each cell sector is different, so there is no hard and fast rule for when to block calls. Without any experience in the specific system, we would recommend setting the blocking criterion to half the maximum power and seeing how often the amplifier gets close to full output. If the amplifier overloads at this blocking criterion, we would lower the call-blocking power threshold. And if the system is actually blocking calls and the amplifier never gets above 80 percent, then we might raise the blocking threshold to 60 percent of full power.

Our time-of-day efficiency is the peak-to-average ratio, the amount of traffic demand we can serve compared with how much we could serve if our system were busy all the time. In Chap. 45 we discuss ways to improve this efficiency.

The traffic engineering job is done when every cell sector is engineered to meet its blocking requirements for the full suite of services with time of day taken into account for the serving sector, its neighboring sectors, and its neighboring cells.

33.6   Migration from FDMA or TDMA to CDMA

Conversion of existing FMDA or TDMA systems to CDMA is an opportunity to increase revenue greatly with existing radio spectrum. We wish that we could throw the switch and instantly transform the entire system and subscriber community to CDMA. However, it is not that simple because there is a substantial community of legacy system users with legacy cell phones who are not going away. Thus we have to form a plan that gradually replaces the old equipment with more efficient CDMA technology. If the transition is managed properly, then a dwindling legacy subscriber community continues to receive good service while a new CDMA community brings in increased revenue.

The specific conditions for frequency migration depend on the technology. Advanced Mobile Phone Service (AMPS), Global System for Mobility (GSM), U.S. TDMA, and so on are different technologies and require different migration plans. Frequency migration also depends, of course, on the CDMA technology being substituted. Guard-band requirements will be different depending on which radio interfaces are being used.

All CDMA transition plans have some things in common. The transition will be a lot easier to manage with multimode user terminals. Having a large community of dual-mode cell phones means that we can introduce CDMA more slowly than the user terminals. The CDMA system may have high blocking, but subscribers will not experience ineffective attempts because their calls will overflow onto the old system. As the population of dual-mode user terminals increases, we can increase the share of CDMA radio spectrum. Over time, the system will be nearly all CDMA, the result of a seamless transition.

Changing a wireless air interface is a complex task. Both the legacy air interface and CDMA have engineering issues, reuse rules, and interference management that have to continue during the migration for a seamless transition.

33.6.1   Greater radio efficiency with CDMA

The principle behind the migration to CDMA is that the FDMA and TDMA air interfaces are far less efficient than CDMA. They support fewer calls and serve less demand per megahertz of bandwidth per cell, making the migration to CDMA economical.

Any FDMA or TDMA system can benefit from the increased efficiency of CDMA. Any of these technologies can allow partial radio spectrum conversion so that we can introduce CDMA in part of the bandwidth. We start by introducing dual-mode CDMA user terminals into the market. Once there is a substantial community of CDMA-ready subscribers, we can begin to convert radio spectrum.

The idea is to take some piece of radio spectrum from the existing system and turn it into a CDMA carrier. The remaining spectrum should be enough to handle the remaining legacy subscribers. As the legacy population decreases, we can convert another piece of spectrum into another CDMA carrier.

Let us take an example, a 50-MHz GSM system with 25 MHz in each direction, 125 two-way GSM carriers with eight channels each. Using $N = 4$ with three-sector cells, we have 10 GSM carriers, 80 channels, serving each cell sector. This means that our GSM system can support about 70 erlangs per sector in 25 MHz each way.

Let us take one-fifth of this radio spectrum and use it for a wideband CDMA (W-CDMA) carrier. A W-CDMA carrier, 5 MHz in each direction, can handle about 130 voice calls. This is more telephone traffic than the entire GSM system could support. Thus, if we have a significant population of dual-mode user terminals in our GSM population, we can serve all of them in a single CDMA carrier and still have 80 percent of the GSM spectrum left over.

We do not want to overlap radio spectrum between GSM and CDMA. Even if the CDMA channels are tough enough to withstand some narrowband interference from GSM, the GSM system was never designed to handle CDMA interference. It is certainly best to keep the two technologies in separate radio spectrum allocations.

We may decide to convert only the busy area to CDMA. We can share the radio spectrum between the two technologies if we leave a sufficient geographic buffer zone around the CDMA radio spectrum before reusing it in FDMA or TDMA cells. The buffer zone may be an area two cells wide between the CDMA part of the system and the full-bandwidth FDMA or TDMA section.

33.6.2 Greater traffic efficiency with overflow

Neither AMPS nor GSM is designed to hand off calls into a CDMA system. The CDMA system, on the other hand, does have the ability to hand off calls to something else. Thus we will plan on setting up calls on the CDMA spectrum and keeping them there until conditions require them to be handed off to the legacy system.

If we plan on setting up calls on CDMA, then we can treat the legacy system as an overflow server for the CDMA system traffic. We start by selling a large supply of dual-mode user terminals and then switch on the CDMA carrier for them. We also make the CDMA carrier their first choice for setting up calls. The legacy system now serves two roles, the overflow for the dual-mode CDMA user terminals and the primary server for the legacy cell phones. If we have sold enough dual-mode user terminals, then this combined system should be more efficient than the legacy technology using the full spectrum.

Knowing that we can overflow traffic, we should engineer the CDMA system for high blocking, perhaps very high blocking. As we move traffic from FDMA and TDMA to CDMA, the legacy system will become lightly loaded, and we can grab another piece of spectrum for another CDMA carrier. We can keep doing this until there is only a small amount of legacy bandwidth left. We want to leave some small piece of FDMA or TDMA bandwidth for those stubborn subscribers who hang onto their old cell phones for many years.[7]

As the market develops for 3G services, the service provider may want to reserve some capacity for data and other services. It may make sense to put more voice traffic on the legacy system so that CDMA capacity is available for these advanced services. The service provider may decide to give priority to data and other services on the CDMA carrier to develop these markets.

---

[7]One of us (Rosenberg) still has the AMPS car phone he bought in 1986 and still uses it occasionally.

33.6.3 Evolution starts with a small CDMA system

It is very expensive to deploy a partial-spectrum CDMA system on every cell of a legacy system. Using our earlier example, a 5-MHz CDMA carrier handles twice the traffic of 25 MHz of GSM. Thus we may want to start the CDMA evolution on only a fraction of the legacy cells, a larger CDMA grid superimposed on the FDMA or TDMA system.

In Fig. 33.3 we show every third cell being marked for CDMA service. The resulting one-in-three grid is shown in Fig. 33.4. Even with only one-third of the cells using CDMA, the radio-efficiency advantage still favors using it.

Thus we start with a large-cell CDMA grid and let the revenue from that investment pay for the conversion of more cells to CDMA. It is expensive to reoutfit hundreds of cells, and we certainly would rather let the CDMA revenue stream pay for the change as the system continues to migrate to CDMA. Spreading the introduction out over more cells means that it will take longer, but the initial outlay of cash will be much smaller.

If the cells are not too large, then we may decide to start with every fourth cell marked for CDMA service, as shown in Fig. 33.5. Several options are available for initial CDMA introduction, and the best choice depends on how much a service provider wants to invest initially, how quickly a service provider wants to switch to CDMA, and how great the pressure for radio efficiency is in the market.

The coordination of cell conversion and frequency conversion is guided by a desire to make better use of radio bandwidth and a desire not to spend too much money up front. The migration plan is ready to go when it takes into account market penetration of dual-mode user terminals, introduction of CDMA carriers to specified cells, increasing the CDMA presence to more cells, and increasing the CDMA spectrum to realize the gains in efficiency.

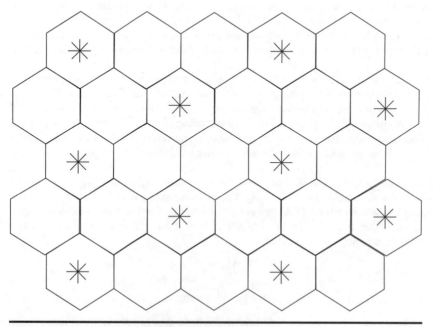

Figure 33.3 Every third cell marked for CDMA service.

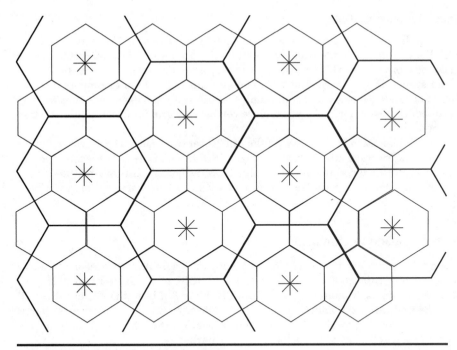

Figure 33.4 Resulting one-in-three CDMA grid.

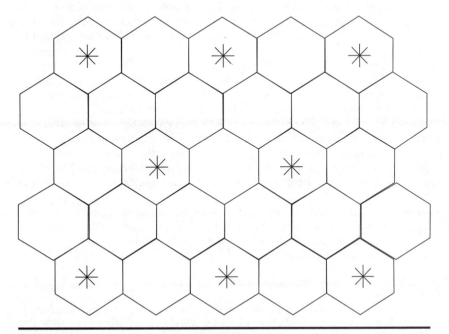

Figure 33.5 Every fourth cell marked for CDMA grid.

33.6.4 Evolution ends with mostly CDMA

When the evolution is complete, the wireless telephone system will be virtually all 3G CDMA.[8] We can do the same capacity analysis for a migrated CDMA system as we would do for a brand-new CDMA system. The migration becomes part of history.

The old system is still there, and we probably want to continue to support it. However, it may be reduced to a very small number of subscribers. This means that we may decide to reduce the old technology to a large-cell overlay grid for lower administration and maintenance.

The full CDMA system is in place with the full 3G service choice. The service provider has one major advantage after a migration from FDMA or TDMA. There is a large collection of cells with a large amount of surplus CDMA capacity. This means that it may be a very long time before the service provider has to spend money for more base stations.

33.7 Migration from cdmaOne to cdma2000

Migration from cdmaOne to cdma2000 is far simpler. There are two reasons to make the change. First, cdma2000 does allow more voice calls in the same radio spectrum. The increase is not as dramatic as going from FDMA or TDMA to CDMA, but it is still welcome. Second, cdma2000 allows data service subscribers to join the wireless party. If there is a burgeoning market for wireless data service, then moving from cdmaOne to cdma2000 is a good way to participate in that market.

cdma2000 does allow frequency superposition with cdmaOne. In fact, the cdma2000 channel directly supports cdmaOne user terminals making cdmaOne cellular calls. The whole issue of managing the transition of subscribers, managing the sale of dual-mode user terminals, and managing the overflow of traffic from one air interface to the other goes away in migrating cdmaOne to cdma2000. These two air interfaces were meant to fit together.

Thus the only decisions for the service provider are when and where to replace cdmaOne base station equipment with cdma2000 equipment. This decision should be driven primarily by the market for 3G services rather than the need to expand voice channel capacity because of the compatibility between the two systems.

There is some improvement in voice capacity with cdma2000, but most or all of it is only realized when subscribers switch to cdma2000 user terminals supporting the reverse pilot. If the only argument for replacing a system were the increased capacity for voice calls, then it probably would not be worth it. However, the compatibility issues and the data-market opportunity both make conversion to cdma2000 a more favorable decision.

What may be the most important issue in changing cdmaOne to cdma2000 is managing the expectations of subscribers expecting 3G services. If marketing efforts for data services run too far ahead of reality, then we can disappoint subscribers anticipating 3G services long before they are available.

---

[8]We are assuming that the CDMA air interface used for the migration is a 3G technology. It makes more sense to use cdma2000 rather than cdmaOne because cdma2000 supports cdmaOne user terminals.

33.8   Quality and Cost Issues in Air-Interface Planning

In our discussion so far we have talked about meeting demand as some kind of absolute. We have so many erlangs of demand for each service offering, and here is what we have to do to satisfy that demand. In fact, we have some broader choices to make—how we are going to manage the bigger tradeoffs of quality versus cost in planning the air interface.

The attitude that quality is an absolute standard runs deep in the telephone culture, from way back in the Bell System days.[9] When the receiver is picked up, anytime or anywhere, there *will* be dial tone. When a telephone number is dialed, there *will* be a connection. Perfect service was not a business case in the Bell System—it was a vision, it was a mandate, it was a commandment, and it was achieved! It was a great culture shock to admit that cellular systems would have an engineered blocking rate and lost-call rate, even if it was a very low 2 percent.

Designing wireless telephone service as a business requires more flexibility than the old telephone company mandate of universal perfect service forever. Increasing quality is increasingly expensive, and the service provider has to weigh each increment of performance against its cost. For readers who used to work for the telephone company,[10] we can console ourselves that evaluating quality and cost is our opportunity to offer good service at better prices.

As we are writing this book in 2002, higher quality of wireless telephone service is a major feature being advertised on television in the United States with commercials making fun of the consequences of cellular static.[11] Lower blocking rates, fewer lost calls, and better sound likely will continue to be used as differentiation in the wireless market. High data rates and high-quality service are really the two things a CDMA 3G provider can offer to attract subscribers. And when 3G is nearly universal, everybody will have the high data rates.

On the other hand, availability of service may be the marketing route of some wireless service providers. By covering more territory or offering lower rates than competitors, a provider may offer the market more benefit by reaching more subscribers. The combination of higher blocking rates and lower monthly rates may be a better business opportunity for some providers. Offering a service of lower cost and lower quality may make the most sense in a large wireless market where there are several providers. Managing subscriber expectations can make them happy to be paying less even if they have to try their calls more often.

Another tactic in the quality-of-service game is to delay service rather than to refuse it. If a busy CDMA sector has 100 calls on it and the average call is 200 seconds long, then the average time until somebody ends a call is 2 seconds. We may decide that subscribers would rather wait 2 seconds than be told that the system is not available and

---

[9]At least, that was how things were here in the United States.

[10]One of us (Rosenberg) worked for Bell Telephone Laboratories in the old Bell System days.

[11]Apparently this has not caused a subscriber shift away from wireless and back to their landline telephones. The airlines, for example, have avoided a similar advertising campaign where one airline would exploit the poor safety record of another. The effect of such advertising probably would be to heighten fear of flying and keep people from buying airline tickets on any airline.

to try again later. Allowing a few seconds delay in placing a call allows a higher occupancy on the CDMA carrier and allows us to operate closer to full utilization. This may be particularly useful in data service, where subscribers are already accustomed to waiting several seconds for a session.

In Sec. 22.3 we discussed the concept of a feasible region of possibilities, all the choices we are able to make in designing a system. On the boundary of that feasible space is a frontier, a region of possibilities where any change we make has to degrade some aspect of the system. The frontier of wireless telephone service is defined by cost, quality, and quantity. In the big picture, these are the three components vying for our attention. In managing the tradeoff of three related components, one effective approach is to constrain one of the variables, reducing its degree of freedom, and then manage the tradeoff between the other two components. Since we can choose to constrain any one of the three variables, this gives us three options.

We can fix the amount of traffic we will serve. Once the number of subscribers and their usage patterns are fixed, we have a tradeoff between cost and quality of service (QoS). We can spend money on more CDMA capacity to offer lower blocking and lower occupancy. This means that subscribers will see fewer ineffective attempts and higher call quality. Their data links will have faster response times and higher average rates over time. The service provider has to determine how willing the subscribers are to pay more for these improvements.

We can fix the budget for the wireless telephone system. A fixed amount of money buys a fixed amount of CDMA capacity, and we have a tradeoff between traffic and quality. We can continue to add subscribers to the system, or we can encourage those subscribers to use more air time or more services. As the usage increases, the system has higher call blocking and more data delays. Since most service providers are loathe to turn away new subscribers, this road usually leads to the lower-quality choice.

Finally, we can fix the QoS. We can decide that the wireless telephone system will meet specified targets of ineffective attempts, lost calls, lousy calls, slow bit rates, and data errors to its voice, data, and other subscribers. This commitment may be a marketing decision, or it may be in a contractual form with service-level agreements (SLAs). Then we have a tradeoff between traffic and cost. As we add subscribers, we add carriers, sectors, and cells to maintain the QoS. This is closer to the mindset of the old telephone company engineers.

These frontiers are not easy to find. It takes first-rate engineering and a lot of hard work to get the most CDMA capacity for the money we spend on a system. It takes keen marketing planning to offer the mix of services that generates the most revenue for the same use of CDMA resources. The financial planners should be made aware that these are difficult decisions that require a thorough understanding of the technology and the market being served.

Once we have decided that there are tradeoffs to manage, it is tempting to use this as an excuse for not making the most revenue we can from a system. At any given point, we could have more subscribers, higher quality, or fewer expenses. The complexity of the problem demands greater attention and vigilance because there is a continuous three-way balancing act going on every day in a wireless telephone system. Any time we add subscribers, change quality, or spend money, we have an obligation to consider the global effects of these changes.

The air-interface design issues are blocking and maintaining adequate signal quality. The subscriber issues resulting from these are ineffective attempts, lost calls, lousy calls, slow bit rates, and data errors. A CDMA carrier allows us to block less and to degrade more, to adjust the nature of the impairments. The initial data market may be very tolerant of delays and errors because wireless data service will be such a novelty in itself. But subscribers will soon decide that they want more than wireless data service, that they want *good* wireless data service. Back in 1985, cellular telephones were a novelty, and subscribers tolerated very poor service just to have service at all. This has changed in the voice call arena, and it will change in the data service arena as well.

The management of a wireless telephone system should form a very clear picture of the product they are selling, the mix of services, how much they are willing to pay for incremental quality, and how much they are willing to pay for more traffic. This picture is going to be the input to the decision process for the wireless engineers building and maintaining the system. It does not matter that the engineering is first rate if nobody knows what kind of system and service we are trying to offer.[12]

33.9   Reliability in Base Station Design and Layout

Base station equipment does not last forever. Any piece of electronic equipment has a maintenance life cycle; it is going to break eventually. To make maintenance matters worse, base stations are often in remote locations. Part of wireless system design is understanding base station maintenance and reliability issues.

A base station failure is a service outage. When an antenna face stops working, the subscribers in that cell sector get degraded service from other antenna faces, or they get no service at all. The outage from a broken base station is not a few seconds or even a few minutes but typically is several hours and maybe days. Unfortunately part of the QoS picture we are offering wireless subscribers includes these outages.

More reliable base stations cost more money for more reliable components. Each component of a base station can be made more rugged, more reliable, and more likely to operate for a long time without failing. When designing the base stations for a wireless telephone system, engineers should do a thorough reliability analysis of each component to form a mean time between failures (MTBF) estimate for the base station. Most likely there will be one or two components that contribute the most to that MTBF, and we have to decide whether we want to spend more money on more reliable equipment for those components.

More reliable base stations cost more money for redundant components. As discussed in Sec. 4.8, we can make a system more reliable by duplicating some of its functions. When one component breaks, its duplicate takes over its function, and the repair facility is notified. But the overall reliability of the system is determined not only by the redundancy but also by how quickly the failed component is repaired. After all, the period between a component failure and its repair is a period of vulnerability during which another failure will result in service outage.

A service provider may decide that partial service is an acceptable condition. If the typical antenna face is down 0.1 percent of the time, 9 hours per year, then we can decide that it is reasonable to let its subscribers be served by neighboring antenna faces during that time. The decision to live with these failures rather than to design more reliable base stations depends on whether or not the cell is adequately covered by neighboring base stations. Will the other base stations be able to cover the entire cell?

---

[12]There is a joke like that: The bad news is we're lost; the good news is we're making really good time.

In making this decision, it is appropriate to consider mean time to repair (MTTR) as well as MTBF. The effect on subscribers of nine 1-hour failures per year is very different from the effect of one 9-hour failure.

Of course, blocking rates will be higher during a base station outage. Not only are we serving k cells of traffic with $k - 1$ base stations, but the missing cell is likely to form a pilot-polluting valley for the other $k - 1$ cells and perhaps one with poor radio signal paths. After all, if the remaining $k - 1$ base stations could cover the missing cell with good coverage and low blocking, then we would not have put a base station there in the first place.

Subscribers probably will tolerate periods of lower performance as long as they do not happen too often. A system with 5 percent blocking most of the time can easily add 0.1 percent base station outage to its overall ineffective call rate. However, the essential difference is that the 0.1 percent base station outage is a continuous period of time, several hours during which service will be poor. Even subscribers used to 5 percent blocking will notice when no calls go in or out for hours.

The strategy we recommend is to have a large-cell grid of more reliable base stations. In Fig. 33.3 we show a one-in-three grid. (In Sec. 33.6.3 we were using this picture for CDMA migration, but it is the same concept of a large-cell grid overlaid on a small-cell grid.) We could ensure that the large-cell grid has high reliability and adequate coverage so that a failure of one of the other base stations still leaves service over the entire area. If one cell in three is too expensive, we can use every fourth cell, as shown in Fig. 33.5.

It is important to make the reliability decisions early. They are part of the system planning process and should not be a reaction to a base station failure. Having broken radio equipment and angry subscribers is not the time to be making decisions about how reliable base stations should be. Decisions made under that kind of duress tend to be short-term patchwork rather than well-thought-out plans.[13]

A complete base station plan includes a reliability plan. Each major component of a base station must be quantified as to its failure patterns, rate of failure, and how long it takes to repair it. If failures of different components tend to be correlated, then this should be noted in the plan. An example would be that one sector failing drives the other transmit amplifiers to nearly full power so that they are more likely to fail as well.

Once the component reliabilities and correlations have been cataloged, the service provider must decide what level of reliability to offer for each base station. Is an occasional base station outage a tolerable event? And how good does the service have to be during an outage? The answers to these questions should be firmly in place in the form of a base station reliability plan as part of the base station planning process. The base station reliability plan is the foundation of the maintenance plan and standard operating procedures for operations, administration, and maintenance (OA&M).

33.10 Conclusion

In this chapter we have discussed the options available for planning the equipment to be placed at base stations to provide coverage for subscribers and meet peak demand with appropriate levels of blocking and acceptable QoS at reasonable cost.

---

[13]We might be able to tell a lot about the planning of a wireless telephone system by the mechanical condition of the cars in the parking lot. If they are maintained only when they break down, then our confidence in the resulting telephone system may be reduced.

Mobile Switching Center (MSC) Planning

The mobile switching center (MSC) is the heart of the wireless network. The wireless network, like an arterial network, is a fan-out network, and the MSC is the single point from which all voice calls and data transmissions fan out for mobile-terminated calls and is also the central processing point for all calls coming in from mobile-originated calls. Like a heart, an MSC must be robust, well supported, and well protected.

In this text we use the term *MSC* to refer to the switch. The term also can refer to the building, the physical facility that houses one or more MSC switches and related equipment. When we count MSCs, we are counting the number of switches needed to support all the base stations on the network. In Sec. 34.2 we will discuss the physical housing of MSCs. In this chapter we will also discuss component planning for MSCs and for auxiliary equipment.

34.1 Estimating MSC Count

Just as we did with base stations, we want to estimate the total number of MSCs required for a wireless telephone system. The issues are less complex for MSCs than for base stations because MSCs do not depend so heavily on the CDMA air interface, but there are more issues.

Switches cost money, and more switches cost more money. To put MSC cost into perspective, if one MSC costs US$5 million and serves 200 base stations that cost $500,000 each, then the MSC cost is 5 percent of the base station cost. Thus MSC cost is a significant fraction of total equipment cost, enough to merit some attention. While some of the MSC cost depends on the number of base stations, there are enough fixed costs of installing an MSC that we want to avoid having extra MSCs in the system.

Each time we add an MSC to a wireless system, we add to the administrative workload. Telephone switches have internal databases and maintenance schedules. There are more connections and more transport pipes as well.

Finally, having more MSCs in a wireless system means having more MSC boundaries and more inter-MSC communication. This affects call originations and terminations because dividing the system into more MSC regions makes more of the wireless calls into roamer calls. They may not look like roamers on the wireless telephone bills, but they are roamers as far as signaling and inter-MSC communication are concerned. And having more MSCs in a system means more inter-MSC handoff activity.

In determining the number of MSCs required, we should think in terms of a fully loaded maximum-configuration switch. Of course, we do not plan to equip every MSC with all the equipment it can support, but we should base our estimate of MSC count on the full capacity of each MSC. Once we have the number of MSCs, we can equip each MSC as needed in Sec. 34.3.

34.1.1 Main processor capacity

The MSC processor handles call processing, short message service (SMS), other signaling, and handoffs, as discussed in Sec. 24.1. The total workload for an MSC is the total sum of these tasks, so it suffices to have a switch processor powerful enough to handle the combined peak load with some headroom to spare. This is not an area for traffic engineering or any kind of aggressive engineering. When a switch processor runs out of capacity, the system stops working. Calls are not completed, short messages are lost, and handoffs are mismanaged. There is some ability for switch-processor tasks to be put into a queue, but a telephone system, wireless or not, has little tolerance for switch-processor delays. Thus we can take the total call processing, SMS, and other signaling loads, add some fraction for inter-MSC handoffs, add some headroom factor, and divide that by the maximum processor capacity of one MSC. This gives us a lower bound for the number of MSCs we need.

The good news for processor capacity is that data services typically involve proportionally less call processing. As demand for third-generation (3G) services grows, the ratio between air-interface demand and processor capacity demand actually will put less pressure on the MSC main processor.

34.1.2 Transport load capacity

The transport load of a telephone switch is the total number of DS-1 or E-1 links from the switch to something else. The MSC has links to base stations, the public switched telephone network (PSTN), the public packet data network (PPDN), private networks, the ANSI-41 signaling network, other MSCs, and itself via loopback trunks. This is a deterministic calculation, the total of all the links compared with the total communication capacity of the switch, as discussed in Sec. 24.2. We can take the total of all of these, with some proportional estimate for inter-MSC transport, and divide that by the maximum transport capacity of one MSC. This gives us another lower bound for the number of MSCs we need.

34.1.3 Circuit switching

The circuit-switching load of a telephone switch is the number of voice call actions described in Sec. 24.3. The total voice call activity divided by the maximum voice call activity for one MSC gives a lower bound for the MSC count. If 3G realizes its promise of an increasing share of data service calls, then the circuit-switching limitation of an MSC should recede in importance. Real-time video may be circuit switched at the MSC, but the high data rates required for it should keep the number of real-time video calls small compared with voice calls. For example, it is the number of calls being switched that taxes the circuit-switching resources of an MSC, not the quantity of data in each call.

34.1.4 Speech coding

The speech-coding load described in Sec. 24.5 is another load we must consider in estimating the number of MSCs required. We divide the total speech-coding load by the maximum number of speech coders at one MSC.[1] We should add a small fraction of that quotient for the Erlang-B traffic engineering required for the speech coders to get a speech-coding lower bound for the MSC count.

34.1.5 Packet switching

The packet-switching load on a 3G wireless switch includes all the base station communications because they are almost certainly all packet pipes, all data services, and most of the other services. The packet handler is a queuing system, so the packet capacity of one MSC is limited not just by the volume of packet data but also by the delay requirements of the data services, as described in Sec. 24.4. Once we have an estimate of the packet capacity of the MSC, we divide that into the total system packet load in the forecast to get a final lower bound for the MSC count.

34.1.6. Total MSC count

Main processor capacity, transport load capacity, circuit switching, speech coding, and packet switching are five factors that limit the capacity of an MSC. The final MSC count estimate is simply the largest of the five lower bounds based on the MSC components.

In second-generation (2G) systems, we have real-world experience with MSCs and base stations. Even here, the MSC count can vary considerably depending on the number of carriers and the distribution of traffic by time of day. However, code division multiple access (CDMA) providers tell us that a rough rule of thumb is one MSC for 200 cells in a system with eight cdmaOne carriers.

In the new generation of 3G systems, we have new services and new markets to consider. The mix of data services in the market is uncertain. A mix favoring high-rate data subscribers will keep the MSC packet-switching components busy while leaving the processor nearly idle. On the other hand, a larger community of low-rate data services, heavy e-mail and SMS activity, for example, will limit the MSC processor first.

As we did with base stations, we should leave some room for initial system growth. If we install three MSCs all operating at 97 percent of their capacity, then it is virtually certain that we will need a fourth MSC almost immediately.

To refine the MSC count estimate, we need to quantify the mix of voice, data, and other services by region and time of day. In this way, we can figure out the true limiting factors that determine how many MSCs will be needed. The result of this study should be a multiple-service, time-of-day analysis that tells us how many MSCs we expect to need at the end of the planning horizon.

---

[1]We are assuming that speech coders are bidirectional.

34.2 Location Planning for MSCs

It is important to find a good place for the MSCs. Because there is no radio link, MSC site selection is far less important than base station site selection, but it is still an important part of the wireless telephone system planning process.

An MSC has to have reliable electrical power. This may sound obvious, but it is an important consideration. The less reliable the source of electrical power, the more important it becomes to have a good power backup.

The real cost issue in MSC location is the availability of telephone transport. In addition to the base stations, the MSC has to communicate with the PSTN, the PPDN, any private networks, the ANSI-41 signaling network, and other MSCs. Clearly, we want the MSC to be in a place where large-volume telephone transport is available and cheap. Most areas, particularly where there is an extensive landline telephone system, have transport backbones, networks of very high speed telephone transport.[2]

In a system with more than one MSC, we can put all of them in one place. This keeps administrative costs down because all the equipment is in one place, so one service visit can attend to all the MSCs. Having all the MSCs together means being able to share the electrical power supply. And any communication between MSCs is going to be cheaper and simpler when the MSCs are all in the same building.

Distributing the MSCs allows them to be closer to their base stations. In the case of a pure fan-out network without a major backbone, this can reduce backhaul costs significantly. Also, having distributed MSC locations avoids having a single point of failure for the entire wireless telephone system. This is only partial consolation for the service provider because having one MSC fail still takes all the base stations that it supports out of service.

For MSC locations, we are looking for places that are central in the transport network for base stations, PSTN, PPDN, private networks, and signaling. Where there is a cheap backbone transport network, this becomes less important than the availability of real estate and electrical power. Where there is no backbone, or where the backbone has no extra transport available for a wireless service provider, we have to do some serious MSC location planning, scouting locations, evaluating transport costs, and finding the locations for MSCs that minimize transport costs.

The planning process also should have an eye toward the future. A system with three MSCs is expected to grow and will require a fourth and a fifth MSC over time. The MSC plan should take into account the space, power, and transport requirements of future switches as well as the present ones.

The MSC location process is done when there is a clear decision where to put the MSCs for the system, whether they should be colocated or separated, and how their power and transport needs will be met. Further, there should be a clear plan as to where future MSCs will be located.

34.3 Equipment Provisioning for MSCs

There are many components of the MSC in a wireless telephone system. The individual vendors and equipment will determine the specific MSC equipment needed, but we can go over some of the basic issues here.

---

[2]In the United States and western Europe, any city or large town has a telephone transport backbone. There may not be such a thing in more remote parts of the world.

MSC equipment should be deployed in sufficient quantity that there is very little blocking at the switch. This is the result of traffic engineering studies based on forecast demand for MSC resources. It is wasteful to spend a lot of money on base stations, antenna faces, and air-interface capacity only to lose calls because the MSC lacks enough capacity. The MSC may seem expensive at US $5 million, but it is a trifling expense compared with all the base stations. We are protecting the base station investment by having enough, and perhaps more than enough, MSC equipment and MSC capacity.

For the main processor, we should evaluate the total load from call processing, SMS, and handoffs. We should make sure that there is enough processor capacity to handle all this activity with enough extra that we do not run out of capacity.

The MSC has to have enough capacity for all its transport needs. We can add up all the transport requirements for base stations, the PSTN, the PPDN, private networks, and the ANSI-41 signaling network to get a minimum figure for transport. We must add some extra transport capacity for inter-MSC transport, as described in Chap. 37. These transport links are required for inter-MSC handoffs and are recommended for cost reduction in roamer terminations and mobile-to-mobile calls. There also should be enough MSC capacity for loopback trunks to support mobile-to-mobile calls where both ends of the call are at the same MSC.

Circuit-switching capacity at the MSC should be enough for all the voice calls and all real-time video calls as well. Even if real-time video becomes very popular, we do not expect it to be a major player compared with the quantity of voice calls in using up circuit-switching capacity at the MSC. There also should be enough speech coders at the MSC to convert speech between the base station packets and the PSTN full-rate pulse code modulation (PCM) links.

Packet switching is going to be harder to engineer because there are so many functions using packet data. All the base station links, all the PPDN traffic, and all the signaling are going to be using packet links. The MSC must have enough capacity to handle all these data flows. We must form an estimate of the total packet data requirement at the MSC based on forecasts of data services.

MSC equipment should include provision for electrical power and backup equipment for reliability. We need the same sort of reliability studies we did for base stations in Sec. 4.8. MSC reliability is very important because hundreds of cells can go out of service with an MSC failure.

We should be prepared for the mix of services to change over time as the market grows for 3G wireless telephone service. As people get used to the idea, they may decide to buy more and more wireless data service over time. As wireless data service becomes part of the technical environment we all live in, the demand may grow much faster than the demand for voice telephone calls. This growth in data service will change the requirements for MSC equipment.

The result of MSC equipment planning is a clear picture of what equipment will be required for each MSC in the wireless telephone system. This picture should be based on forecasts of traffic and the mix of services. There also should be provision in the equipment plan for growing wireless traffic and shifts in the mix of services subscribers use.

34.4   Assigning Base Stations to MSCs

For systems large enough to have multiple MSCs, the assignment of base stations to MSCs can be a complicated process. The basic idea is to distribute the workload relatively evenly among the MSCs and to control the amount of inter-MSC communication.

The first job of the base station assignment is to balance the workload. If we assume that each base station generates the same amount of MSC activity, then we start the assignment by assuming that we will divide the base stations evenly among the MSCs.

It is not necessarily true that each base station uses the same amount of MSC resources. Some have more CDMA carriers than others, and some may have more antenna faces than others, so these base stations will generate more call activity. Also, base stations with a high fraction of data-service subscribers will use different MSC resources than those with mostly voice subscribers. In considering the demand that a base station places on an MSC, it is best to focus on the resources that are most scarce, as determined by the lower bounds given in Sec. 34.3, where we define the one resource driving the MSC count. Base stations requiring more of that resource can be said to require more of the MSC's most critical resource. For example, if processor capacity is the resource driving MSC count, then a base station that is busy with 3G data traffic is actually using relatively little of the MSCs most critical resource. However, it can be valuable to identify base stations that are heavy users of resources that are nearing capacity, even if they were not the resource defined as critical during planning. Base stations serving primarily voice calls will have more call processing and less packet data than those serving primarily data service.

The result of the base station assignment is an even distribution of workload. This workload should be calculated in terms of whichever MSC components were the limiting factor in deciding how many MSCs to use. The assignment also should manage the inter-MSC borders discussed in the next section.

34.5   Inter-MSC Borders

In a multiple-MSC wireless telephone system, we can think of the MSC regions as giant cells. Just as the service area is divided into cells, base stations are divided into MSC regions. In the CDMA air interface, the borders between cells are the problem areas from both a capacity and a communication perspective. In the same way, the borders between MSC regions are the problem areas for inter-MSC communications.

We want to assign base stations to MSCs so that the MSC giant cells have little interaction, less demand for inter-MSC handoffs and inter-MSC handoff requests, which are generated when the signal strength is close to equal at base stations served by different MSCs. Calls from inter-MSC border areas will have inter-MSC handoffs. If the border area involves three or four MSCs, then inter-MSC handoffs become more complicated. Those are the areas where the ANSI-41 signaling system uses all the complicated messages in Sec. 15.1.

This means that we want to keep inter-MSC borders away from busy areas. Like regular cells, the giant cells of MSC regions have to have borders, and somewhere, they have to have triple-points. Thus one goal of base station assignment is to ensure that those MSC triple-points are areas with low wireless telephone demand, and another is to have few zones involving four or more MSCs and to place those in areas of low demand as well.

If the giant cells have borders in high-traffic areas, then we may need to change the assignment of base stations to MSC so that the borders are in low-traffic areas. We may have to do some base station assignment gerrymandering.[3] There are inter-MSC communication and confusion issues during call setup as well as handoffs.

The job of assigning base stations to MSCs is done when the base stations generate an even workload among the MSCs and the borders between the MSC regions are not high-traffic areas. Keeping the inter-MSC borders in low-traffic regions is probably more important than maintaining a convex shape of a given MSC region. It certainly would be better to have an inter-MSC boundary follow a winding river with few bridges than to have it follow a straight line near the river. For a more complete discussion of inter-MSC border areas, see Sec. 36.1.

34.6 Auxiliary Component Planning

Subscriber service depends on the home location registers (HLRs) and the visitor location registers (VLRs) having enough capacity to do their jobs. Also, SMS and voice mail are important features to wireless subscribers, so the short message service centers (SMSCs) and voice mailboxes have to have enough capacity.

These auxiliary components may not be glamorous to radio engineers who are used to solving highly mathematical optimizations of CDMA carrier capacity, but a shortfall in the capacity of an auxiliary component is going to cause a loss of service just as certainly as a poorly designed radio antenna.

Part of the auxiliary component design is deciding how to distribute their services. While each MSC is assigned an HLR, a VLR, an SMSC, and a voice mailbox, these functions may be aggregated into common equipment. We may have a single machine to be the HLRs and VLRs for an entire wireless system, or we may decide that each MSC has a colocated machine that does all four of these auxiliary functions.

When we are done planning the auxiliary systems, we will have an equipment and capacity plan for the HLRs, VLRs, SMSCs, and voice mailboxes.

34.6.1 Home location register (HLR) planning

The HLR should have a database large enough for all the subscribers. There should be no surprises in the HLR database because we know the number of subscribers associated with each MSC.

The processing capacity of the HLR depends on the amount of activity per subscriber. This depends on the number of registrations, calls, and parameter downloads. This means that a 30-second voice call and a stationary 30-minute high-speed data session use the same amount of HLR processing time. The mix of services tells us a lot about how busy the HLR is going to be.

34.6.2 Visitor location register (VLR) planning

The VLR database size depends on where calls are actually happening. Some areas will have surges of roaming activity, whereas others should be steadier. We expect airports

---

[3]In the United States, congressional districts are assigned to decide which communities are represented by which seats in Congress. Politicians have chosen some very strange shapes for these regions to change congressional representation on key issues. This political process is called *gerrymandering.*

to have large surges of roaming activity when flights are late, we expect office buildings to have medium-sized fluctuations at lunchtime or the end of the work day, and we expect shopping malls to have a fairly continuous stream of users from different places.

Most of the VLR activity we envision is voice call activity, but a VLR may end up concentrating on SMS or even data service. We have to look at the mix of services in the demand forecasts to form our best estimates of how much processing capacity a VLR needs.

The VLR should be traffic engineered to serve its demand with very low blocking. When a VLR is slow, wireless service is slow.

34.6.3 Signal transfer point (STP) planning

The signal transfer point (STP) required by Signaling System 7 (SS7) handles all signal traffic from the public switched telephone network (PSTN) SS7 network, plus all traffic from all SMSCs and multimedia message switching centers (MMSCs) sending short messages, enhanced short messages, and multimedia messages. We need to estimate this total signal traffic in order to plan STP capacity.

To meet SS7 redundancy requirements, there must be two STPs, and the total capacity at peak demand must equal no more than 80 percent of the total capacity of one device. This allows for the peak demand to be met with each STP running at only 40 percent capacity. If one should fail, then the other can take over the whole load while still operating at only 80 percent of capacity so that it has room to spare.

The SS7 requirements apply whether the STP is a separate physical device, as is the case with Telekelic Eagle STPs, or are integrated with the MSC switch itself, as is the case with switches from Nortel.

34.6.4 Short message service center (SMSC) planning

The size of the database for the SMSC is the size of the home subscriber community. Like the HLR, there should be no surprises in the number of short message entities (SMEs) served by an SMSC.

We have to plan the SMSC to handle the short message activity of those subscribers, and this can vary considerably. The estimate of SMS activity per subscriber depends on the service mix of the subscriber base. We might expect data service subscribers to send fewer short messages than voice subscribers.

The good news for SMSC planning is that the SMSC is associated with the home MSC rather than with the visitor MSC of a user terminal. This means that a late flight at the airport generates a lot of short messages, but those short messages are distributed over the home SMSCs of the passengers. Unless the airline sold its tickets to a lot of people who live in the same place, this surge of SMS messages will be distributed over many SMSCs.

The traffic engineering job for an SMSC is less critical than the VLR because the wireless system will still operate when the SMSC is slow. Short messages are a feature, a service, but wireless telephone life can go on during a few minutes of slow SMSC performance. This is in contrast to the VLR, where a slowdown affects all roaming traffic.

34.6.5 Voice-mail planning

The size of the voice-mail database is the size of the voice-mail subscriber list. Like the HLR and SMSC, there should be no surprises in the number of subscribers who are paying for voice-mail service.

The number of voice links to the voice-mail system should be traffic engineered based on two sources of demand. There are callers leaving messages, and there are voice-mail subscribers receiving their messages. Both these groups require voice connections from the MSC to the voice-mail system, and neither of these likes being blocked. We recommend a very low blocking rate for voice mail because subscribers remember for a long time when they cannot get through.

Another dimension of voice-mail planning is how long voice messages are stored. Both subscribers and voice-mail planners have a say in the average message storage time. The subscribers may get their messages quickly, or they may let messages sit for several days. And the voice-mail system can have a maximum storage time limit after which voice messages are deleted automatically. The size of the voice-message database depends on how many messages need to be stored, how long the messages are, and how long they need to be stored. There is marketing experience in voice mail for different groups of subscribers, and the voice-mail planners for a new wireless system should use their forecasts of subscriber demographics to design a voice-mail system with adequate message storage space.

When things go wrong, there is a surge in voice-mail activity. Late flights and traffic accidents generate more calls, and some of those calls roll over to voice mail. Also, when the sudden increase in call traffic uses up all the air-interface capacity, calls will roll over to voice mail. Thus a traffic accident not only generates more voice-mail traffic from drivers sitting in their cars, but it also generates voice mail to those drivers when the air interface is blocking wireless call terminations.

34.7 Anticipating System Growth

We can design a wireless telephone system to serve forecast traffic at a specified quality of service (QoS). We can do our financial planning based on these designs and form a clear picture for investors and stockholders.

We expect our wireless telephone system to grow, however, to serve more traffic over time. We expect voice call traffic to increase and new 3G services to increase even faster. The financial plan of a system should include not only the expected initial traffic but also the growth of that traffic over time. Our base station plan already should be designed for growth in demand and evolution of service mix.

Adding a new MSC is a big deal. It costs a lot of money, of course, but it also involves a lot of engineering work. The new MSC requires transport links to the PSTN, the PPDN, private networks, signaling, and other MSCs. New MSC regions have to be formed, and new inter-MSC borders have to be managed. If MSC regions are thought of as giant cells, then adding a new MSC is giant-cell splitting.

If the system design with m MSCs is close to requiring $m + 1$ MSCs, then the initial deployment probably should install the extra MSC up front. Just as we did with base stations in Chap. 31, we should plan the initial system for a deployment time horizon. Anticipating growth in demand for a time horizon gives us a breathing spell to get the system working well before we have to start changing it to satisfy more subscribers.

34.8   Conclusion

Following the methods in this chapter, it should be possible for us to plan the number of MSCs, assignment of base stations to MSCs, and capacity for each component of each MSC, as well as capacity for auxiliary components. Now we will turn our attention to the planning of the backhaul transport network.

Backhaul Planning

The primary function of backhaul planning is to reduce the cost of transport from all base stations to the mobile switching center (MSC). The capacity requirements of the backhaul network were calculated in Chap. 26. Now we need to plan the routing of the pipes. Our budget process needs to take into account both up-front installation costs and ongoing maintenance and leasing costs. Backhaul is communication between base stations and MSCs, and the actual link is between proprietary equipment from the vendors for our particular network. Data compression, efficiency of packet pipes, and other factors may vary on different proprietary equipment. Therefore, estimation of backhaul capacity should be based on the specifications of the communication requirements of the proprietary equipment selected for the specific cellular network. In addition to cost reduction, our planning should consider reliability.

35.1   Cost Issues in Backhaul Planning

All the backhaul optimization models rely on the notion of the *cost* of a transport link as a function of its capacity. If the wireless service provider is leasing its transport from the local telephone company through the public switched telephone network (PSTN), then transport-cost calculation may be a simple exercise in looking up tariffs. Rates are published based on zones and distance for single and multiple DS-1 or E-1 links. On the other hand, having more options can make the cost calculations far more complicated.

Some pricing is based on one rate for cost of reaching the backbone and another rate for transport on the backbone. The implications of this difference are discussed in Sec. 35.3.

Different vendors have different plans for leasing transport. It is important to compare rates among vendors and to make sure that the same rate rules are being applied in comparison. It may be as simple as one vendor including sales and excise taxes in the rate, but it may be more complicated, with one vendor offering a better rate than another but for a less reliable technology.

Transport can be priced in bundles. We get a good rate if we buy 100 DS-1 links or if we sign a long-term transport contract. We rely on separable pricing for our heuristic optimization, that the price of one transport link did not depend on using another. In the face of vendors bundling services and requiring long-term service contracts, we have to be careful that we are really getting the service and prices we put into the computational model. Ideally, engineers who are aware of alternative backhaul technologies could be consulted in advance of contract negotiations. Bringing this level of expertise to the table may make it possible to negotiate contracts with flexibility that will allow reduced costs within the contract but still give the service provider options outside the contract for parts of the network backhaul.

We could lease dark fiber and provide our own terminal equipment. Or we could use point-to-point microwave links. These options involve paying more in initial cost to save some or all of the recurring expenses down the line. The value of such a tradeoff depends on the time value of money as seen by the financial planners. From the point of view of backhaul planners trying to save a buck, it makes the equation for cost more complicated. Each transport link has several options, some involving purchasing equipment up front and others involving leasing facilities from vendors. Some of the vendors are going to offer combinations that make it hard to evaluate the separate cost of each link in the system.

In terms of the tradeoff between buying equipment up front and leasing facilities, we should consult the financial planners and ask them how to evaluate this tradeoff. Financial planners have a set of tools, such as *net present value* (NPV) calculations. These tools allow them to make decisions that are better for the business. They have their priorities, and we want to make sure that we build the system that is most cost-effective according to the needs of those who are paying the bills. There is another reason to consult the money people before engineering backhaul with these kinds of tradeoffs. If we are going to make decisions about ongoing expenses, then consulting the executives will encourage them to be part of the process. At least they will feel more comfortable with a decision when they have had a chance to participate.

The result of a transport-cost study is a list of all the transport options reduced to costs we can compare. Once corporate priorities are established, fixed costs can be compared with recurring expenses, and long-term contracts can be compared with shorter-term commitments. Each transport link option in our calculations should have a clearly defined cost as a function of how many DS-1s or E-1s it is going to carry.

35.2 Fan-out Backhaul Networks

Designing an efficient fan-out backhaul network starts with evaluating the transport options from base station to base station. The mathematical model in Sec. 26.1 starts with a cost for every possible connection in a fully-connected, complete graph, so we can start the backhaul engineering process by considering the transport options between every pair of base stations in the entire system. Then we can feed all those costs into a mathematical program and get an optimal solution for fan-out backhaul, at least one optimal for the specific base station layout being considered. Once the system grows and adds base stations, the solution needs to be reevaluated.

We probably can afford to be a little more relaxed and consider only the more eligible hops. Connections between neighboring base stations should be on the fan-out candidate list along with connections two and three cells away. The list of fan-out backhaul connections also should include a direct connection from every base station to the MSC.

Once all the possible hops are cataloged, we need to find out how much each hop costs. Because we expect to be bundling the packet links together as we get closer to the MSC, we need to know the link costs as a function of their capacity. As we discussed in Sec. 35.1, getting well-defined cost values for transport may be a difficult problem all by itself.

The idea is to form a daisy chain of packet links from base stations to the MSC. We can visualize a fan-out network as streams and tributaries merging into rivers, and we can visualize the transport network as packet flows from the base stations to the MSC. These are two-way packet pipes, of course, but we can visualize a one-way flow from base stations to MSC to see the design process more clearly. Each base station node in the fan-out network has a next node where all its packets are sent on their way to the MSC.

We can approximate the mathematical optimum using a heuristic approach to solving the fan-out problem. We tend to call a solution technique a *heuristic* instead of an algorithm when it is more like an approximation that gradually improves rather than a full solution for a large problem. In the case of a fan-out network, we can look for good solutions locally at each base station to get a pretty good global solution.

We can start our backhaul heuristic with a feasible solution of having each base station send all its packets directly to the MSC with no daisy chains. Some service providers may stop right here and use the star graph solution described in Sec. 26.1 as their backhaul network.

Now let us consider each base station and see if there is a cheaper way to get the packets home to the MSC. We only need to consider all the transport links on the list from that base station. For each link on the list, we add the cost of using that link and the incremental cost of carrying those packets on the downstream links, and we subtract the cost of the transport we were using before.

Let us put this in more mathematical notation. For every base station node b, we are going to consider changing its next node toward the MSC in the fan-out network. Suppose that the current next node for b is node c and we are considering changing that to base station node d. Then the cost added by the move from c to d is the cost of all the b packets (including any packets upstream of b) being sent to d plus the extra cost of adding all the b packets to whatever links d is using to get to the MSC. The cost saved by the move from c to d is the cost of all the b packets being sent to c and the cost saved by removing all the b packets from whatever links c is using to get to the MSC. We do this for all the transport links on our list from b to anywhere and then move on to the next base station to look for improvement.

This iteration continues for all the base stations and all the links on our list until no improvement is found. Is this interactive, heuristic solution as good as the full-system-view optimal solution of the network flow optimization in Sec. 26.1.3? Probably not, but it is close to optimal, and it is a lot less work. Furthermore, an incremental solution technique, such as this iterative heuristic, is well suited to the continuously changing backhaul environment of a growing wireless telephone system.

The fan-out backhaul network design is complete when we have evaluated all the reasonable transport links, assigned them cost values taking into account the fixed cost and recurring expenses discussed in Sec. 35.1, and completed this heuristic until there are no improvements found in the iterative search.

35.3   Backbone Backhaul Networks

In areas where there is a backbone of cheap and plentiful telephone transport, the backhaul problem is easier.[1] The transport backbone makes itself visible at specific

---

[1]Planners in major industrialized cities can take this kind of transport for granted, but the future of wireless telephone is worldwide, with much of its development expected in places where telephone transport is neither readily available nor cheap.

places called *points of presence* (POPs). Transport from POP to POP is enormously cheaper than transport from anywhere else to anywhere else. There can be a single line of POPs, or the backbone can be a complicated network of transport with POPs all over the city.

The backhaul design from the base stations to the MSC becomes a backhaul design from the base stations to the nearest and cheapest POP. It is like driving in a town with a few major highways: Everybody gives directions from the nearest major highway, and most trips involve getting to the highway, a few minutes of high speed motoring, and getting from the highway to the destination. Because of its transport requirements, the MSC will be on the backbone, so it suffices to find the least expensive way to get each base station's packets to the nearest POP.

We expect the backhaul solution to consist of daisy-chained packet links as before. This means that one large fan-out network of packet links is going to become a collection of smaller fan-out networks gathering at several POPs. Generally, network planning is more than proportionally simpler for smaller networks: It takes less than half the effort to design a network half the size. Thus we expect the planning effort for a piecemeal network with a backbone to be simpler, overall, than the full fan-out backhaul problem we saw earlier.

We should make sure, however, that there is room on the backbone network for our wireless telephone transport. This may sound obvious, but it is possible to plan on using the backbone for transport, to design the network with this expectation, and to find ourselves scrambling to find alternative transport from POP to POP because the backbone is completely sold out. There also can be problems if the terminal equipment at the POP nearest our base station is completely full and does not have the ability to accept new pipes. We know of one case where this actually happened during a major upgrade of an urban backhaul network.

The backhaul network design with a transport backbone is still a fan-out network. The backbone gives us more structure and we can use that structure, to find simpler and faster heuristics. In this case, most of the simplicity is that the smaller subnetworks are easier to solve than one big fan-out network.

We can use a heuristic for a backbone backhaul network. From each base station, we find the cheapest link to a POP. As before, what is cheap depends on how we evaluate costs, as we discussed in Sec. 35.1. We add transport hops to the list of links for neighboring base stations and hops of one or two cells as before. We certainly do not need to consider any link from a base station that costs more than the hop to a POP.

We start our heuristic with the feasible solution of having every base station connected to the cheapest POP. As before, this is a reasonable solution by itself, and it is easy to administer. Some providers may stop right here and decide that this is a perfectly fine backhaul network. We recommend continuing the process to reduce the cost of backhaul.

For each base station, we look for a cheaper hop. We consider each transport link on the base station list and evaluate the cost impact of switching the downstream link from its current link to this one. Base station b currently sends its packets downstream to node c on its way to the MSC. Node c could be another base station, it could be a POP, or it could be the MSC itself. We are considering changing that path from c to d. In our thought experiment, the total packet traffic from b and any base stations upstream from b is going to be sent along the link to d, so we add the cost of that link. We also add the cost of increasing the flow downstream from d by the packets from b and upstream. The cost saved is the cost of the link from b to c and any cost saved by removing all the b packets from c downstream to the POP. We do this for all the transport links on our list from *b* to anywhere and then move on to the next base station to look for improvement.

As before, we continue this iterative investigation until no improvement is found. When we reach this local optimum, we have a good backhaul network using the transport backbone. The backbone network for backhaul is complete when this optimization is finished.

35.4   Data Direct from the Base Station

Another possibility is to have packet traffic for subscriber data services go directly from the base station to the public packet data network (PPDN). This makes sense when there is a large packet data market in the subscriber community served by the base station. Having data services going directly from the base station reduces the need for backhaul transport from that base station to the MSC.

As we described in Sec. 21.2.1, we might be able to install the equivalent of a digital subscriber line access multiplexer (DSLAM) at the base station. In this way, the base station itself becomes an Internet Protocol (IP) node. Not only might this reduce backhaul transport costs, it also might reduce data service delay times because there are fewer links in the packet path.

This idea may not make sense if the user terminal can hand off to another base station. Then we could have the extra administration of managing the flow of packets from one base station to another after a handoff is complete. Even if we manage hard handoffs by clever use of IP address assignment, soft handoff could make such a scheme into an administrative nightmare.

The third-generation (3G) code division multiple access (CDMA) standards provide for shared packet data channels deliberately designed to offer efficient data service for stationary subscribers who do not hand off anywhere. If there is a large enough community of stationary data service subscribers to warrant the use of a shared-packet CDMA channel, then there also might be a good case for making the base station an IP node and siphoning those packets off into the PPDN without hauling them back to the MSC. There is no reason stationary data users need to send their packets through the MSC. The MSC can handle the call processing, signaling, and billing, while the local IP network handles the subscriber's data packet stream.

Short message service (SMS) is being extended to form enhanced message service (EMS) and multimedia message service (MMS), as described in Sec. 20.5. MMS is designed to send audio and video information, cellular snapshots, to other wireless subscribers. While the signaling information for MMS messages should go through the usual channels of MSC and ANSI-41, a large image file can be sent through the IP network directly from base station to base station. If the 3G subscriber community starts sending a large number of MMS images, then this could be a major saving in backhaul expense.

Turning a base station into an IP node still means connecting it to the nearest POP on the IP network. This may be easier and cheaper than using a bigger pipe all the way back to the MSC.

35.5   Reliability Engineering

A fan-out network, with or without a backbone, has only one route from each base station to the MSC. This makes the backhaul link a single point of failure, raising a reliability issue. We have given some thought to reliability in base station design and MSC equipment planning. The backhaul network deserves some reliability planning as well.

The backbone network itself may have the same lack of redundancy. Part of a service-level agreement (SLA) for any leased transport should include the level of reliability and the level of redundant facilities. Any facilities of our own design, dark fiber or microwave links, for example, should be evaluated for their reliability. If all the transport facilities are rock solid with negligible failure rates, then we have done our reliability homework for the backhaul network. If, however, there are some questionable links, then we have to think about the consequences if those links stop working.

The basic scheme to make a transport network more reliable is to have diverse routing, redundant paths from one place to another. Not only is transport diversity insurance against equipment failure, it is also protection against environmental damage. The environmental damage could be a flood or a tornado, but it also could be the result of human actions. More than one T-carrier has been disconnected by a road crew digging a ditch, and underwater cables also have been severed.[2] A letter of apology is small consolation to subscribers who are left without service.

The bad news is that diverse routing costs more money. A truly diverse network has enough surplus capacity to handle the full demand even with a link missing. We may decide to divide the backhaul load along geographically distinct paths so that we get some of our transport when there is a failure, but even this is usually more expensive than a simple fan-out network. Almost any telephone transport network has increasing economy of scale, so dividing the load between two paths means operating each path at higher cost per unit transport.

Transport facilities do not fail often. We should be able to treat transport failures as rare events rather than routine outages. This allows us to decide which base stations require redundant transport and which cells we are going to protect against these rare failures. The best strategy is probably to have a skeleton network of base stations with geographically diverse transport and to let the rest of the network rely on single transport links. This skeleton system should provide good coverage for most of the service area. If a link fails, then there will still be service on the skeleton system, so no large areas will be left without service. No matter what the outcome, the backhaul planning process is only complete when reliability, redundancy, and diverse routing options have been explored.

35.6   Anticipating Growth

Backhaul growth is a funny thing. Wireless telephone systems grow by adding cells rather than by growing cells, so the individual serving nodes in the backhaul network

---

[2]The industry term for this is *backhaul fade*. It is the most common source of service failure in the transport network. The consequences, both for PSTN service and for users of leased lines, can be quite severe. A few years ago, New York City lost half of its capacity for calls leaving Manhattan due to a dredging error made by the phone company itself. The leased line carrying radar data among the three New York area airports was severed, and all the airports were forced to shut down for several hours. There was a redundant fiberoptic link for the radar data from an alternate service provider. However, no one knew, until too late, that the alternate service provider leased space on the same fiberoptic cable bundle under the Hudson River that held the primary service pipe. It pays to investigate the details of the plans that your service providers offer.

become more numerous rather than more capacious. Most communication networks grow by having small nodes get bigger, but our communication network is growing by having more and more small nodes. Our pipes, however, will grow as the fan-out backhaul network concentrates more traffic from more cells on the same transport links. When cells split, the new base stations have to send their packets somewhere, and the best place to send them is probably one of the existing base stations nearby. Thus the backhaul transport pipes are likely to get bigger as well as more numerous. It is important to anticipate growth in planning the backhaul network for the initial system.

35.7   Other Backhaul Technologies

In certain areas, there may be alternative technologies for backhaul networks. In Sec. 32.5 we discussed leasing space or colocating transport facilities with cable TV networks. Using the cable TV network would be similar to using PSTN backbone facilities, as we discussed in Sec. 35.3.

Point-to-point microwave is frequently used in rural networks, especially in areas with rocky terrain that makes underground cable installation expensive. It also can be used in urban areas if there is some difficulty reaching the network.[3] In addition to installing our own point-to-point microwave, we might find innovative alternatives. For example, if a company has built its own point-to-point microwave link for internal use, we might be able to colocate base stations on their towers and lease space on their microwave pipe at low cost.

One service that has been defined and prototyped but not fully deployed is worth mentioning as a possible backhaul alternative. It is the multichannel multipoint distribution system (MMDS), which is designed to function as a high-capacity radio network operating on a cellular model. It can offer digital television and high-speed data services. Cells have a radius of up to 35 miles. Where MMDS exists, it might be possible to use it as backhaul for one or more base stations. MMDS deployment has run into financial difficulty, so in cities where prototypes exist the owners of the prototype systems might welcome the revenue they would receive by providing backhaul for cellular telephone networks.

35.8   Conclusion

Following the steps in this chapter, it should be possible to develop a reliable, cost-effective backhaul network that provides sufficient capacity to support all the voice and data services through each base station of our cellular system. We now turn our attention to planning our networks so that they have sufficient capacity to support signaling.

---

[3]One of us (Kemp) worked for a university that used microwave transmission to deliver cable TV programs to the TV studio for broadcast but was unable to receive cable TV in the same building for a number of years. This was in central Manhattan, where it cost over $1 million to lay a pipe with cable across a small crosstown street.

Signaling Capacity Planning

In planning, we must ensure that our network has the capacity to handle the signal load at peak capacity. This includes the signaling load required by the ANSI-41 standard (including short message service support), as defined in Chap. 15 and discussed in Chap. 25. It also includes signaling required between base stations served by a single mobile switching center (MSC), which are not included in the ANSI-41 standard, and any other signaling features available on proprietary equipment that exceed the ANSI-41 standard, such as support for inter-MSC soft handoff.

Although we are able to define all the types of signaling messages, and we have proposed methods for estimating them, it is difficult to develop precise statistical measures of signaling demand, such as the number of handoff measurements per minute per call for a particular geographic area. Due to the lack of precise, measured input, it is probably better to err on the side of caution and allow some extra capacity for signaling. This is not excessively expensive because each signal is small, so adding a significant margin for error does not require a major investment in equipment or transport.

36.1 Handoffs

Handoff activity generates communication messages between base stations. When the base stations share a common MSC, their handoff messages go through the MSC. When the base stations involved in a handoff are served by different MSCs, the MSCs use ANSI-41 messages to determine the need for a handoff, to establish the handoff, to make the required connections, and to end the handoff. The ANSI-41 specifications to date do not require soft-handoff capability between base stations on different MSCs, but several vendors support inter-MSC soft handoffs.

The ANSI-41 signaling network is involved with any inter-MSC handoff, whether or not it is on the same wireless telephone system. In the jargon of ANSI-41, any change of MSC is a change of system. In estimating the ANSI-41 signaling load for signaling capacity planning, we care about inter-MSC handoff activity whether the MSCs are on the same service provider's system or on two different wireless telephone systems next to each other.

The signaling workload for handoffs is generated by the subscriber traffic at the borders of MSC regions, as described in Sec. 34.5. It is the call activity in the border zone that determines how much inter-MSC signaling has to be done. The calls do not have to hand off to generate signaling messages. Once they get close enough, the serving MSC sends handoff measurement messages to any candidate MSCs for the serving base station. If the signal at the candidate MSC is good enough, then the MSCs coordinate a handoff using more ANSI-41 signaling messages.

We want to know how much handoff traffic, or handoff candidate traffic, to expect. The best indication is to look at a 500-m-wide band (one-third of a mile) between MSC serving areas. Calls in this zone are likely to have good radio signals in each MSC area often enough to be generating handoff messages. If we get close to a triple-point where three MSC regions meet, then we expect a circle about 500 m across where all three MSCs are actively sending handoff messages for each call.

Planning tools can help us refine the estimate of how much area is involved in handoff activity. Not only might the activity band be wider or narrower than our 500-m estimate, but local hot spots of high radio signal path gain can produce areas of inter-MSC handoff activity well inside an MSC service area. The signal reaching a base station served by a different MSC signal may never be strong enough for an actual handoff to take place, but the ANSI-41 signaling goes on as long as the radio path gain is high enough on different base stations served by different MSCS.

Moving calls generate more handoff activity. At the most basic level, this is obvious. Calls that move are likely to be calls that change cells and change MSCs while they change cells. However, moving calls are also calls with rapidly changing radio paths. The handoff itself does not have to happen to keep the signaling network busy sending messages; it suffices that the radio signal levels are close. Moving calls with their greater radio path variation are more likely to trigger handoff measurement messages.

As the number of MSCs increases, the amount of border territory increases as well. The more MSC regions there are in the system, the more lines appear on the map separating them from each other, more border zones for more handoff activity. Thus, adding an MSC to the system adds handoff signaling workload as well.

The border zone handoff activity increases faster than the number of MSCs. We say that the border area increases at a rate more than linear in the number of MSCs. Going from two to three MSCs adds more border area than going from one to two. And going from three to four MSCs adds more than going from two to three. Each new MSC adds more new border area than the prior addition does.

The more-than-linear increase happens because the MSC regions themselves are getting smaller as the system serves more traffic. Smaller MSC regions have a greater ratio of boundary to area, a simple fact of mathematics that boundary increases with size and area increases with size squared. Subdividing the MSC region into smaller regions keeps the total area the same but creates new border zones.

The inter-MSC border zone, that 500-m band we described a few paragraphs back, does not get thinner just because the MSC regions are getting smaller. If the border zone goes through a part of the wireless telephone system with smaller cells, then the handoff zone will be smaller in area. Since we only add small cells where the subscriber traffic density is greater, the total border-zone traffic will be at least as great in small-cell areas as in large-cell areas.

The border activity affects call setup as well as handoffs, but with a lower volume of signaling. When a call starts near an MSC border so that it is detected by cells on multiple MSCs, the MSCs all contact the home location register (HLR) for a decision about who is going to serve the call. This border-zone call-setup signaling activity is also in proportion to the traffic in the border zone.

To estimate the amount of inter-MSC handoff signaling traffic, we need a good estimate of the total subscriber traffic in the inter-MSC border zones. This estimate can come from a careful population study with maps or it can come from planning tools, but the estimate of inter-MSC handoff signaling starts with an estimate of how much traffic is in the MSC border zones. This estimate should be calculated for each MSC and, preferably, for each pair of MSCs sharing a common border.

Once we have the estimate of how much border-zone traffic is being served, we need an estimate of how much handoff signaling activity is being generated. In Sec. 15.1 we go over the various handoff activities and how many ANSI-41 signaling messages each generates. The rate of these activities for border-zone traffic has to be estimated based on radio conditions and code division multiple access (CDMA) experience, including local experience if there has ever been a CDMA system in the area.

The result of this calculation is the best estimate we can form, given what we know about the system borders, for how much inter-MSC handoff activity there will be and how much ANSI-41 signaling that handoff activity will generate.

In addition to planning MSC boundaries within our own system, we also must plan the boundaries between our system and neighboring systems operated by different service providers. Although the calculation of capacity requirements is the same, the management of the boundary with another company can raise some additional issues worth mentioning.

- The exact placement of the boundary line may be set by governmental regulation or may need to be negotiated with the other carrier.

- Our planning for inter-MSC handoff may have been based on proprietary features of the equipment in use on our network or on parameters we set. Assumptions derived from this situation may not hold true when planning, negotiating, or managing a boundary with another service provider.

- Communication between any two companies typically takes longer than communication within one service provider, so it makes sense to start planning our external borders early.

- At the border, the actual radio environment depends on the operational quality of the other service provider's base stations and is not under our control. An outage of the other provider's base station at the border will increase traffic on our network.

These issues should be taken into account when planning and managing the boundaries between service providers.

36.2 Call Processing

The connection between call processing and signaling messages is based on subscriber activity, as described in Sec. 25.3. The capacity requirements for call processing and signaling messages are not correlated with borders zones or the locations of subscribers, unlike the requirements for handoffs and roaming. Call processing is well understood and happens everywhere in the wireless telephone system.

Every time a user terminal is turned on, it registers with the nearest base station, and the serving MSC contacts the HLR. When the user terminal moves from one MSC region to another, it registers again and updates its database record in the HLR. In addition, base stations periodically will check to see if registered user terminals are still within range, adding a bit more demand to the signaling load. The registration load on the signaling network is the number of user terminals being turned on or off plus a bit extra for confirmation of registration.

The rest of the call-processing signaling workload is from actual call activity. Every time a call starts or stops, messages are sent to the HLR. Every time a subscriber uses call waiting or three-way calling, messages are sent to the HLR. We should have good forecasts of the amount of call processing we expect in a wireless telephone system.

For voice calls, we know the average call duration, and we have forecasts of the total call demand. We needed those forecasts to decide how many base stations and MSCs to build and how much backhaul to provide. Data service also uses call processing to set up sessions and to end them, but we do not have as good a feel for the average duration of a wireless data session as we do for the average voice telephone call. We have to guess, however, and base our signaling capacity on our forecast call-processing load for the mix of services we expect on a third-generation (3G) CDMA system.

The call-processing estimate is the best estimate of the call-processing activity based on the mix of services. The signaling workload estimate for call processing is the number of ANSI-41 messages generated by that activity.

36.3   Roamers

Virtually any user terminal activity sends ANSI-41 messages to and from the HLR. Registration, call activity, and short messages all involve the HLR no matter where the user terminal is. When the user terminal is served by an MSC other than its home MSC, then most activity also generates messages to the visitor location register (VLR).

When we do our calculations to estimate roaming traffic, we think of the subscribers far from home using another wireless telephone system. As far as ANSI-41 signaling is concerned, any user terminal served by an MSC other than its home MSC is a roamer. This population of roamers depends on the number of MSC regions and how they are laid out. As we have more MSCs and more MSC regions, it is more likely for a user terminal to register or place a call as a roamer.

Once we have an estimate of how many roamers our system will be supporting, we can use the message counts from Sec. 25.2 to estimate the total signaling load from roamers. The signaling load occurs in three main areas, MSC to HLR, MSC to VLR, and VLR to HLR. Each of these routes has message flows when roamers register, call, and send short messages.

We should spend some time analyzing the patterns of subscriber roaming. People have different calling patterns and use their wireless telephones at different times of day. Commuters, travelers, and shoppers have different calling patterns, and we should take some time to form a clear idea about how that affects roamer activity levels. If an MSC region contains a major airport or train station, then we expect it to perform differently than an MSC region serving mostly shopping malls. The roamer signaling estimate is a total of the signaling needs for roamers from other wireless telephone systems and roamers in another MSC region in their home system.

36.4   Short Message Service (SMS)

Short message service (SMS) is a subscriber feature handled by the ANSI-41 signaling system. Other signaling functions are internal to the wireless telephone system, messages that support subscriber activity. In the case of SMS, the subscriber's message is sent right on the signaling channels.

SMS messages are sent from the sender's serving MSC to the sender's home short message service center (SMSC). If there is a concentration of short messages, then the serving MSC sees that concentration. After a message leaves the serving MSC, the sender's home SMSC takes care of it. There are also signaling messages between the sender's SMSC and HLR.

Once the message is stored safely at the sender's SMSC, the message recipient is involved. The recipient's HLR is contacted to see whether the receiving user terminal is turned on. If the receiving user terminal is able to take the message, then the sender's SMSC sends the message to the recipient's MSC on the ANSI-41 signaling network. These functions generate signaling workload.

Enhanced message service (EMS) allows subscribers to send and receive ring tones, icons, and graphics. The system uses concatenated, binary-encoded SMS messages to send these larger messages. Since EMS messages involve more data than normal SMS messages, EMS subscribers place a heavier load on the signaling system than SMS subscribers. If EMS becomes popular enough, then it can be a major component of signaling activity.

Multimedia message service (MMS) sounds scarier because it is sending full graphic images from one wireless user terminal to another. MMS uses an Internet Protocol (IP)–based multimedia message switching center (MMSC) separate from the SMSC. Also, MMS sends its images as data files over packet data channels, so they do not go over the ANSI-41 signaling network. Active MMS subscribers may use a lot of CDMA resources, and we hope that they will pay for them, but their images do not use signaling resources.

As we discussed in Sec. 25.4, SMS and EMS are special cases not only because they send subscriber data through the signaling network but also because there is no other, larger bit stream associated with these messages. For example, a sudden surge in cell phone voice calls caused by a major traffic accident will quickly fill the CDMA carrier's capacity with calls. Each of these calls generates some signaling activity, but the calls themselves use thousands of times more bits than the signaling messages. A similar surge in SMS and EMS messages also may fill the CDMA carrier, but now all the bits in those messages will be using the signaling channels. Heavy use of SMS and EMS will create one-for-one heavy use of the ANSI-41 signaling channels that we rely on for our system to operate. This means that SMS and EMS are signaling wildcards.

The system operator can control the quantity of SMS and EMS messages by setting rates. If the signaling channel is getting full of messages, then we can raise the price to keep usage down. If demand elasticity favors a lower price with more SMS messages than the ANSI-41 signaling network can support, then it is probably time to increase signaling capacity. Different markets in different parts of the world have dramatically different SMS activity levels. As of the writing of this book, Japan is well known for the high volume of short messages on its wireless telephone systems, but the fad could catch on in other places as well.

It is important for wireless system planners to have good estimates or at least good upper bounds of SMS and EMS traffic levels. The signaling consequences of this activity should be reviewed carefully so that we can forecast the signaling loads for MSCs, SMSCs, HLRs, and VLRs resulting from short messages.

36.5 Other Signaling Functions

There are other signaling functions in a wireless telephone system. Many of these do not use much signaling capacity because they are not used often. The operations, administration, and maintenance (OA&M) in Sec. 15.4 and the over-the-air service provisioning (OTASP) and over-the-air parameter administration (OTAPA) in Sec. 15.5 are all very important signaling functions, but they do not use much signaling capacity.

Fraud prevention is an extra overhead function. The extra messages we need to send to protect subscribers against fraudulent use of their telephone numbers likely will increase over time as crooks get cleverer. The obvious case is when a real user terminal and a phony cell phone with the same identification number place calls at the same time. There are going to be fraud-specific messages to and from the HLR to deal with this kind of problem, and such messages may get more complicated.

We need an estimate of the amount of other signaling activity. If it is small, then there is no point in sweating to make it accurate. If there is a significant signaling load from these other functions, then they should be added to the total estimated ANSI-41 signaling load.

36.6   Conclusion

In this chapter we have provided a process for estimating the signaling requirements of the cellular network so that the network plan can include equipment sufficient to meet peak signaling demand. We also have discussed ways of reducing signaling demand by careful planning of inter-MSC borders. Now we will turn our attention to planning the transport between any given MSC and the other networks and systems with which it communicates.

37

MSC Transport Planning

It would be a shame to build a code division multiple access (CDMA) network with high capacity and quality on the air interface and then lose the calls on the way to the public switched telephone network (PSTN), or to another mobile network. We need to do traffic engineering and plan the transport between mobile switching centers (MSCs), as well as the transport from each MSC to the PSTN. We also need to plan transport for data to the public packet data network (PPDN) and private networks.

37.1 MSC-to-MSC Transport Planning

Inter-MSC handoffs and some roamer terminations require links between MSCs, and we can reduce costs by carrying mobile-to-mobile calls between MSCs without passing the calls through the PSTN.

37.1.1 Handoffs

When a call is handed off from one MSC to another, a connection is formed for that call between the two MSCs. The initial ANSI-41 signaling for the inter-MSC handoff is to establish radio levels and carrier assignments, but much of that signaling is to establish the specific MSC-to-MSC connection that brings the call to its new serving MSC. The packet link for the call is maintained until the call ends or goes back to the original MSC.[1] The ANSI-41 planners did not build soft handoffs into the inter-MSC handoff specification, but some vendors do allow inter-MSC soft handoffs.

Inter-MSC handoffs are handled by packet links over inter-MSC facilities dedicated to handoff traffic. These facilities may themselves be leased from the PSTN, but the handoff links are not set up as calls in the PSTN. MSCs do not hand off calls to each other without these facilities in place.

The inter-MSC handoff workload increases as more base stations are involved. This can happen because more cells are in border zones or because the cells in the border

---

[1]The handoff link also could go away if the handoff path gets complicated enough that a path minimization is done and removes the link.

zones are smaller. Smaller cells do not always have more handoff activity, but they can be more sensitive to the variations in radio propagation, and thus we can expect more handoffs.

As more MSCs are added, there are more MSC border cells, and this means that there are more inter-MSC handoff opportunities. As we pointed out in Sec. 36.1, the increase in border areas is more than proportional to the increase in MSCs. Each new MSC added to a system causes a greater increase in border zone traffic. Therefore, we can expect a greater proportion of inter-MSC handoffs as MSCs are added.

When more than two MSCs are involved in one wireless telephone call, things can get complicated. A call can bounce around from A to B to C and back to A and then to C. The people writing the ANSI-41 specifications for inter-MSC handoff thought of these problems and dealt with them, as discussed in Sec. 15.1. When more than two MSCs are involved, each handoff pair has to have dedicated facilities.

A rough rule of thumb for estimating the amount of inter-MSC handoffs can be created based on the traffic in a 150-m-wide band (one-tenth of a mile) between the two MSC regions. Not all the traffic in this band is going to be handed off from one MSC to the other, but this band is a reasonable place to expect inter-MSC handoffs. We suggest a narrower band here for actual handoffs than the band for handoff measurements in Sec. 36.1 because most handoff measurements will not result in a handoff. Planning tools can provide more accurate propagation models that may help refine the estimate of the handoff traffic between each pair of MSCs.

Handoff traffic between MSCs must use the links dedicated for that purpose. There is no overflow to the PSTN, so it is important to maintain an adequate supply of inter-MSC transport. Once these facilities are full, there are no inter-MSC handoffs even when CDMA radio conditions warrant them.

We need to form a confident estimate of the peak-hour amount of inter-MSC handoffs for each pair of MSCs in the system and plan for transport to meet those needs. It suffices to deal with MSCs in a pairwise fashion because three-MSC and four-MSC handoffs still happen two MSCs at a time.

37.1.2 Roamer terminations

When a roamer termination appears, an incoming call for a user terminal that is registered somewhere else, the home MSC forwards the call to the serving MSC, where the user terminal can be paged for the call. We can send these roamer terminations through the PSTN to their serving MSCs, but we have the option of using dedicated transport to save the PSTN charges for these calls. The cost saving can be substantial for pairs of MSCs with a lot of roamers between them.

In the case of inter-MSC handoffs above, the engineering issue is having enough capacity to support the handoff activity between switches. We certainly want to do this in the most cost-effective manner, but the primary motivation is having the system function, ensuring sufficient capacity so that it will work well. In the case of using dedicated facilities for terminating roamer calls, the engineering issue is cost reduction because the PSTN is available as an alternate route for overflow.

In Sec. 26.3.1 we discussed the design criteria for engineering the dedicated transport for roamer terminations. The idea is to use dedicated facilities when the recurring cost is less than the PSTN usage costs of traffic it is carrying. Only the busiest MSC-to-MSC routes are going to benefit from dedicated facilities for incoming roamer calls.

As the MSC count increases, the inter-MSC roamer termination count also will increase. More specifically, the fraction of incoming calls at each MSC that are roamer terminations at other MSCs will increase. As a system has more MSCs, the MSC regions get smaller, so the likelihood becomes greater than a subscriber is in a neighboring MSC region when a call comes in.

The time-of-day picture is also very important for this kind of analysis. It is not the peak-hour traffic load that matters for cost savings but the amount of inter-MSC traffic that is sustained over enough hours of the day to justify the expense of a dedicated link.

The roamer termination traffic can be combined with the mobile-to-mobile traffic (see below) in taking advantage of the cost-reduction opportunity of dedicated transport. The handoff links, however, are separate.

37.1.3 Mobile-to-mobile calls

Just as in terminating roamer calls above, sending mobile-to-mobile calls through dedicated transport instead of the PSTN is an opportunity for cost savings. We can run these calls through the PSTN on a call-by-call basis, or we can lease our own transport for them. In Sec. 26.3.2 we discussed the criteria for saving cost with dedicated inter-MSC transport.

The first step in determining where to save cost in mobile-to-mobile calls with dedicated inter-MSC transport is figuring out which MSC pairs would benefit. The best candidates have a lot of mobile-to-mobile traffic between them distributed fairly evenly throughout the day, so a leased DS-1 or E-1 link would be kept fairly busy for several hours per day.

The best candidates for inter-MSC transport, for roamer terminations as well as for mobile-to-mobile calls, are going to be MSC pairs in the same system serving the same city. Some of these pairs of switches are going to have a lot of traffic between them. Any city with two major wireless calling areas in two separate MSC regions is likely to have a lot of calls between the two areas.

We need to form a clear picture of the traffic loads as a function of the time of day. It is not the peak traffic but high daily average traffic that makes dedicated transport attractive.

We should combine our best hour-by-hour estimates of roamer terminations and mobile-to-mobile calls between each pair of MSCs to decide where to use dedicated transport and how much to use. The analysis in Sec. 26.3 can be used to determine how much transport, how many DS-1 or E-1 links, to put between each pair of MSCs for this cost-saving opportunity.

Estimates of mobile-to-mobile traffic should take marketing plans and promotional offers into account. For example, a marketing plan that promotes several cell phones on one plan for a family and offers free calling among those phones might well increase the inter-MSC mobile-to-mobile traffic among MSCs that support the business district, residential neighborhoods, schools, and shopping malls of one city.

37.2   MSC-to-Network Transport Planning

In planning the links from the MSC to the PSTN and to data service connection points, our first goal should be a very low blocking rate. Our second goal should be to reduce the cost of transport facilities.

37.2.1 Voice transport to the PSTN

The connection between the MSC and the PSTN carries almost all the voice call traffic. As described in Chap. 36, we may decide to save some money by siphoning off some mobile-to-mobile calls, but this is a small share of the MSC-to-PSTN voice traffic. Mobile-to-mobile calls within our own system are handled through direct MSC-to-MSC links. We can save additional cost by establishing direct links to the MSCs of other carriers in the same region, avoiding the PSTN. To do this, we need to program our MSCs with the mobile identification number (MIN) ranges used by the other carrier, and the other carrier needs to do the same with the MIN telephone numbers assigned to our customer's user terminals. Least-cost routing inside the MSC switch will use the direct lines first and route the call through the PSTN for overflow.

Except for private voice networks and mobile-to-mobile calls handled separately, voice call originations from a wireless telephone in the MSC region go directly from the MSC to the PSTN.[2] Call terminations to a user terminal's home MSC come directly from the PSTN to the MSC.

Calls to roaming user terminals come into their home MSCs from the PSTN, routed by the telephone number of the user terminal. Roamer terminations for calls coming in to a local access transport area (LATA) where the user terminal is roaming to a different MSC but is in the same region should be routed on inter-MSC trunks. We traffic engineer these trunks to reduce costs, but overflow will add to the MSC-to-PSTN traffic load. If we do not transport these calls separately to their visitor MSCs, then they go back out to the PSTN to be sent to their destination, where they come back in from PSTN to the visiting MSC. The cost savings we realize by using our own MSC-to-MSC transport for roamer terminations and mobile-to-mobile calls as well is not the incremental cost of MSC-to-PSTN transport. It is the PSTN charge for each call that we hope to save by carrying our own calls on our own facilities from one MSC to another.

The situation for long distance (inter-LATA) roaming is different. The latest version of ANSI-41, ANSI-41D, supports a method that prevents tromboning. When the Signaling System 7 (SS7) network contacts the home MSC and home location register (HLR) prior to establishing the call, the HLR can identify the current location of the call in a different LATA. The HLR provides the SS7 network with the temporary listed directory number (TLDN) of the user terminal in a different LATA. The SS7 network can then contact the MSC serving the user terminal through its TLDN, and the call can be connected directly from the callING party's local office to the serving MSC without routing through the LATA of the home MSC.

The voice connections from the MSC to the PSTN are full-rate links, 64-kbps pulse code modulation (PCM) links. This is the format the PSTN uses for voice calls at this time, so the conversion from our low-bit-rate packetized voice channel to a full-rate PCM channel takes place in the MSC. The links may be μ-law encoded DS-0 links or A-law encoded E-0 links; they are in the form expected by the landline telephone network.

The bad news is that all these voice links are full-rate trunks. The good news is that the MSC can be centrally located to have cheap transport available. There may be a tradeoff between backhaul expense and other MSC transport, but the MSC can be placed close to a point of presence (POP) in the PSTN to reduce cost.

---

[2]We will be casual and refer to a user terminal being in an MSC region. The user terminal is served by a base station in the MSC region, but we can skip the base station in our discussion of MSC transport.

The first job of MSC-to-PSTN facilities provisioning is to figure out the total voice demand for the busy-hour for all base stations served by the MSC. We may need to analyze different groups of subscribers and evaluate their behavior and calling patterns to determine what the busy-hour traffic looks like. The mall traffic may be 20,000 erlangs and the commuter traffic may be another 20,000 erlangs, but the total can be a lot less than 40,000 erlangs if these communities reach their peak traffic at different times.

This peak voice demand should be traffic engineered for a very low blocking rate. It does not make sense to invest in all the CDMA air-interface capacity only to be blocking those calls for lack of a few transport links to the PSTN. With traffic quantities of tens of thousands of erlangs, we can offer very low blocking and still maintain high occupancy on these trunks. The engineering of voice transport from the MSC to the PSTN is done when each MSC has a well-thought-out plan for the amount of transport needed for initial service and for the planning horizon.

So far we have discussed the link from MSC to PSTN as if it were a single pipe. However, it is possible to reduce costs by traffic engineering pipes between the MSC and individual destinations on the PSTN. The first priority is always to maintain a low blocking rate, but in those cases where alternate routing is available, the particular traffic engineering job for the direct link can be optimized for cost savings. For local calls, the traffic engineering is from the MSC to the local LATA. For long-distance calls, we set up alternate routes based on lowest-cost options, and the MSC makes a least-cost-routing determination on a per-call basis. A wireless service provider is likely to make arrangements with multiple long-distance carriers to reduce cost.

Regulations also require us to arrange for an appropriate link from the MSC to the local emergency call center for emergency calls (911 in North America). This destination is called a *public service answering point* (PSAP) and is usually reached through a type 1 connection to the PSTN. In engineering this link, our goal would be to prevent blocking the reporting of multiple simultaneous emergencies within one MSC serving area. A single emergency, such as a traffic accident on the freeway at rush hour, might generate a large number of calls. In such a case, what matters is that at least one of those calls gets through. However, there is a possibility that, simultaneously, there will be another emergency in the same area, perhaps a medical emergency inside a house, where only one caller is dialing 911. That call also must get through. In planning for capacity for emergency access, we also need to take two other factors into account: We must support all cellular telephone calls, even from unregistered user terminals, and we must ensure enough capacity to allow real calls to get through even while allowing for the possibility of prank or nuisance calls to emergency services.

37.2.2   Data transport to the PPDN

The total data service traffic from the MSC is a little harder to predict than voice traffic because we do not have a 20-year history of wireless data service or other services, as we do with cellular voice service. However, the estimates we made earlier for counting cells and the assignment we made of base stations to MSCs should give us a good idea of the volume and quality-of-service (QoS) requirements of packet data services.

We start with the real-time packet data services and add up their total rates. This list of services includes two-way video calling and any data service with a very low delay requirement. The low delay requirement may be in a service-level agreement (SLA), or it may be a planning decision to offer high-quality service. The top tier of packet services is defined by the services that require the lowest amount of delay. We want to count the services going from the MSC to the PPDN so that we can eliminate from consideration those services which we can offer directly from an Internet Protocol (IP) node in the base station, as described in Sec. 35.4.

The next step is to chart demand for these services on the clock so that we can determine the hourly requirement for these services. This may involve evaluating the time-of-day usage of a mix of services, perhaps services used in different base stations. It does not matter which base stations are involved because we are planning transport from MSC-to-PPDN transport, one big packet pipe to the outside world.

We then find the busy hour of this mix of services and do the traffic engineering to ensure low delay for the maximum packet load. This is the amount of transport required for these services. The packet pipe from the MSC to the PPDN will have to have at least this much capacity.

Now we can go to the next level of delay requirement. The next level may be a lower level of the same service, perhaps a Web browsing service that allows 3 seconds of delay when there is a premium service with a half-second maximum wait. Whatever the second tier of services may be, these are the services we add to the list. Again, we evaluate the demand for these services by time of day to get a full usage picture.

Now we take the total packet data demand of the first two tiers of service and the delay requirement of the second tier and do another traffic engineering exercise. If this calculation gives a bigger capacity requirement than the first tier alone, then we expand the transport requirement to this amount.

We repeat this process for each delay requirement until all the data and other services are represented. In each cycle of the traffic engineering process, we use all the data requirements for this level and all previous levels, and we use the delay requirements for this level. There will be enough capacity to carry all the top-level data with top-level delay, all the top-two-level data with second-level delay, all the top-three-level data with third-level delay, and so on. Any one of these levels could produce the maximum capacity requirement for the MSC-to-PPDN packet pipe, so we need to check them all.

The important feature in ranking these data and other services is not their general QoS, not their bit error rate (BER), but specifically their delay. The other factors are taken into account in their E_b/N_0 calculations in the air interface. When we are done, each MSC will have a packet pipe traffic engineered with enough capacity for all the levels of data service delay.

37.2.3   Private wireless networks

We have discussed public wireless voice, data, and other services, but we expect third-generation (3G) systems to develop a growing market for private networks. Part of this demand is simply a result of the enormous number of subscribers we expect 3G CDMA to be serving, whereas part of this demand is a result of the flexibility of 3G CDMA, especially in the data services area. Private network customers will arrange their own transport for voice or data.

When customers decide to use a wireless telephone system for their own private voice networks, they will get mobile telephones for their users, and they typically will get their own transport from the MSC to their own telephone network. They may lease their facilities from the PSTN, but those facilities are not public telephone trunks. These services would generally be within a LATA and therefore would be available as metropolitan area network (MAN) services from the local access provider.

Private wireless voice networks have been around for a while, but private wireless data networks are as new as wireless data service. Private data network customers will equip their users with data terminals of whatever sort suits their needs. Perhaps their data terminals will be wireless laptops, but they may be devices designed by the customer, custom tailored to a specific task. The private data customers may rely on the PPDN, or they may arrange their own transport.

Having their own packet pipes gives private data network customers direct control over the performance of their networks. Sharing their IP packet traffic with the Internet leaves them out of control of their own data. Not only does the public network leave them at the mercy of PPDN delays, but it means that their data may be intercepted more easily by third parties.[3] They will have their own QoS standard and may even have an SLA with the wireless service provider.

The SLA for data service performance defines the amount of data delay that the private packet links can have. From the delay and traffic estimates, we can do a traffic engineering exercise to figure out the capacity requirement for the packet pipe.

If the private network is concentrated in a cell, then one base station may serve the entire private data network for a customer. Then we may decide to have the dedicated packet pipe for that customer's network go directly from the base station to the customer.

37.3 Conclusion

Careful planning of the transport from each MSC to other MSCs and to various termination points on the PSTN, the PPDN, and private data services guarantees a low blocking rate while keeping costs reasonable. In Chap. 38 we will turn our attention to unusual circumstances that require special planning, completing Part 6.

---

[3]The air interface is hardly private, but it is difficult to intercept data for a particular user who may be anywhere in the coverage area. Encryption helps, but customers may not even want their competitors to know how *much* data traffic they are sending.

Chapter

38

Special Situations

To wrap up our discussion of planning code division multiple access (CDMA) networks, we need to examine two more issues. In telephony, there is a traditional assumption that the demand is the same in both directions. However, with the advent of third-generation (3G) data services and future video services, we will encounter asymmetric demand that will require planning for different capacity requirements in each direction. The second special situation is the case where a CDMA service provider is entering a new market. The combination of startup costs, the need to gain and maintain competitive advantage, and the unpredictability of demand creates a real challenge for making the business case for a new network and then implementing that network profitably.

38.1 Asymmetric Demand

Cellular voice traffic is symmetric traffic on the air interface. The number of bits in the forward and reverse directions should be nearly the same for a voice conversation.[1]

Data services may not have the same symmetry. Consider a wireless Web surfer, for example. The reverse link sends a Uniform Resource Locator (URL), and the forward link sends a Web page. Eighty bytes are sent in the reverse direction, and 80 kilobytes come back in the forward direction—and then the process repeats itself. The flow of bits for subscribers using the Internet in this fashion is highly asymmetric, favoring the forward direction.

The asymmetry can go the other way, however. Suppose that we have a community of wireless multimedia message service (MMS) subscribers sending pictures to their landline "cyberfriends" on the Internet. Or suppose that we have a Web server that is itself a wireless data terminal. Then the flow of bits would be similarly skewed, but

---

[1] We can scratch our heads and come up with some exceptions. For example, subscribers calling in for recorded messages of some sort would punch two or three Touch Tones and listen to a minute or more of forward-direction speech. But we have a hard time believing that this sort of traffic represents a serious fraction of wireless voice traffic. Of course, many of us experience listening to people who talk a lot on the other end of the phone. Statistically, however, this balances out because there are as many talkative people on each end of each telephone call, landline or cellular.

favoring the reverse direction. Video broadcast services create a similar asymmetric demand. If a roving reporter sends a live video feed using four cellular carriers in one direction plus one two-way voice link, then the demand is much greater in the reverse direction.

While it is tempting to evade the issue by suggesting that the two asymmetries will cancel each other out and leave a perfectly balanced data channel, we believe that the imbalance issue deserves a little more scrutiny than this. The decision of how much capacity to allocate in each direction depends on the estimated market for various kinds of data services. Time division synchronized CDMA (TD-SCMDA) offers the advantage that it can allocate forward and reverse data capacity dynamically so that the service provider can engineer the total two-way traffic without having to make the decision in advance of how much traffic will go in each direction.

38.1.1 Carrier asymmetry

cdma2000 and wideband CDMA (W-CDMA) allow the service provider to use more CDMA carriers in one direction than the other. This gives one-time control to a service provider, who can estimate the service mix and the traffic in both directions.

cdma2000 offers a variety of carrier options, 1x, 3x, 6x, and so on. The service provider is allowed to use a different carrier in the forward and reverse directions, so forward can use 3x while reverse uses 1x. We believe that it is more likely that cdma2000 service providers will be using multicarrier systems with three 1x carriers forward rather than 3x because the multicarrier option is compatible with cdmaOne. If the market has a leaning toward forward data traffic, then the service provider can allocate more carriers in that direction.

The cdma2000 1x EV-DO standard (1x carrier bandwidth, evolution data only) allows for very high rates at close range in the forward direction. This allows a provider to select a target audience for high-performance data downloads using the fact that the forward direction can be time multiplexed (TDMA) as well as code multiplexed (CDMA). This technology will provide excellent support for data services that lean toward the forward direction, such as our Web server example.

W-CDMA allows for asymmetry as well. A provider with 15 MHz may decide to use two 5-MHz carriers in the forward direction and only one in reverse. The W-CDMA system has no requirement for equal numbers of carriers in the two directions.

38.1.2 Shared packet channels

Third-generation (3G) CDMA has shared packet channels in addition to dedicated CDMA data service channels. The provider can set aside CDMA capacity for these channels and allow many wireless data terminals to have access to them. The shared channels are high-speed data links, so one terminal can get or send a burst of data at high speed without having to set up a high-speed data channel. The number and capacity of the shared packet channels can be set independently for forward and reverse directions. The shared packet channels set aside capacity for data services so that other subscriber activity does not squeeze them out.

Use of the forward shared packet data channel is controlled by the system. The data terminal is told which time intervals on the shared channel contain its data. It is the reverse use of the shared channel that requires coordination through the Medium Access Control (MAC) protocol. In cdma2000, the MAC is sent on a dedicated control channel, and in W-CDMA, the MAC is sent using slotted Aloha. Even with the extra overhead of coordination, the reverse shared packet channel allows more efficient high-speed data service than the system would have if we set up high-speed links for short bursts of data.

38.1.3 Time division duplex (TDD)

Time division duplex (TDD) allows flexible distribution of the CDMA resource between forward and reverse directions. The TD-SCDMA provider has to estimate the total CDMA traffic in both directions and allocate enough TD-SCDMA carriers for the total traffic. The TD-SCMDA system balances forward and reverse on its own.

The TD-SCDMA carrier has seven time slots in its time-division cycle. The system is required to use at least one of them in the forward direction and at least one in the reverse direction. Use of the five time slots other than the two required in each direction is allocated according to subscriber demand.

This allows the TD-SCDMA channel to be highly skewed in either direction. A ratio as high as six-to-one can be supported or a ratio as low as one-to-six (a high ratio in the other direction). If the system has at least two-sevenths of its resource utilization in balanced services such as voice, then the TD-SCDMA channel can be nearly perfectly efficient in allocating forward and reverse resources to subscribers.

Unlike the cdma2000 and W-CDMA providers, the TD-SCDMA service provider does not have to decide up front how much capacity to allocate in each direction. This is a tremendous advantage in flexibility because these usage patterns can change. As an example, a community of workday Web surfers may go on holiday and start snapping MMS pictures of their family activities. Such a system would have mostly forward-direction data use on work days and mostly reverse-direction image transmission on weekends. The flexible TD-SCDMA carrier allows the service provider simply to allocate enough capacity for the total traffic at any particular time and let the system take care of the balance.

38.2 Market-Entry Planning Issues

The history of the cellular market in the United States has been a tough road. The Federal Communications Commission (FCC) allocated spectrum to two service providers in each market so that subscribers would have a choice and there would be competition. The service providers who got to market first found a swarm of subscribers waiting eagerly for cellular service. The second providers had the harder job of establishing their markets. Typically, they offered very favorable deals with extremely generous offerings of cellular air time. The effect of this was to overload their systems and, because the first providers would match the deal, to overload both cellular systems.

A 3G CDMA service provider is investing a lot of time and money in a large system designed to serve millions of subscribers. If we are building a 500-cell system at US$500,000 per cell with three mobile switching centers (MSCs) at $5 million per MSC to serve 1 million user terminals at $300 each, we are spending $565 million before spending a dime on telephone transport, maintenance, or marketing. Investors spending this amount of money want to see a return on their investment, a revenue stream that will earn them a profit.

Most of the new 3G CDMA systems are going into markets already well served by existing cellular technologies. The voice market is well established but also well served. The data market is not established and will have to be created. Subscribers will have to buy cordless laptop computers or other types of wireless data terminals, and they will have to subscribe to a new service. The service provider will have to be aggressive in its promotions to entice people to use this new 3G capability. It will have to be easy, available, and cheap. In the initial phase, the price will be a critical component for subscribers to decide to buy the service.

This means that we will be offering a costly service that uses a lot of capacity at a very attractive price. Deals will be offered that virtually give data air time away to get people used to the idea of wireless data service. We hope that they will sign up, we hope that they will use the service, and we hope that they will like the service.

The problem is that subscribers may sign up for a big promotion in large numbers and may use large amounts of CDMA capacity. If we are not careful, then we may sell the entire air-interface capacity at a promotional rate. This leaves us the unfortunate choice of continuing to lose money selling high-quality service or trying to generate enough sales to make money by overselling the service and hoping subscribers do not notice the decline in quality.

The marketing plan for a new technology will have to walk the fine line between promoting the technology and maintaining a sufficient revenue stream. It is important to charge enough money for data service to pay for the investment. One possible approach is to limit the quantity of the promotion deals. Once half the capacity of the system is being given away or sold below cost, it may be time to stop promoting and to plan for a profitable future.

We have a problem in managing subscriber expectations in wireless data service. Many of us have digital subscriber line (DSL) service that offers unlimited access for a fixed monthly fee.[2] The most promising cost figures we have seen in 2002 are a few cents (U.S.) per megabyte of data.[3] If we can achieve 30 megabytes per dollar, this still means paying $20 for 600 megabytes (a compact disk) of data.

Is the market going to tolerate paying incrementally for data? Will subscribers be willing to have a tollbooth on the information highway, which, up until now, has been free of usage charges? This is going to depend on how well wireless data service is marketed, how much value it actually adds to its subscribers, and the ability of wireless service providers to offer high-quality and high-capacity wireless data service.

Acceptance for volume-based usage charges for data and more expensive data rates also depends on the alternatives. Cellular data services are competing not only with each other but also with a number of other rapidly growing alternatives. There is some evidence that the road warriors who are seen as key data customers actually may prefer high-speed DSL-based landline services at hotels, even if it means that they are not connected all day long. And the IEEE 802.11b wireless local area network (LAN) stan-

---

[2]Never mind that some Internet service providers (ISPs) are selling a big pipe to a tiny spigot by overselling their server capacity. Customer expectation in landline data is a modest monthly charge for all the bits you can use.

[3]While these figures came from marketing hype, technology prices tend to come down fairly fast once a market is established. We will wave our hands and assume that the baloney factor of marketing hype will be balanced by cost reduction and technology development. Let us therefore assume that this is attainable cost.

dard, popularly called *wireless fidelity* (Wi-Fi), is being deployed rapidly, including, in some cases, for public access that will compete with cellular data services in so-called hot spots, where there is high demand for data services.

38.3 Conclusion

This concludes Part 6. In these last seven chapters we have addressed the challenge of building a new network or performing a major upgrade to an existing network to expand capacity and provide new services, particularly 3G data services.

Building or upgrading an entire network is exciting, but the humdrum daily operations, administration, and maintenance (OA&M) tasks of an existing network, with its repairs and minor upgrades for new technology and increased capacity, are equally important to maintaining a high-quality, high-capacity, profitable CDMA network. We will now turn our attention to the job of increasing capacity on existing networks.

Increasing Capacity

Part 7, "Increasing Capacity," is written for engineers and
technicians who are providing operations, administration and
maintenance (OA&M) support for growing cellular networks. In
Chap. 39, "Measuring System Performance for Growth," we look at
the key measures of capacity for all parts of the cellular network: the
air interface, base stations, backhaul, and system components. We
help planners identify where the limits to growth are and recommend
solutions.

In Chap. 40, "Turning User Complaints into Useful Data," we
discuss how to move from the symptom perceived by the user to the
underlying problem on the network and how to solve that problem.
Chapter 41, "Increasing Capacity of a Base Station," discusses how
to add power, to add carriers, to add sectors, and to install
repeaters.

Sometimes base stations in a particular area have reached their
growth limit, and we need to increase capacity by adding new cells.
This is the topic of Chap. 42, "Adding Cells to a CDMA System." We
discuss cell splitting, keeping the grid, and downtilting antennas of
nearby base stations to reduce intercell interference. We also cover
microcells, picocells, handoff management, and reliability.

In Chap. 43, "Mobile Switching Center (MSC) Growth," we tackle
the challenges of upgrading the MSC switch and auxiliary
equipment, adding new MSC switches (either at an existing MSC site
or at a new site), reassigning base stations, and setting up new MSC
borders. We follow this with Chap. 44, "Adding Transport," where we
discuss adding backhaul and adding transport between MSCs and
other network components and networks.

We move out of the territory of engineering into other parts of the
cellular business in the last chapter of this part. In Chap. 45,
"Regional Growth in a Specific Area," we discuss ways of increasing
net revenue by adding subscribers who will use services that are less
costly to add to our network than our most expensive peak services.
We bring ideas from cellular engineering, economics, and marketing

together in an effort to address how a company can synergize its marketing effort with its cellular network growth to improve profitability.

In essence, Part 7 gives a wireless system planner or engineer just about everything needed to grow a high-quality, cost-efficient cellular network.

Measuring System Performance for Growth

How well is our code division multiple access (CDMA) system performing? Is it serving all customers, or are calls being dropped? Is there room for growth, or is it reaching its limits? Are problems at local spots of poor coverage, cellwide, or systemwide? We want to be able to answer these questions before customers detect problems and before our system becomes overloaded.

Periodically, we will want to take radio signal path measurements throughout the coverage area. Also, we will want to monitor a number of crucial statistics in an ongoing way and to set alert levels to warn us of problems. Some of the key items to monitor are forward and reverse power levels, handoff activity, backhaul congestion, mobile switching center (MSC) performance, signaling capacity, and transport utilization. We will explore each of these in turn, discussing how they are monitored, what the key measures are, and what kind of problems are implied by measurement results that show that a particular measure is outside its normal operating range.

39.1 Radio Signal Path Measurements

Knowing the radio environment is an important input to CDMA system design. In the earliest planning stage we start with computer predictions of radio. Once a wireless telephone system is up and running, we can supplement these predictions with measurements from the system itself. These measurements help plan for the growth and improvement of the system.

We want to know the geographic layout of every cell sector. To do this, we need to know the best radio path, the base station with the highest signal path gain to every (x, y) location in the system. Once we have the best radio path for every place, we can derive the coverage areas for each cell and each sector of each cell.

The first step toward complete knowledge of radio propagation in a wireless telephone system is running the computer models described in Chap. 47. The computer gives us maps showing us the geographic areas of each cell, the geographic regions served best by each base station.

Once we have the best base station for each place, we want the second best, third best, fourth best, and so on. These calculations help us understand how efficiently

CDMA can serve the subscriber demand. The fourth-pilot delta described in Sec. 32.1 is determined by comparing the best radio path and the fourth-best path. The CDMA carrier performs best when there is a single best-serving antenna face and all the others have far lower radio path gain. When the first and fourth radio paths are close, the CDMA carrier experiences pilot pollution and does not perform at its best.

For each (x, y) location in the service area, we want to know the best radio path to a base station, but we also want to know any radio path to a base station that is close, within 10 dB. We can find this out from the live system by driving or walking around and making radio measurements. We use specialized measurement equipment designed to find the strongest CDMA signals. These measurement terminals have the same kinds of radio receivers as user terminals, but they probably do not make telephone calls.[1] It is important that we measure areas representative of actual traffic. If our subscribers are shopping in the stores, then we do not want to make our mall measurements in the parking lot. And we will need to get inside the office buildings that represent the bulk of our business traffic.

For each location we reach as we move around, the measurement terminal reports its location, using the Global Positioning System (GPS), and the best serving base station. In this way, we actually measure what cell each location is in. We also want the base stations that are not the best radio path but are close. These are the sources of pilot pollution in our CDMA planning process.

An ongoing process of radio measurement gives us an accurate picture of radio performance, including pilot pollution, cell by cell and place by place. The end result of the measurement process is a map nearly as complete as the computer prediction, but with the authenticity of actual measurements. Where the prediction is a guess, the measurement is an observation.

39.2 Forward Power Levels

The measure of forward-link performance in a CDMA system is how often the base station transmit amplifier is close to its maximum power. When the CDMA carrier reaches saturation, the power-control loop will keep raising power levels on the channels until the amplifier runs out of power. What we want to get out of transmit power measurement is the busy-hour performance of the system, so we need to measure the amplifier power during the busy hour.

Watching the power levels at the base station amplifier is our way of knowing how often the CDMA carrier is at or near its breaking point. Amplifier overload is the system telling us about CDMA overload. The curve in Fig. 28.2 shows us how the system reacts to increasing traffic. When the amplifier reaches half power, we can figure that traffic is getting close to the maximum. It does not matter if the traffic is a lot of voice calls, some real-time video, or a few really big data users. It also does not matter if the event that pushes the CDMA carrier near its limit is a new call, a handoff, or a change on a neighboring sector. The amplifier power is the indicator we are looking for.

---

[1]Back in the Advanced Mobile Phone Service (AMPS) days, we used to drive around in "road rallies" making actual calls on cell phones with signal-level meters and channel displays. We would write our location, channel, and signal level as we made telephone calls. Newer technology has specialized equipment for making these measurements.

The fraction of the time the amplifier is at or above half power is an approximate measure of the forward call blocking rate. The actual threshold to use may depend on the cell itself. A coverage cell may spend quite a bit of time around half power without being in trouble, whereas a very small cell in a dense area may reach 95 percent of its maximum capacity while still at one-third power. The last 5 percent of the traffic sends a small cell scurrying up the steep part of the curve in Fig. 28.2. If forward amplifier power is being used as a criterion for blocking new calls, then we should use the system blocking level as an approximation in this section. If the busiest period of the day has this power threshold crossed 5 percent of the time, then we can presume that we have a peak-time blocking rate of 5 percent.

In a conventional-reuse system (FDMA or TDMA), there is a fixed number of radio channels, and once they are full, there are no more. We can measure blocking by seeing how often they are full, or we can measure their occupancy. The CDMA carrier capacity varies with environment and call activity in nearby cells. The only way to get a firm handle on the capacity is to measure it when the system is reaching its limit.

We should be keeping track of all the sectors of all the cells, if not every day, then on a regular basis. A CDMA carrier often gives little warning that it is reaching saturation, and the expected rapid growth of high-rate data service makes this even more likely. We recommend forward transmit power vigilance.

39.3 Reverse Power Levels

Just as we can watch forward transmit power, we can keep track of reverse transmit power. The transmit power of the user terminals tells us how close the reverse direction is to CDMA saturation. As in the forward direction, we want to use user terminal power levels to determine the busy-hour blocking performance of the system.

If some equipment from some vendor makes it hard to catalog the user terminal power levels, then we can monitor the power-control messages going out on the forward channels. If they are consistently more often up than down, then there are user terminals that have reached their limits. This is not a gentle cry for help; this is a CDMA system demanding capacity relief *right now*.

There is an easier measurement we can make of reverse congestion on the CDMA carrier. We can look at the total received power at the base station over time. The curve in Fig. 28.1 shows us how the received power starts to climb as we get close to CDMA capacity. We can monitor the base station receiver power level over time to determine reverse-link performance.

So what levels do we use? In the forward direction, we could relate the transmit power to the maximum amplifier power, and in the reverse direction, we can compare the received signal to the receiver noise. If the total signal power is close to the noise level, then we are getting close to overload. And when the signal power is two or three times the noise level, we are critically close.

The carrier is getting overloaded when the total signal of all the channels rises to the point of being just equal to the noise level. This sounds a bit counterintuitive. Shouldn't we be happy when our signal-to-noise (S/N) ratio is high? The answer is that the CDMA channel is designed to operate with signal less than noise, typically far less than noise. CDMA gets high S/N ratios through processing gain, but the physical channel is mostly noise. This is true even when there are many users, many channels on the same carrier.

The fraction of the time the base station receiver is getting more CDMA signal power than noise is an approximate measure of the reverse call blocking rate. The actual threshold to use can vary, but in the opposite way from the forward direction. A coverage cell may find itself with one-fifth signal and four-fifths noise when it is nearly overloaded, whereas a small cell in a dense area may reach two-thirds signal and one-third noise before it gets into capacity trouble.

We should be watching the received power levels, either their transmit power levels or the total received power, and looking for signs of saturation. If the busiest period of the day has the reverse power threshold crossed 5 percent of the time, then we can presume that we have a peak-time reverse blocking rate of 5 percent.

As in the forward direction, we must keep a watch on reverse-direction performance on a regular basis. CDMA's lack of warning and the growth of new services can combine to create surprise situations with a short fuse.

39.4   Handoff Activity

The handoff activity in a CDMA system is another way to monitor system performance. On the surface, high handoff activity means a high workload for the signaling channel if the handoffs are from one mobile switching center (MSC) to another. However, on the air-interface level, the rate of handoffs tells us a great deal about how well our traffic is using the CDMA air interface.

The areas of nearly equal radio propagation are the least-efficient service areas in CDMA. These are the areas where one CDMA channel creates interference on two CDMA carriers. If there are three or four sectors with nearly equal radio performance, then we have pilot pollution, as discussed in Sec. 32.1. It does not take a lot of demand in those areas to use up all the CDMA capacity we have available.

An easy measure of the amount of traffic in multiple-cell zones is the amount of handoff activity. After all, what makes a call hand off from one sector to another is that those two sectors have nearly equal radio paths at the point of the handoff. If there are frequent handoff patterns from A to B to C or any other order of the same three sectors, then there is some area of significant population where all three sectors have nearly the same radio path gain. Getting this information from the system means that we have to measure not only the individual handoffs but also the sequences of handoffs so that we know which groups of cells are nearly equal in radio path gain.

Measuring soft-handoff activity gives us an excellent handle on the nearly equal radio zones. While it appears that soft handoff reduces forward-link capacity by having multiple CDMA channels for one call, it mitigates the even greater problem of having a hard-handoff process serving the call on the wrong sector while trying to catch up. Three-way, four-way, and five-way soft handoffs are the CDMA system's way of telling us which areas are pilot-pollution zones.

Once we know which cell and sector boundaries have a lot of handoff activity, we know which areas are using the CDMA carrier less efficiently than we would like. We should be measuring the handoff activity levels as part of our CDMA system capacity monitoring process.

39.4.1   Soft handoff

Hard handoffs are events that happen, and we can record these events. Soft handoff is an ongoing process, a call can be in a soft handoff for an extended period of time. The events that define soft handoff are the adds and drops of cell sectors to the soft-hand-off list. It is the soft-handoff add and drop events that we should be tracking so that we know the level of soft-handoff activity.

Once we have the soft-handoff additions and deletions, we can infer the soft-handoff states in between these events. If a call starts on sector A and we see sector B added, then we have a soft handoff between A and B. When we see sector C added, we have a three-way soft handoff among A, B, and C. When sector B is dropped, we have a soft handoff between A and C. We can calculate the soft-handoff states over time from the add and drop events that we observe.

The result of tracking soft-handoff states from soft-handoff add and drop events is to find the time calls spent in various states of soft-handoff. We can add up all the soft-handoff state times during the busy hour and determine which sector combinations have high soft-handoff activity levels. We should do this calculation for sector pairs, triplets, and combinations of four or more.

When we see a combination of four or more sectors with high soft-handoff activity, this is a loud warning sign that there is a capacity problem. The geographic area where those calls are coming from is a pilot-pollution zone that is a good candidate for a new cell.

39.4.2 Inter-MSC handoffs

We should measure handoffs between MSC areas as well. The wireless telephone system should give us counts of inter-MSC hard handoffs and inter-MSC soft-handoff events if the vendor's equipment supports inter-MSC soft handoffs.

We can measure signaling activity by counting not only the inter-MSC handoffs but also all the inter-MSC handoff measurement messages. The total message count for each MSC is part of the signaling load and should be used to make sure that we maintain enough signaling capacity. If the handoff message load is higher than we predicted in the system planning stage, then we may need to add some signaling facilities.

The other side of inter-MSC handoffs is the inter-MSC transport to support them. Here we care how much time calls are spending in inter-MSC handoff because this determines how much inter-MSC transport capacity they are using. High handoff rates that were not foreseen in our forecasts can use up our inter-MSC facilities. The consequence of this is the loss of ability to do handoffs from one MSC to another, a serious consequence.

We can watch the global handoff rates to and from one MSC, between each pair of MSCs, or among combinations of three or more MSCs, but we can be more specific in our observations and note which cell combinations are causing high handoff rates. The solution may lie in the radio engineering department rather than in the signaling and facilities areas. If there is a zone with too many handoffs, then we should go back and diagnose the radio problems. Handoff add and drop parameters may need to be reset, or antenna patterns may need to be shifted by changing antennas or downtilting the existing antennas.

39.4.3 Intercarrier handoffs

The system also should be telling us how many intercarrier handoffs we have in our wireless telephone system. These are handoffs from one CDMA carrier to another as a call is handed off from one sector to another. The receiving sector may have a lot more available CDMA resources on a different carrier, so it makes sense to switch the call from one carrier to another as it changes cell sector.

Some vendor products use semisoft handoff to handle intercarrier handoffs seamlessly, as described in Sec. 11.4.7. They make both antenna faces receptive to the call so that the system misses nothing as the user terminal switches frequency. It looks like a soft handoff to the system, whereas the user terminal performs a hard handoff. cdmaOne and cdma2000 require hard handoffs to change CDMA carrier frequencies, whereas wideband CDMA (W-CDMA) may allow a soft handoff by having the user terminal receive two different carrier frequencies from the two cell sectors. If we have to use hard handoffs to switch carriers, then this is a less efficient utilization of the CDMA resource.

High levels of intercarrier handoffs tell us two things. First, they tell us where carrier boundaries are causing high levels of switchover and poor utilization of the CDMA air interface. Second, they tell us where carriers are getting so congested that the system has to move traffic from one carrier to another.

In either case, high rates of intercarrier handoffs are telling us that we need to add carriers. If a carrier boundary zone has a lot of calls falling off the edge of the carrier area by handing down to another carrier, then we should add that carrier to some neighboring cells to extend its coverage. If a carrier boundary zone has a lot of calls moving into a cell with more carriers and being pushed up to another carrier by congestion on the lower carrier, then the neighboring cells also should have more carriers. When adding carriers to cells, it is important to maintain the convex shape of the carrier boundary.

We should develop a list of intercarrier handoff rates by individual cells and sectors so that we can see carrier coverage problems developing before they cause subscriber service-level problems.

39.4.4 Intermode handoffs

While frequency division multiple access (FDMA) and time division multiple access (TDMA) systems typically do not allow handoffs to CDMA systems, most CDMA systems do allow intermode handoffs to older technology. These can happen when the user terminal moves into an area without CDMA coverage or into an area where the CDMA system has no capacity left to receive the handoff.

Handoffs from CDMA to other technologies move traffic from a more efficient to a less efficient air interface. This is a bad thing to do just from a radio utilization standpoint, but there are other consequences as well. The CDMA subscriber is paying good money for a suite of services. It is unlikely that the older FDMA or TDMA system is going to have all those services available. The third-generation (3G) services are certainly not going to be found on older systems used for traffic overflow.

A high rate of intermode handoffs could be a sign of poor CDMA coverage. If we see a steady stream of intermode handoffs from a particular cell sector, then we should inspect the rate of intermode handoffs when the cell sector is not busy. If there is still significant intermode handoff activity when the sector is lightly loaded, then we should conclude that there is a pocket of poor coverage subscribers are moving into and thus losing their CDMA signal. The high intermode handoff rate is telling us to improve the radio coverage of a CDMA base station.

A high rate of intermode handoffs could be a sign of CDMA congestion. If we see a high rate of intermode handoffs only when the CDMA carrier is busy, when power levels are high, then we suspect that the older technology is acting as an overflow medium for an overloaded CDMA air interface. The high intermode handoff rate is telling us to improve the radio capacity of a CDMA base station.

We should be keeping records by cell sector and by time of day of the rate of intermode handoff activity. In particular, we should be watching that rate as a function of how busy the CDMA system is when those intermode handoffs occur.

39.5 Backhaul Congestion

The backhaul network should be designed and engineered for very low blocking. The packet streams from base stations to MSC should see very low delay almost all the time because the MSC is carrying real-time voice traffic and may be carrying real-time video as well. Because the backhaul network may be complicated not only by a multiple-hop backhaul configuration but also by redundancy in the network with some alternative packet routing, it is important to monitor backhaul links and to ensure high-quality packet performance.

The system should be measuring packet delay on the backhaul links. As in any packet system, we can measure delay in two ways. The global delay can be determined by knowing when a packet is sent and when it is received. When the time stamps are far apart, the packet has been delayed. The other method is to measure the sizes of backhaul packet queues at the MSC and the base stations. In a multiple-hop link, the delay is the total delay time of all the individual hops.

Different packet protocols have different measuring tools for delay and latency. Time stamps on data packets can be monitored to determine delay on any hop or path of multiple hops. Also, packets can be sent and returned for a loopback test, which not only ensures that the link is running but also performs timing benchmarks. As quality-of-service (QoS) protocols are improved and deployed, monitoring methods and intelligent network response will continue to improve.

Backhaul delay should be monitored routinely. Alert levels should be set so that we are automatically notified of any delay that might be noticeable to users or even a bit tighter than that at, say, 80 percent of the limit of acceptable performance. If alerts happen very seldom, then we may decide to live with occasional bursts of slow backhaul performance. If backhaul delay happens too often, then it will degrade subscriber service with noticeable dropouts in voice calls.

39.6 MSC Performance

The MSC has several components that can become bottlenecks in a wireless telephone system. Each of these components should be watched so that congestion can be dealt with before it becomes a customer service issue.

The main processor can be too slow for the call processing, short messages service (SMS), and handoff activity. For the same CDMA resource, voice calls tend to use a lot more call processing than data calls, so a provider counting on a migration from voice to data may be surprised when voice service remains dominant. SMS traffic follows social trends that can change dramatically. And a few pilot-pollution zones that escape the air-interface design process can increase handoff activity beyond our expectations. Any of these can place loads on the MSC processor.

The circuit-switching resources may not be adequate. This will happen if there is an increase in voice calls instead of the forecast growth in data service. The circuit-switching components of the MSC that were engineered for the data-heavy market will be underengineered if voice traffic grows instead. An unexpected increase in voice call demand also will bring about a similar demand for speech-coding equipment at the MSC.

The packet-switching resources may not be adequate. If data service takes off faster than expected, then we will find ourselves short of packet capacity. Customer interest in real-time video or multimedia message service (MMS) could bring packet data levels to a higher level than we originally planned for.

We do not have to measure the total transport connected to the MSC. We can count the DS-1 or E-1 links to the base stations, other MSCs, the public switched telephone network (PSTN), the public packet data network (PPDN), private networks, and the Signaling System 7 (SS7) network. This is the total transport. (At least there is one switch component whose usage is obvious from looking at it.)

The system should monitor the performance of the MSC components over time: the main processor, circuit switching, speech coders, and packet data. Congestion in any of these areas should be addressed by the wireless system planners before it affects the quality of system service.

39.7   Signaling Capacity

Signaling congestion can do major damage to system performance. ANSI-41 signaling over the SS7 network is a critical part of every registration, every call, and every other subscriber service in the wireless system. For this reason, SS7 procedures tell us to keep utilization of its fully redundant facilities below 40 percent.

All SS7 links in the wireless system should be monitored closely for traffic levels at all times of the day. This includes all the A links from signal transport points (STPs) to MSCs, home location registers (HLRs), visitor location registers (VLRs), short message service centers (SMSCs), and service control point (SCP) databases. It also includes F links that carry signaling messages directly from one MSC to another. There may be a chance to watch the other links in the SS7 network as well.

These links should stay less than half full all the time. The SS7 rules say that we cannot go above 40 percent, so any occupancy over half is a warning sign that things are getting congested in the ANSI-41 signaling network. We should be watching as much of the SS7 network as we can get reports for.

Any SS7 link in our wireless signaling network reaching 50 percent of its capacity should send up a red flag for us. Either it is getting too full of ANSI-41 messages, which is a capacity engineering concern, or its redundant SS7 link is down, which is a maintenance concern. In either case, we should respond when a signaling facility is half full.

The dual SS7 links also should stay balanced. If we see more than two-thirds of the traffic on one link, then we should investigate to see why the signaling traffic is not staying balanced.

39.8   Transport Utilization and Congestion

Transport should not be a limiting factor in wireless telephone system performance. It is a pity to spend all the money we are spending on radio equipment and other base station equipment only to lose calls because we did not provision enough transport facilities.

There are some transport links where overflow is available. Dedicated transport for mobile-to-mobile calls can overflow to the PSTN. Links from the MSC to private networks may be able to overflow to the PSTN if their facilities are full.

However, most transport in our wireless telephone network is critical to system service. When backhaul links are full, or when links to the PSTN and PPDN are full, service is denied. If these trunks are too busy, then the system may need more facilities, or it may need service to some of the transport equipment.

39.8.1 MSC-to-PSTN links

The voice links from the MSC to the PSTN are critical for voice traffic. When these links are full, voice call traffic is denied all over the wireless system. Blocking voice calls is bad for revenue because it turns away money-making calls and annoys subscribers who are paying for wireless service. These voice trunks should be traffic engineered for very low blocking.

The voice links are full-rate 64-kbps pulse code modulation (PCM), so they can be traffic engineered according to Sec. 23.1. With the very high quantities of these voice trunks, we expect to be able to maintain low blocking and high occupancy during the busy hour.

Because of the importance of the voice links to the PSTN, we should be watching their performance. We should have a plan in place if the voice facilities become full. On the other hand, if the voice links from the MSC to the PSTN are never more than 80 percent full, then perhaps we have more facilities than we need. Under these circumstances, we can consider retaining the current level of voice facilities for a while as the system grows.

39.8.2 MSC-to-PPDN links

The packet pipes from the MSCs to the PPDN are just as critical for data service as the voice trunks are for voice calls. Unlike the single service of voice calling, wireless data calls come in a variety of rates and service levels.

The packet links should be engineered for very low delay on all real-time services, including video calls. These services should be checked for any excess delay on the MSC-to-PPDN packet pipe.

All services should have acceptable delay on these packet pipes. Even the background services, those which can tolerate high delays, should have enough capacity on these links. Each packet service, whether for voice (VoIP), video, still images, or data, should have low enough delay from the MSC to the PPDN.

Every packet service has a delay requirement, and the fact that some services allow higher delays is not a license for huge delays from the MSC to the PPDN. After all, the other links in the service chain are going to have delays engineered around the service requirements, and they are probably not expecting an extra delay here.

39.8.3 MSC-to-MSC links

Links between MSCs are used for inter-MSC handoffs. Only pairs of MSCs that have dedicated transport links are able to have inter-MSC handoffs. Once these links are full, that pair of MSCs can have no more handoffs. It is important to measure the load on these links because their congestion has a direct negative impact on inter-MSC handoffs. If these links are getting full, then it is time to diagnose the reasons and fix them. If cells are handing off more than they should, then perhaps some parameters need to be adjusted. Or perhaps some antennas need to be downtilted. If the cells are handing off properly, then there may be more traffic moving across the MSC-region border than we had anticipated in the system design. In such a case, it is time to get more facilities there.

We also can have dedicated links between MSCs to save PSTN costs for some roamer terminations and for mobile-to-mobile calls. These facilities are used to reduce costs rather than to provide a vital function, so we should be measuring their occupancy. If they are not carrying enough traffic to pay their own way, then it is time to reduce the size of these dedicated inter-MSC packet pipes. The traffic actually on the facility determines how much money it is saving us in PSTN charges.

39.8.4 Backhaul links

Backhaul link engineering should be easier than most other transport because we know the capacity of the air interface. Thus we should have few surprises from the base station to the MSC. With few exceptions, packets over the air become packets to the MSC. Different vendors handle the channel data differently on their packet links. Most use compression schemes, so the backhaul links are not perfect data mirrors of the CDMA air interface.

The CDMA carrier does vary in capacity due to changes in the environment. Calls from areas of low radio signal path gain consume more of the CDMA radio resource, but this is not going to overflow the backhaul link. A CDMA carrier will have more capacity if neighboring cells have less traffic than usual due to some cause or just due to statistical randomness. When this happens, the backhaul link might be the bottleneck.

We should measure backhaul packet link delay. If this delay becomes significant, then we should act. Most likely, the action we will have to take to fix a backhaul delay problem is to call the maintenance technician and have him or her repair a failed DS-1 or E-1 link.

39.9 Conclusion

Monitoring the status of our entire cellular network, especially these crucial measures of performance, allows us to identify potential problems early. We can then maintain and grow a healthy network, offering high-quality service to the maximum number of users we can support. We also can detect early signs of growth in usage or changes in usage patterns and respond before QoS is affected. This lead time allows us to get changes in place before customers experience poor service, or at least before they experience a lot of poor service. The more we know about exact network usage patterns and capacity performance, the more we can tune our network rather than overengineer it. And a finely tuned network is more cost-effective and therefore more profitable than an overengineered network.

A network operations center (NOC) can monitor and record the status of all network equipment and links across a continent. The NOC monitors the performance levels discussed in this chapter and also monitors the network for outages and equipment or link failures, allowing prompt dispatch when problems do occur.

Sometimes, however, rapid subscriber growth can lead to problems before we are aware of them through monitoring. In such a case, we need to be able to turn user complaints into useful data. This is the topic of Chap. 40.

40

Turning User Complaints into Useful Data

How many times have you had a cellular telephone call drop or given up and hung up due to static but not called to report the problem? Most users suffer poor service and do not complain until one day they change vendors, and we lose their business. The complaints we do get each represent many problems we never hear about at all. Therefore, it is essential that we take every complaint seriously and use it as information that can help us improve our systems.

Service complaints describe the user's experience, the symptom. Unfortunately, a single symptom, such as static or a dropped call, may have many different causes. The cause of the event could be an area of poor coverage, a cell operating at capacity, or an event outside our control, such as a failing user terminal. We need to gather all the information we can about each complaint and look at each complaint both individually and as part of a larger pattern of complaints and performance statistics to understand what is really going on and what, if anything, we can do about it.

The journey from symptom to problem comes first, and it is followed by the journey from problem to solution. In this chapter we will look at how to turn user complaints about voice and data service into useful information that helps us identify problems. We also will look at how changes in subscriber usage patterns can be monitored so that we can provide the best mix of voice, data, and video services.

40.1 Symptoms and Causes

Diagnosing technical problems from user experience is sometimes easy but at other times can be very difficult. When a user terminal experiences a rising bit error rate (BER) because of interference from a neighboring cell, the subscriber is only going to complain that the call was lost. The serving sector, the carrier, and the interfering cell are not going to be mentioned in the user complaint. If we are lucky, then the subscriber will remember when and where the call was made.

If we know the mobile telephone number and the time, then we can often track the location because third-generation (3G) code division multiple access (CDMA) user terminals use the assisted Global Positioning System (GPS) to determine their position.

The combination of position and time can give engineers critical information in diagnosing the wireless system problems that are causing customer complaints.

When a call fails in a radio system, it could be a lack of signal strength or it could be a lack of capacity. The CDMA air interface blurs the distinction between coverage and capacity. The CDMA carrier can lose a call because some other call was made, perhaps on a different cell.

The CDMA capacity resource is a link budget, and more of that capacity budget is consumed as a user terminal moves away from the base station during a call. Lower radio-signal path-gain calls consume transmit power, and the carrier runs out of capacity when that transmit power runs out. This is true in both the forward and reverse directions.

We expect to see user complaints when the CDMA carrier is close to its full capacity. As it overloads, calls are blocked, voice calls are degraded, and calls are dropped. Especially in a small-cell scenario, a carrier operating at very low power can suddenly find itself in an overloaded state with less than enough bits and E_b/N_0 to go around. Data users may have a different perception of degraded performance from voice callers. Data services can become slow or can develop errors.

40.2 Ineffective-Attempt Complaints

Ineffective attempts are calls attempted by subscribers that do not get started. Typically, the reason a call does not happen after a subscriber presses the SEND button or after somebody calls the user terminal is a lack of room on the air interface for another call. We should keep in mind, however, that call-attempt failures can be caused by equipment failures or a lack of resources elsewhere in the system.

Especially in a small-cell, high-density area, air-interface blocking is usually due to the fact that there are already so many bits that there is not room for another call. We are climbing up the steep part of the curves of Figs. 28.1 or 28.2, and it does not matter how much transmit power we find in the forward or reverse direction. This is the case most like conventional reuse (FDMA or TDMA).

Ineffective attempts also can be caused by low radio signal path gain, an impossibly weak signal. A user terminal tries to set up a call and fails because the paging messages are lost in the radio noise or the user terminal lacks enough power to set up a call. This, too, looks just like a conventional-reuse system not setting up a call because it is in such a weak radio area that a call cannot exist there.

Ineffective attempts also can be caused by poor coverage in a manner unique to CDMA. The call would have been fine if it had been closer to the base station, and the call would have been fine if there were fewer users on the CDMA carrier, but the combination of the two was enough to keep the call from being started. We can think of these calls as casualties of poor coverage because their signal path is good enough to make a call, but their low radio signal path gain makes them use more CDMA resource than we have available.

If we know where the user terminals were located when they failed to start their calls, then we can use propagation maps to evaluate the cause of the ineffective attempts. We can think of the system in three parts:

- Green areas where any call can be set up as long as the CDMA carrier has enough bits left, enough room for the call between its current load and the theoretical CDMA maximum

- Red areas where no call can be set up no matter what the traffic because the radio path gain is too low

- Yellow areas where there is some ability to set up a CDMA call, but it requires more CDMA capacity than its bit rate and E_b/N_0 suggest because of low radio signal path gain

Even if they do not use this particular color scheme, the radio maps can help us understand which of these three categories should contain the ineffective attempts our wireless telephone system is experiencing.

The clue that tells us which kind of air-interface congestion is blocking calls is the distribution of the ineffective attempts:

- If ineffective attempts are distributed fairly evenly throughout the sector or are distributed as the overall traffic is distributed, then we have a green area as described above.

- If ineffective attempts are concentrated in specific areas and happen whether or not the sector is busy, then we have a red area as described above.

- If the ineffective attempts concentrate in some areas and seem to spread out to a larger area as the traffic load increases, then we have a yellow area.

From these clues, we can deduce what radio conditions are causing our CDMA carrier to block calls.

40.3 Lost-Call Complaints

Lost calls are a major embarrassment to a wireless service provider. When subscribers are cut off in the middle of their calls, they are reminded that they are not using the reliable landline telephone network. Lost calls create the impression for subscribers that this is still a toy telephone system, not a medium for serious or professional communication. It is usually best to block calls rather than to lose them in progress.

The usual reason for a lost call is a good radio signal becoming bad. Unless a piece of equipment actually fails, this is the only reason a call should ever stop once it has started. Radio signals can go from strong to weak because a user terminal moves from a high radio path gain area to low path gain. We think of a subscriber driving into a valley or walking into a building.

But CDMA can lose calls by running out of capacity as well. When the antenna face serves a call over the limit, the CDMA carrier becomes overloaded. It is not necessarily the new call, the extra call, that is going to be lost. Even if our design is smart enough not to take on new calls that overload the CDMA carrier, a handoff can come along and add to the CDMA carrier load. Turning down a handoff request and leaving the call on a neighboring cell sector on the same carrier is actually worse for CDMA capacity than accepting it when radio conditions warrant a handoff.

Usually it is the call with the worst radio conditions, the weakest signal from the lowest radio path gain, that is going to be lost. The overloaded CDMA carrier is going to experience high BER on all its channels and is going to send all its user terminals power-up messages. In the reverse direction, one of them is going to run out of transmit power before the others and is going to be drowned out by the others as they climb up the power curve. In the forward direction, the base station amplifier is going to run out of transmit power, and one of the calls is going to have more noise or external sources of interference than the others and is going to be lost first.

The cause of the CDMA carrier overload does not have to be local. In conventional reuse, a call several cells away can cause cochannel interference. In CDMA, the interference is mostly the same cell and neighboring cells, but a call appearing on a neighboring cell can send a CDMA carrier into overload and cause it to drop one of its calls. A user terminal can move from high to low path gain, and somebody else loses a call, perhaps on the same cell sector but possibly on a neighboring sector or cell. Our CDMA carrier frequencies are low enough to be resistant to rain drops, but there is some attenuation in a heavy storm, and this may be enough to drop one or two calls.

40.4   Lousy-Call Complaints

Poor sound quality is usually caused by channel degradation less severe than a lost call. The interference and noise go up faster than the signal can, the E_b/N_0 goes down, the BER goes up, and the speech quality deteriorates. Depending on the speech coder, the subscriber may hear lower quality or short periods of dropouts, but either way, the performance drop is audible.

Customer perception of low sound quality has a lot to do with the time intervals involved. A listener facing 30 seconds of steady bad sound in a 10-minute call is going to notice the loss of performance after a few refrains of "Could you say that again?" The same listener experiencing ten 3-second intervals of bad sound may miss a word here and there but can fill in the details from context. And 100 dropouts of 300 ms during the same 10-minute call will lose syllables here and there, and the user may not notice any problem at all. All three of these examples are 5 percent signal outage, but the listener's perception of the outage is quite different.

Brief periods of CDMA carrier overload can come from changing radio conditions. When subscribers change their own radio paths by walking into buildings or driving into tunnels, they expect their mobile calls to take a few seconds to adapt, about the same amount of time it takes their eyes to adapt to the changing light. However, in CDMA, every other user terminal on the same cell or on a nearby cell is a significant part of the radio environment of a given user's call and contributes to the quality of the call. When somebody else walks out of a building, however, that user terminal has a much higher path gain and will create excessive interference on the CDMA carrier until its power is reduced. There also will be changes in radio conditions as vehicular traffic blocks the radio path between user terminals and base stations. These environmental changes can cause brief periods of high CDMA interference and low speech quality.

Another cause of brief periods of high interference are bursts of activity from high-speed data terminals. A wireless Web surfer downloads an image in the forward direction or a multimedia message service (MMS) subscriber uploads a snapshot in the reverse direction, and there is a short period of high-volume data activity. If the bursts are short, then there is not even time for power control to react, just a moment of reduced E_b/N_0 and higher BER. There will be more of these brief hiccups as 3G services become more popular, but the interruptions should be so brief that voice callers barely detect them or do not notice them at all.

When an extra call appears on the CDMA carrier, the effect may be longer than a data burst. The power-control process will try to compensate, but sometimes there will be too much traffic on the CDMA carrier for everybody to have a good call. Then how long will it take, on average, for some call to end and to relieve the overload? Well, consider a fully loaded cdmaOne carrier with 20 calls.[1] And consider the average voice call direction on a cell sector being about 1 minute before somebody hangs up or the call is handed off.[2] This means that the average time until some call goes away is 3 seconds. This is longer than a syllable, long enough that somebody listening for it might hear it, but not long enough to create a bad impression unless it happens frequently. A wideband CDMA (W-CDMA) carrier with 130 voice calls brings the average call ending time down below half a second, not a problem unless it is very frequent.

Calls that are lousy and stay lousy are a warning that something other than an occasional brief CDMA carrier overload is happening. Sustained poor sound quality is a sign that something more is wrong. It is these calls that should get our highest level of attention.

40.5   Slow-Data-Link Complaints

Slow response is a data link's equivalent to poor speech quality in voice calls. A slow data link simply could be a slow server at the other end of the line, but let us assume that we have narrowed the complaint down to the wireless telephone system.[3] There are several links in the data chain, several opportunities for delay. Before we launch an investigation into air-interface issues, we should make sure that the packet links in our transport network are not overloading and causing delay. These links are one or more backhaul hops, the mobile switching center (MSC) packet switch, and the links from the MSC to the public packet data network (PPDN).

Once we have narrowed the search for data delay down to the air interface, we know that the CDMA data channel is not sending and receiving subscriber bits quickly enough. Because the CDMA channel is a packet data link, reduced capacity means that packets have to wait in a queue before they are sent.

A short period of slow data service could be a short-term radio performance issue. The wireless data terminal could be moving into an area of low radio gain, or some other subscriber could be changing radio conditions, perhaps even in another sector or another cell. The CDMA carrier may take a second or two to reach a new equilibrium.

The cause of slow data service for one subscriber may be a burst of data activity from another data subscriber. If two bursty data subscribers are using the same CDMA carrier, then there may be brief periods when they will both be requesting data and there is only room for one. Then the other will have to wait a fraction of a second. If this happens only occasionally, then a subscriber might not even notice that the delay is in the CDMA system and not the data server at the other end.

Sustained slow data rates are a more urgent concern. Subscribers will notice them and, over time, will become sensitive to them. Initial 3G service will be exciting and novel, and people will be anxious just to get wireless data at high speeds. However, frequent periods of slow service will dilute that excitement and generate a community of

---

[1]This is the worst case because all the 3G CDMA carriers handle more than 20 calls.

[2]Handoffs are a little tricky here because a CDMA channel interferes with its neighbors even after a handoff is complete. The migration of a call from one cell sector to another is made even more gentle by soft handoff.

[3]It has been a long tradition to blame the telephone company for problems at the other end of the line. We see no reason for this tradition to change for wireless data service.

subscribers who feel they are getting less than they are paying for. Significant periods of slow data are a sign of an overloaded CDMA carrier. It may be that there are enough bursty data subscribers so that the traffic engineering parameters have to be adjusted. This means that the CDMA carrier may have to operate at a lower occupancy than originally planned, and this means sending fewer revenue-producing bits over the same air interface.

It is important to respond to slow data problems because of subscriber perceptions, as well as system performance. The marketing success of 3G depends on its users feeling good about it, but its technical success depends on its ability to deliver high data rates on demand. When this starts to fail, especially to the point where subscribers are complaining about it, we must find the root causes.

40.6   Data-Error Complaints

When subscribers complain about data errors, we first make sure that the equipment path is solid. The subscribers' equipment may be causing the errors, or the wireless service provider's packet data equipment may be in need of maintenance. Once the equipment issues are put to rest, we can focus our attention on the air interface as the source of subscriber data errors.

Poor radio conditions are an obvious place to look for wireless data errors. The path gain is low, or the signal is weak, or the receiver cannot resolve the errors even with forward error correction (FEC) in place. If the radio conditions change too rapidly, then the power-control loops can get caught by surprise and leave some errors in the bit stream.

Data errors may be a symptom of parameters set with too much optimism. If a data service is supposed to have a BER of 10^{-5}, then we should add some margin to this BER for CDMA channel environmental factors, particularly changes in the channel environment. The inner and outer power-control loops set E_b/N_0 based on BER and set power to achieve E_b/N_0. If the power-control setting for BER is too high, even if it is lower than the true target BER, then the CDMA channel will exceed the true target BER from time to time. We need some headroom, some extra BER margin, to ensure adequate error rates in a changing CDMA radio environment.

A wireless service provider in 3G CDMA is a data service provider. And that data service comes with subscriber expectations of performance. We may have to spend some more money, and therefore charge some more money, to meet those expectations. However, this probably makes more sense than selling inferior service that falls short of subscriber expectations. If we fall short of subscriber expectations, no matter how low the price, people will talk about our system, and what they say is not likely to be very nice.

40.7   Reasons for Change and Growth

There are several reasons to modify and expand a wireless telephone system. We can fix quality and capacity problems, anticipate growth in demand, provide new services, anticipate change in service mix of demand, or upgrade equipment. These activities are distinct from repairing equipment that breaks, but repair and improvement can be coordinated to work well together.

The easiest reason to change a wireless telephone system is to fix something that is wrong with it. In these discussions we are not writing about repairing equipment that does not work, but we are writing about perfectly good, working equipment that is not delivering the service we want to offer.

The system may not be delivering the quality of service (QoS) that is required for the subscriber community. For voice traffic, we can have too many blocked calls, too many lost calls, or too many lousy calls. For data traffic, we can have too many blocked calls, too much delay, or too many errors. These may be concentrated in specific geographic areas that require improved coverage. These are the red areas we described at the beginning of this chapter, areas of such low radio path gain that they cannot be served well with the antenna faces currently in place. We can adjust, modify, or add radio equipment to provide service to these hard-to-reach areas.

The QoS problems may be associated with high-demand times of day. If entire cell sectors are affected, then we have a green area as described earlier with a CDMA carrier whose fundamental, mathematical capacity limit is too low for subscriber demand. If partial cell sectors are affected and the QoS problems expand geographically as the traffic increases, then we have a yellow area, a CDMA carrier reaching its limit on the coverage-and-capacity frontier. Either of these cases is relieved by adding more CDMA capacity, more carriers, more sectors, and more cells. The yellow-area case also can be relieved by improving the radio signal paths with more or better radio equipment such as higher-gain antennas or repeaters.

The system may be working fine today, but growth in demand makes it likely that it will not be working well fairly soon. This is a good problem to have, more subscribers wanting to pay for our wireless telephone service. However, it means that we need to add capacity to our system even though it is working well at present. The new capacity starts with new sectors and perhaps new cells and works its way back to the landline world. This means more backhaul, more MSC equipment, more signaling resources, and more transport to the landline networks. This capacity expansion may require more MSCs and auxiliary equipment, as described in Secs. 34.6 and 43.6.

The system also may need to change, to serve a changing subscriber community, using a different mix of services. We hope that this is associated with a growing subscriber community so that we can add new equipment rather than changing old equipment. The change can be a shift from voice to data service, but it also can be growing demand for cellular snapshots in the form of MMS or even real-time video calls. Whatever shift there may be in subscriber demand, we have to make sure that the wireless telephone system is able to serve the changing demands well enough to ensure a revenue stream.

Classically, the distinction between repair and improvement is this: *Repair* returns a system to its prior operating status or brings it to the QoS level committed to in a service-level agreement (SLA) or similar specification. *Improvement* is change that results in performance better than what was previously achieved or exceeding the previously defined standard or specification.

Sometimes the line between repair and improvement can get blurry. We call a vendor to fix a software problem in the packet data switching area, and we are told that the repair requires us to buy a new software release that also will increase packet data switching capacity. Are we fixing a broken piece of equipment, or are we buying an enhancement to serve subscribers with more packet data?

We are constantly updating parameters, upgrading software, modifying configurations, and adding equipment. Each of these changes should be justified by a need in the wireless telephone system, fixing performance or capacity, adding capacity for growth, or following trends in the service mix.

40.8 Conclusion

It is possible to take the troubleshooting process one step deeper, beyond cause to root cause. If a certain type of problem recurs even after it is fixed, it may be the case that there is a deeper underlying problem that needs to be resolved. For example, if we find ourselves perpetually fighting pilot pollution, then it may not be enough to adjust antennas and add new base stations. We may need to review our planning methods or tools, asking: Is there some parameter we can change, or different approach we can take, that will reduce the pilot pollution problem overall? Perhaps we have been using three-to-one cell splitting, and we should change to four-to-one cell splitting. Or perhaps the radio propagation measurements we took were taken in winter, and summer foliage is creating interference that we did not include in our planning model.

This process, called *root-cause analysis,* and the subsequent action, called *permanent prevention,* can increase system capacity and QoS a great deal, improving profitability and also improving the system's reputation with customers.

Increasing Capacity of a Base Station

We need to monitor code division multiple access (CDMA) systems closely because, unlike conventional-reuse systems, they can show few signs of trouble even when operating very close to capacity limits. When our measurements of forward or reverse power levels indicate a need to improve base station capacity, we can increase amplifier power, add CDMA carriers, add sectors, or add repeaters. However, these solutions are not interchangeable. Each solves a different set of problems and works in different situations. We explain when to use each one in this chapter. The primary tool we use to describe the status of a cell in need of added capacity is the system of green, red, and yellow areas described in Sec. 40.2.

41.1 Determining When to Add Capacity

Most communication systems offer a clear way to tell when they are getting busy. There are a fixed number of channels, and the system is blocked when those channels are full. Or a packet pipe has a fixed capacity, and the system is too busy when there are more packets than the pipe can carry. We can get a rough estimate of the blocking performance of the system with 10 percent more traffic than it has by observing how often the system is 90 percent full. The CDMA carrier does not have the same notion of a fixed capacity, so we cannot simply count channels or packet volume. There is no obvious point where a CDMA carrier is 90 percent full, a point where another 10 percent will overload it. We have to look for more subtle signs.

The air interface does give some clues of impending traffic overload, however. CDMA systems should have systems in place to measure air-interface performance, and planners should remain vigilant. The CDMA carrier often looks just fine at 95 percent of its capacity, so we may get very little warning before the CDMA channel reaches its limit.

As traffic gets denser and cells get smaller, the thermal noise floor of the receiver becomes less important compared with the interference from our own system. In Figs. 28.1 and 28.2, small cells reach saturation far up along the steep vertical asymptote. This means that the system may be operating at one-third of its power or less when it is 95 percent full. The system may be giving us a warm sense of complacency that everything is just fine when it is only a few percentage points away from carrier saturation.

There are two different directions, forward and reverse, with two different radio environments. The forward direction has shared radio paths that aggravate multipath losses and has to use multiple transmissions for soft handoff. The reverse direction has a much smaller power amplifier available and has to use more rapid power control to get the same balance. While we can speculate as to which direction will saturate first, as we did in Sec. 30.7, it is best to keep a close watch on both the forward and reverse CDMA carriers.

In the forward direction, we are looking for a base station amplifier frequently reaching one-third or one-half power. The rate of reaching significant power levels now will be the rate of blocking when the system grows. The warning power threshold should be set higher for larger cells that are mostly for coverage, around half power. The warning threshold should be set lower for smaller cells that are mostly interference-limited, 20 to 30 percent of full power. When the amplifier reaches these warning levels for more than 5 percent of the busy hour, we are looking at a base station that will soon need relief as the system grows.

In the reverse direction, we are looking for total received CDMA signal power frequently exceeding the receiver noise floor. The rate of matching the noise floor now will be the rate of blocking when the system grows. The warning threshold for CDMA signal power should be set lower for larger-coverage cells, around one-third of the noise power. The warning threshold should be set higher for smaller cells, around two-thirds of the noise power. When the received CDMA signal reaches these warning levels for more than 5 percent of the busy hour, we are looking at a base station that will need relief as the system grows.

The other sign to look for is the blocking level itself. If the system is reaching saturation 1 percent of the time, then we can look at its traffic levels and deduce the CDMA carrier capacity from the traffic engineering principles in Chap. 23. If 10 percent more traffic will raise blocking rates to unacceptable levels with that capacity, then we know it is time to grow the system.

41.2 Adding Power

If a base station is running out of power in the forward direction, then a bigger power amplifier may offer capacity relief. Clearly, this is only an option when the reverse direction is showing no signs of stress and no signs of impending overload. It does not make sense to spend a lot of money for a shiny new power amplifier only to find the user terminals running out of their transmit power and losing calls anyway.

We can only consider adding power in the forward direction if more power is available. We can increase power in two ways, buying a bigger amplifier or using more amplifiers. When smaller cells are added to a system, we often use smaller power amplifiers because their coverage range seems less important. Even for a larger cell, we may have decided to use amplifiers of one average power level, with the expectation that a few base stations might require more power and need a bigger amplifier. The other approach is to divide and conquer, to use separate amplifiers for separate carriers. If a base station's architecture allows it, then we will often put multiple carriers on a single power amplifier because it saves a lot of money. When that amplifier is no longer powerful enough, we can separate the carriers onto their own amplifiers. Depending on the combining technology available, we may be able to keep the same antenna with multiple-transmit-power amplifiers.

We can consider using higher-gain antennas. This has the advantage of increasing the radio signal path gain in both forward and reverse directions. However, the horizontal radiation pattern is already determined by the sector angle. Making the angle narrower means adding more sectors (described in Sec. 41.4), which is a more costly proposition. Most of the initial base station designs probably use the highest-gain antennas consistent with CDMA air-interface performance, so using higher-gain antennas is unlikely to be a major design option to improve the radio path.

Once we have decided that the system is limited in the forward direction, and once we have determined that more power is an option we can use, we look at the distribution of amplifier power to see if more power will help. If our capacity crunch is in a green area, where the limit is available bits, then adding power will not help at all. More power will not increase capacity, and user terminals in this area already have more than adequate coverage. While more power will reach more subscribers in red areas and improve system coverage, it will not relieve a capacity problem in doing so. It is only a capacity problem in a yellow area that will be relieved by adding more forward transmit power, and only if there is sufficient user terminal transmit power in the reverse direction. We can learn about the distribution of subscribers from the distribution of forward transmit power.

If the forward amplifier is spending time operating at over half power, then we have a need for capacity increase. It is the distribution of power during the rest of the time that tells us if the capacity need is in green areas, where more power will not help, or in yellow areas, where more power will directly increase the CDMA capacity of the base station. If the amplifier is spending most of its time at very low power levels, then we conclude that the demand we are trying to serve is green-area demand that already has high radio signal path gain between the user terminals and the antenna face.

When the forward transmit amplifier is spending significant time in the 20 to 40 percent power range, then we conclude that there is significant subscriber traffic in yellow areas, where increasing radio power will directly increase CDMA capacity. These are the sectors where we want to consider adding power.

41.3 Adding Carriers

When we put a cell in a CDMA system, we do not automatically install every CDMA carrier on every antenna face. The reason is simply a matter of equipment cost. Adding carriers is expensive. Even if multiple carriers can share transmit amplifiers, there is other base station equipment that has to be present for each CDMA carrier.

We should only consider adding a CDMA carrier to a cell next to another cell that already uses the same carrier. Having a gap in a CDMA carrier does not reduce the interference by keeping the reuse further away because calls using that carrier will extend into the middle area and create either a coverage problem or a pilot-pollution problem. Each carrier is a separate CDMA planning problem, and each carrier has to maintain a solid grid that avoids pilot pollution. We want to avoid multiple small islands of carrier coverage, and we want to maintain a convex coverage area, as described in Sec. 32.3. By keeping a tight grid of cells in a convex shape, we can avoid four- and five-way zones, as described in Sec. 32.1.

There are two circumstances where we are comfortable having a new area of a CDMA carrier. The first is when we expand the radio spectrum to include a new carrier. If we are going to introduce a new carrier to a wireless system, then we have to start somewhere. We can get a new carrier because a government agency gives us more radio spectrum, but we also can add CDMA carriers as we migrate a frequency division multiple access (FDMA) or time division multiple access (TDMA) system over to CDMA, as described in Sec. 33.6. The second case is when we have a large service area with two distinct population centers. Here in the United States, we have several service areas with pairs of adjacent cities, for example, Minneapolis and Saint Paul or Dallas and Fort Worth. If they were separate wireless telephone systems, then we would expect each city to have its own high-density, maximum-carrier zone, as described in Sec. 33.4. And defining one large service area with both cities in it does not change the fact that each city has its own high-traffic area. We expect such a dual city to have two maximum-carrier regions isolated from each other. If traffic increases, then those isolated maximum-carrier areas might grow into each other and merge into one large high-traffic CDMA service area.

How we add a carrier depends on the base station architecture. Some equipment vendors may have us buy a lot of equipment to add a new carrier, and others may have a simpler scheme. If the existing forward transmit amplifier is already close to its maximum power limits, then we will need to add another amplifier as a part of this process.

Sometimes the need for a new carrier does not occur in a nice convex-grid cell. It may be necessary to expand carrier coverage by adding the new carrier to transition cells (described in Sec. 33.4) so that the carrier region stays convex as it grows. The alternative is loss of capacity due to pilot pollution.

41.4   Adding Sectors

CDMA sectors are independent servers, so adding sectors to a cell adds capacity. A three-sector CDMA cell does not divide up the traffic into thirds but instead offers three separate CDMA servers. When a cell runs out of capacity, we can divide it into more sectors by putting more antenna faces at the base station. This is attractive as long as new sectors serve new traffic more cheaply than a new cell.

Each antenna face at a base station adds less capacity than the one before. There is a decreasing capacity gain as a cell has more sectors. This happens because the same amount of sector-to-sector radio interference is a greater proportion of a smaller sector angle. The result of this sector interference is that a three-sector cell generally serves 2.5 times the traffic that an omnidirectional cell would serve, and a six-sector cell serves about 4 times the omnidirectional traffic. This means that twice the equipment generates 1.5 times the revenue, but we save the cost of an extra building and tower going from three to six sectors.

We want to keep sector boundaries away from cell boundaries, especially three- and four-cell zones. A sector boundary produces just as much pilot pollution as a cell boundary. While cell boundaries depend on terrain, buildings, and even vegetation, sector boundaries are clean, straight lines from the base station. We have to be careful about relying on cookie-cutter hexagons while splitting cells as described in Sec. 32.3. However, we can rely on cookie-cutter lines on a map for sector boundaries that depend on the antenna radiation patterns and their orientation and not on the terrain between the base station and the user terminal.

Sectors can be whatever shape satisfies the subscriber demand. We do not have to have three or six sectors. We do not have to have evenly spaced sectors. If a 30-degree wedge sector with a super-high-gain antenna is the best solution to serve a particular region, then we do not need 11 other sectors in that cell. The rest of the cell can be three 110-degree sectors so that it otherwise looks like a normal three-sector cell.

All the cells do not have to use the same sector plan either. A CDMA system can mix three- and six-sector cells and even five- and nine-sector cells. Whatever configuration best uses the CDMA air interface to serve the demand is what we should put into the system. There may be an administrative hassle in having 50 different sector plans in a system of 500 cells, but this has to be weighed against the foolish consistency of insisting that all the cells look the same.

An alternative to adding more sectors with narrower fixed-beam antennas is to change a base station to adaptive phased arrays (smart antennas). Time division synchronized CDMA (TD-SCDMA) requires smart antennas as part of the specification, but other CDMA systems might use them as well. They may be more expensive and complicated than fixed-beam antennas today in 2002, but this is the kind of technology that comes down in price as it becomes more widespread.

Base stations with smart antennas can be the equivalent of six-sector cells. The versions being used in TD-SCDMA generate a 60-degree beam width. With larger and more complex antenna arrays, we might be able to get radio path isolation equivalent to 40 degrees or even 20 degrees, the equivalent of 10 or 20 sectors per cell.

We add antenna faces or smart antennas for capacity when adding the equipment to an existing base station is more cost-effective than adding more base stations to the wireless system. We want to make sure that new sector boundaries do not create a capacity problem in pilot pollution that mitigates the capacity advantage of having more CDMA sectors in one cell.

41.5 Installing Repeaters

Repeaters are often thought of as a stop-gap coverage measure, a quick fix for areas with low radio signal path gain to their base stations. In a growing CDMA system, repeaters can fix poor-coverage red areas, as described in Sec. 40.2, but they also can support yellow areas and improve the capacity of their base stations.

Repeaters are radio retransmitters, as described in Sec. 4.6. Without any telephone transport of their own, they enhance radio communication by breaking a single radio path into two hops. When a weak radio area is supported by a repeater to a base station, the cell changes shape to include that weak area. Coverage is increased when subscribers in red areas can get service that was hard to get at any time without the repeater. Capacity is increased when subscribers in yellow areas can get service using less CDMA radio resource with higher effective radio path gain because of the repeater.

The best candidates for coverage repeaters are places with poor radio access to base stations that have convenient electrical power but no convenient telephone transport. This is how we discussed repeaters in Sec. 32.4.

The candidates for capacity-enhancing repeaters are areas of significant subscriber density where there is acceptable CDMA service when the system is lightly loaded. This indicates that the radio path gain is low, and calls from this area cannot compete in transmit power with calls in better areas. A base station might not be cost-effective or it might be hard to get telephone transport, so we use a repeater. Also, if our system has a four- or five-way zone of pilot pollution, then a repeater might be a cost-effective way to break the tie and have one dominant server in that geographic area.

Repeaters are not base stations. They do not do as much as base stations, but they do not cost as much either. We can use repeaters as cost-effective ways to add capacity as well as coverage.

41.6 Conclusion

When we add capacity to the air interface of a base station, we need to monitor the backhaul from the base station to the mobile switching center (MSC) to make sure that we have not exceeded the capacity of the packet pipe. In fact, if the pipe is close to capacity, we should plan to upgrade it on the same visit to the base station.

Although new base stations are expensive, the entire principle of cellular telephony is that we can grow our system by adding cells. Let us turn our attention to adding new base stations and new cells in the next chapter.

42

Adding Cells to a CDMA System

We grow cellular networks by adding cells and, in doing so, increase capacity and revenue. If we plan carefully, the new cells arrive before customer complaints about blocking or poor call quality but not before they are needed. Base stations are expensive, so adding cells too soon decreases net revenue. In addition to adding cells at the right time, we want to add the right type of cell at the right place and engineer the network with the new cell for minimal intercell interference.

42.1 Determining When to Add Cells

Ultimately, a cellular telephone system grows by adding cells. We can use more powerful amplifiers and higher-gain antennas, we can use all the code division multiple access (CDMA) carriers that are available, and we can add sectors, but there is a point where a base station is carrying as much traffic and earning as much money as it can. If we want to increase the traffic and revenue of a system, then we have to use more base stations. Once we have our base stations set up for their maximum load, we need to know when to add more of them.

A cell-splitting criterion we do not want to use is waiting until we have high blocking and subscriber complaints. Good planning is the art of staying ahead of trouble rather than waiting for it to happen and then reacting to it. As we discussed in Chap. 41, the primary warning of impending overload in a CDMA system is power levels: high transmit power in either direction, too many power-up messages for the reverse link, or more signal than noise at the base station receiver. In a small-cell, high-density area, these symptoms may not appear until the base station is very close to saturation.

Therefore, planning ahead is essential. We can use the planning models described in Chap. 48 to estimate how much traffic each sector is supporting and how close it is to its maximum. The planning tools take into account such factors as the effects of terrain on propagation, intercell and intracell interference, and pilot pollution. We have to use some judgment because the computer program may not know the usage patterns and the mix of services of our subscribers.

In a growing system, we can use earlier experience to calibrate our computer predictions. As our system has been growing, we have been adding cells, and we know how much traffic the old cells could handle before they ran out of capacity. That traffic level is going to be a good estimate for how much traffic the current cells can handle. If our own system has been consistently worse or better than the computer prediction, then it may be that we have a tougher or easier subscriber community. Perhaps we have more bursty data users or fewer multimedia messages than we assumed in the computer models.

Adding base stations means adding backhaul. A new base station needs its own transport back to its mobile switching center (MSC). This will be discussed in Sec. 44.1, but we should keep in mind that available cheap transport is an important component in selecting new base station locations for growth, just as it was in initial system base station placement.

42.2   Cell Splitting

As demand grows, each cell sector is carrying more traffic. We can represent the peak-hour traffic on each cell sector as a statistical distribution based on the average demand level and the mix of services. Some part of that demand distribution will be above the capacity of the cell sector. If we have engineered the system well, then it will respond to too much demand by blocking calls. As the demand grows, the fraction of that statistical distribution on the high side of the maximum capacity will increase, and the blocking rate will rise.

The way cell splitting relieves capacity is simple. The new cell is placed amid the old cells so that it serves some of the region of the old cells and therefore serves some of the traffic. The old cells are now smaller in area and have reduced subscriber demand. The distribution of traffic is now lower as a result, and the old cells should return to an acceptable blocking rate. If two neighboring cells are overloaded, then a new cell can be placed midway between them. In some ideal sense, we would like the three cells to share equally the burden of the original two. This would reduce the average demand on the older two cells by one-third. In actual practice, the new cell initially serves less than this ideal one-third share of the total traffic.

The first step of the cell-splitting process is to identify high- and medium-urgency candidates for traffic relief. These are base stations that are currently experiencing high peak-hour blocking or are going to in the near future.

Like any other coverage picture in a real place with real terrain and real buildings, the new cell will be some kind of odd shape. We use a geometric construction with hexagons and straight lines to get a conceptual picture of what the new cell can do for capacity relief.

The cell sectors that have high blocking during the busy hour are in need of traffic relief. They are already telling us that our wireless telephone system needs to increase its capacity to provide adequate service. Once they are at their maximum configuration, we should identify cells with high blocking rates as high-urgency candidates for cell splitting.

We should anticipate growth as well. If the distribution of traffic driven by subscriber demand has a large fraction close to the maximum capacity, then it is only a matter of some more growth before that cell sector is going to have high peak-hour blocking. Our power-level observations are telling us which cells are medium-urgency candidates for cell splitting.

We can use the planning models described in Sec. 48.1 to estimate the demand loads and traffic distributions on our existing cells. These forecast demands can tell us when we expect to reach overload on our existing cells. The predictions identify cells where we expect high blocking in the near future, and these cells are also medium-urgency candidates for cell splitting.

42.3 Keeping the Grid

We have identified those cells and cell sectors in greatest need of capacity relief, and we have equipped their base stations with as much capacity-enhancing equipment as we can. We are going to add new cells to the system to serve traffic so that these identified cells will have less traffic. The new cells will take big bites from the service regions of the old cells. Because sectors of a cell interfere with each other, reduction of any sector traffic on a cell causes significant relief to the entire cell.

We want to select new base station sites to provide the maximum relief for the old cells. As we compile a list of candidate locations for new base stations, we want these new cells to take the biggest bites they can from the old-cell areas to provide the greatest level of traffic relief. The more we relieve and reduce the traffic on a busy cell, the longer it will be before continuing growth compels us to relieve it again.

It is important, however, to keep system issues in mind when splitting cells for traffic relief. In particular, having four- and five-way zones of nearly equal radio paths causes pilot pollution and degrades CDMA capacity, as discussed in Sec. 32.1. Since we are adding cells specifically to improve CDMA capacity, we do not want to compromise this mission by adding cells that create low-capacity zones, as discussed in Sec. 32.3.

The best way to maintain a CDMA system relatively free of pilot pollution is to maintain a regular cell grid. Adding cells in hot spots without regard for the grid structure causes four-way zones of pilot pollution to appear, as shown in Fig. 32.3. In Fig. 32.1 we showed how the grid has no areas with more than three cells competing for CDMA capacity. The regular hexagonal grid keeps the three-way zones far apart so that variation in radio propagation does not often bring them together into four-way zones.

We can add a cell in the triple-point of three existing cells to form a geometric picture like Fig. 32.5. The reality of a three-to-one split, however, looks more like Fig. 32.6 with three new four-way zones of pilot pollution. Eventually, the new grid fills in and forms a new hexagonal grid with three times as many cells handling three times the traffic, as shown in Fig. 27.15, but the boundary of the new grid is bristling with pilot-pollution zones. Even though the three-to-one split looks tempting, we recommend using the four-to-one split to maintain the grid and to keep pilot pollution to a minimum.

The four-to-one split adds a cell midway between two cells. The new cell, as drawn in Fig. 32.7, is one-quarter the size of the old cells. As we continue to split cells in the four-to-one configuration, we end up with a denser grid of cells one-fourth the area handling four times the traffic, as shown in Fig. 27.14.

The initial new cell is actually a more rectangular shape, as shown in Fig. 32.8. Notice that the rectangle does not have the four-way zones at its corners that we saw in Fig. 32.6. Not only is the eventual four-to-one splitting grid reasonably free of pilot-polluting four-way zones, but the transition and boundary areas of the four-to-one grid are also low in pilot pollution.

We have to remember to keep the new grid of small cells in a convex shape, as described in Sec. 32.3. Any U-shaped regions in the boundary will be four-way zones with some pilot pollution. Sometimes we need to add transition cells, as described in Sec. 33.4, to keep the region convex.

Getting base stations in the right places is an ongoing tough battle. We should not allow difficulties in site acquisition to thwart our growth plans. If a site is hard to get, then we should consider sites nearby.[1] If a base station for a transition cell is taking longer to build, then we probably should deploy the other growth cells and live with some pilot pollution for a while.

Many factors make site acquisition a challenge. It is hard enough to engineer a CDMA system with hundreds of cells without the extra challenge of not being able to put base stations where we want them, but this is part of the real engineering world. We can put an X on the map where the base station should be, and then there is the challenge of putting the base station close to that X. Here in the United States, local zoning boards often prohibit towers in some areas, usually areas where a lot of rich people live, areas where we want good cellular coverage. Getting the land for a base station can be hard as well. A tower that is already built and is not too far away may be a far better choice than starting from scratch. Or a location closer to a transport node, a point of presence (POP) in the public switched telephone network (PSTN), may have major cost advantages.

A growing system needs a continuous plan for new cells, new base stations, and new places to put those base stations. The growth plan should keep the hexagon cellular grid as well as it can within the reality constraints of zoning, property, and transport. This includes continuously developing a list of good base station locations for the future.

42.4 Downtilting Antennas

The original mission of our earliest base stations in a wireless telephone system is coverage. We built our original cell grid with the objective of reaching every place where a subscriber might be willing to pay money for wireless service. The link budgets have a line item for radio signal path gain, and the capacity of a coverage cell is higher when the radio path gain is higher.

Thus we specifically designed our cells with the highest possible radio path gain so that we could reach the hard-to-reach areas. We put our antennas on high towers to get better propagation performance over terrain, vegetation, and buildings. We used powerful amplifiers in the base station transmitters and low-noise amplifiers (LNAs) at the receive antennas to get a strong radio signal as far as we could. And we used high-gain antennas with a narrow vertical beam width to increase the end-to-end radio signal path gain.

A growing system has priorities other than long-range radio coverage. The base station design that worked well for a large cell may be counterproductive when the cell shrinks because neighboring cells are sharing its load. A neighboring cell is a significant interferer in a CDMA system. The radio overlap between two cells forms the zone where those two cells interfere most with each other. Where two cells have nearly equal radio paths, calls in either cell consume CDMA radio resources in both, more in the forward direction because of soft handoff.

---

[1]The grid mentality can become inflexible. A carrier deploying an 8-mile (13-km) cell grid fought a long zoning battle for a site in a resistant neighborhood when there was an industrial park less than 2 km away where a site would have been easy to get.

As neighboring cells get closer to a base station, we want to reduce the coverage so that there is less overlap with neighboring cells. We want to find a way to keep the radio signal path gain high throughout each cell-sector area and to have it drop off sharply outside the area.

There is not much we can do about sector-to-sector interference as a system grows. Even in its large-cell coverage form, a base station has greater capacity if its antenna radiation patterns have well-defined edges so that the boundaries are sharp between sectors of the same cell.

Downtilting the antennas, however, seems like the best way to reduce a cell radius. As shown in Fig. 33.1, downtilting sector antennas simultaneously increases the path gain for near calls and decreases the path gain for more distant calls.[2] Furthermore, the radius of the cell is adjustable simply by changing the tilt angle of the antennas.

We can evaluate the positive effect of downtilting antennas on CDMA capacity using the predictive planning models described in Sec. 48.1. These computer programs should quantify the positive effects of changing the vertical-beam radiation pattern of the sector antennas.

The result of a downtilting study is a plan that coordinates splitting cells for CDMA growth and downtilting antennas to get the most capacity from both the old cells and the new cells. We can lower antennas on the tower to reduce the effective cell size. This is an alternative to downtilting, but it is usually more expensive and less effective. The path-gain reduction from a lower antenna is greater at longer distances, so lowering the antennas gives a relative advantage for calls close to the base station.

42.5   Microcells for High Traffic Density

Sometimes system growth is concentrated in a small area so that a full-sized macrocell would spread the CDMA interference over a larger area than the concentrated traffic. And sometimes the concentration is so great that there is enough wireless traffic in a small area to keep one or more CDMA carriers busy.

In the case of concentrated demand, a microcell may be a better solution than a macrocell. A microcell is a smaller base station designed for short radio range of a few hundred meters. A microcell has a low antenna height and a lower-power amplifier than a macrocell. Since its coverage area is deliberately small, a microcell may be placed off the grid pattern without producing the interference and pilot pollution of a regular macrocell in that same off-grid location.

We can put microcells on the grid positions for cell splitting as well. If the traffic continues to grow, then we can continue to split cells in the grid so that the microcell grows in function to be a regular cell. We may need to raise the antennas higher and downtilt them to get the local coverage needed for macrocell function. The plan for microcells is to identify concentrated areas of growth in small geographic areas that are easily reached with a shorter-range radio system.

---

[2]Obviously, downtilting only works at a base station with multiple antenna faces. But any cell with only a single omnidirectional sector is almost certainly going to gain CDMA capacity by installing more sectors.

42.6   Picocells for Specific Local Areas

The concentration of traffic can be so extreme that even a microcell is too large. We think of a single office building with a lot of traffic, enough to justify one or more CDMA carriers. While such a load seems like an unusually large building when we think of voice calls, third-generation (3G) data services may generate this kind of demand in many office buildings in a large city.

Picocells are designed to serve traffic concentrated in a single building. The ranges are limited to 10 or 20 m mostly by placing the picocell inside the walls of the building. Since a picocell is already isolated from the rest of the wireless telephone system by being inside a building, issues of other-cell interference and pilot pollution should be attenuated because the radio signal is attenuated by the walls of the building being served.

Picocells offer close-range service, so we can use the very high data rates of 3G that were designed for short distances and high radio path gains. The cdma2000 1x EV-DO offers a data service of 2.45 Mbps in only 1.25 MHz of radio spectrum using more complex constellations than the normal CDMA air interface. This technology is available in the forward direction only when the radio path is very short and very clean. Having a picocell right in the building gives us the opportunity to create such a clean link and offer these high data rates to subscribers.

The high-rate data service offered by picocells may be part of the main wireless telephone system, but it also may be offered on a private data network. The picocell could be an Internet Protocol (IP) node by itself, as we described in Sec. 35.4. A private community of data subscribers would arrange to be able to use this one picocell as a private data node in its own data network. Data packets could go directly from the picocell to their internal network without taking a detour through the mobile switching center (MSC) or the public packet data network (PPDN). The private network wireless data terminals might be configured to use the MSC and PPDN for their packet routing if they are not being served by their private picocell.

A picocell also can be used for a private voice network in a large building. Special cell phones could be configured so that their voice calls would use the private voice network from inside the building. The picocell would send the voice traffic from these user terminals directly to a private branch exchange (PBX) inside the building on a private network. These special cell phones might be allowed to use the MSC and PSTN when being served by another base station outside the building.

We can use picocells for very small spot service but also for specific service in a building. Such specific service can be high-rate data service using the most advanced 3G technology, or it can support private networks for wireless data or voice subscribers.

42.7   Handoff Management

New cells mean new neighbor lists for handoffs. When a growth cell is added next to an old cell, the growth cell should be put on the handoff neighbor list for the old cell. Perhaps there is an old cell that used to be a neighbor that now has an entire growth cell in between, so it should be taken off the handoff neighbor list. And the new cell itself needs a list of handoff candidates.

The first step in creating new handoff neighbor lists is to draw a picture of the cell layout. The base stations are dots on the map, and we can draw cookie-cutter lines halfway in between the dots for cell boundaries. Then we can draw sector boundaries radiating from the base station dots. The resulting shapes on the map are the cells and sectors represented geometrically. We are using distance to represent radio signal path gain. We expect to have significant handoff activity between any pair of cells sectors sharing a line on this cookie-cutter map, no matter how short that line is.

We can start building a handoff neighbor list from this map by going around each cell sector and noting all the other cell sectors that share a common border. The list of neighbors on the map is a starting list of handoff neighbors for the system. As a rough estimate, we expect to see more handoff activity between two sectors when the border line on the map is longer.

The question arises what to do if the line on the map between two sectors is very short or if two sectors meet in just a corner. The first thing to do when this happens is to rethink the cell layout and sector plan because this only happens when there are four or more sectors with nearly equal paths. A four-way zone is a candidate for pilot pollution and inefficient use of the CDMA radio resource. If we cannot fix the four-way zone in the layout process, then we probably should put any pair of sectors sharing a line in common on each other's handoff neighbor lists.

The map process relies on distance to estimate radio signal path gain. Of course, there are many other factors to consider, such as terrain, vegetation, and buildings. These other factors can create a more favorable radio path environment for one cell over another and move the boundary between two cells. However, this effect is not likely to change the picture so much that two cells that look like neighbors on the map are not neighbors in the radio environment. However, the opposite can happen. Two cells that do not look like neighbors on the map may share radio traffic as a result of propagation irregularities.

We can use planning tools to add to the sector handoff neighbor lists. We are loathe to remove a handoff neighbor candidate that looks good on the cookie-cutter map, but we are certainly anxious to include a candidate suggested by a planning tool, even if it is not visible on the map.

It is important to maintain the neighbor lists as new base stations are added to the CDMA wireless telephone system. It is also important to manage the add and drop parameters for soft handoffs.

42.8 Reliability Issues for Growth Cells

When we built the original cell layout, we decided on a level of reliability for our wireless telephone system. This reliability requirement was the input to the decision of what kind of equipment and how much redundancy to put in each base station we were building. Now we are adding base stations to create growth cells to serve more traffic. These new base stations are required to serve the increasing demand for wireless telephone service while keeping the quality of service (QoS) high.

If a growth cell base station fails, then the system will experience a decrease in its QoS below the level required. If this were not the case, then we would not have added the new cell. However, the system worked without the growth cell when there was less traffic. We may decide that serving less traffic is acceptable for the relatively rare and brief periods that a growth-cell base station is out of service. The loss of the growth cell would bring us back to the system as it was before the growth cell was added and the system had full coverage of the service area.

The result of this reasoning is that we should consider the cost savings of adding growth cells without extra equipment for redundancy. Over time, the subscriber demand may be great enough that we will change our minds and decide that these cells are too important to be left without redundant equipment. However, the initial installation of a growth cell may be a lot less expensive if we can live without duplicate systems.

This cost-saving opportunity may go away if the neighboring old cells no longer have adequate coverage of the new-cell area. This would happen if we downtilted the antennas on the old cells to restrict their interference with the new cells. In such a case, the decision not to employ redundant base stations for growth cells is a decision to risk loss of system coverage and not just higher blocking during a base station outage.

The wireless service provider has to make a decision, one of those cost-versus-quality tradeoffs. The consequence of a growth-cell outage is a much higher blocking rate for a period of time. We have already made the decision in our design process of what level of QoS we are willing to pay for in buying equipment and adding base stations. We now have to decide what QoS is going to have the extra level of reliability that a more expensive, redundant base station provides.

It is reasonable to decide that all the base stations should be fully redundant and reliable. If we marketed the 3G CDMA product as being rugged, tough-as-nails dependable, then an outage that causes poor service is not going to please subscribers who are paying a premium for our superior product. Or we may have large business contracts with service-level agreements (SLAs) that we can only meet when all the base stations have high reliability.

It is also reasonable to decide that occasional periods of high blocking are acceptable. If we marketed our product to a casual voice call community on the basis of price, then it may be perfectly reasonable to have two or three days per year when it is very hard to make a call. The base station equipment cost saving can be passed on to the subscribers in the form of lower, more competitive rates.

In any case, equipment failure in a base station should be a rare event. If base station equipment fails more than twice in a year, then we probably should be switching equipment vendors. And a key component to the redundancy decision is how quickly we can repair a base station once it has an equipment failure. If the base station is at a busy shopping mall, then we should be able to dispatch a repair truck in a few minutes. If the base station is on a hill at the end of 1200 m of dirt road outside town, then its repair is going to take longer simply because it takes longer to get there.

Transport failures in backhaul should be similarly rare, but we should think about failures there as well. We may have alternate routing in place for the large-cell grid, and we may decide to connect each growth cell to one large cell in the fan-out network. The failure of that one link would bring down the growth cell and cause an outage. If this is sufficiently rare or we can live with the growth-cell outage, then it makes sense not to spend the extra money on diverse transport routing in growth-cell backhaul.

The result of a reliability analysis should be a firm plan for redundancy in growth-cell equipment and transport. Such a plan should be based on the frequency and duration of outages and the QoS consequences of those outages.

42.9   Conclusion

In this chapter we have addressed when, where, and how to add base stations, creating new cells to improve coverage or to provide enhanced 3G data services. When adding base stations, it is important to pay attention to the allotment of base stations to MSCs and inter-MSC borders, particularly those which are also intersystem borders. For example, we will need to notify neighboring service providers of the need to update their handoff lists, and we may want to cooperate with them when planning border base stations to reduce pilot pollution.

43

Mobile Switching Center (MSC) Growth

Our mobile switching centers (MSCs) and auxiliary equipment must grow to meet the capacity demands of our ever-growing cellular network. In this chapter we will examine when to upgrade MSC equipment, how to assign users and base stations to new MSCs and deal with MSC borders, and options for MSC placement, equipment provisioning, and auxiliary equipment.

43.1 Determining When to Upgrade MSC Equipment

As the subscriber traffic grows, the MSC will run out of critical resources to serve that traffic. As this happens, we will have to buy more MSC equipment to keep up with subscriber demand. As in the case of base station equipment planning, this is an ongoing process. While adding MSC equipment is a simpler process than site planning for new cells, increasing subscriber demand for an evolving mix of services can keep planners busy figuring out which MSC components need to expand to meet the growing demand for service.

It is important that the MSC maintain enough capacity for peak-hour demand. Having the switch run out of capacity, even for 5 minutes during the busy hour, means noticeably degraded service over a large area with many subscribers. Each component of the MSC has its own limitations and its own time-of-day demand cycle. Of course, the time-of-day demands on MSC components will be highly correlated with each other, but it is possible that the mix of services will be different at different periods during the day.

Peak-load processing increases with call processing, short message service (SMS), other signaling, and handoffs, as discussed in Sec. 24.1. Growth in call volume, increasing popularity of short messages, and more mobile subscribers will drive us to increase the MSC processing power. A trend toward longer calls or data services may have the opposite effect because they generate less call-processing activity. Increasing use of multimedia message service (MMS) ties up a lot of code division multiple access (CDMA) air-interface resource and comparatively little MSC processing, so an increase in MMS might relieve the MSC processor for the voice call traffic it is displacing.

The total DS-1 or E-1 connections on the MSC are increasing as we add backhaul for more cells. Growing activity also increases the transport needs to the public switched telephone network (PSTN) for voice traffic and to the public packet data network (PPDN) for data traffic. It is also important to remember that higher traffic levels beget higher inter-MSC handoffs levels, and the transport links between MSCs also may have to expand to meet this need. Finally, we may want to increase inter-MSC transport for cost reasons when there is an increase in mobile-to-mobile calls or roamer terminations within the system, as described in Sec. 37.1.2. As we said in Sec. 34.1.2, transport is a relatively deterministic item. Once we have done the traffic engineering and determined how much transport we need, we can make sure that the MSC has the required equipment to support that transport.

As the number of voice calls increases, so must the amount of circuit switching in the MSC. The introduction of third-generation (3G) services can push subscriber demand away from voice and actually might reduce the voice calling load. It is more likely that voice and data service will both grow, and it is up to us as planners to make sure that the MSC circuit-switching equipment can handle the voice calls. The number of speech coders in the MSC also increases with the amount of voice calling. There is a subtle difference that could be affected by changing call patterns. The circuit-switching activity depends on the total *number* of calls, whereas the speech-coding usage depends on the total *duration* of calls. If there is a trend toward shorter voice calls, then the circuit-switching components of the MSC will have to increase. If the trend is toward longer voice calls, then we have to add more speech-coding equipment.

The packet-switching component of the MSC serves the backhaul packet pipes, data services, and most of the other services. If there is steady growth in 3G services, then we expect an increasing share of the MSC workload to be packet switching. This means that the packet-switching equipment has to keep up with an increasing share of increasing traffic.

In expanding the MSC, it is important to think about not only adding equipment but also making equipment better. Technology gains can make upgrades attractive because they increase capacity at lower cost. If the new version of a piece of equipment saves money, then this is good. If it means that the total capacity of an MSC is greater so that we can postpone purchasing a whole new switch, then this is even better.

Upgrading equipment offers more than cost and capacity advantages. Often new equipment offers capabilities and features not available earlier. For example, if a new speech coder is developed that provides high-quality speech in 4800 bits per second, then we can install these speech coders in the MSC, sell user terminals with them, and realize an air-interface capacity increase for those subscribers with the new cell phones. As another example, a more efficient backhaul compression scheme for the packet link could be installed at the MSC and at the busiest base stations.

Many of these upgrades are not capacity-driven but vendor-driven. Our equipment supplier makes us aware that there is a better piece of equipment that will save us money or will serve more subscribers or, perhaps, will be more reliable and easier to service. There is typically a window of upgrade opportunity between the first introduction of new equipment and the end of support for the old equipment.[1] Early in the up-

---

[1] We all have sorry stories in the software arena, where we buy upgrades to inferior versions of products simply to maintain our support relationship with a vendor. Fortunately, most telephony equipment upgrades are to genuinely better versions.

grade window, we have the advantage of getting the benefit of the new technology sooner, but we may end up being the beta-test guinea pigs for the upgrade. In the middle of the window, we are buying a more stable product but are delaying the benefits of upgrading to a later time. If we wait until the last minute, then we can rely on the experience of other customers who have purchased the upgrade, and we might be able to find training materials and resources. However, we may end up on a rush deadline to buy the upgrade because our old equipment is fading into obsolescence, and we are delaying the benefits of the improvement even further.

In the case of any upgrade, we must be sure that the upgraded equipment works with every MSC and network component we have and with our network monitoring and management tools. This testing process is ongoing as the MSC and network evolve over time. Two separate upgrades may both be good products, but they may not work with each other. It is important to perform thorough unit and integration tests for each upgrade.

We must continue to monitor the MSC for performance, and we must maintain its ability to serve subscriber traffic. This requires making sure that each component in the MSC has enough capacity to do its job and making sure that we properly configure and use the features and capabilities of new equipment and maintain it appropriately.

43.2   Assigning Base Stations to a New MSC

At some point along the MSC growth road there will not be enough capacity in one switch to do the job. We will run out of main processor capacity, transport capacity, circuit switching, speech coding, or packet switching to the point where we can no longer just add more equipment and make a bigger switch. At this point we need to add a switch.

We have to decide how to divide the base stations among the new MSCs. We had m MSCs before the addition, and we have $m + 1$ MSCs afterward, so we have to redistribute hundreds of base stations.

There are two separate but related issues in MSC assignment. One is the assignment of base stations to MSCs, and the other is the assignment of user home MSCs in the home location register (HLR). The management issues in HLR reassignment depend on the logical and physical relationships between the HLR or HLRs and the MSC switches, and this varies from implementation to implementation. In cases where MSC switches are located at different physical sites, there is an additional issue: Variations in backhaul cost become a significant component in deciding which base stations to assign to the new MSC.

We could do a global reassignment of base stations to MSCs. Such a reassignment could end up changing the MSC assignment of many base stations. This could mean new HLRs and short message service centers (SMSCs) for a huge number of user terminals depending on the logical and physical relationships between HLRs and MSCs and how closely we have assigned subscriber home service to a particular base station. At the very least, reassigning a lot of base stations is a lot of administrative work and raises the risk that most of the system might not work if something goes wrong.

When adding MSCs to a growing system, we can think of the MSC regions as giant cells, as we did in Sec. 34.5. The base stations are assigned to the MSC giant cells, and we have to make changes in that assignment when we add another MSC and another giant cell.

The size of the administrative job of global reassignment favors a simpler solution: treating the new MSC base station assignment as giant-cell splitting. We take the busiest pair of adjacent MSCs and create the new MSC region in between the two regions by taking a bite from both of them. In this way, the only base stations reassigned to a different MSC are the base stations being assigned to the new MSC.

The other argument for minimizing the base station assignment changes is that the backhaul links have to be rerouted to the new MSC. Even if the MSCs are all in the same place, there is a significant amount of reconfiguration in changing the backhaul link of a base station from one MSC to another. If the MSCs are not all in the same place, and especially if the MSCs are not all on the same transport backbone, then transport cost is a significant factor in the decision of how to assign base stations to MSCs.

If we are optimizing backhaul cost for the reassignment of base stations to MSCs, then we should formulate the decision model carefully. If we simply decide to serve each base station where it costs the least, the transition cost of moving the backhaul transport from one MSC to another will keep us using the old MSC as much as possible. We can ignore the transition costs, but this also ignores the possibility that changing one base station backhaul link may be much more expensive than changing another. We also can insist that the number of base stations on the MSC be fixed at an appropriate level and look for cost-effective base station swaps.

Let us start with a solution that has the desired number of base stations served by the new MSC. We want to consider changes in the new MSC region that keep the same number of cells but reduce the backhaul cost. A simple heuristic that will do a good job is to make incremental changes that reduce cost. We look at the border of the new MSC area and consider all the combinations of one base station in the new MSC area and one base station in an old MSC area so that exchanging the two still leaves contiguous MSC regions. We consider exchanging these two base stations and ask if the exchange reduces the backhaul cost. If it does, then we repeat the process until we find no improvement in the calculation. This generally will find good assignments, but it can make a good change in one step that prevents a better change later on. We can paint ourselves into a corner.

A more sophisticated heuristic looks further ahead in the assignment-exchange process. We can look at *all* the pairs of border cells where we can exchange their MSCs and keep the MSC regions contiguous. We calculate the backhaul cost difference of each exchange of base stations and pick the one that reduces cost the most. We repeat this process until no further improvement is found. Even this heuristic might not reach the true optimal minimum-cost solution that a full-blown mathematical program would find, but the difference in cost should be very small. As the system grows and adds base stations, it is even more unlikely that the result over time will be much different.

43.3 Dealing with New MSC Borders

The job of assigning base stations to MSCs should take into account inter-MSC border signaling for call processing and transport for handoffs. The MSC border area is a busy area. There are inter-MSC handoffs, the transport required to support these handoffs, and all their associated signaling messages for measurements and actual handoffs. In addition to handoffs, we have call registrations and setup activity in the border zones creating inter-MSC communication. The other component of inter-MSC communication is roaming traffic that has little to do with the border areas.

The MSC border areas are similar to pilot pollution for giant cells. There is no air-interface capacity loss when four or five MSC regions come together, but there will be a greater load on the signaling and handoff resources. We want to manage the MSC regions so that the giant cells have their borders in places with less activity. It is also in our interest to put those borders in places where subscribers move infrequently and vehicles move less rapidly because this reduces handoff activity.

Regular air-interface cells are at the mercy of the radio frequency (RF) environment. We can aim or tilt an antenna so that it covers a desired area, but we do not have direct control over base station coverage. MSC giant cells, on the other hand, are directly in our control in the assignment of base stations to MSCs. We can shape the MSC regions the way we want, and we can move their borders to stay away from high-density or high-mobility areas.

43.4 New MSC Placement

Some wireless service providers put their MSCs in different locations to save backhaul transport costs or to ensure diverse geographic locations for reliability. Other vendors colocate all the MSCs in one complex so that they can use the common location to get lower costs for transport, power, maintenance, and other resources. We can place the new MSC with one or more existing MSCs, or we can put it someplace new.

Transport cost is a major input factor in MSC placement. Depending on the local telephony environment, it can make a tremendous difference in cost to have the MSC in one place or another. The MSC has its entire backhaul network to all its base stations and all the transport it requires to the public networks, the PSTN and PPDN. In addition to all that transport, we can add any inter-MSC transport we need, the Signaling System 7 (SS7) signaling transport we require, and any private wireless networks connected directly to the MSC.

If there is no transport backbone, or if there is no available room on the backbone, then we have to consider the mileage costs of backhaul. This leads us to placing the MSC near the geographic center of the service region so that it is at the minimum total distance from all its base stations. Of course, the availability of cheaper transport from some base stations rather than others will tend to bias MSC placement toward the more expensive transport base stations to minimize their backhaul costs.

The transport costs to the public networks (PSTN and PPDN) are also an important factor. The backhaul links tend to be expensive because they are small pipes of a few DS-1 or E-1 links. The public network transport comes in larger quantities, DS-4 and E-4 levels, so we should get a better price per unit on PSTN and PPDN transport. However, the voice traffic to the PSTN is uncompressed 64-kbps pulse code modulation (PCM), so the total volume of voice transport is higher from the MSC to the PSTN than it is from the MSC to the base stations.[2]

If we have a very cheap transport backbone, then all the MSCs will be placed along that backbone, and the backhaul problem becomes the backbone access problem

---

[2]There is a possible exception to this. It may be possible to carry compressed signals all the way to the PSTN switching office and install our voice speech coders at the far end. If this option is available, we choose by comparing the reduced transport cost to the cost of leasing space for the speech coders and maintaining equipment at a separate site, as well as a risk calculation for the chance of equipment failure at the separate location.

described in Sec. 35.3. Also, the transport costs between the new MSC and the public networks (PSTN and PPDN) are likely to be low along the backbone. In such a case, MSC placement depends mostly on the rental cost of the physical location because telephone switches are physically large and take up a lot of floor space.

Diverse placement of MSCs can protect the wireless telephone system so that some part of it will continue to operate even after a catastrophic local event. However, it does not seem likely that one MSC will be easily able to take over the base stations of another in the event of such a failure. Thus, trying to maintain reliability through diverse locations may generate a false sense of security.

For the system to be reliable, it has to have diverse backhaul routing as well as reliable MSC equipment. We could exercise the option of having every base station connected to two geographically separate MSCs, but this is going to raise the price of service significantly and is not likely to produce any noticeable improvement for subscribers. Having one MSC with diverse backhaul routing or at least redundant equipment on the same route is going to produce an improvement in reliability that improves service to our subscribers and does so at lower cost.

Rather than figure out how to make the MSC network more reliable, it may make more sense to invest in a more reliable configuration for each MSC. Equipment can be duplicated using the models described in Sec. 4.8, or we might have the option of buying better equipment, probably more expensive equipment, that breaks down less often.

43.5   Equipment Provisioning for a New MSC

When we install a new MSC in a wireless telephone system, we need to make sure that it has enough equipment to do its job for the system. These are the MSC capacity requirements we discussed in Sec. 34.3. Once we have decided on the base station assignment for a new MSC, we should be able to calculate the equipment required.

The main MSC processor has to have enough capacity for all the call processing, SMS, other signaling, and handoffs. Since we are adding a new MSC to serve base stations already in place, we can measure the components of the MSC processor load. We know how many calls per hour these base stations are generating, so we know the call-processing activity. We know how many short messages are being sent from each base station, so we can add them up for a total SMS load. We know how many handoffs these base stations are doing, and we can use that handoff activity. Thus we have a good handle on the amount of MSC processing power we will need.

Backhaul is also easy. We can look at the current backhaul requirements of the base stations in the new MSC region. The backhaul requirement of the base stations does not change with a change in MSC, so we can add up the total backhaul for the new MSC region.

Transport to public and private networks requires more than just a total of existing resources. We have to estimate the amount of PSTN, PPDN, and private network traffic for the base stations being served by the MSC. If the service mix across the wireless telephone system is fairly uniform, then we can use the number of base stations to estimate the voice traffic for the PSTN and the data traffic for the PPDN. We should be able to calculate the private network traffic based on contracts. We can do traffic engineering exercises for voice and packet data to figure out how much transport we need.

We need to estimate the number of loopback trunks for mobile-to-mobile calls within the new MSC region. Since these calls can overflow to the PSTN, admittedly at higher cost, we can start the new MSC out with a good estimate and see if we have enough loopback trunks. We recommend taking the amount of loopback traffic on the old MSCs, figuring out the proportional share for the new MSC region, which may be smaller than the old MSC regions, and doing a traffic engineering exercise.

For inter-MSC transport, we need to figure out how much handoff activity we expect and how much mobile-to-mobile and roamer-termination traffic we expect as well. These calculations are the same as those in Sec. 37.1. The inter-MSC transport for handoffs must be adequate or else inter-MSC handoffs will fail, and the CDMA air interface will lose capacity as a result. The mobile-to-mobile and roamer-termination links can overflow to the PSTN. This may cost more money, but it will not lose system capacity.

We can take the combined voice traffic for the base stations in the new MSC area to get the requirement for how much circuit-switching equipment and speech-coding equipment to use. Similarly, we can total the packet-switching requirements of the base stations to get the requirement for the packet-data-switching capacity.

The mix of services can change over time. When we are buying equipment for a new switch, we should be prepared for our subscriber community to change its preferences. Third-generation (3G) services may increase dramatically over time, and we should be prepared for this. We also should be prepared for a system that has a steadily growing voice demand, as cellular telephone services has had for two decades.

The decision to add a new MSC to a wireless telephone system also should include a complete description of the equipment required so that MSC can serve its traffic. As we continue to expand MSCs, both old and new, we will find ourselves upgrading equipment, as we described in Sec. 43.1. A new MSC will have all new equipment, and that equipment, has to work with our existing base stations. It also has to communicate over inter-MSC links with the other MSCs that may not yet have all the latest equipment. We also must make sure that the systems that monitor and operate a new MSC are compatible with the new equipment. We have to remember that a new MSC needs electrical power and environmental support to operate and that it should be designed for enough redundancy to meet the quality-of-service (QoS) requirement for our switching systems.

43.6 Auxiliary Equipment Growth

We are using the term *auxiliary equipment* for the home location registers (HLRs), visitor location registers (VLRs), signal transfer point (STP), short message service centers (SMSCs), and voice mailbox systems. The growth issues for these machines are similar to the capacity issues in their initial configuration. We discussed those issues in Sec. 34.6. The growth of these items should be planned and managed, whether they are standalone equipment or integrated components of the MSC.

As the MSC gains subscribers, the HLR database has to grow as well. This is simply a matter of maintaining the database capacity to follow the increase in subscriber count. The HLR processor will be busier because more subscribers will have more registrations, calls, and parameter downloads. However, if the service mix moves away from voice calls and short messages to high-rate data service, then the HLR could find itself doing less work to serve fewer actual calls. On the other side of the same coin, an increasing trend toward short messages will increase HLR activity.

The VLR traffic is going to grow as the roamer community grows, so adding MSCs to a system will increase the number of roamers. Calls that were *intra*-MSC become *inter*-MSC as the system capacity grows and MSC regions become geographically smaller. A shift of subscriber service mix can change the VLR activity level just as it affects the HLR. We have to plan for both the throughput and the total database capacity of the VLR.

The STP required by SS7 handles signal traffic from the PSTN SS7 network plus short message traffic. The two STPs for each MSC and the SS7 transport links must remain below 40 percent occupancy so that the SS7 network remains fully redundant, as required by SS7. We can add STP capacity and SS7 transport to maintain enough capacity to meet this objective.

The SMSC will have to expand to meet increased demand for SMS. This may increase in proportion to other services, or it may increase more rapidly as SMS becomes more popular in the subscriber community.

Voice-mail growth is simply a matter of tracking usage. As subscribers use more voice mail, we need more voice links between the MSC and the voice-mail system, and we need more voice mailbox capacity. An increase in demand may increase the volume during voice-mail activity surges when things go wrong, as described in Sec. 34.6.5, but we do not expect system growth to increase the number of events that drive up voice-mail activity. We are not going to have more late flights because we sell more cell phones.

New services will create new demands on auxiliary equipment. Enhanced message service (EMS) sends longer messages and will use more message center storage than SMS. Multimedia message service (MMS) allows subscribers to send large graphic images from their cell phones, and this could create large demands for intermediate storage. Network-based voice commands for cell phone subscribers could increase message activity as well.

When we increase the capacity of an auxiliary box, we should keep in mind that these are computers with computer-upgrade choices. We may decide to buy bigger disks, but it may make more sense to get a new computer instead. While we are busy making sure that our base stations and MSCs are keeping up with subscriber demand, let us not forget the less glamorous auxiliary equipment that manages our calls message by message.

43.7   Conclusion

In planning MSC growth, we have a major effect on the quality of the services we offer, the reliability of our network, and the profitability of our company. Careful work does more than keep the network running well. It reduces customer complaints and loss of customers to our competitors. Careful planning and implementation also reduces the need for frequent, rushed, poorly planned, expensive upgrades followed by high maintenance costs.

44

Adding Transport

As our cellular network grows, we need to add transport capacity, both in backhaul and in the connections from the MSC to other network components and other networks. In addition, we may be able to reduce network costs by reorganizing the configuration of the backhaul network periodically.

44.1 Adding Backhaul

The normal way a telephone system grows is by adding capacity through the use of bigger pipes. As more telephone users served by a local exchange office make more calls, the telephone company adds more trunks between that local exchange office and other local exchange offices and interexchange carriers. When a business expands its local telephone network, it upgrades its private branch exchange (PBX) and adds capacity by adding trunks to the existing links to the public switched telephone network (PSTN).

Cellular growth is different because of the limiting nature of the air-interface capacity. Capacity is added to a cellular system by adding cells. This means that a wireless system grows by adding base stations, and the backhaul network grows by adding more transport pipes rather than by making the existing pipes bigger. The backhaul requirement for a base station can be no more than the maximum capacity of the air interface for all the sectors using all the code division multiple access (CDMA) carriers.

Sometimes a backhaul pipe from one base station gets bigger. When a cell gets more CDMA carriers or more sectors, its base station requires more transport to the mobile switching center (MSC), and we enlarge the backhaul network to provide this increase in transport. Once the base station reaches its maximum configuration, the growth of the backhaul network is continuing growth in the number of backhaul network nodes rather than a continuing increase in the capacity of the existing backhaul links.

We have discussed how to set up a backhaul network for a wireless telephone system. We wrote about mathematical global optimization techniques in Sec. 26.1 and then explained in Sec. 35.2 how an incremental-improvement heuristic can get a near-optimal calculation for a minimum-cost backhaul network. The heuristic has us form a list of candidate backhaul routes from each base station and has us repeatedly evaluate each base station's choices until no improvement is found. Since the backhaul network grows by adding nodes, we have to think of its cost minimization as an ongoing optimization in the backhaul planning process.

Each new cell has subscriber demand to be served, so each new base station has a backhaul requirement. The new base station is a new node in the network flow optimization. Thus we can continue the heuristic by forming a candidate list for backhaul hops from the new base station to the MSC and to all the nearby base stations within three cells. Let us examine some of the issues that make backhaul growth different from the initial cost minimization we described earlier.

Incremental growth on the existing transport links is almost certainly cheaper than forming new links because transport costs are typically concave. This means that they have decreasing incremental cost as the quantity increases. The exception to this rule is when incremental demand pushes the transport requirement past a small number of big units so that the twenty-fifth DS-0 requires another DS-1 and the twenty-ninth DS-1 may require another DS-3.[1] If these backhaul networks are continuing to grow, then we probably should ignore these exceptions. Otherwise, we may end up with a backhaul fan-out network with fully loaded DS-3 or E-3 links that is not well designed for growth. The next new base stations are going to push many of those over their limits and make us pay dearly for taking the short-term view on transport costs.

For a new base station, we can expect to add a link to a nearby old cell and then to add the new cell's backhaul requirement from the old cell to the MSC. This is usually the best way to grow the backhaul network because it involves the fewest changes and usually keeps transport costs down.

We could decide to use the new base station as an intermediate node, a hub in the backhaul network. If it saves enough transport cost, then it makes sense to change the routing from old cells so that their backhaul goes in daisy chains through the new cell. Since this involves significant costs for reconfiguration, the reduction in transport cost has to be enough to pay for the administrative costs and headaches.

As a wireless telephone system grows over time, we should evaluate the entire backhaul network periodically from a systemwide perspective. Each new cell adds a new backhaul node, and we incrementally add transport in a locally cost-effective way. However, it is possible for a sequence of good short-term decisions to add up poorly in the long run. If a system of 150 cells grew to 500 cells by adding a few cells at a time, it is possible that the backhaul network can be reconfigured to save a lot of money.

Growth and change in our own network are not the only reasons to consider a backhaul overhaul. Transport technologies are changing, and costs are coming down rapidly. Thus a 2-year-old system built on a 2-year-old backhaul plan based on 2-year-old transport prices may be ripe for a new backhaul plan based on current transport technology and prices. A new backhaul plan may be an opportunity to renegotiate some old transport lease contracts.

Of course, reconfiguring a backhaul network of 500 cells is an administrative headache and a costly operation. The transport savings has to be significant enough to justify the cost of making the change. We certainly do not recommend doing a global backhaul optimization every time a few cells are added to the system. Rather, revisiting the backhaul plan from time to time may uncover cost-saving opportunities.

---

[1]Similarly, the thirty-first E-0 requires another E-1, and the seventeenth E-1 may require another E-3.

44.2 Adding MSC-to-MSC Capacity

As MSCs serve more traffic, they communicate more with their base stations via a backhaul network, they communicate more with the outside world, and they communicate more with each other. The increase in inter-MSC communication means an increase in inter-MSC transport requirements.

The critical inter-MSC transport need is inter-MSC handoffs. Handoffs require a dedicated link at the DS-1 or E-1 level. If there is no capacity on the link, then there will be no handoffs, and the CDMA air interface will suffer. As we monitor the inter-MSC handoff links, we will see them getting full, and we will know when more capacity is required.

Mobile-to-mobile calls and roamer terminations are more complicated because they are cost-saving links rather than required transport. If the inter-MSC handoff links are completely full for half an hour each day, then our system is in trouble. However, the most economical solution for mobile-to-mobile calls may be to have significant periods during the day where we are overflowing traffic to the PSTN and paying the per-call charges. It is a cost tradeoff we are managing, the cost of dedicated transport against the cost of sending those calls back out.

As we would expect, a new MSC increases the amount of inter-MSC communication. We might not expect the reality that the increase in communication for each new MSC is *more* than the one before. Another way to think of this more-than-linear increase is to realize that *each* MSC in the system has an increasing amount of inter-MSC communication when a new MSC is added. In Sec. 37.1 we explained how this increases the inter-MSC transport requirement.

More MSC regions mean more borders, so we expect more inter-MSC handoffs. And more MSC regions mean smaller regions, so calls that would be handled within a single large MSC region are now going to be inter-MSC events.

If the MSCs in our system are all in one place, then the transport is a few meters of cable from one switch to another. However, we should not forget that even a few meters of transport cable requires terminating equipment at both ends. This terminating equipment is part of the transport cost, so even a 10-m cable run incurs some expense. In addition, the MSC has resource limitations as to how many DS-1 or E-1 links it can support, and using more of these for inter-MSC transport means that there are fewer links available for other MSC communication.

If the MSCs are not colocated but are on the a common transport backbone, then the transport overhead should remain low. We expect that most wireless system providers in developed areas will have access to a high-capacity transport backbone at the Synchronous Optical Network (SONET) level. The MSCs will be located at one or more points of presence (POPs) along the transport backbone. We must make sure that backbone transport is available. Just because the local telephone company has a lot of transport does not mean that it can lease it to a wireless service provider. All of it already may be leased or committed to other telephone transport customers.

Worldwide, wireless telephone service is becoming more popular in places without a large telephony infrastructure. This means that service providers are setting up wireless telephone systems, often as primary telephone service for subscribers, where there is no backbone transport network. This means that each MSC is probably going to be located near the geographic center of its base stations to keep its backhaul costs down. In such a case, we have to minimize inter-MSC transport cost. With only a few MSCs, it is reasonable to enumerate all the possible networks of inter-MSC transport and select the least-expensive option. Even a huge wireless telephone system with 5 MSCs has only 10 pairs of MSCs to connect.

As the wireless system grows and the MSCs grow, the inter-MSC transport must grow along with it. We should be prepared for the inter-MSC transport to grow proportionally faster than the growth of the rest of the system.

44.3   Adding MSC-to-PSTN Capacity

Voice traffic in cellular systems has been growing steadily for two decades. During the last decade of the twentieth century, cellular telephony doubled every 2 years, a 40 percent per year increase in subscriber traffic. While the rate of growth may be declining, we expect voice traffic to continue to grow.[2] This means that we expect to see a continued increase in the amount of transport required between the MSC and the PSTN.

One of the two reasons for third-generation (3G) CDMA is to provide a large increase in wireless capacity at a low cost. As 3G CDMA brings lower cost cellular service to the world market, we may see strong growth in voice traffic. This means that the MSC-to-PSTN links will have to increase to meet the growing demand.

The other reason for 3G CDMA is the introduction of wireless data and other advanced services. It may be that the growth opportunity is not in voice traffic but in these advanced service offerings. In many countries, such as Finland, cellular voice traffic has already reached almost everybody, and there may be a leveling off of subscriber voice demand in favor of other uses of wireless communications. While an increase in data service or subscriber interest in multimedia message service (MMS) is good for the wireless provider's business, these increases do not require the addition of more transport from the MSC to the PSTN.

We believe that there will be sustained growth in the voice market. Even as we read about cellular market saturation, more people are buying their first cell phones, and others are buying more than one. Developing countries are developing markets for voice, and wireless systems will see large gains in voice traffic over time as subscribers begin to rely on their cell phones as their primary telephones.

The bottom line for MSC-to-PSTN transport is that it will continue to increase as voice traffic continues to increase. Our monitor systems will generate reports on voice call activity and transport utilization, and 3G CDMA planners should continue to read these reports, no matter how boring they may seem. Voice call service may not be glamorous in the high-tech, 3G picture, but it is the bread and butter of the wireless telephony industry, it is likely to stay that way, and we should make sure that our transport facilities are supporting that voice call traffic.

44.4   Adding MSC-to-PPDN Capacity

The big unknown in the wireless market at the time we are writing this book (2002) is the size of the wireless data market. The popularity of other new services is also in question. Multimedia message service (MMS) allows subscribers to send graphic im-

---

[2]We read sad reports that cellular service may grow at a rate of *only* 10 or 20 percent per year. It is easy to forget that adding 10 percent to the world's current cellular community is an addition of 100 million subscribers.

ages at the press of a button. These services send their packets along a route from the base stations through the MSC to the public packet data network (PPDN). The size of the MSC-to-PPDN link depends on the sizes of the markets for these features.

The big selling point of 3G is data service, particularly high-rate data service. If the market develops as the 3G promoters say it will, then the packet data flow from the MSC to the PPDN is going to grow very rapidly as subscribers sign up for data service. Along with an interest in wireless data, the current interest in short message service (SMS) could migrate into a similar interest in MMS, and the packet data link could start carrying a lot of cell phone snapshot image files.

It is not at all clear that there is a big market for these features. Landline data popularity has relied on the low cost of data service, particularly the low incremental cost. The home data user typically pays a flat monthly fee and no extra money for more data activity. People pay a premium for wireless telephone service so that they can make and receive telephone calls in a lot of different places. We will have to wait to see if people have a similar desire to use data service in enough different places to pay a premium for wireless. For wireless data service to succeed, it must offer its subscribers a high enough quality of service (QoS) that they continue to use it.

If the data market develops, then we want to make sure that the MSC-to-PPDN packet pipe has enough capacity. Few things dampen subscriber enthusiasm for a new, premium service than not being able to use it. The rush can come fast as people find out their neighbors have wireless laptops and they run out to buy their own. The wireless service provider has to watch the market and be prepared to add packet capacity on short notice.

As explained in Sec. 23.3, the traffic engineering of packet data is based on what amount of delay is considered too long and how often we can tolerate delays over the limit. In Sec. 37.2.2 we went over a nested-delay approach that enables the MSC-to-PPDN packet data link to have enough speed and capacity for each category of data delay tolerance.

It is important to evaluate the mix of data services and to engineer the packet data link so that each level of delay receives adequate QoS. The same half-second hiccup that looks really lame on a real-time video link is unlikely to be noticed by a computer user. And the 15-second pause that would annoy a Web surfer is not going to bother somebody waiting for e-mail. Each level has its own requirements, and we can engineer the packet data pipe to meet those individual requirement without buying more capacity than we need. Our ability to engineer QoS on packet pipes will grow as new standards and technologies for managing QoS on packet pipes and networks are deployed.

44.5   Conclusion

We add transport to our cellular network to keep up with growth and with changes in usage of different types of service, and we modify our backhaul network to reduce cost; these are routine, mundane activities, especially in parts of the world where the PSTN provides a solid backbone. The routine nature of the work should not make us think that the work is of low value; the opposite is true. Continually maintaining quality and lowering cost is the basis for increased net revenue. This increases our network's growth potential both by providing financial resources for growth and also providing a robust network that is easier to modify with new technology.

Regional Growth in a Specific Area

In our wireless planning analysis, we have written about meeting the demand. We traffic engineered to meet the demand in Chap. 23, we located base stations to meet the demand in Chap. 32, we planned base station equipment to meet the demand in Chap. 33, and we spent the rest of Part 7 designing the rest of the system to meet the demand. In Chaps. 39 and 40 we discussed how to use system measurements and subscriber feedback to measure the demand and to calculate the mix of services so that we could design our wireless telephone system better and more cheaply to meet the demand.

The demand for our wireless telephone service is not a given fact of nature like the weather. We can exercise some control over subscriber demand for wireless service to reduce the cost of meeting that demand. We are already familiar with using different rates for different times of day as a way of encouraging off-peak usage. In this chapter we introduce the concept of choosing geographic locations for subscriber activity that reduce the cost of meeting that demand. We will discuss the kinds of places where we want the demand to be and how we can encourage our subscribers to be there when they use the system.

The goal of this chapter is to show ways to increase net revenue most effectively by getting more new subscribers who will use their cellular telephones in areas that can support increased capacity inexpensively. Doing this requires a combination of engineering capacity analysis, economic analysis, and market analysis.

There are a number of ideas that tie these disparate fields together:

- New subscribers increase revenue. However, with most voice call packages operating primarily on flat-fee usage, additional use of voice services by existing subscribers does not increase revenue. On the other hand, selling additional minutes or vertical services, such as data services, does bring revenue.

- The key question is: Can we add new subscribers who will use their cellular telephones in the areas where we can grow our networks inexpensively? People use their cellular telephones wherever they are, especially when they are away from home, so location of residence is not necessarily a good predictor of location of use.

- We can target services to the business market, the consumer market, or both, and we can shift that balance in our marketing plan.

- Marketing ourselves as a national service provider may be a barrier to marketing particular regions intensively.

- Marketing through franchises limits our flexibility in marketing by region.

We will begin by determining where to encourage demand. We will then discuss how to target growth, returning to the issues discussed here.

45.1   Determining Where to Encourage Demand

Not all subscriber demand is equally expensive to serve. We certainly can calculate the cost per unit usage of our wireless telephone system by dividing the entire cost of the system by the total amount of usage. We can use a weighting scheme so that high-rate data users are recognized as more expensive to serve than voice call subscribers. When the total cost is divided up this way and assigned to each unit of usage, we call that share the *allocated cost* of the usage. While an allocated-cost model gives a good insight into the fair share of each user of a system, it misses a very important component of system cost analysis, the concept that some usage is far more expensive to provide than other usage.

What we want to get a handle on is the extra cost of serving a particular component of the subscriber demand. We ask how much less the system would cost if we did not serve some particular usage. That cost is the *incremental cost* of subscriber activity. An incremental-cost model does a poor job of reflecting the total cost of a system, but it does an excellent job of highlighting the demand that directly drives up total cost.

Peak-hour calls are the expensive calls for a cellular provider because it is the busy-hour call volume that determines how much equipment we need. Roughly speaking, the incremental cost of each busy-hour call is the total cost of the system divided by the total number of busy-hour calls.[1] The rest of the calls at the other times have an incremental cost of zero because the system is designed to have enough capacity for the peak time.

The concepts of peak-time and off-peak usage in telephony go back to long before cellular was invented. Most cellular service in the United States has a cost structure that favors off-peak usage because the incremental cost of that usage is small.

However, we do not increase our revenue by getting users to use the off-peak minutes they have already paid for. Our goal would be to find people who would go counter to the trend and tend to use their cell phones late at night and on weekends but not much during the day. This might be hard to do.[2]

What is less obvious is that some geographic areas of service are costing the wireless service provider far more than others, and we can market to potential subscribers who will tend to use their cell phones in those areas. Just as we have high incremental costs of peak-hour calling, we have high incremental costs of peak-area calling. While we may or may not be inclined to have different rates for different areas, we can encourage growth in the areas that cost the least to serve.

---

[1]Those who have a background in economics will please forgive this oversimplification. We are ignoring the time value of money, the fact that equipment has a life cycle, or even the difference between fixed costs and continuing costs. The point we are making here is simply that the fraction of calls at the busy hour are the only calls with a significant incremental cost to the service provider.

[2]We could consider a targeted campaign that aired during *Buffy*: Vampires, don't get caught without a cell phone.

Mobile telephone service is particularly popular on major routes of travel. These would include highways and commuter rail routes. Some of these routes go through areas where it is difficult to get base stations to provide adequate coverage. Sometimes it is the lack of telephone transport or even available electrical power that makes it difficult to reach these areas. And sometimes it is that communities try to put noisy roads and rails in sound-protected areas so that their residential neighbors do not have to listen to the noise. Those sound-protected areas are often well protected from radio signals as well.

We want to spend some effort figuring out where demand is cheap to serve. Those are the areas where we want to promote subscriber demand so that we get a greater revenue stream while paying less for it. This is the kind of engineering that makes financial investors very happy.[3] The best areas to promote are those where we have excess capacity in a base station that is already operating. The next best areas are those where transport is cheap and easy to get. And the third best areas are those where we have plenty of capacity in the mobile switching center (MSC).

45.1.1 Existing base stations

While we have been writing about base stations being installed for capacity expansion, even a large and populous wireless telephone system has a lot of base stations whose purpose is coverage. Without these base stations in the system, areas that are currently green areas would have been red or yellow, as described in Sec. 40.2. Many of these cells have only a subset of the total code division multple access (CDMA) carrier set because there is not enough traffic volume to justify the entire set.

A one-CDMA-carrier base station is a prime candidate for low-cost growth. Adding a carrier to a base station is typically less expensive than adding sectors and a lot less expensive than adding new cells. And many of these one-carrier base stations are not even using the full capacity of that one CDMA carrier.

Other base stations may be less extreme cases of underutilization, but they also represent an opportunity to increase their capacity at lower cost than cell splitting. These base stations may not be using all the CDMA carriers, or they may be able to support more sectors.

We want to encourage demand in these areas. As more subscribers use wireless services in these places, the service provider earns more revenue without paying for more equipment. This increase in revenue can be an opportunity to lower rates to a more competitive level, or it can be a chance to make some profit on the enormous investment that a large CDMA system represents.

45.1.2 Existing transport backbone

Some areas have low-cost, readily available telephone transport. One of the major issues in base station placement is installing adequate backhaul facilities. If we can find base station locations where backhaul is cheap and easy, then these are good places to put base stations.

---

[3]We recall a Dilbert comic strip where the boss was enthusiastically promoting Project BIFF. What did BIFF stand for? Big Improvements For Free. Well-targeted growth is the closest we can come to Project BIFF in the wireless telephony world.

The most obvious place for transport availability is a point of presence (POP) in the public switched telephone network (PSTN). The transport backbone of the telephone company usually has enormous resources at rates far lower than transport off the backbone.

Another good opportunity for base station transport is a local cable television network. Cable networks often have high-bandwidth wires running down every street in a neighborhood and they often have bandwidth to spare at the low end of the spectrum below the very high frequency (VHF) band. We might be able to use that bandwidth for backhaul transport. While neighborhood locations may not be prime candidates for regular macrocell base stations, they may be ideal locations for microcells that cover a particular neighborhood for wireless service.

Major office complexes are another opportunity for readily available transport. Business buildings today are built with plenty of telephone transport in anticipation of high volumes of voice calls and high-speed data networks. We can put picocells in these buildings and use some of that transport for wireless services inside the building.

An example of a marketing approach that could be linked to picocells would be an effort to get a large business with its own office building or a hospital or other organization with a campus to sign an exclusive contract. We provide the company with discounted rates, cellular telephones for all qualified employees, and a picocell at their facility, and they pay for cellular telephones and service for several thousand employees. Companies that have merged recently are good candidates because they will be looking to consolidate vendors and save money and because they will have many employees traveling to other facilities within the company. Giving every employee a cellular telephone is easier than trying to track each one down through the private branch exchange (PBX).

These are areas where we want to encourage subscriber demand to grow because we can add base stations in these locations without the burden of acquiring or building adequate backhaul transport. We have discussed backhaul under the premise that we have to find a way to get transport to base stations that follow subscriber demand. It is often more cost-effective to find a way to get that demand to follow the transport opportunities we already have. However, we need to be careful when we do this. Continuing with the example of adding a picocell for a major corporation, we may find ourselves in trouble when all these employees head for home at five o'clock and get stuck in the same traffic jam at peak time on our most overburdened cell. We should not turn down the contract because of this problem, but we should not be blindsided by it either.

45.1.3 Existing MSC capacity

No matter how carefully we engineer a wireless telephone system, some of its MSCs are busier than others. More specifically, some of the MSCs are closer to reaching their capacity saturation limit, where they will run out of a critical resource. The MSCs with extra capacity are better places to add base stations because they do not require the addition of a new MSC.

Encouraging growth in the regions with extra MSC capacity makes better use of the existing MSC resources than adding subscribers in places where the MSC is near capacity limits. Even if we are buying more equipment for an existing MSC, this is still a lot less expensive than installing a whole new switch. We should identify areas where the MSC has a lot of extra capacity and encourage growth in these areas.

45.2   Targeting Growth

In the preceding section we determined where growth of subscriber traffic can be served at minimum cost. This is where we would like the system to add new subscribers. A wireless service provider is in the business of making money by selling wireless service. More subscribers buying more service should be good for business no matter where they are located. It is the job of the engineering and planning staff to ensure that the system is well engineered so that it continues to make money for its owners while it grows to meet increased demand. We certainly do not intend to turn away new subscribers because they are located in the wrong part of town.

We do have the option, however, of encouraging new business in the areas where it costs us the least to serve that new business. We target growth in the preferred areas by advertising, pricing, managing the mix of services, and maintaining a marketing presence. The result of the analysis in this chapter is a regional growth plan that targets specific areas for new subscribers based on the economic efficiency of serving them.

45.2.1   Advertising

While we like to think of ourselves as cold, rational decision makers, the decision to buy wireless service is often an impulse decision. People pass by a storefront that says "CELLULAR" or "WIRELESS" in big letters, and they decide to sign up for service.

Some people have wireless service that they are just not happy with. It has high blocking, calls are noisy and get dropped, and customer service is slow. These subscribers often tolerate poor service until something stimulates them to change to another wireless service provider. While the reasons to change may have been rational, the change itself is an impulse decision.

Local advertising targets local subscribers. We can use billboard advertising in areas where we want more subscribers to sign up. We can put more storefronts in the neighborhoods where it costs us less to add capacity. We can even run telephone sales campaigns selling service in areas where we want increased growth. Newspaper advertising allows specificity of location that television and radio do not.

We want to try to reach subscribers based on where they will use their cell phones. A neighborhood advertising campaign should be aimed at people who will make calls in the neighborhood. One approach is to market extra cellular telephones on family plans in areas where schools and local malls have room for growth. This region-based targeting should attract more local shoppers than commuters. On the other hand, if we are trying to boost demand on the commuting route, a different targeting strategy would be required.

45.2.2   Price promotions

We can do more than advertise usage where we want it to happen. We can promote usage by selling it for a lower price. It is like those gaudy commercials we have in the United States, "We pass the savings on to you."

By selling airtime for less in target areas, we encourage subscribers to use the service in those areas. However, we also present a more attractive price, encouraging new subscribers to sign up for service in those areas. In a highly competitive market for wireless service, a small price difference can stimulate a lot of growth. Where this is true, we can use price differences to control geographic usage patterns. This will not be feasible for some service providers who offer identical service nationwide. On the other hand, it is excellent for service providers who have specifically sought regional or rural licenses and have identified their companies with the area.

We are used to this kind of pricing pressure to encourage subscribers to use their cell phones during off-peak times. Since all the cellular providers are likely to have the same busy hour, there is little differentiation among them in time-of-day pricing. However, not all wireless systems have the same geographic cost issues because their cell layouts are different and their transport networks are different. Thus a mild price incentive on the part of one provider for using wireless service in targeted areas may produce significant increases in subscriber demand in those areas.

We can use a lower price to encourage usage in targeted areas. We also can use other promotions similar to how we treat off-peak usage. We can give away free minutes as part of a promotion, or we can sell minutes in targeted areas as part of a subscriber package. In designing these packages, we can seek business or residential customers by customizing the packages to offer services that meet different needs at acceptable prices. Sometimes it is possible to charge a business substantially more money per month for a service similar but not identical to residential service. For example, we might allow more daytime minutes, or we might allow people in different households to share the same plan at a surcharge. We also might be able to tie third-generation (3G) services, wireless data, and multimedia service (MMS) into the promotion package.

45.2.3   Mix of services

The capacity of a 3G CDMA system is highly dependent on the mix of services. A CDMA carrier may support hundreds of voice callers and only a few high-rate wireless data users. We would hope that higher-usage services would be able to command an appropriately higher price in a competitive market. It is not always obvious which services are using the most capacity resource.

Some services use more resources than others, but some of those resources are more critical. A bursty data user may use the same number of bits as a lower-rate, steady-stream data user, but maintaining enough spare capacity on the CDMA channel for high-speed data bursts may displace more voice calls than the steady-stream user. The higher variability of a bursty usage pattern reduces the capacity of the CDMA air interface.

Real-time video and Web surfers both use high bit rates, but the Web surfer may tolerate delays of a second or two that would be unacceptable in a real-time video call. Thus, on a busy packet channel, the low-delay video caller may be displacing more voice calls so that we can maintain the high packet rates necessary for low delay.

These high-rate services displace other revenue. How much revenue they displace depends on the mix of services being used and the quality of service (QoS) we are trying to provide. How much revenue they earn depends on the demand for these services and the competition in the market.

High-rate 3G services cost the least to provide at off-peak times, of course, and in the targeted areas. If we are going to promote 3G services, especially services that use a lot of CDMA air-interface resources, then we would like to encourage that growth where we already have capacity or where we can easily get backhaul transport.

We can control the mix of services through price promotions and product packages. In Sec. 38.2 we discussed the problem of entering a market for new services. We explained that raising consumer interest in a new product often requires selling that product very cheaply and making promises that are hard to keep in the long run. Here is a case where we can promote a new service that costs us very little in a specific region by selling it cheaply in that region. As subscribers get used to using the new services, they will want to use those same services in the more expensive areas and may become comfortable paying a premium for those areas.

The plan for targeting specific regions based on the cost of capacity should take into account the mix of services we are selling in those areas. The low cost of incremental capacity may be an opportunity that we can use to promote 3G services in the subscriber community. In marketing data services, it is particularly important to define the different types of demand and peak-demand times created by business users and consumers and the different rates each is willing to pay.

45.2.4 Marketing presence

People buy wireless service for many reasons. There are rational factors and there are impulse factors in the selection of a wireless service provider. In the 3G wireless world, not only are people deciding to buy wireless service, but they are also deciding what specific voice, data, and other services to buy.

The big-picture view of the consumer purchase decision process is vendor recognition. People feel comfortable buying service from companies whose names they know and associate with high quality.[4] Price and availability of service are major concerns in consumer selection. And advertising is often the catalyst that stimulates people into taking action and buying the service they want to buy.

Marketing presence is related to reputation not only in the sense of good or bad reputation but also in the sense of the association people make with the company. Dell Computer Company offered personal computers that were as reliable as and price competitive with IBM, Compaq, and Hewlett-Packard, but it took over a decade for Dell to gain a reputation as a provider of personal computers for businesses rather than home users. Dell succeeded in changing its reputation by establishing product lines that solved business problems, such as repair cost. Companies can tailor their marketing plans to change their image and get customers of a particular type well suited to cost-effective network growth. As the cellular market begins to shift from a growth market to a competitive market, we can expect to see this kind of image making. Service providers will succeed by developing their networks and their image synergistically.

It is important to have market presence where we want to sell our product. This applies not only to the large service area of the wireless system but also to the specific areas that we can serve more cheaply. Superbowl advertisements on television are an example of very broad-based advertising for a national market, whereas local television commercials may have coverage comparable with that of our wireless telephone system. Supermarket fliers, on the other hand, allow targeting of very specific areas. The right mix of broad and narrow promotion gives a service provider control of the growth of subscriber demand.

Since people often buy wireless service from a storefront sales office, we can position these stores where we want the growth to occur. We may be limited in managing

---

[4]We have to be a little careful because the association has to be in the same general area. Maytag has a good reputation for dishwashers but might have a hard time selling computers. Xerox had a tough time bringing its excellent reputation in photocopiers into the computer marketplace.

growth through storefronts if we sell largely through franchises rather than through our own name-brand stores. Some franchises put restrictions on targeted marketing opportunities in an effort to ensure fairness to all independent managers. As mentioned earlier, we also must be cautious about nearby areas where our network is congested. Selling lots of service to an underserved mall near several underserved new housing developments sounds excellent. However, we may discover that the freeway from the housing developments to the mall has high sound barriers and no access to electrical power.

Business marketing also can be targeted. Small businesses in areas with extra base station capacity can be offered special deals. Larger businesses in areas with extra transport capacity are prime targets for picocells in their buildings that take advantage of the capacity we have available.

A growth plan for a wireless telephone provider should take into account the preferred areas for growth and a marketing plan that generates demand in those areas. It makes sense for engineering and marketing staffs to work together to do cost estimates and growth studies so that the provider can make the most economical decision for both supply and demand of its product.

45.3   Worldwide Marketing Plans

We have focused on competitive marketing plans for nations with well-developed cellular service. However, it is appropriate to mention that the ideas discussed here are equally valid anywhere in the world. In nations where cellular use is growing rapidly and the transportation and power infrastructures are relatively poor, it is reasonable to focus on developing mobile telephony urban service first.

However, this is not the only approach. Whether through wireless local loop or through innovative marketing plans, we can explore the possibility of developing services and marketing them synergistically as long as we take economic and cultural factors into account and seek to understand our potential subscribers and their needs.

45.4   Conclusion

We have moved far away from capacity engineering, but it is crucial to realize that competitive success depends on cellular engineers being able to understand market and economic issues and present technical cost considerations in the service provider's process of defining itself in the marketplace. Much of American business revenue over the last 100 years has been based on an engineer's new idea of what is possible joining with a businessperson's sense of what customers want. We may not be in a position to invent the light bulb or the cellular telephone, but we are in a position to understand the real relationship between what our system can do and what our potential subscribers want. Closing this gap is the key to long-term success as the cellular telephony market becomes increasingly competitive.

Modeling for CDMA

Theoretical models of how code division multiple access (CDMA) subsystems should work allow us to build simulations to test ideas about our systems. A simulation is a model of a real-world system that can be used in an effort to predict what will happen in the real world or to assist with troubleshooting problems on the real-world system it models. In addition to simulations, we also can perform analysis to solve problems. Each approach is effective in different situations, as we will discuss here in Part 7, "Modeling for CDMA."

In Chap. 46, "Business-Case Models," we look at approaches to estimating the cost of a cellular system plan and ways to estimate revenue. Combining revenue estimates with cost estimates, we can generate cost-benefit analyses, return-on-investment (ROI) calculations, and other useful business analysis factors, such as revenue per customer.

In Chap. 47, "Propagation Models," we present four different types of models that help us simulate coverage in CDMA cells: distance-based models, terrain-based models, in-building models, and ray-tracing models. In Chap. 48, "Subscriber Traffic Modeling," we add in our customers, the subscribers who use cellular services. This addition allows us to create simulations of our planned and existing cellular networks in use. We can run computer-based simulations that will show us a picture of likely call usage, capacity requirements, blocking rates, and capacity and quality problems on our networks.

46

Business-Case Models

Business-case models are designed to validate investment choices. They model costs per some functional element, such as time or number of subscribers. They allow us to compare alternative systems and determine the cost per functional element for each one, permitting us to make the choice most appropriate to the forecast of the demand for services. We can then apply these cost models to our investment through return-on-investment (ROI) models by calculating the expected income in relation to the cost investment.

46.1 Fixed and Variable Costs

The cost of something sounds basic and simple, but defining what a component costs as part of a system can be very complicated. There are a lot of different contributions to the cost of a system part—what we pay up front, what we pay over time, and what else we have to pay for.

In general, we try to identify linear cost functions, how much something costs per unit. The units may have dimensions of some number of things, some amount of time, or something else. Sometimes we have to use estimation to get a linear function that reflects our cost per unit.

46.1.1 Time

Most things we buy have a one-time cost, a purchase price. When we own equipment, the purchase price is usually the largest component of the cost. Even a lease can have a substantial cost at the beginning for equipment we have to buy or for some kind of up-front service, an installation fee. This is consistent with our experience buying and leasing consumer products. If I buy a car, then I pay the dealer a lot of money, and the car is mine. I still have expenses associated with the car as I drive it, but the up-front cost is major. Even if I lease the car, however, I still have to pay some one-time fees that do not recur each month of the lease.

There is also a cost per unit time. Telephone transport is typically leased from a facilities provider, often the public switched telephone network (PSTN). We give them the A-to-B routing and the size of the pipe, and they tell us how much to pay each month if we want to lease the facilities. The payment continues as long as we have the equipment or the service. We can have leasing agreements that specify when we can stop paying, but the payment is a linear time function, a cost per time for each item or service.

This is not the same as a monthly payment on equipment we own. Even though we are paying a certain amount per month in installment payments, we own the equipment as an asset, and we carry an accounting liability on our books. The distinction between one-time and ongoing payments is not about cash-flow but about a truly different relationship, buying something or leasing it.

The ongoing cost of a piece of equipment does include operating and maintaining it. Lower power requirements or less frequent repairs can justify a higher price in the purchase decision. Equipment life expectancy also fits into the equation because the cost of more frequent replacement is also an ongoing cost of ownership. Cost components are sometimes broken down into capital expense (CAPEX) and operating expense (OPEX).

How we compare up-front and ongoing costs depends on how we value money over time. If our company has a lot of loans at 10 percent per year, then we can equate a $1000 one-time cost with a $100 per year expense. Of course, the time value of money is one of many factors affecting the comparison of one-time and ongoing costs. Other factors are how much money we have on hand, how much we can borrow, and how quickly loans have to be repaid.

The typical scenario for comparing up-front and ongoing costs is the decision to purchase equipment or to lease it. In addition to the time-value issues of money, we also have the time value of equipment and the element of risk associated with that. The obvious risk is that something will break down and leave the owner responsible, but most of our wireless telephone equipment is sophisticated enough to have service contracts and maintenance schedules. The more subtle risk is that equipment will become obsolete, that a year or two from now there will be much less expensive alternatives that provide the same or better service.

An example of an own-versus-lease decision is backhaul transport from a base station to the mobile switching center (MSC). We can pay the monthly tariff for the required number of DS-1 or E-1 links for the packet pipe, or we can purchase the equipment for a microwave link between the base station and the MSC. Of course, the cost and feasibility of the microwave link depend on the height of the base station and the terrain between the base station and the MSC. Once we determine that we can use a microwave link, and once we determine the up-front cost of the microwave equipment required, we can compare that one-time cost with the monthly tariff of leasing the DS-1 or E-1 facilities over time. However, if DS-1 or E-1 lease rates drop, we may regret our investment in the microwave link.

46.1.2 Equipment

The normal price of a piece of equipment is the price per unit item. We expect each amplifier to cost some amount, and we expect 20 amplifiers to cost twice as much as 10. While evaluating the cost per unit of a piece of equipment, we should consider how modular it is. A machine that does twice as much for less than twice the price may not be a bargain if the smaller unit already meets our needs.[1] A wireless telephone system

---

[1]In the consumer arena, a shameful example of this kind of marketing is the selling of a 20-minute long distance call for one dollar with no discount for shorter duration. Since most calls are much shorter than 20 minutes, most calls have a far higher unit price per minute than the advertised 5 cents.

with hundreds of base stations may not offer the option of concentrating its equipment into larger and more economical units.

Buying equipment often has a fixed cost as well. If we are buying amplifiers from one vendor and decide to buy amplifiers from another vendor, then we may have to buy a whole new suite of test equipment and train our repair staff on new procedures. These costs are fixed costs of owning *any* amount of a product, no matter how little. To make purchase decisions more complicated, the fixed cost of owning a line of equipment often depends on whether or not we already own equipment from a particular vendor.

We do not have to count every piece of equipment separately in a cost model. If every antenna requires a cable and mounting equipment, then all of these can be lumped together in estimating cost. Whatever need made us buy an antenna is going to make us buy all the associated equipment as well.

Another issue in evaluating equipment cost is the cost of multiple subcomponents. A base stations is divided into sectors, and a sector may have several code division multiple access (CDMA) carriers. Depending on the level of detail we need, we can aggregate some of the subcomponents

For example, an MSC handles 200 base stations, each base stations has three sectors, and each sector has up to five carriers. When we are estimating the cost of a particular layout and deciding among several plans, we consider each of these separate components and calculate the total cost of each plan.

If we are calculating the cost of the system at a high level, however, then we may add up all the components of an entire cell and call that the base station cost. We can say that each cell has three sectors, the average sector has four of the five carriers, and the per-cell cost of the MSC is $\frac{1}{200}$ of the total MSC cost. As we shall see in Sec. 46.2, this level of aggregation gives us cost estimates per subscriber for financial planning, but it does not give us the level of detail needed to compare two system plans.

Equipment has variable costs as well. Most equipment has time component of cost, a monthly cost of operation and maintenance. Of course, each piece of equipment also has a life cycle, time period after which we expect to replace it. A $3000 piece of equipment with a 6-year life has a $500 per year cost associated with it for its periodic replacement.

And some equipment has a cost of usage. An antenna does not cost any more money because we are using it for more wireless telephone calls, but an amplifier can use more electricity and run up a power bill. While few components of a base station have a usage-sensitive life expectancy, user terminals can wear out faster with increased use. Even if the electronics last forever, the buttons on the keypad wear out with use.

46.1.3 Proportional-cost model

Another way we can simplify cost calculations is to distribute costs of shared equipment proportionally. For example, if all the cells have three sectors, then we can take the per-cell costs of a base station controller, tower, antennas, power, and building and count one-third of that as a per-sector cost. This is a convenient way to combine costs for a simpler calculation.

Let us take this example one step further. If the average sector has four CDMA carriers, then each carrier on a sector inherits one-quarter of the per-sector cost. If all the carriers on one sector can share an antenna and power amplifier, then the per-carrier cost includes one-quarter of the antenna and power amplifier cost.

Each layer in the equipment hierarchy inherits the proportional cost of shared equipment at higher levels. Since a base station supports three sectors and a sector supports an average of four carriers, we can assign each carrier one-third of the per-sector cost and one-twelfth of the per-cell cost because each cell in this example has 12 carriers.

We can extend this further up the equipment hierarchy to include the MSC cost as well. If each MSC can serve 200 cells, then the per-cell cost includes $\frac{1}{200}$ of the cost of an MSC. If we carry this down to the sector level, then the per-cell cost of an MSC is $\frac{1}{600}$ of the cost of the MSC.

And we can extend this further down the equipment hierarchy to the individual channel level. If each cdmaOne carrier has 24 channel cards installed, then each of these cards inherits $\frac{1}{24}$ of the per-carrier cost, which includes a share of the per-sector, per-cell, and per-MSC costs.

This is a good method for calculating average costs as long as the decision we are making with it does not involve changing the ratios. If we are considering adding three-sector cells to a system, then dividing the per-cell cost by three for each sector makes sense. If, however, we are considering going from three to six sectors for some of the cells, then we have to separate the per-cell and per-sector costs.

46.1.4   The postage-stamp model

Sometimes the proportional-cost model does not have enough detail. A base station requires an MSC so that it can operate, but up to 200 base stations can share the same MSC. If we deploy base station number 201 in our example, then we are going to buy another MSC for the system. Thus, if the cost of one MSC is M, then we would like to use a proportional-cost model and to say that the MSC cost *per cell* is $M/200$. However, there is an offset factor, a fixed cost of MSCs, that is analogous to mailing letters.

When you mail a letter in most countries, there is some charge for the first unit of weight, and then there is a subsequent charge for each unit above the first unit. Let us assume a rate of 30 cents for the first ounce and 20 cents for each ounce beyond that.[2] We can graph the postage charge as a function of weight as in Fig. 46.1. We pay the full cost of each ounce even when we use only a fraction of that ounce.

Sometimes we can use a linear approximation of the postage function, as shown in Fig. 46.2. The diagonal line is a linear representation that smoothes out the step function of the postage rate. The linear representation charges us 20 cents to mail a letter and 20 cents per ounce. If we have a collection of letters all the same weight, then we use the exact step function to calculate the postage for them. However, if we have a collection of letters of varying weight, then we can estimate the cost by weighing the entire pile and using 20 cents per letter plus 20 cents per ounce as a good estimate of the total postage cost.

If one MSC serves 200 base stations, then we can use a similar approximation for the MSC cost of a growing system. The MSC cost step function is M for the first 200 cells and M for each 200 cells after that. The linear approximation is $M/2$ plus $M/200$ for each cell. In fact, the actual cost of an MSC increases with the number of cells it is serving, so the MSC cost function has steps that are rising up to the maximum of 200 cells. Figure 46.3 shows this step function and its linear approximation.

This linear approximation to a step function is only useful when we are using multiple steps. If we are planning a system that is never going to grow beyond a single MSC,

---

[2]The United States Postal Service still uses ounces for weight, where 1 ounce is 28.3 grams. The example is easier to follow with round numbers of 30 and 20 than with the actual rates of 37 and 23 cents in use as we are writing this book.

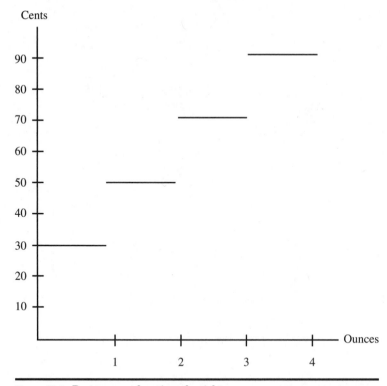

Figure 46.1 Postage as a function of weight.

then it does not make sense to use a linear approximation that takes multiple MSCs into account. The linear function that makes sense for a single-MSC wireless telephone system is the first step of the function in Fig. 46.3 with a higher fixed cost and lower per-cell cost.

Another example is at a more detailed level of base station design. Suppose that a base station manufacturer has a design in which a channel backplane in a base station supports 32 channel cards and has a cost of B. The base station has to have one channel backplane, and it requires another one at 33 channel cards and another one after that at 65 channel cards. We have a step function of channel backplane cost for a base station. If base stations vary considerably in the number of channel cards, then it makes sense to use the approximation that the channel backplane cost is a per-cell cost of $B/2$ and a per-channel-card cost of $B/32$.

We can use the postage model and its linear approximation when there is a step-function cost that we want to estimate over several steps.

46.2 Cost-per-Subscriber Models

We can put all this together to form a per-subscriber-cost model for a wireless telephone system. This model is generic. The different CDMA air interfaces have different capacities, and vendors vary in their equipment architectures. Thus the most we can hope to

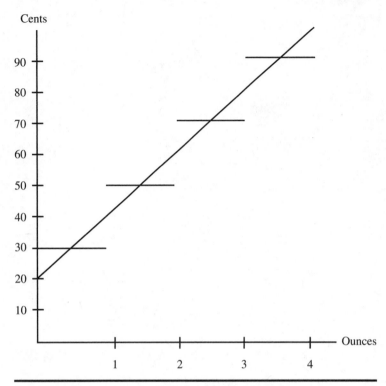

Figure 46.2 Linear approximation of postage function.

accomplish here is to provide an illustrative example of the cost-calculation process. In this example we are calculating the cost per subscriber so that our financial backers can feel comfortable with the money they are investing in our wireless telephone system.

Let us start by considering a cdmaOne system with five carriers, as summarized in Table 46.1. All costs are in U.S. dollars. We will go through each level of the hierarchy. A full-blown cost study would include operations, administration, and maintenance (OA&M), along with a more complete and detailed list of equipment components at each level.

We start with the subscriber who has a user terminal. The cdmaOne user terminal is a mobile telephone that costs about $300.[3] Each subscriber generates 0.02 erlang of peak-time demand. In other words, the busy-hour load is 0.02 times the number of subscribers.

Each equipped channel has a $200 channel card plugged into a $1600 channel backplane that has enough slots for 32 channel cards. In our example, each DS-1 of backhaul transport can carry 100 voice calls. To our investors, the average tariff rate $200 per month for a backhaul DS-1 is worth $40,000 in up-front money. We are using the postage-stamp model described earlier to add 1/32 of a channel backplane and 1/100 of backhaul DS-1 to the cost of a channel card. These are marked with (P) in Table 46.1.

---

[3]The subscriber probably sees a much lower price for the mobile telephone. This is so because much of the cost of the user terminal is built into the monthly rate.

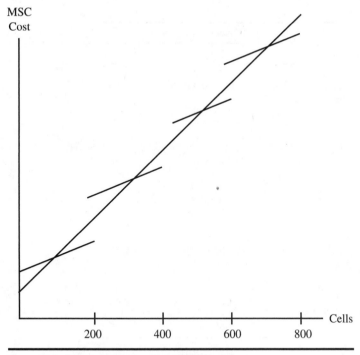

MSC
Cost

Cells

200   400   600   800

Figure 46.3 Linear approximation of MSC cost.

We are going to equip each carrier of each sector with 24 voice channels. With an average load on neighboring cells and sectors, the cdmaOne carrier supports about 21 users, but we will keep three extra channels to carry more traffic when neighbors are more lightly loaded. We are engineering our system for a busy-hour blocking rate of between 2 and 5 percent, high enough to serve 19 erlangs of busy-hour demand. This means that each equipped channel can serve 40 subscribers and that each cdmaOne carrier with 24 equipped channels can serve 960 subscribers.

Each equipped carrier has a low-power combiner that costs $3600. Our base station design allows us to use one big power amplifier and one set of antennas for all the carriers serving one sector. The low-power combiner adds up all the modulated signals from the channel cards and sends the combined signal to the power amplifier for the forward direction.

Our wireless telephone system has five cdmaOne carriers, but not all cells have all five carriers. In our example, the average number of carriers per sector is four. Thus each sector serves four carriers at 960 subscribers per carrier, which works out to 3840 subscribers per sector.

Each equipped sector has the full array of radio equipment. The three antennas, one transmit and two receive, are $3000 each; there are $2000 worth of cables and connectors; and the power amplifier is $13,000. Each sector also gets charged for half a channel backplane using the postage-stamp model because each sector has to have a channel backplane for each 32 channel cards. The half a backplane is marked with (P) in Table 46.1.

TABLE 46.1   CDMA Component Costs

Item	Ratio	Cost	Per Subscriber
		Cost per subscriber summary	
Per-subscriber costs			
User terminal	50 subscribers/erlang	$ 300.00	$300.00
Per-channel costs	0.8 erlangs/channel		
Channel card		200.00	5.00
1/32 channel backplane (P)		1600.00	1.25
1/100 backhaul DS-1 (P)		40000.00	10.00
Per-carrier costs	24 channels/carrier		
Low-power combiners		3600.00	3.75
Per-sector costs	4 carriers/sector		
Three antennas		9000.00	2.34
Cable and connectors		2000.00	0.52
Power amplifier		13000.00	3.39
1/2 channel backplane (P)		1600.00	0.21
Per-cell costs	3 sectors/cell		
Power supply		25000.00	2.17
Base station controller		50000.00	4.34
Building and climate control		120000.00	10.42
Tower		40000.00	3.47
1/2 backhaul DS-1 (P)		10000.00	0.43
Per-MSC costs	200 cells/MSC		
Maximum-configuration MSC		5000000.00	2.17

In our example, all the cells have three sectors. Thus each cell serves three sectors at 3840 subscribers per sector, for a total of 11,520 subscribers per cell.

The per-cell costs are the big-ticket items: the $25,000 power supply, the $50,000 base station controller, $120,000 for a building and climate control, and $40,000 for the radio tower. We also have the postage-stamp component, marked with a (P) in Table 46.1, for the backhaul DS-1 that every cell requires.

Our MSC costs $5,000,000 for a switch with enough capacity for 200 cells. In this example, the MSC is modular enough that the fixed-cost component is insignificant, so an MSC equipped for 100 cells is half the price, $2,500,000.

The grand total of this analysis comes to $349.46 per subscriber, 86 percent of which is the user terminal. We can consider just the system cost of $49.46 per subscriber and realize that this cost-per-subscriber estimate depends heavily on the assumption that each subscriber uses 0.02 erlang during the busy hour. If our subscribers are significantly more or less active than this, then the system cost per subscriber will change in proportion to their busy-hour usage.

In this example we dealt only with the fixed cost of equipment. We aggregated the monthly costs of backhaul transport and ignored the other repeating costs, such as OA&M and equipment replacement. In a more complete cost analysis, we would consider the initial expense for equipment and the continuing costs as separate items.

In the third-generation (3G) environment there is a large selection of services, and there is a cost per subscriber for each of them. We would like a cost per subscriber for each service. As described in Sec. 45.1, we can compute an allocated cost, the fair share of system cost per subscriber, or an incremental cost, the extra system cost for each subscriber added. The incremental cost is very important for marketing decisions about what services to promote.

46.3 Return-on-Investment (ROI) Models

Cost per subscriber, even with a time analysis, is only half the money equation. We want to make sure that our system is earning its keep, and this means that we need to analyze the relationship between revenue and cost. In particular, we want to ensure that the revenue is more than the cost.

Investors in any business venture, and wireless telephony is no exception, want a positive return on their investment. In particular, investors have a time horizon during which they want to see that positive return. The time window for a positive return on invested capital depends, in general, on the risks involved in the investment and, in particular, on the specific industry.

Let us consider the preceding example, where our investment in the wireless telephone service is $349.46 per subscriber. This includes the user terminal, channels, carriers, sectors, cells, and switches.

Let us suppose that the subscribers in this system are paying an average rate of $20 per month. For this discussion, we really do not care if they are paying a flat $20 per month or they are paying $10 per month and another $10 for busy-hour air-time charges and other services.

If we take a simple ratio of system cost and monthly revenue per subscriber, then we get $349.46 divided by $20 per month, and we see a payback period of 17 and a half months. If investors in our example want to see their money back in a year and a half, then this study shows that we will make it with almost two weeks to spare.

We know that equipment has continuing expenses associated with it even after it is paid for. For this example, let us assume that the ongoing cost of the equipment is 10 percent of its purchase price per year. For the user terminals, the ongoing expense consists of mobile telephones breaking down or being lost under the provider's warranty. For the system equipment, the ongoing expense is regular maintenance. Adding 10 percent per year, $2.85 per subscriber month, to the cost side of the equation pushes the breakeven point to 20 months and 11 days.

Another cost on the balance sheet is the cost of growth. As subscriber demand increases, a wireless service provider typically will add equipment to satisfy that demand. This creates more revenue than not serving the demand, but it does mean spending more money to expand the capacity of the system.

Since the cost of a wireless telephone system has a large up-front component with continuing expenses after that, we are counting on the revenue stream to catch up with the cost. Investors are counting on having their up-front cost paid back not too long after the system is running. Once the payback period has passed, the revenue stream continues, and the costs, we hope, continue to be significantly less. This leaves the wireless telephone system operating with a profitable revenue stream.

It is not enough for a system simply to repay the up-front cost. It has to earn a sufficient rate of return; it has to pay interest to its investors. The wireless system payback has to be enough to make it a worthwhile investment.

The measure of success of a financial investment is its return on investment (ROI), the interest rate it pays. We select a time window, calculate the costs and the revenue, and determine the rate of return the wireless telephone system earns during that time. We should compare the ROI of our system with other financial opportunities because we can be sure that our investors will be making such comparisons.

Once the initial capital outlay has been made, there is typically a period of time where equipment is new and requires less repair and replacement than it does in the long term. Initial low repair costs make the payback period shorter, a good thing for investor confidence.

As demand for wireless service grows, we want to grow our system to serve the demand. This is good for business in the future because we are configuring the wireless telephone system to earn the most revenue from the subscriber demand opportunity. Paying for a growing system, however, makes the paypack period longer and makes the ROI figures less favorable for short-term gain. If we continue to grow our system, then we continue to spend money on new equipment. The new equipment eventually will earn revenue, but that revenue is used to pay for even more new equipment as growth continues.

In the early days of Advanced Mobile Phone System (AMPS) cellular, service providers were caught by surprise by the astonishing growth rate of cellular and had growing concerns about the continuously increasing costs, whose revenue was continuously being consumed to provide for yet more growth.

Wireless service providers faced with growth in subscriber demand have to make a decision about how to manage their costs and revenues. Paying for growth without realizing a profit is tough on investors. On the other hand, making a profit-taking decision to spend less on equipment and not to satisfy all the demand rubs engineers the wrong way. However, deciding how to manage costs during extended periods of rapid growth is a good problem to have.

46.4 Conclusion

As the market for CDMA services continues to evolve, business models will need to be adapted. Subscriber growth is slowing, and the increased churn and the ups and downs of a competitive market need to be taken into account. Multiple service offerings compete for the same air-interface resources. Each offers a different rate of return and has a different and as yet unknown popularity with subscribers. Business and engineering teams are most likely to succeed if they cooperate closely with one another to determine the best investment of available funds in the relative growth of different service offerings.

47

Propagation Models

Modeling propagation allows us to plan locations and design of our base stations for better cell coverage with less interference. The basic approaches to modeling and implementing code division multiple access (CDMA) cells contain some interesting contrasts from methods used for conventional-reuse systems. The problems that arise are not the same. There are different models of radio propagation, each requiring different inputs. The best model for our system depends on a number of factors, such as the type of terrain over which we are providing coverage and the size of the cells.

47.1 CDMA Issues and Challenges

As students of CDMA, we can scoff at the trials and toils of those who manage conventional-reuse systems with their intricate channel-set-reuse paradigms and the complexities involved with multiple cell sizes. While they pore over their "most likely interferer" maps trying to track down a problem in North Lake caused by too strong a signal across town in Springfield, we sit quietly knowing that our $N = 1$ CDMA system has no channel assignment issues and balances power second by second.

As students of capacity, we might be smug about the coverage problems of initial system configurations. It is the ability to add subscribers to an existing system at minimum cost that excites the growth planner. Getting the wireless telephone radio signal to the millionaires in Sweetwood Canyon is something they better do right in early deployment because those people are the corporate executives who decide which wireless carriers are getting the big contracts.

However, before we get too comfortable, as students of CDMA capacity, we have a few issues and challenges of our own. The conventional-reuse channel sets with their distant cochannel interferers protect us from the immediate threat of interference from neighboring cells and sectors. In a CDMA call, about half the interference comes from same-sector calls, and about half comes from immediate neighboring sectors and cells. We are being a bit loose here because "about half" could be anywhere from one-quarter to three-quarters, but the important point is that little interference is coming from anywhere *else*. A large area of nearly equal radio signal path gain to four or more cell sectors is a CDMA capacity nightmare. Thus our conventional-reuse friends can relax while we worry about the intricacies of the radio environment at cell boundaries that concern them far less.

In conventional reuse, capacity planning is an issue for mature systems with a lot of cells. During the early stages, there are plenty of radio channels to spread around the few cells being constructed. The distinction between coverage and capacity engineering is not so clear-cut in CDMA. Weak signal paths in CDMA lead not always to poor calls but often to low capacity. The link budget of a CDMA carrier that could carry 60 calls under ideal circumstances can be used up by a single call with low enough path gain. Since poor coverage is a capacity-limiting phenomenon, even the early stage of CDMA system development is a capacity engineering phase.

We also would point out that those people who engineer conventional reuse and those people who design initial system layouts come from the same community of wireless telephony engineers as long-term CDMA capacity planners. We should not be too surprised if "those people" turned out to be *us*.

What is our point? That the automatic radio reuse and automatic power control of CDMA may make our lives easier in some important ways, but we still had better learn about radio propagation models and how to use them if we want our CDMA systems to carry the quantity of calls the financial planners promised the investors who ultimately pay us.

47.2 Variation in Radio Propagation

The cellular principle depends on radio path gain decreasing considerably as a cellular telephone gets further away from a base station. This is true for the most part, but real-world radio has some surprises.

Terrain has a dramatic effect on radio propagation. Higher-frequency radio waves are less likely to find their way around hilltops into the valleys beyond. And increasing demand creates increasing pressure to move wireless telephony into higher frequencies. The 900-MHz band was bold in 1977, whereas 1800 MHz is routine for cellular service in 2002. Cellular engineers, who had to pay some attention to orography[1] in planning Advanced Mobile Phone System (AMPS) networks, are faced with more serious engineering challenges in hilly terrain and at higher frequencies.

The less obvious impairment is the texture of the ground. Pavement is highly reflective, whereas shrubbery is not. Big leafy trees not only impair microwave transmission, but their impairment is seasonal.

Weather is less of an issue for us wireless telephone folks than it is for higher-frequency microwave engineers. The rain factor at 30 GHz is serious, several decibels at a typical cell radius of 5 km (3 miles), but down at 2 or 3 GHz, we can save our worry for other issues.

Buildings create three kinds of radio degradation:

- They create radio shadows when the base station and user terminal are on opposite sides.

- Their vertical reflective surfaces create alternative radio paths and add to the multipath propagation problem.

- Subscribers *inside* a building have walls, windows, wiring, and other impairments, and they still want to use their wireless telephones.

---

[1] *Orography* is the fancy scientific term for hills and mountains.

The urban environment is hostile to wireless telephony, again in three major ways:

- Urban structures create regions of considerably lower path gain (more path loss from more obstructions) than is typical in more rural areas.

- The urban environment subjects the radio link to more variation and more *rapid* variation.

- The multipath component of the radio path is greater because of building surfaces, broad, smooth streets, and metal vehicles on those streets.

Regions of low path gain are an obvious concern. In conventional reuse, low path gain means lousy call quality ("Hello? Hello? Are you still there?"). The CDMA power-control algorithm spreads the ill-effects of a weak radio path onto the calls for an entire sector. The signal weakens while thermal noise and interference from other sources stay the same, power-up messages are sent, and the transmit power increases in both directions. This creates more interference to other CDMA calls, and their power levels have to increase, too. Let us assure readers that this is not a runaway power increase; the system soon reaches a new equilibrium with higher power levels all around. Of course, if the path impairment is large enough, then there may not be enough power to compensate, and some calls can be lost.

The increase in variability is more subtle. In conventional reuse and in spread spectrum, we *rely* on the radio signal path decreasing as distance increases. The greater the "random" component of path gain, the less we can rely on distance to discriminate between near and far signals.[2] This means not only that are cell boundaries fuzzy but also that cells can even have noncontiguous components, little pockets of isolated real estate in their service areas.

The CDMA equations in Eq. (28.17) tell us that the amount of radio resource consumed by a radio link is 1.0 for the serving cell sector plus the path gain ratios to other sectors. Highly variable and irregular radio paths contribute to the off-diagonal cross terms in these equations and contribute to other-cell-sector interference.

The variation in urban radio paths is not just greater but more rapid. A suburban subscriber walking in a shopping mall does not have the rapid signal fluctuations of a city subscriber coming around the corner and out from behind a building's radio shadow. This keeps both power-control and handoff processes busy, and the service provider has to leave larger performance margins for the same overall performance level.

With hundreds of power adjustments each second, it is tempting to think of CDMA power control as some kind of iron control of received power, but this is not the case. It takes a few tenths of a second for power control to compensate for a large signal path change. CDMA service providers have to leave some margin for out-of-balance received power levels from imperfect power control, and areas with larger and more rapid variation require greater margins.

The urban environment also has many smooth surfaces, paved roads, and building faces, creating more radio paths with less attenuation. As we discussed in Sec. 8.4, the

---

[2]These extra components are not truly random at all, but they are independent of distance. This makes them "random" from this point of view. Just as one call's signal is another call's interference, one mathematical model's subject matter is another model's random variation. A distance-propagation model sees urban effects as "random."

CDMA rake filter mitigates much of the multipath damage, but we are considerably better off with only one radio path than with multiple paths and a rake filter, even if the rake filter is doing a perfect job. When the delay-spread picture changes rapidly, the rake filter has to work hard to keep up, and multipath takes a greater toll.

Calls inside office buildings aggravate all these factors. Walls and windows create rapid changes in the radio path, both in intensity and in multiplicity. Inadequate facilities based on optimistic estimates lead to high blocking, lost calls, and poor voice quality. The mad scramble to repair the damage by adding sectors and cells leads to blame, finger pointing, and recrimination between the engineers and the financial planners. A realistic plan will take these factors into account so that there are few surprises later on. Good tools help us form a realistic plan.

47.3   Distance Models

Back in Sec. 1.7 we discussed the inverse fourth power propagation rule. Radio in free space decreases in power with the square (second power) of distance, but radio on the surface of the earth decreases at a faster rate than this.[3] Based on measurements made around 1973, Bell Telephone Laboratories estimated an exponent of 3.84, which we feel can be rounded off safely to the fourth power in Eq. (47.1):

$$p = \frac{\alpha}{r^4} \tag{47.1}$$

This was for one frequency, 900 MHz, in one city, Philadelphia. This formula is adequate for some cellular planning, but the next level of sophistication is specific to frequency and environment.

47.3.1   The Hata-Okumura model

The next level of sophistication in distance models is the Hata-Okumura model, derived from measurements in Tokyo. These estimates are a closer fit to measured data than the simpler inverse fourth power rule.

The Hata-Okumura model divides the radio environment into "large city," "medium-small city," "suburban," and "open."[4]

The basic Hata-Okumura formula is Eq. (47.2):

$$L_0 = 69.55 + 26.16 \log f - 13.82 \log h_b - a(h_m) + (44.9 - 6.55 \log h_b) \log r \tag{47.2}$$

---

[3]This is good news for cellular radio because the decrease of radio power with distance is what discriminates between the nearby signal cell and the more distant interfering cells.

[4]It has been pointed out that the building types in Tokyo are different enough from those in the United States that the model is less valid here. Our suburban areas are somewhere between Okumura's suburban and open areas, and Okumura's suburbs are like our metropolitan areas with large groups of row houses.

where L = path loss in decibels
  f = carrier frequency in megahertz
  h_b = base station height in meters
  h_m = mobile telephone height in meters
  r = distance between them in kilometers

All the logarithms are base 10. L_0 is the radio path loss in decibels from the base station to the mobile telephone. In the large-city environment,[5]

$$a_{\text{large}}\,(h_m) = 8.29(\log 1.54 h_m)^2 - 1.1 \qquad \text{for } f \leq 200 \text{ MHz}$$

$$a_{\text{large}}\,(h_m) = 3.2(\log 11.75 h_m)^2 - 4.97 \qquad \text{for } f \geq 400 \text{ MHz}$$

and elsewhere

$$a(h_m) = (1.1 \log f - 0.7)h_m - (1.56 \log f - 0.8)$$

There are also correction terms for suburban and open environments. The Hata-Okumura equations for four environments are shown in Eq. (47.3):

$$L_{\text{large city}} = L_0$$

$$L_{\text{medium-small city}} = L_0$$

$$L_{\text{suburban}} = L_0 + 2[\log(f/28)]^2 - 5.4 \qquad (47.3)$$

$$L_{\text{open}} = L_0 + 4.78(\log f)^2 - 18.33 \log f + 40.94$$

- The frequency range for f is 150 to 1500 MHz.

- The base station height h_b is from 30 to 200 m.

- The mobile telephone height h_m is 1 to 10 m.

- The distance r is 1 to 20 km.

These ranges are neither guarantees of validity nor absolute boundaries. The model will be valid to the extent that your neighborhood looks like the parts of Tokyo where these measurements were taken. And these predictions are not instantly useless at 1501 MHz. Using this model at 1800 MHz, for example, will be less valid than using it at 900 MHz, but it may be better than a simple inverse fourth law rule.

This is a statistical-fit model. The various terms and coefficients chosen here do not represent components of radio signal path reality. We tend to prefer models where each term represents some physical aspect or relationship, but the Hata-Okumura model is not such a formula. It represents a good mathematical estimate of a great deal of diligent measurement effort and should be taken and used in this light. For example, we would be pleasantly surprised if the term $\log^2 (f/28)$ represents some component of the radio signal path process. It is simply a term in a statistical formula fit.

---

[5]The astute reader will notice that there is a gap in the large-city model from 200 to 400 MHz. This is not a typographic error; rather, this model "does not know what to do" around 300 MHz in large cities.

47.3.2 The Walfisch-Ikegami model

The next level of radio signal path model sophistication is the Walfisch-Ikegami model, often called the *COST 231 model*.[6] This model takes into account several factors not included in the Hata-Okumura model.

The path loss is broken up into three components:

L_f = the free-space loss in decibels

L_{rts} = the rooftop-to-street diffraction and scatter loss in decibels

L_{ms} = multiscreen loss in decibels.

The average path loss in decibels L is the total in Eq. (47.4):

$$L = L_f + L_{rts} + L_{ms} \qquad (47.4)$$

or $\qquad\qquad L = L_f$ when $L_{rts} + L_{ms} \le 0$

The free-space loss is the easiest, the inverse-square rule with a frequency component:

$$L_f = 32.4 + 20 \log r + 20 \log f \qquad (47.5)$$

where r = distance in kilometers
$\quad f$ = carrier frequency in megahertz

This equation has a physical reality associated with it. The $20 \log r$ term is the inverse square of distance, the $20 \log f$ is an inverse-square rule in terms of frequency, and the 32.4 is the constant term that scales the result. In free space, we already know that twice as far means four times the path loss, and this formula tells us that twice the frequency also means four times the path loss.

The rooftop-to-street diffraction component L_{rts} adds the mobile telephone environment to the radio path loss, as shown in Fig. 47.1.

$$L_{rts} = -16.9 - 10 \log W + 10 \log f + 20 \log \Delta h_m + L_0 \qquad (47.6)$$

where W = street width in meters
$\quad \Delta h_m = h_r - h_m$
$\quad h_m$ = mobile telephone height in meters
$\quad h_r$ = rooftop height in meters

The L_0 term describes the effect of the radio path orientation relative to the street, as shown in Fig. 47.2.

---

[6]COST is the European Cooperation in the field of Scientific and Technical Research (COoperation européenne dans le domaine de la recherche Scientifique et Technique), which has several committees. Their Committee 231 finished its work in land mobile radio communications on April 4, 1996. The Walfisch-Ikegami model is the result of one of their studies.

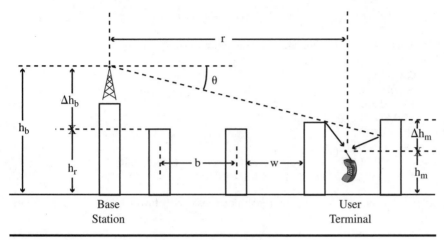

Figure 47.1 Side view of environment for Walfisch-Ikegami model.

$$L_0 = -9.646 \qquad\qquad 0 \le \phi \le 35$$

$$L_0 = 2.5 + 0.0075(\phi - 35) \qquad 35 \le \phi \le 55 \qquad\qquad (47.7)$$

$$L_0 = 4 - 0.0014(\phi - 55) \qquad 55 \le \phi \le 90$$

where ϕ is the incident angle in degrees relative to the street.

The multiscreen loss adds the base station and path environments to the total radio path loss, as shown in Fig. 47.1.

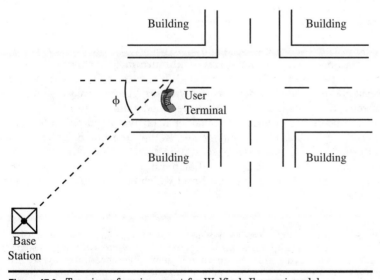

Figure 47.2 Top view of environment for Walfisch-Ikegami model.

$$L_{ms} = L_{bsh} + k_a + k_d \log r + k_f \log f - 9 \log b \qquad (47.8)$$

where b = distance between buildings along radio path

$L_{bsh} = -18 \log(1 + \Delta h_b),\ h_b > h_r$
$L_{bsh} = 0,\ h_b \le h_r$
$k_a = 54,\ h_b > h_r$
$k_a = 54 - 0.8h_b,\ r \ge 0.5,\ h_b \le h_r$
$k_a = 54.0 - 1.6\Delta h_b r,\ r < 0.5,\ h_b \le h_r$
$k_d = 18,\ h_b < h_r$
$k_d = 18 - (15\Delta h_b/\Delta h_m),\ h_b \ge h_r$
$k_f = 4 + 0.7(f/925 - 1)$, midsized city, moderate tree density
$k_f = 4 + 1.5(f/925 - 1)$, metropolitan center

Note that two terms, L_{bsh} and k_a, increase path loss with lower base station antenna heights.

- The frequency range for f is 800 to 2000 MHz.

- The base station height h_b is from 4 to 50 m.

- The mobile telephone height h_m is from 1 to 3 m.

- The distance r is 0.02 to 5 km.

As with the Hata-Okumura model, these ranges are not absolute barriers, nor are they guarantees of accuracy. If one does not have actual measurements of all the input parameters, then the following default values can be used:

- Base station height h_b, 20 to 50 m

- Street width W, half the base station height

- Street angle ϕ, 90 degrees

- Roof height h_r, 3 m per floor plus 3 m if the roof is pitched

The Walfisch-Ikegami model takes rooftop heights, street width, and street orientation into account, factors not included in the Hata-Okumura model. This makes it a more detailed model but not necessarily a *better* model. If the extra input parameters are available, then the more detailed model should give more precise forecasts and allow the engineer to see differences that the less detailed model might miss.[7]

As an example of the modeling differences, consider a fairly spread out suburban neighborhood compared with a concentrated city center. In the suburban neighborhood, with winding streets, low houses, and high base station towers, the Hata-Okumura model might do just fine. In city center, on the other hand, it could become very important which radio paths go right down the road and which ones are at a 45-degree angle to the block pattern.

---

[7]The difference between accuracy and precision becomes important when discussing mathematical models. Let us summarize the difference this way: A model is *accurate* when it gets something close to the right answer, and a model is *precise* when it shows differences correctly. We are relying on model accuracy when we want absolute radio path gain, and we are relying on precision when we explore the effect of environmental changes on the signal path.

The Hata-Okumura model tends to predict lower path losses than the Walfisch-Ikegami model. There are several parameters in Walfisch-Ikegami that are not in Hata-Okumura, and making different assumptions about these parameters affects the relative values of the two models. Comparing two models with such different levels of input-data detail is tricky for precisely this reason.

The measurements made in developing the Hata-Okumura model were made under conditions in and around Tokyo that are, in some sense, internal assumptions in the model. The Walfisch-Ikegami model specifies these parameters explicitly. Thus we can think of the distribution of these conditions as the parameter assumptions for street orientations and rooftop heights in the Hata-Okumura model.

The good news for us CDMA growth planners is that the absolute path gain levels are less important than the relative levels. It sounds cavalier to wave away what might be a 15-dB difference between two propagation models, but most of the CDMA capacity issues involve small cells where user terminals have high path gains (low path losses) between themselves and their base stations. As a wireless telephone system grows denser, the cells get smaller, the serving radius gets smaller, and the path gains get larger. The CDMA equations in Sec. 28.1 make it clear that the relative path gains between user terminals and base stations determine the capacity of a CDMA system. It is good for CDMA capacity when each serving area has one cell sector with much higher path gain than the others, and it is bad when there are large areas of nearly equal radio path gains. The absolute levels of these path gains become less important as the system grows.

47.4 Terrain-Based Models

The Hata-Okumura and Walfisch-Ikegami models are excellent propagation models that depend on the distance from the user terminal to the base station. The next level of refinement is to add terrain data so that we can form path-elevation contours to predict radio propagation.

We divide the service area into small regions called *bins*. These are laid out in a rectangular (x, y) grid. The terrain database starts with an elevation, typically height above sea level, for each (x, y) bin. Some terrain databases have more detail, including the vegetation and building density (what the weather radar reports sometimes call "ground clutter"), in addition to the altitude. Once we have the bins laid out, we establish the locations of the base stations we are studying. We need not only their (x, y) coordinates on the grid but also their heights, so we have three-dimensional (x, y, z) coordinates for each base station antenna set. The combination of (x, y) bins and base station locations is shown in Fig. 47.3.

The terrain database contains a lot of information. Smaller bins give greater precision in propagation studies, but a smaller bin size means a lot more bin data. Consider a service area that is 100 km by 100 km, a typical U.S. urban area and its suburbs. If we use a fairly course bin size of 200 m, then we have to keep track of 250,000 elevation values. This is not a large database by computer memory standards, but it can be a large amount of elevation data to obtain. If we move to a finer grid of 50 m, then we have 4,000,000 bins.

Elevation data are comparatively easy to get in the United States, where we have the U.S. National Geological Survey (USGS). The USGS has a complete digitized record of elevation data for the entire United States. Worldwide, terrain elevation data can be harder to get. Some governments are concerned about letting private companies have

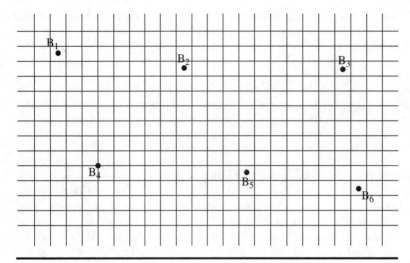

Figure 47.3 Dividing the service area into (x, y) bins.

access to their (x, y) altitude data for reasons of national security. For those countries, we sometimes have to resort to satellite imagery to construct a three-dimensional picture of a wireless telephone service area.

For each (x, y) bin, we generate a path to each base station, as shown in Fig. 47.4. Each of these paths represents a radio path between a base station and a bin. Once we understand the radio conditions from every base station to every bin, we can begin to analyze the propagation of the system as a whole, including its terrain environment. Each bin represents a possible place for a user terminal to be on a wireless telephone call.

Each of these terrain profiles is constructed using the elevations of the bins between the base station location and the (x, y) bin being studied. An example of a radio path

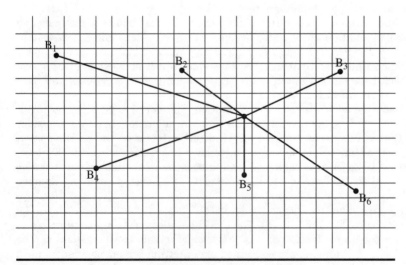

Figure 47.4 Each bin has a radio path to each base station.

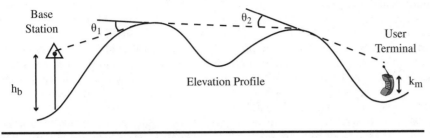

Figure 47.5 Radio path over terrain.

over a terrain profile is shown in Fig. 47.5. Each bin represents a piece of the service area where a user terminal could be engaged in a wireless telephone call. Thus we have to choose a realistic height for each bin as an end point of a call radio path.

Once we form the geometric radio path over hills and through valleys, we need to calculate the radio signal path gain between the base station and a user terminal in the bin. We treat the hilltops as knife-edge diffraction points and use rules of propagation to determine the signal path gain. The loss at each hilltop is based on radio frequency and the angle of the path change over the hilltop. If we have information about vegetation or density of buildings, then that information also goes into the propagation model.

The terrain-based propagation calculation is done for each base station. Sector plans come later because all the antenna faces are within a few meters of each other. If the antennas faces are mounted at different heights for some reason, then their propagation would have to be calculated separately.

The result of a terrain-based propagation model is a calculation for each base station for each bin. The full-blown propagation database can be very large if every bin has a path gain value stored for every base station. Storage space can be saved by realizing that there is a range beyond which the radio contribution of a base station is insignificant.

47.5 In-Building Models

The terrain-based models described earlier work well for open space but are not helpful when an entire subscriber community is entirely within one building in one (x, y) bin. Adding the third dimension of height to create (x, y, z) bins is not going to address the radio propagation issues of a single, large office building. When considering one or more picocells in a building, we need to model the radio propagation within the building structure.[8]

For each area we want to serve, we can do a direct ray calculation from the picocell antenna to the user terminal location. We draw the straight-line radio path and calculate the attenuation of whatever walls and floors are between the picocell and user terminal.

Normal office building floors start at 10 dB of attenuation for floors of concrete and ceilings of acoustic tile with pipes above. If there are steel panels between the ceiling and the floor, then the attenuation may be much greater. The choice of office furniture makes a difference in the radio path gain between floors of an office building.

---

[8]We are going to give a brief overview of in-building and ray-tracing propagation models. For a much more involved description of these models and other radio propagation issues, we direct the reader to Bertoni (2000).

Attenuation of a radio signal through a wall depends on the radio frequency, the materials used in the wall, and the furniture on either side. The signal loss through a fabric cubical barrier is small, and even a concrete interior wall only adds about −2 dB in path gain at 900 MHz. These same concrete walls lose 10 to 15 dB at 6 GHz. If the walls and floors are made with steel reinforcing rods spaced 20 cm or more apart, then the effect on microwave radio is small. If the builders used corrugated steel pans that were left in place as casting molds for the concrete, then the attenuation is much greater, and there is a significant reflected signal.

The next level of sophistication in radio path modeling is to go beyond the direct point-to-point path to a ray-tracing model that takes into account multiple paths between the two end points.

47.6   Ray-Tracing Models

A ray-tracing model uses a database of all the relevant surfaces between and around the radio path. Not only must we know where the surfaces are, we also must know their radio characteristics at the frequencies we are considering. For microcells in a business district with large buildings, a two-dimensional ray-tracing model may give excellent attenuation and multipath predictions. For a picocell inside the building with its subscriber community, we may need a three-dimensional model to get the level of detail required to ensure adequate wireless telephone service.

The ray-tracing program calculates all the paths it can find from source A to target B. Radio waves can do three things at a change of medium:

- They can *refract* as they go from the old medium into the new medium and change direction according to Snell's law.[9]

- They can *reflect* as they bounce back into the old medium.

- They can *diffract* in a new radiation pattern from the medium boundary.

The ray-tracing program analyzes all the paths that contribute to the total signal.

We only need to do this calculation in one direction. The radio paths from A to B are the same as the radio paths from B to A.[10] All the refractions, reflections, and diffractions are going to be the same coming and going. The difference in frequency between the forward and reverse directions is too small to matter in a ray-tracing computation.

This is a seriously large computation to do. Recent computer animation has produced astonishingly realistic images in movies using ray-tracing models, and the computation effort has been enormous. A single wall or floor can have multiple combinations of reflection and refraction at the two sides where concrete meets air. Radio waves can bounce around corners in many ways, and all of these have to be calculated and added up.

---

[9]Snell's law says that the angle of incidence θ equals the angle of reflection θ_R and that the angle of transmission θ_T satisfies the equation $\sin \theta = \sqrt{\epsilon} \sin \theta_T$, where ϵ is the change in the speed of transmission called the *index of refraction*. The index of refraction can depend on the frequency of the radio or light wave.

[10]The fading characteristic in a wireless system may be different in the forward and reverse directions. This is so because the same collection of radio paths may have phase differences that combine at one frequency and cancel at another. However, the combination of multiple paths in the two directions is the same.

To conduct a full three-dimensional ray-tracing study of an office building requires a huge database of every floor and wall. To do it right requires having all the internal structures of the building with their three-dimensional coordinates. This includes elevator shafts, metal reinforcements for concrete walls and floors, pipes, and electrical fittings.

Because of the complexity of the models and the vast data requirements for accurate prediction of radio propagation, we only use ray tracing for critical evaluations. Microcells used in areas with a lot of tall buildings are a good candidate for ray tracing. Picocells delivering cdma2000 1x EV-DO service require high radio path gain with minimum multipath to achieve very high data rates. It may be worthwhile to do a full ray-tracing analysis to ensure that these radio conditions will exist so that the promised data rates can be delivered.

47.7   The Statistical View

The important trend in these models is increasing precision from increasing information. More sophisticated models are not necessarily *better;* rather, they resolve differences that simpler models do not. We prefer to think of these models as tools for different purposes rather than a list of models rank-ordered from inferior to superior.

The statistical view starts with an average path gain for the community of subscribers in a cell sector, the horizontal line in Fig. 47.6.[11] The distribution of path gains in decibels has some standard deviation, 11 dB in our Advanced Mobile Phone System (AMPS) measurements.

---

[11]We may loosely refer to the average path gain for the cell sector, the path gain for a subscriber, or the path gain for a location. In each case, we mean the radio signal path gain (not including any antenna gain) between user terminals and their serving base station antenna faces.

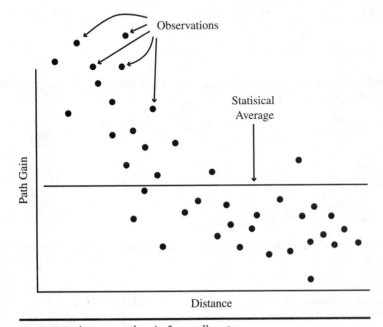

Figure 47.6   Average path gain for a cell sector.

Playing a little fast and loose with statistics, let us express that signal spread as a variance budget. Variance is the square of the standard deviation, so, statistically speaking, halving the standard deviation reduces our variance fourfold. Thus our initial variance budget for a log-normal standard deviation of 11 dB is 121 dB^2.

Now, the object of increasing model precision is to reduce the variance yet unexplained by the model. We think of this as adding knowledge about the signal path. The most obvious thing we can know about the radio path is its length. In general, the further it goes, the weaker it gets. The best statistical fit to measured signal data is our inverse-fourth-power rule shown in Fig. 47.7. The *residual* standard deviation of path gain, once distance is removed, is about 8 dB. We can tell ourselves that about half the variability of wireless radio path gain is "explained" by distance, leaving about 64 dB^2.

We can increase our knowledge about the path by finding out the terrain path (see Fig. 47.5) and using that information in a more refined model. In the vehicular world, this removed about half the remaining variability, so the standard deviation came down to about 6 dB, with variance about 36 dB^2.

Knowing about streets and rooftops might bring the variability down to 4 or 5 dB (about 20 dB^2) with tools such as the Walfisch-Ikegami model. The more we know about the base-station-to-user-terminal signal path, the more precisely we can estimate its path gain.

The fundamental limit of this process is not zero. Rayleigh fading varies the radio path on a microscopic scale not within the scope of these models. This Rayleigh variation, while not log-normal, has a standard deviation of about 3 dB, so there is a variance 10 dB^2 in a narrowband channel that propagation models will not explain. Spread-spectrum systems have less variation due to fading, depending on the delay spread, as discussed in Sec. 2.1.4.

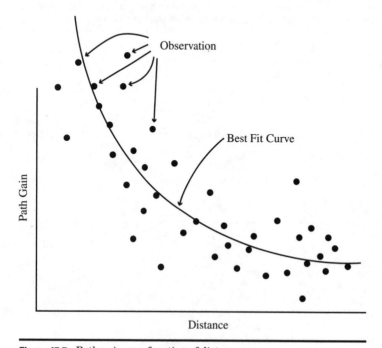

Figure 47.7   Path gain as a function of distance.

The portable telephones that dominate today's wireless market have more serious issues and more sophisticated solutions. The variance budget with subscribers going from parking basements to high-rise office buildings is considerably higher than the AMPS 121 dB^2. To help us, we have in-building models that reduce this higher variance and get better predictions of path gain.

As we move from macrocells and microcells to picocells and in-building cells, even these models are inadequate, and we need to build a more detailed model of the wireless subscriber environment. We believe that urban ray-tracing models that sound like science fiction today will be used in a few years to plan high-density wireless service within dense high-rise workplaces.

47.8   The Signal Matrix

A good summary of sector-to-sector interference is the *signal matrix*. In conventional reuse, it expresses the relationship between the serving signal and potential cochannel interferers. In CDMA it expresses the interference from nearby sectors that degrades CDMA capacity. The signal matrix is typically computed from propagation models, but it can be derived from system measurements.

We put all the server groups across the top of the page and all the base station antenna faces along the left side to form a matrix structure.[12]

In Table 47.1 we show a simple signal matrix for three cells, A, B, and C, in a row, as shown in Fig. 47.8. The signal path gain from each base station to its own subscribers is -100 dB, on average. The direct neighbors enjoy about 15 dB less path gain, and the further reach from A to C has about 25 dB less than the serving signal. The columns represent communities of user terminals, so the first column is three sets of signal paths from three cells to subscribers being *served* by cell A. The rows represent the antenna faces, so the first row is the three sets of signal paths from all the user terminals to cell A.

Another way of looking at this signal matrix is the columns represent the forward link and the rows represent the reverse link. That the A column has higher interfering path gain than the A row suggests that cell A has a tougher time on its forward link than on its reverse link. The forward direction has 1 dB more interference from cell B and 2 dB more from cell C. The asymmetry in the directions could be from elevation differences or different obstructions, but it appears in the signal matrix in any case.

---

[12]This distinction between server group and serving face is not important in CDMA, where a server group is completely defined by an antenna face, but in conventional reuse with overlaid cells, we may serve a large cell and a smaller overlaid cell with the same physical cell sector antenna.

TABLE 47.1   Signal Matrix for Three Cells

	A	B	C
A	-101 dB	-115 dB	-126 dB
B	-114 dB	-100 dB	-116 dB
C	-124 dB	-117 dB	-99 dB

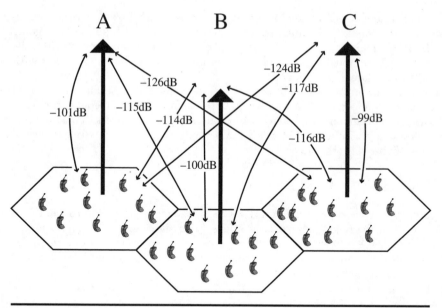

Figure 47.8 Three cells in a row.

The average path gain values shown here are just part of the signal matrix picture. We certainly want to know the log-normal standard deviations of the signal and inter-fering path distributions in the complete signal matrix picture.

The next level of refinement of the signal matrix would be to consider correlation fac-tors between radio paths. The path gain from A subscribers to cell A tends to be lower when the path from A subscribers to cell B tends to be higher. Path gains of neighbor-ing cells tend to be negatively correlated because subscribers close to the neighbor tend to be further from their own serving cells. While this is statistically present, it is not terribly important in conventional reuse. It may be more important in CDMA, where neighboring cells are interfering cells.

The signal matrix tells an important story. In the compact form of a numerical ma-trix, it tells the cellular engineer the source and approximate amount of interference from other cell sectors. Where these other-cell interferers are strong, the CDMA ca-pacity will be reduced for two reasons. First, there will be more interfering signal from the interfering cell's subscriber traffic. Second, there will be more soft handoffs going on when signal paths are close to equal. A soft handoff is essentially two radio links from two cells for one call.

The signal matrix describes the radio environment the way the CDMA equations in Sec. 28.1.2 described the CDMA radio interference. The rows and columns of the two matrices are similar in structure. We can think of the signal matrix as the foundation environment that the CDMA equations are built on.

47.9 Radio Propagation Maps

The result of a propagation study is typically a collection of visual maps telling cellu-lar engineers what is going on in their systems. These maps can tell us about existing systems, or they can predict what a system is going to look like before we build it.

The most basic map shows the most likely server for each (x, y) location in the service area. Each base station is assigned a color, and the blobs of color on the map correspond to the regions served. We picture these maps as having regular-shaped coverage areas with smooth boundaries between them. When the most likely server map has a fuzzy boundary between two base stations or, worse, areas specked with different colors because the path-prediction algorithms come up with different base stations from bin to bin, the cellular engineer should be looking closely at that area to determine if the system will provide adequate CDMA service.

A radio path gain map shows strong and weak coverage areas. If we use green for good coverage, yellow for marginal coverage, and red for no coverage as we did in Sec. 40.2, then we can see a big green circle around each base station with some yellow areas between base stations. If there are significant red areas, then we have areas where the system might provide no service at all.

Our final example of CDMA planning maps is the pilot-pollution map showing areas of nearly equal radio signal paths from multiple base stations and antenna faces. These are areas where CDMA coverage may be adequate but capacity is compromised because multiple servers are competing for the CDMA signal and interfering with each other. We will have more to say about pilot pollution maps in Sec. 48.5.

47.10 Conclusion

Radio propagation models use four basic approaches:

- Distance and general environment
- Detailed terrain models
- In-building analysis
- Ray tracing

In the order presented in this chapter, the models grow more precise, providing a greater level of detail, but do not necessarily grow more accurate. Assumptions inherent in the models may not match our terrain or equipment in important ways. Nonetheless, propagation calculations are of great value. We can build or purchase computer systems that apply these tools to our wireless networks. They work both as planning guides, as well as troubleshooting tools. Once networks are operational, it is best to supplement these tools with a program of regular measurement. Mathematical models combined with actual measurements give us the best solutions.

Having an analysis of radio coverage for our network is useful, but it is not enough. We also need to know where the demand is. Which of the locations on the map have lots of cellular users? We turn our attention to this now in Chap. 48.

48

Subscriber Traffic Modeling

Wireless is an unusual industry not only because our customers are mobile while they are using our services but also because we can serve them without knowing quite where they are. We can identify their location within the system, that is, the cell (or cells) that is serving each subscriber's user terminal. However, at least until recently, we have not identified the exact physical location of each subscriber over the length of each call. Yet, to design and grow good service, we need to know where our customers are. We once again turn to statistical modeling, applying a combination of common sense and knowledge of our market to picture what is happening and what is likely to happen.

48.1 Traffic Tally Calculations

One reason for adding subscribers to our network model is to ensure that we have enough capacity for current use and to meet our growth forecasts. We do this by entering the peak demand we have (or expect) to our model. We should remember that peak-demand times differ for different locations (such as commuter routes, offices, and shopping malls). We can vary the expected growth of demand based on forecasting from different marketing plans, coordinating cellular growth (described in Chaps. 31 and 45) with our simulations. We also can run the models with average or aggregate usage data to calculate the financial return on the investment in growth in specific parts of our network. We can work to answer such questions as

- Which of several buildouts will generate the most return on investment?

- What do we need to do to create hot spots to serve localized demand for wireless data?

- What will it cost to build an indoor picocell for a customer who wants a bid on providing that service?

As we integrate subscribers into the model of our cellular network, we become able to address both business and engineering questions more completely.

Once we have an adequate picture of radio propagation, we can move from there to subscriber traffic modeling. It is not enough to identify geographic areas of poor radio conditions, we also must identify congested areas so that we can plan for system growth. We also want to use the traffic data to decide which areas of radio problems to work on. It makes sense to put the most effort into the areas with the most revenue opportunity.

This means that our (x,y) bin database needs more information; it needs the subscriber demand for that location. In the second-generation (2G) cellular environment, this could be a single number for the busy-hour demand in erlangs per square kilometer. As we move into the third-generation (3G) world, we want to know the demand across a mix of services. Some areas will be busy voice calling regions, some will have heavy wireless data service traffic, and others will serve large volumes of multimedia message service (MMS) messages. Different services have different radio requirements and create different kinds of interference for other users.

We start with demand estimates and forecasts. If we have wireless service in place, then we can augment the forecasts with measured traffic levels to estimate subscriber demand. We should remember that measured traffic levels can be truncated by system capacity and that we may need to do some reverse traffic engineering to figure out what the demand would have been to create the traffic and occupancy observed. Once we have the demand by geographic area, we divide it as well as we can into (x, y) bins for our computer models to use.

This division into bins may be as simple as dividing the total demand by the total area and assigning an equal share to each bin. We recommend using a little more local knowledge than this. We know where the roads and railroads are for voice commuter traffic, we know where the parks and shopping malls are for recreational users, and we know where the major office complexes are for business users. This gives us a far more refined picture of traffic distribution. This greater level of detail will be more revealing when we consider splitting cells.

We use the propagation models to find out which bins are in each sector and then add up the total demand for each sector. This may be done on a service-by-service basis or for the total code division multiple access (CDMA) air-interface resources. Once we have the total demand for each cell sector, we can look at how the sector is equipped, how many CDMA carriers it has, and estimate its blocking rate. If the blocking rate is too high, then we can add carriers, add sectors, or even split cells, as described in Chaps. 41 and 42.

It is important to remember that 3G CDMA systems serve a variety of services. It may not be enough that the total CDMA air interface has enough capacity for the total demand. There may be specific requirements for one service that are not met by a total capacity analysis. A simple example is a wireless laptop user buying a premium wireless data service that guarantees fast access to the Internet. Even if there are more than enough bits available on the CDMA carrier, there may not be enough at a given time to meet this subscriber's quality-of-service (QoS) requirement.

48.2   CDMA-Specific Factors

CDMA has its own specific issues with both radio propagation and subscriber demand that we can deal with in predictive models. The more we can analyze these issues in the planning stages with computer models, the less we have to deal with them in the field in a working system.

In conventional reuse (FDMA and TDMA), the same-channel interference is coming from a cochannel cell several cells away. In CDMA, the neighboring cells are using the same carrier. This means that the planning model has to be able to add up the interference from neighboring cells. These cells are close enough to require summing their interference on a bin-by-bin basis. These sums are reflected in the CDMA equations we saw in Sec. 28.1.

Not only do we have to consider interference from neighboring cells, we have to consider interference from neighboring sectors on the same cell. Because the interferers in these cases are so close, we have to consider each bin separately rather than treat the entire interfering cell or sector as a single source of radio power, as we could do in conventional reuse.

The blocking-rate calculation in CDMA is also different from conventional reuse. The CDMA air interface works best when each user terminal is served by the antenna face with the highest radio path gain, so there is no need for complex traffic overflow models. However, CDMA has soft blocking (described in Sec. 30.6), where the capacity of each CDMA carrier depends on the utilization of its neighbors. This is a kind of load sharing that changes the traffic engineering calculations in favor of using higher occupancy for the same blocking rate.

A single server with 20 channels can support 13.2 erlangs of traffic at 2 percent blocking. However, a cdmaOne system, limited to 20 channels per carrier by intercell and intersector interference, can support 15.6 erlangs of traffic per carrier at the same 2 percent blocking. If a single carrier had all its neighboring cells and sectors continuously serving 20 users, then it would be limited to 13.2 erlangs for 2 percent blocking. However, having an average load of 15.6 erlangs on all the cell sectors, the system is essentially sharing the CDMA air interface. A CDMA traffic-planning tool must take soft blocking into account.

Another important CDMA traffic issue is pilot pollution, where nearly equal radio paths degrade CDMA capacity, as described in Sec. 32.1. The CDMA planning tool must calculate the loss of capacity from having multiple cell sectors competing with each other while sharing a CDMA carrier.

Time division duplex (TDD), as used in time division synchronized CDMA (TD-SCDMA), is far more complicated than the frequency division duplex (FDD) used in other CDMA standards. In an FDD system, the interfering paths in the forward direction are from other antenna faces to the user terminal, and the interfering paths in the reverse direction are from other user terminals to the antenna face. There are no interfering radio paths from one base station to another base station or from one user terminal to another user terminal.

In a TDD system, there can be interference from one base station to another base station. One antenna face may be transmitting in its forward direction, whereas another antenna face is receiving in its reverse direction. This adds additional paths to the interference calculations, but the total number of base-station-to-base-station paths is not very great.

The interesting part of TDD interference analysis is when one user terminal is transmitting in its reverse direction while another is receiving in its forward direction. Now we have a radio environment where any user terminal can interfere with any other user terminal in a different cell. If the sectors of each cell are not synchronized, then interference may be possible between user terminals in different sectors of the same cell. This is an enormous increase in the amount of calculation required because now any (x, y) bin can interfere with any other (x, y) bin.

48.3 Simulation Models

Computer simulations are powerful tools for diagnosing system problems. We break a system down into basic functions and reproduce those functions in a computer program. The simulation technology allows us two liberties we cannot use in a real setting. We can monitor internal aspects in a computer program that would be hard to measure in real life, and we can control the environment of a simulated system and do experiments that would not be practical, or even possible, on a real system.

We start with a collection of base stations. Each base station has a physical location, an (x, y) location, and a tower height. A base station has a collection of antenna faces for sectors, and each antenna face has its own radiation pattern and antenna gain. Each antenna face has a collection of CDMA carriers using whatever air-interface technology we are simulating in our system.

The physical environment is represented by (x, y) bins covering the entire service area. For each of these bins we have a radio path gain that comes from terrain-driven propagation calculations. If there is a subscriber community that makes calls from a significantly higher or lower elevation, such as an elevated highway, the upper floors of an office complex, or an underground shopping mall, then we may need another set of bins with another set of radio gains for the other elevation.

The subscriber demand is represented by a call volume for each bin, some number of erlangs. Each different service in the system has its own demand profile, its own erlang count, for each bin.

Now we have a representation of the entire system in our computer memory. We have cells, sectors, terrain, propagation, and subscriber demand for each service. The simulation program is our way of putting the system into motion.

Simulated subscribers start their calls in proportion to the erlang counts of their bins. These calls are created by random numbers generated by the computer program. Each service type has its own distribution of call duration. Once calls appear in the system, we simulate their access and paging messages, and once the calls are active, we simulate the forward and reverse CDMA channel, including power control.

As the call continues, the user terminal can move, and it may change radio conditions by doing so. This leads to power changes in the CDMA channel that may lead to handoffs, which may be soft, softer, semisoft, or hard as needed. The messages associated with handoffs are simulated, and under marginal radio conditions, these messages can be lost. Calls can be lost because of poor signal quality, or finally, they can end because their callers have nothing left to say; that is, they end normally.

What we can do in the simulation environment is keep detailed statistics of what happens to each call. We do not need to wait for a subscriber to tell us there was some static on the line because our simulation will detect lower performance for that period of time. We know when and where every moment of substandard call quality is occurring. More important, we know the particular interferers that are causing lower call quality or other impairment each time it occurs.

Computer simulation is a fantastic tool for investigating system problems. Periods of high blocking, services with reduced call quality, or regions with a lot of lost calls can be simulated and diagnosed. Not only can we determine the root cause of a system problem, but we also can test solution strategies on the computer before we install equipment in the actual system.

Simulations are generally not a good way to ensure overall QoS in a wireless telephone system. When a system problem is bad enough to be noticed, then a simulation can explore its cause, but telling the difference between good enough and not quite good enough is a hard thing to do with a random process. The finer the line between good enough and not good enough, the longer it takes to see the difference.

Let's take a simple example from statistics. A coin is fair if it is equally likely to flip heads or tails. If the coin is actually fair, then it takes about 200 tosses to have a 95 percent probability that the fraction of heads will be between 40 and 60 percent. To get this 95 percent confidence interval down to a range of 48 to 52 percent, it takes 20,000 tosses. Once we observe the fraction of heads and tails, we have to apply statistical confidence bands to the observation because we do not know the true probability, so the confidence interval on the fairness of the coin is about twice as wide as the probability range of its performance. This means that in 19 times out of 20 trials, 20,000 tosses of a truly fair coin is enough to give us 95 percent confidence that the coin is within 4 percent of fair.

Consider simulating a 500-cell system engineered for 2 percent blocking during the busy hour, and let us suppose that the system actually meets the universal 2 percent blocking objective. We want to simulate the system to verify that the blocking is less than 3 percent, a comfortable margin over 2 percent.

Suppose that we have a system with universal 2 percent blocking, and we want to simulate it for a long enough period of time to be 99 confident that the blocking rate is less than 3 percent. This means that each base station has to experience thousands of calls for that base station to give us that 99 percent confidence. By the time each base station has accumulated a few thousand calls, the simulation has performed millions of calls. And this still only gives us 99 percent assurance on a cell-by-cell basis, for an expected 495 of 500 cells. If we want comfort that all 500 cells are delivering adequate blocking, then we would have to let the simulation run a lot longer.

Analytical tools are better for this kind of analysis, calculation of small probabilities at the tail of the probability distribution curves. We can do the mathematical computations at the tail of the curve without dragging the computer through simulation of millions of perfectly fine calls.

48.4 Analytical Models

An analytical performance tool for a wireless telephone system looks directly at the statistical distributions themselves rather than sampling them. We can use the computer to integrate the small numbers to come up with precise estimates based on the propagation and traffic data.

Let us go back to our coin-flipping example and ask about the probability of flipping a fair coin and getting 10 heads in a row. We could flip the coin hundreds of thousands of times to get a pretty good estimate, or we could simply do the arithmetic of taking one-half to the tenth power. The probability of 10 heads in a row is $\frac{1}{1024}$, and it is a lot easier to do the analysis than to simulate all those events with all those small probabilities.

An analytical tool uses the same propagation and demand data for each bin, but instead of simulating calls, it adds up the statistical properties of each bin for each base station. We can develop mathematical calculations to integrate the CDMA equations over all the bins in the system for all the base stations and for all the services. The result is a precise estimate of blocked calls and poor radio performance for each bin and for each service.

This approach is particularly advantageous for the low-density areas of the system. If our aim is global coverage of an entire geographic service area, then an analytical tool can do its calculations for every bin, including bins where the subscriber traffic level is very low.

A simulation is a better tool for finding out message rates, handoff rates, and gross event counts. It deals directly with the dynamics of subscriber activity, placing calls, moving from bin to bin, and ending calls. The analytical approach is better for finding small probabilities and ensuring QoS over a geographic area and not just for a population of subscribers.

48.5   Pilot Delta Maps

For CDMA capacity planning, the CDMA equations tell us that more capacity depends on less pilot pollution, fewer areas with multiple sectors of nearly equal radio signal path gain. We need a map that makes it visibly clear where there are areas of high pilot pollution and significant traffic.

Areas of pilot pollution are only a capacity problem when there are subscribers using the system in those areas. Thus we want to know where the worst combination of high interference and high traffic occurs.

We use our propagation models to calculate the radio signal path gain from each base station to each bin. We can use the antenna radiation patterns to figure out the path gain from each antenna face to each bin. Then, for each bin, we can rank-order the antenna faces in order: first, second, third, and fourth. We look at the ratio, or the difference in decibels, between the first and fourth best path gains for each bin. This difference is the fourth-pilot delta described in Sec. 32.1.

The areas where the fourth-pilot delta is low are areas of pilot pollution, areas where CDMA capacity is compromised. These are the areas we want to see geographically on a map so that we can figure out what to do about them. The pilot delta map shows these areas superimposed on a layout of the CDMA system.

When these areas are also high-traffic-density areas, we have a loss of system capacity as a result. This is why we need a map that shows areas with both pilot pollution and heavy call volume. It is these areas that require attention if we want to get the most capacity from our wireless telephone system. It is these areas that we want highlighted on the pilot delta map.

48.6   Measurement-Based Decisions

A CDMA system analysis starts with predictive tools. The computer programs give us a preview of system performance and a diagnosis of system problems. However, the ongoing operation of a CDMA system should incorporate measurements into the engineering process.

The predictions of a computer program are no match for observing the system first-hand with measurement terminals, as described in Chap. 39. An ongoing measurement program, in concert with computer prediction, fine-tunes the mathematical models and generates a clear and correct picture of the system.

We can improve the predictive process by using measurement results to refine the predictive databases. The easiest example is subscriber traffic levels. The system reports its usage, and we can use these reports to adjust the (x, y) bin demands. If a cell sector is consistently serving more traffic than the computer models predict, then we should increase the bin demands in that sector. It may be worth some effort to look at a map and estimate where the extra traffic is located. If there is a major highway, a large shopping area, or a big business complex, then maybe bins in that area should be getting their subscriber demand values boosted.

The next step is to use measured radio signal levels to revise computer-generated propagation predictions. We drive around, we walk around, we take the elevators up the office buildings, and we measure the radio signals. If the predicted propagation is consistently lower or higher than the measured value, then we should adjust the database to match reality.

Our last area for comparing prediction to reality is the area of pilot pollution. As we are moving around the service area, our measurement terminals should be reporting areas where four or more pilots are competing for CDMA service. These areas should match the pilot-pollution zones on the fourth-pilot delta maps. When the measurements and the maps disagree, then it is worth significant effort to figure out why they are coming out differently. We should determine which propagation prediction is not matching the CDMA wireless telephone system.

48.7   Conclusion

With a model of our system and a sense of the locations of our subscribers, we can troubleshoot problems using both simulation and analysis. We also can use these systems to plan growth. Growth is becoming more complicated as we begin to provide a variety of services over the air interface. CDMA is not just for telephone calls any more, but it remains to be seen what data services will be popular. The level of demand will be a key factor for income, and the ability to serve that demand when and where it is located will be key to success in the increasingly competitive cellular market. A business with more accurate models of its systems and customers and better ability to link engineering plans and requirements with business requirements and plans has a much better chance of succeeding. Such a company is more likely to allocate scarce resources for growth to the most important new services and to end up with more profitable business operations.

Soon we may be able to leverage systems that locate our subscribers during calls. This would allow us to populate our simulations with even more accurate data and derive more accurate results. New and upcoming technologies like this, as well as the challenges facing the cellular industry as a business, are the topic of our final chapter.

Conclusion

Engineering is only one essential part of a successful code division multiple access (CDMA) wireless network. As engineers, we serve our companies best by understanding a variety of business issues and applying our engineering knowledge to create solutions. Part 9, "Conclusion," looks at the current picture and future possibilities and challenges of CDMA-based cellular telephone service to the world.

CDMA Now and in the Future

In August 2002, Sprint PCS activated the first third-generation (3G) cdma2000 1x network across the United States. Yesterday, service was briefly unavailable in San Antonio, Texas, probably as the base station was switched over. For the first time, 3G code division multiple access (CDMA) is a reality across an entire large national network. Technically, it appears sound. Will it demonstrate the 50 to 100 percent increase in call capacity expected? How much of that call capacity improvement will be realized before customers are using cdma2000 cellular telephones? Will customers have enough interest in data services to make them profitable?

These major questions face the industry as a whole and face each company as the market reaches saturation and the industry moves out of its initial growth phase and into its mature, competitive phase. Growth will become more a matter of market share than a matter of attracting new customers to the industry. There are a host of minor questions as well. In the first section of this chapter, "CDMA Now," we will look at these issues across the globe. In the second section, we will look at some of the challenges and opportunities CDMA faces in the future.

49.1   CDMA Now

CDMA faces different market challenges and different technical challenges in each part of the world.

49.1.1   North America

At the same time that Sprint has opened the first 3G nationwide service spanning a continent, Cingular and Verizon have local 3G data services available in some areas. However, they are difficult to obtain in the consumer market because demo systems and good staff training are not readily available in stores. This may change rapidly.

The most disturbing question arises from the confused and conflicting views of the market value of data services. Are they primarily for businesses or for the consumer? Is the hot application personal digital assistants (PDAs) and laptops with e-mail or custom business applications or Internet access or cell phones with built-in cameras? Or both business and consumer services? Or neither?

Let us take a look at business services first. There is no question that it is valuable for businesspeople, especially salespeople and executives who travel, always to be in touch. In 1998, a typical executive in any industry spent perhaps one-third of his or her time on e-mail. Sales force automation, including customer relationship management (CRM) systems, are of real value to business. Enabling these through wireless data will put the mobile work force in an always-on state. If this connectivity helps increase sales or reduces inventory-management errors so that customers get the right thing more often and get it sooner, then the use of wireless services will help the business and might be worth the cost.

However, three questions remain to be answered:

- How important is this in the current economy?
- How soon will it happen?
- Will businesses use the cellular network to do it?

For the last 5 to 7 years, companies have succeeded by implementing more effective technology than their competitors and getting it working sooner. The rush to enterprise resource planning (ERP) and CRM systems, plus the movement of both business-to-customer (B-to-C) and business-to-business (B-to-B) transactions to the Internet, was valuable to those companies which made sure the new information technology (IT) systems benefited their businesses. In general, money was not the constraining factor. Time, in relation to the speed of the competition, was. Even more so, quality was a driver. Systems that did the right thing and worked reliably helped businesses grow.

In the current economy, it is not clear what is best for business. It is certain that there will be fewer efforts to push into the use of new technology. It is likely that these efforts will be smaller and that the successful ones will do a lot of early work to try to ensure the business value of the chosen application of technology. One consultant reports that requests for proposals for IT development work are coming in months earlier than usual. He believes that internal managers are being more diligent to ensure successful use of funds in the current economy.

Surveys from the first half of 2002 suggest that always-on e-mail may be the crucial business application for 3G wireless services. However, the same studies also indicate that businesses are not very interested, at least not yet. Most think that high-speed landline access is more valuable. Road warriors will meet during the day and take care of their e-mail in the evening at the hotel, appreciating the fact that the hotel moved from modem-speed connections to high-speed Internet access just recently.

As many businesses take the wait-and-see or the wait-until-we-have-more-money approach, competition is growing. The Wi-Fi IEEE 802.11b wireless ethernet standard is trying to offer high-speed Internet access in key public *hot spots,* business and travel centers that call for high usage. An innovative alternative to cellular recently demonstrated an ability to provide high-speed data access in the tens of megabits-per-second range over a distance of 20 miles. Even if there is a demand for wireless connectivity for business, it may not go to the cellular industry.

What can CDMA-based cellular companies do to encourage business use of wireless data services? More and more, companies are seeking services with predefined business solutions built in. Connectivity is no longer the selling point. The point is useful business value based on connectivity. E-mail may be a starting point, but it probably will not be enough to drive the market to success. Successful service providers will team up with user terminal providers and developers of Internet-based business software to sell CRM and other knowledge-management solutions that happen to ride on their service. For engineers, this will mean

- Successful service providers will need to provide service based on contractual service-level agreements (SLAs) with guaranteed quality of service (QoS). The customer can wait or take its scarce dollars elsewhere, so it is a buyers' market. What kind of pressure will engineers and maintenance technicians face when every outage means a refund to the customer?

- There will be less money available for development and deployment. Cellular service providers have less money and are not a very popular investment right now. Since the money that they have will be shared among divisions, service development, business alliances, and marketing will take a higher priority than they did before.

Wireless service providers face many of the same issues when planning 3G data services for the consumer. Poor service quality increases the risk of churn as customers seek better service from other vendors. This applies to both those services based on good engineering, such as voice clarity and a low call blocking rate, and those services largely independent of engineering, such as good customer service and acceptable plan offerings and billing. Therefore, the wireless provider must decide how to balance its money across engineering for growth, operations, administration, and maintenance (OA&M), customer service, and marketing.

However, what might make 3G data services interesting to the customer? What will make people want to pay for wireless data access? For this, we need to distinguish between data services and video services. Although, at present, both use the same transmission technology over the air interface, they are very different products for consumers. At this point, our nonbusiness customers have very little use for high-speed mobile data per se. Most demand for data use can be satisfied through short message service (SMS) and enhanced message service (EMS) without even using a voice or data channel.

What may be valuable to customers is the ability to transmit images.[1] The picture phone, as such, is of little value and has never become popular. Why? Because it adds little information to a phone call. It might have some value in a conference-call setting because it allows people to see who's talking so that they do not have to interrupt to ask, "Who said that?" High-quality full-speed video phones might be valuable for medical, therapy, and coaching sessions, where facial expressions and body language could add information valuable to the purpose of the communication, but such phones are beyond the bandwidth capacity of current cellular technology.

Pictures are interesting and useful not when they show the speaker's face but when they capture images from the speaker's surroundings. The first significant use of cellular data for images has been in the use of the cellular phone as a camera to show people other things. As we look at the development of cellular services across the world and probe into the future, we will return to this issue of how people are finding new ways to use cellular services that change the way we live.

---

[1]Thanks to Steven M. Kemp, Ph.D., a research psychologist (and the coauthor's brother), for these insights about the value of new media, based on Marshall McLuhan's distinction between hot and cool media.

It is important to make the distinction between fads, which provide short-term revenue and introduce items into the consumer market, and changes in the way we live, which last a long time. A fad cannot last long enough to pay off the debt created in deploying a major new technology across an entire cellular network. Cellular service providers will see long-term benefits to net revenue not from fads but from responding to people who integrate wireless communications into our lives in new ways.

Before we look to other parts of the world, we should take a look at the status of cellular telephony voice services in North America. The major CDMA carriers are moving to cdma2000, as discussed earlier. Some Global System for Mobility (GSM) and time division multiple access (TDMA) carriers are converting to 3G CDMA as well; others are opting for GPRS/EDGE-based systems. However, there is one strong pressure pushing CDMA ahead of TDMA-based technology. The U.S. government is requiring that cellular systems be able to locate customers by geographic location within 100 m and not merely to the nearest cellular sector. This is part of the enhanced 911 (E-911) system that ensures effective emergency response to 911 calls. The CDMA system's locator service (based on passing the Global Positioning System timing signal from each serving cellular tower to the user terminal and receiving a reply that is then used for precise time-based distance calculation and triangulation) has passed initial approval. This system allows a user terminal within range of only two base stations to be located with sufficient precision. In North America, almost all locations are within range of two towers. The proof-of-concept system for locating GSM and other TDMA user terminals was not approved. Without the timing signal, location can only be determined with sufficient accuracy if the user terminal is within range of four cellular towers. The government has determined that there are too many rural areas where a user terminal is not within range of four towers for this to be acceptable.[2] It is unlikely that a technical solution for this limitation of the TDMA-based location system will be developed in time to be tested and approved for current government deadlines. Either the deadlines will be extended, or companies will be forced to migrate to CDMA. Extensions of deadlines are common in such circumstances. Nonetheless, the situation with relation to compliance for E-911 significantly increases pressure for conversion to CDMA in North America.

There are a number of other technical issues facing CDMA engineers. Let us take a brief look at them. There is insufficient radio spectrum to support projected growth of voice services in North America. New spectrum allocations were announced in summer 2002 but will not be available until 2008. In the meantime, conversion to cdma2000 will provide some increased capacity, but how much remains to be seen. Also, cellular providers that have or can obtain spectrum in the original Advanced Mobile Phone System (AMPS) range (900 MHz) will continue to convert that spectrum to CDMA or TDMA. This will place additional pressure on planners and maintenance engineers to tweak systems for optimal performance.

Smart antennas will be deployed more widely, and prices probably will drop. It would be valuable to define the exact effect of smart antennas in our cellular simulations. Rather than merely adding some capacity factor to a cell, it would be good to be able to

---

[2]In the language of the U.S. Federal Communications Commission (FCC), all areas are either metropolitan or rural. A significant part of rural areas consists of suburbs with relatively large residential populations. And, of course, emergency services want to be able to reach travelers on highways and back roads as well.

get some answers (or at least some good guesswork) on specific questions. Here is one example. If we identify two separate locations with high demand within one cell, can we determine how much additional capacity we will get on that cell by deploying smart antennas as compared with either the current system of 120-degree sectors or a less-expensive upgrade using narrow-beam fixed directional antennas with a spread of 60 degrees or less? Such a determination would be a valuable addition to a cost-benefit analysis before making a decision on what antennas to upgrade, what technologies to use, and when and how to deploy them.

There is another possible improvement in modeling. The cellular system models we use now are based on measurements taken at specific times in specific cities and rural locations, as discussed in Chap. 47. However, we are constantly taking measurements on our networks. It should be possible to integrate these measurements not only as part of the data sets in our tools but also as part of the modeling equations. Statistical methods possibly could be applied to take networkwide data and use them for new and more accurate incremental improvements to the calculations in our propagation models.

49.1.2 Europe

European 3G systems face a number of challenges. Let us look at the economic ones first. Even more than in North America, the market for cellular phones has reached saturation. This has created challenging times for the manufacturers of cellular user terminals, who are often also the manufacturers of cellular base stations and network equipment.

On an engineering level, existing GSM systems are well established and cannot share spectrum with CDMA systems because the CDMA will interfere with TDMA broadcasts. This leaves only three options: new spectrum allocations, spectrum conversion channel by channel, or total conversion of a whole network all in one day. The first option requires government allocation of spectrum, which usually takes a number of years. The second is an engineering challenge, especially since wideband CDMA (W-CDMA) has no guard bands. If a GSM channel borders on a W-CDMA channel, it will be difficult to ensure that there is no significant interference. To the best of our knowledge, such a solution has not been developed and tested. Channel-by-channel conversion is difficult for capacity planners, who must gauge the demand for two types of frequency, predict the change in that demand in each sector or region, and be prepared to switch frequencies over quickly so that neither the old system nor the new one is overloaded. The third solution is not feasible unless dual-mode telephones are made available to every cell phone owner in Europe. International interoperability across the continent and to many other places across the world is a key feature of GSM. Multi-frequency GSM telephones are common, allowing international travelers to use their own phones in other nations with services on different frequencies. A multimode cellular telephone, though, is much more expensive because it has to include both TDMA and CDMA transceivers. Meanwhile, the effort to deploy systems that meet the W-CDMA standard has encountered a number of technical barriers, as discussed in Sec. 9.2.

What will happen in the long run remains to be seen. CDMA has significant capacity advantages over TDMA, but a huge existing network does not change quickly. Coaxial and fiber-optic cable have tremendous capacity advantages over copper twisted pair for landline services. However, even as coaxial cable has come into wide use for cable television, the local access for telephones remains largely copper, and fiber-optic home telephones are prototypes that have not gone into production. In current economic conditions, and in a market that has reached saturation of cellular telephones, if not of cellular service usage for voice, video, and data, slow growth may be a better choice. And whether that growth will be toward CDMA, with its greater capacity and potential, or a more conservative, but easier step toward TDMA with improved data services remains to be seen.

49.1.3 Mainland China

Two major telephone companies are competing to deploy cellular telephone networks in mainland China, and the larger one has chosen time division synchronized CDMA (TD-SCDMA) as its preferred technology. The Chinese government is allocating spectrum to different companies province by province, and it is not clear whether any single province will have two competing services or not. The mere notion of this level of competition in mainland China is astounding. The reversion of Hong Kong to Chinese control in 1999 probably has helped, because cellular service providers can use Hong Kong as a basis for financial operations, getting investment from international sources.

The equipment probably will be largely manufactured on the mainland, where the capacity to produce electronics is growing rapidly. The factories may be owned all or in part by major Western cellular equipment providers, but the production and use of the equipment will remain largely local. To date, no other nation has adopted the TD-SCDMA standard, but companies could use it in the future. If Chinese factories succeed in making equipment at the capacity needed for even reasonably rapid deployment across the world's most populous nation, it will be interesting to see what happens when internal deployment slows and these factories have capacity to support export. It is impossible to say whether this will take 5 years or 15, but it seems reasonable to speculate that China will be a major source of cellular telephones and cellular network equipment in the future. This equipment may go to nations that are growing their first cellular networks and, especially with new spectrum allocations, could be used in North America and Europe as well.

However, there are many challenges, including the mainland Chinese government's ambivalence about the Western model of a market economy, that remain to be faced both for deployment across the Asian continent and for mainland Chinese participation in the world cellular industry.

49.1.4 Japan and Korea

Japan and Korea have the most advanced digital cellular networks in the world. Some recent interesting developments include

- The first deployment of cdma2000 1x EV DO in Korea.

- The popularity of short message service (SMS)–based text messaging in Japan and the new ways of using cellular phones it is introducing, which we will discuss further in the next section.

- The use of CDMA-based phones that accept GSM-type user ID cards for international operability. These phones were rented to customers from Japan attending the World Cup Soccer Finals in 2002.

■ The popularity of the cellular phone with built-in digital camera that helped Japan's second largest cellular provider add a million users to its 3G network in 3 months.

Success in new ways of using technology in Japan is not necessarily a harbinger of worldwide success. There are two major reasons for this. The first is that Japanese culture remains much more tightly knit than the culture of other first-world nations. In all consumer industries in Japan, fads grow very quickly, and people are willing to pay a lot of money to join the trend. Thus, catching a fad can pay for a significant investment in technology in a small nation to a much greater degree than would be true in a geographically larger nation with a higher cost for national deployment and a more culturally diverse population.

Secondly, quality, that which is seen as being of value, depends on culture to a significant degree. This can be seen in the different interpretations of total quality management (TQM) methods in Japan, the United States, and Germany, which have all been leaders in the field. In Japanese culture, quality is sometimes expressed through an effort to give people what they want even before they know they want it. For example, the ideal restaurant waiter or waitress removes the last dish and serves the next one without ever being noticed. In contrast, the idea of quality in the United States includes offering the customer many choices and interactions with service providers. We want large menus and then ask for advice from the wait staff. Quality customer service in the United States depends on offering diverse options to a diverse group of customers. Interaction starting with "How can I help you?" or "What would you like?" is the mark of quality service.

The approach to quality in the United States favors openness in business as well as open dialog in customer relations. In TQM-based organizations, we will hear forthright discussions of problems and solutions. This is very different from the German approach to quality. In Germany, engineers and companies express quality by quietly and quickly solving the problem without ever publicizing it. After the *Hindenburg* exploded, the manufacturer did some tests and discovered that the distinctive orange flames were a crucial clue. Hydrogen burns clear or maybe blue. However, the orange flame came from a mix of paint on the outside of the *Hindenburg* that turned out to be highly flammable and could be ignited easily by the kind of spark an airborne vehicle encounters when it gets near a ground-based tower. The chemical formula of the paint was similar to what was used soon thereafter for solid rocket fuel. The company saw this and changed the paint on its dirigibles. The company buried the records of their discovery, leaving the world to speculate about terrorist conspiracies for over 50 years. Recently, American engineers working to prove the safety of hydrogen first reproduced their work and then uncovered the records that dated from shortly after the explosion.

The point is that particularly when it comes to fads, we cannot count on what is popular in one country moving to another. And even if it does become popular in another country, it is a very open question whether it will be popular enough to be profitable. Adding a new feature to a nationwide cellular network in Japan is a much smaller venture than doing the same in the United States. And the same is true if we want to make the service available only in major metropolitan areas.

49.1.5 The rest of the world

The economic challenges that most cellular equipment providers are facing probably make them cautious about investment in countries with weaker economies and less stable governments. The development of cellular services, whether for mobile use or for wireless local loop (WLL), probably will be slow. It may well be driven more by innovative entrepreneurs who understand the local culture and tap Western engineering expertise than we might expect. Despite strong doubts from financial backers, an innovative thinker has launched a successful cellular telephone service in rural Pakistan. The key to success was an innovative approach to billing and marketing. The owner of the phone can allow use by others on a call-by-call basis and charge them for it. As a result, the owner can subsidize personal use by income obtained through short-term rental of the service to others. This initial success relied on an entrepreneur who knew the culture and got an advanced engineering and business education in the United States. However, now that the method is succeeding, it will be easier to reproduce elsewhere. We should expect and be open to various innovative approaches to novel technology. Such approaches have a better chance of succeeding than does the simple export of first-world systems to a world of diverse cultures.

Although we cannot predict what will happen, we can say that it is vital to the world that wireless communications continue to grow. In under 60 years, Europe has moved from a small continent torn by the most destructive war the world has ever seen to a cooperative union of nations with shared economic goals, a common currency, and increasing respect for one another's languages and cultures. Communications has played a major part in that transformation, and we need the same to happen across the entire globe.

49.2   The Future of CDMA

It is hard to talk about the future without being speculative. However, others are speculating and sometimes backing their speculations with thousands of millions of dollars of investment in research and development. We can report on these speculations and investments to give a sense of the possibilities—and the problems—that lie ahead.

49.2.1   3G and 4G technologies

In the early 1990s, long before third-generations (3G) technologies were standardized, the fourth-generation (4G) initiative began. The current proposed standards for 4G systems have two distinctive components: bandwidth of over 100 Mbps and use of the Internet Protocol version 6 (IPv6), which has been defined for a number of years but has not entered production.

If CDMA systems can reach the higher end of the bandwidth suggested by the 3G standards, then they will be entering the bottom edge of the 4G data rate capabilities, and some people are already referring to such technologies as 3.5G.

A prototype 4G system already exists, but the standards changed after the prototype was built. The prototypes used Asynchronous Transfer Mode (ATM), but the data standard has changed to IPv6. The transmission method was time division multiple access (TDMA), but the new transmission method is proposed as orthogonal frequency division multiplexing (OFDM). The pilot system achieved the impressive speed of 34 Mbps, matching the level of service provided by E-3 lines. The goal of the standard is to reach 155 Mbps, comparable with OC-3 rates. This is over a thousand times faster than wireless data rates available today.

Despite the changing technologies and goals, some say that 4G systems may be available as early as 2006. Japan has set a requirement that all Internet service providers (ISPs) support IPv6 by 2006, and this, presumably, would include wireless service providers. It remains to be seen if other nations will follow suit.

IPv6 solves many problems of today's networks. It offers a single stack of communications protocols that provides these benefits, overcoming limitations of IP version 4 (IPv4, the current version), ATM, and other services:

- Enough addresses for the foreseeable future, whereas IPv4 is running out of addresses

- Addressing systems that support virtual addresses and mobility, making it easier to prevent tromboning and to deploy flexible networks

- Built-in security

- Support for independent definitions of quality of service (QoS) for voice, video, and various types of data services sharing the same packet network

The downside of IPv6 is simply that it will require the replacement of every router and many other pieces of equipment on the Internet. If backward compatibility works, computers running IPv4 will work on IPv6 networks. However, backward compatibility, even when designed into the new system, always brings engineering challenges.

Estimates for when 4G services will arrive vary from 2006 to 2020. The initiative got several boosts in late 2001 and early 2002, including a major commitment of investors in Japan, who see it as the logical next step, given that 3G data services are popular there.

However, by summer 2002, new developments, at least those reported in the news, ceased. In the United States, two 4G wireless initiatives were canceled. The standard for the multichannel multipoint distribution system (MMDS) is being reengineered. At the same time, Wi-Fi is showing some startling progress, even in mobile and cellular tests. As a result, the future of 4G may be slow or fast in coming, but it also may well not be based on any current CDMA telephony technology.

The 4G concept includes some exciting possibilities, such as all-wireless backhaul and even mobile base stations mounted on trains or trucks. We do not mean base stations that can be transported for emergency use or quick installation; we mean base stations on trains, providing active service to passengers and using a wireless uplink to fixed mobile switching centers even while the base station itself is in motion. Speculation about the use of such service is even wilder, and we will not report it in an engineering text.

49.2.2 Spectrum allocation

As we discussed earlier, North American spectrum allocation for cellular telephone service is already too small to meet demand easily and will be overcrowded before some relief arrives in 2008. This relief brings its own challenges, because engineers may face trouble developing cellular systems that operate well at 4 GHz. We also will have to reengineer our modeling tools, which make assumptions based on the transmission frequency. It is almost certain that more expensive multiband telephones will be needed.

Meanwhile, Wi-Fi and 4G technologies are exploring uses of unlicensed spectrum. There is some thought of developing services at 40 or 60 GHz, well up in the microwave spectrum. The 60-GHz spectrum has a quality that is both an advantage and a disadvantage for cellular services. Radiation is limited to about 1 mile due to absorption of this frequency by atmospheric oxygen. As a result, there would be less intercellular interference. However, coverage at this frequency would require a lot of very expensive base stations. For initial buildout, 40 GHz has the advantages of both lower cost per base station and a larger cell radius, reducing the total number of base stations needed.

Available frequency and plans to use frequencies that are regulated in different ways or that are unregulated will have a significant effect on the speed of deployment and the total cost of each of these competing technologies.

49.2.3 Competition

The future of cellular telephone technology, including CDMA, is very dependent on the development of competition using Wi-Fi, and there are some major players in the game. Intel has announced Project Rainbow, an initiative to build a cooperative venture among major IT industry companies to deploy a public Wi-Fi network nationwide.

One counterproposal has been floated by Andrew Seybold. He suggests that Qualcomm should form a consortium to expand 3G CDMA data service deployment with multiple cellular providers. In an unproven market for data services, both would be risky ventures. At this point, we would surmise that if wireless high-speed Internet access becomes popular for use by stationary customers (a sort of Internet WLL), then it is more likely to be Wi-Fi than cellular because Wi-Fi has a higher bandwidth capacity for this purpose. The best way to provide mobile users with high-speed data access in the 4G range remains to be seen.

Sprint, even while rolling out its national 3G network, is hesitant to invest in 4G data services because the average revenue per user of $30 is too low to make the service profitable. Voice service averages $40 per user and is much cheaper to provide.

In the current economy and competitive situation, it is difficult to know what direction the leadership of cellular service providers might take that will succeed. It may not work to be conservative and wait for customer interest because customers do not know what they will do with a new service until it is in their hands. On the other hand, investing large amounts of money in a new service is risky. Even if there are customers out there, an alternative technology could take the market for any of several reasons. The competing service might arrive sooner, it might be less expensive, or it might have more successful marketing.

It is interesting to note that the 4G standard does not mention voice services at all. We have not heard of any effort to address the unresolved telephony issues from the limited success of the 3G initiative. Two things telephony folks might like to see include standardization of frequency allocation around the world and a truly unified telephony standard.[3]

49.2.4 Beyond the cell phone

Although WLL for data probably will go to Wi-Fi or other technologies, there may still be value in deploying WLL using cellular network technology. If nations that do not have robust, widespread current landline networks want to build out their national telephone systems, cellular telephony WLL is a sound technology that could become cost-effective. If not, then simple cellular networks supporting mobile telephones may

---

[3]Of course, the authors would prefer a standardized *CDMA* telephony standard.

serve, with entire countries skipping the era of the landline telephone and going straight to wireless or perhaps satellite service.

We also foresee changes to the user terminal. The first changes are small but essential. We need better ways to connect laptops to the Internet through mobile wireless and to use PDAs with wireless Internet access. Both the human interface of cellular telephones and the air interface of PDAs and laptops need improvement. But what is most needed is packaged solutions that work when we turn them on. Wireless data will not sell well as long as customers need to order special cables or install anything more complex than a card that shoves into a slot (and even this will be too much for many users). We are in a chicken-and-egg situation where the market is not proven enough for low-cost production of easy-to-use solutions and the hard-to-use solutions are slowing market acceptance. This probably will pass, although whether it will take a couple of months or a few years depends largely on the state of the economy and the quality of new interface design offerings.

What does the future hold? We should look at two areas: uses of cellular technology that do not involve people and new ways of living a wireless life. Some industry prognosticators have realized that there is a fundamental limit to the number of cellular telephones that can be sold and that it is closely tied to the number of people on planet earth. A significant fraction of the people on the planet already have cellular telephones. We may expand the market a bit by having people each own two or even a few user terminals. Probably, though, the personal area network (PAN), with a wireless radius of about 10 m, will allow our telephones, laptops, electronic wallets, and who knows what else to link together and share one cellular interface device.

No, the real market growth, some say, is to sell wireless devices that have nothing to do with people at all. This is an example of the notion of ubiquitous computing, and one proposed term is *invisible mobile*. More simply, we are talking about very inexpensive widespread telemetry. Wireless devices could provide unique identification showing where something is located at any time. One proposal suggests that if the devices were cheap enough, one could be attached to each soda bottle during manufacture. This would allow the manufacturer to track inventory in every store and soda machine automatically and dispense new inventory promptly. Some market studies show that the cost reduction and reduced cycle time offered by such a system actually would pay for its deployment and maintenance. Allowing ourselves to be truly speculative for a moment, we would point out that telemetry provides more than just location. Why not include a thermometer so that the soda bottle (or, more important, the ice cream bar) can call for help when the refrigeration shuts down?

More realistically, such systems will be useful in more expensive pieces of equipment. They also will be valuable to society as we focus more and more on security. A new system for vehicular security at airports has passed its proof-of-concept stage and is being prototyped at a major airport. Fuel trucks and smaller vehicles are potential weapons for terrorists who might use them to attack commercial aircraft. In the prototype system, ignition keys are replaced with an electronic key. If a key is reported lost, then it is locked out of the system. A wireless airport network is notified any time anyone tries to start a vehicle. The vehicle transmits information on the key to the network by a wireless system. The network confirms the key owner's identity, work status (whether or not the employee should be working at this time), and right to use that particular vehicle. If that worker should not be driving that vehicle at that time, the vehicle is shut down, and security is notified. If all is okay, then the vehicle starts. Going one step further, the system tracks each vehicle's location with a fair degree of precision and can shut a vehicle down automatically if it enters an unauthorized area. This would prevent even an authorized employee from using the vehicle to ram a moving airplane.

Far more modestly, water meters and other utility meters that are now read by hand are already being converted to wireless technology. Integrating these networks with wireless telephony, where a wireless call, once a day or once a month, will replace a visit by the meter reader, may not be far off.

In all likelihood, ubiquitous wireless technology will be deployed. It will go first to expensive items or to solve problems where human life or expensive losses are possible. As it is proven and manufacturing costs go down, it will become more widespread. However, to what extent these systems will use CDMA remains unknown.

49.2.5   Conclusion: New ways of living a wireless life

In the end, the wow factor of wireless technology and its revenue-generating value are only small parts of the real contribution of CDMA and other wireless technologies. The real value lies not in the realm of the mind, nor in the realm of business, but in the human heart. We do not primarily mean the opportunities for helping people find mates (although this is always popular and profitable), but the deeper sense of heart that refers to courage and to people's endless ability to find innovative responses to their situations, creating a new life from changing circumstances.

The PAN and the wireless home network will become seamless environments that change the way we interact with our possessions. We will call out for lost keys, and they will answer us.[4] More important, cellular telephones are already changing the way we collaborate, make decisions, and take action.

In Japan, people routinely share photographs of where they are with friends rather than just chatting. When shopping, we can call a family member or housemate, show what is on the shelves, and make collaborative decisions about what to buy if our favorite brands are out of stock. Instead of trying to describe products, we show others what we are seeing, allowing us to collaborate over distance.

With cellular telephones, people have more flexibility in both space and time. We can make appointments approximate and use our cell phones to call again when the time and place are closer, deciding exactly what we want to do. We can call our destinations when we are lost. Both radio location and being talked in for a landing used to be a process used only for airplanes. Now they are used routinely by drivers of delivery trucks and friends who cannot find our houses.[5]

One clear illustration of this occurred on September 11, 2001. While cellular telephones are probably useful to terrorists, their use was certainly key to the courageous decision of the passengers and crew of the flight near Pittsburgh that foiled terrorists. They used their cellular phones to call family from the hijacked flight. The news of the World Trade Center had already reached some of the people they were calling, and people put information together from the news available at home or work and the situation on the airplane allowed a collection of citizens to surmise that this was not an ordinary hijacking. Supported not only by the information but by the human contact with

---

[4]One of the authors (Kemp), thanks his wife, Kris Lindbeck, for this insight.

[5]The authors again thank Steven M. Kemp, Ph.D., for collaborating over distance in developing these ideas.

friends and family, the passengers and remaining crew apparently decided to prevent the hijackers from flying the plane into yet another building. Evidence indicates that they used a luggage cart to attack the cabin door. Whatever the hijackers' intentions were, they succeeded only in crashing the plane and not in destroying their target.

This is one spectacular example of a phenomenon that is happening every day. A senior IT executive of a state police department has informed us that the use of cellular telephones has radically changed the job of the state police. Prior to the cellular telephone being in widespread use, the most frequent job of state police officers was to render assistance to stranded motorists. Officers were instructed to remain with stranded motorists to ensure their safety until a tow truck arrived. Not only were these jobs more frequent than any other, they also took longer than many. Now that stranded motorists have cellular telephones, the tow truck often arrives before the police cruise by. Even more significantly, the stranded passengers are safer when they are alone because they have cellular telephones. Criminals, knowing that stranded motorists can call for help, may well be more reluctant to attack. With cellular telephones, we are less isolated, and criminal activity is more difficult. Of course, criminals and terrorists can and do use wireless technology for their own purposes as well. In the long run, however, there is good reason to believe that the balance will run in favor of safety and peace.

We are changing our lives with wireless technology. The final outcome is uncertain. However, there is a core principle of general systems theory that says that living systems have a property of being self-healing. This means that if we can improve communication among all the parts of the system, then the parts and the system will be restored to balance, become more robust, and be able to grow. CDMA engineers play a part in that evolution by doing a good job ensuring ever-growing quality of service in support of the true quality of life.

Appendices

BER   bit error rate

BSC   base station controller

CBCH   cell broadcast control channel

CBR   constant bit rate

CDMA   code division multiple access

CDPD   cellular digital packet data

DSSS   direct sequence spread spectrum

ERP   effective radiated power *or* enterprise resource planning

ESN   electronic serial number

EVRC   enhanced variable-rate codec

FACCH   fast associated control channel

FEC   forward error correction

FER   frame error rate

HLR   home location register

IMT   International Mobile Telecommunications

IWF   interworking function

MDN   mobile directory number

MIN   mobile identification number

MMS   multimedia message service

MRI   mobile reported interference

MSC   mobile switching center

OCQPSK   orthogonal complex quadraphase phase shift keying

OTAPA   over-the-air parameter administration

OTASP   over-the-air service provisioning

OVSF   orthogonal variable spreading factor

PAGCH   paging and access grant channel

PCH   paging channel

PCS   Personal Communications Services

PDC   Personal Digital Cellular

PMRM   power measurement report message

PN   pseudo-noise

QCELP   Qualcomm code-excited linear prediction

QoS   quality of service

RACH   random access channel

RPE-LTP   regular pulse excitation–long-term prediction

RSSI   received signal strength indicator

S/I   signal-to-interference

S/N   signal-to-noise

SACCH   slow associated control channel

SDCCH   stand-alone dedicated control channel

SDMA   space division multiple access

SID   silence descriptor

SMS   short message service

TD-SCDMA   time division synchronized code division multiple access

TLDN   temporary local directory number

UBR   unspecified bit rate

UMTS   universal mobile telephony system

UWC   Universal Wireless Communications

VAD   voice activity detection

VBR-NRT   variable bit rate non-real time

VLR   visitor location register

W-CDMA   wideband code division multiple access

WLL   wireless local loop

bels   Ratios of powers of 10. We almost always use *decibels* (see *dB*).

bits   Single one/zero, YES/NO, ON/OFF pieces of information.

bits per sample   The number of bits used in a speech coder (or any digital audio system) for each discrete time sample.

bits per second (bps)   The rate of data transmission, the data in bits divided by the transmission time in seconds.

bits per symbol   The number of bits in each sample of modulation in a phase shift keying. If the constellation has 2^k points, then the modulation scheme should be sending k bits per symbol.

blocks per second   The rate of data transmission in 8-bit blocks per second. A *block* is an old telephony term for an 8-bit sample.

bytes   A unit of data that is 8 bits long, also called an *octet*. Typically, a byte is a single character of text, although some languages use 2 bytes for each character.

cents   Hundredths of a U.S. dollar.

characters   A single letter in a written language. Typically, one character is 1 byte, but some languages use 2 bytes for each character.

chips per bit   The ratio between the individual chips and the information bits in a direct-sequence spread-spectrum system (DSSS). The chips per bit ratio is the processing gain.

chips per second   The rate of chip transmission in a direct-sequence spread-spectrum system (DSSS).

centimeter (cm)   Units of one-hundredth (1/100) of a meter.

decibel (dB)   One-tenth of a bel, a ratio of $10^{1/10}$, about 1.26.

dB^2   Decibels squared, the measure of statistical variance in a log-normal distribution.

dB per decade   The relationship between power level and distance. For example, an inverse-square relationship would have a power loss of 20 dB per decade.

dB SPL   Acoustic volume in decibels relative to a very quiet level of 0.0002 dynes per square centimeter, 0 dB SPL.

dBm   Radio power in decibels relative to a level of 1 mW, 0 dBm.

degrees   An arc of a circle. A full rotation is 360 degrees.

erlangs   Units of telephone usage or demand, where the number of erlangs is the average number of calls.

erlangs per MHz per cell   The efficiency of a wireless telephone system, the amount of traffic in erlangs each cell can carry divided by the radio spectrum usage in megahertz (MHz).

frames per second   The rate of data transmission, the data in frames divided by the transmission time in seconds. (See *frame* in the Glossary.)

femtowatts (fW)   Units of 10^{-15} W.

gigahertz (GHz)   Units of 10^9 Hz.

hertz (Hz)   The frequency in cycles per second of a radio or sound wave.

inches   Units of 2.54 cm used in the United States.

kilobits (kb)   Units of 1000 bits.

kilobits per second (kbps)   Units of 1000 bits per second.

kiloblocks per second   Units of 1000 blocks per second.

kilobytes (kb)   Abbreviation for *kilobytes,* units of 1024 (2^{10}) bytes. While the *kilo-* prefix normally is used for 1000, computer geeks have been in the habit of using even powers of 2 rather than powers of 10.

kiloframes per second   Units of 1000 frames per second.

kilohertz (kHz)   Units of 1000 Hz.

kilometers (km)   Units of 1000 m.

kilometers per hour (km/h)   Speed of motion.

kilosamples per second   Units of 1000 samples per second.

megabits per second (Mbps)   Units of 1 million (10^6) bits per second.

megabytes (Mb)   Units of 1,048,576 (2^{20}) bytes. While the *mega-* prefix normally is used for 1,000,000, computer geeks have been in the habit of using even powers of 2 rather than powers of 10.

megabytes per dollar   The price of data service.

megachips per second   Units of one million (10^6) chips per second.

meters (m)   Units of distance. Officially, 1 m has been defined to be 1,650,763.73 wavelengths of the orange-red radiation of krypton-86 atoms. More recently, it is defined as 1/299,792,458 of the distance light travels in 1 second in a vacuum.

megahertz (MHz)   Units of 1 million (10^6) Hz.

miles   Units of distance. Originally the distance of one thousand paces in ancient times, 1 mile is 1,609.334 m.

milliseconds (ms)   Units of 1/1000 of a second.

microseconds (μs)   Units of 10^{-6} s.

milliwatts (mW)   Units of 10^{-6} W.

nanowatts (ns)   Units of 10^{-9} W.

octets   Same as *bytes*.

RSSI   Short for *received signal strength indicator,* one unit of 0.78125 dB. This unit came about because early cellular systems had 7 bits to store signal strength values (from 0 to 127) over a range of 100 dB.

samples per second   The rate of discrete time samples in a speech coder (or any digital audio system).

seconds (s)   Units of time. Officially, 1 second is 9,192,631,770 cycles of a specified radiation of cesium-133 atoms.

terahertz (THz)   Units of 10^{12} hertz.

volts (v)   Units of electrical potential. A volt is not energy or power by itself. Having electrical charge or current change its voltage level becomes energy or power.

watts (w)   Units of power. Power is energy per unit time.

Glossary

access attempt   An attempt to set up a call with the correct transmit power levels in the cdmaOne air interface.

access probe   A single user terminal message in setting up a call with the correct transmit power levels in the cdmaOne air interface.

adaptive matched filter   Another term for *rake filter.*

adaptive multirate (AMR)   A speech-coding scheme used in W-CDMA that uses different coding rates depending on radio conditions.

adaptive phased array   A *phased array* that changes its direction based on changing radio conditions. Also called a *smart antenna.*

air interface   The standard that defines the modulation, demodulation, power control, and signaling for a radio channel.

algebraic code-excited linear prediction (ACELP)   A speech-coding scheme used in W-CDMA that divides the bits according to their subjective importance. The *forward error correction* (FEC) is designed to give the most error protection to the most sensitive bits.

antenna face   A set of transmit and receive antennas on a base station that serves a range of directions, a pie-slice-shaped *sector* of coverage.

autonomous registration   A wireless user terminal *registration* that occurs because a system-specified amount of time has elapsed since its last activity.

backhaul   Transport between base stations and their mobile switching centers (MSCs).

backlobe   The directions of a directional antenna not in the arc of its intended coverage.

backlobe attenuation   The reduction of radio power in the backlobe compared with the coverage area of a directional antenna.

base station   A radio transmitting and receiving station that provides wireless telephone service by sending the forward (downlink) signal and receiving the reverse (uplink) transmission.

base station controller (BSC)   The central processor of a base station that administers call processing, messages, and handoffs. Some systems allow one base station controller to control more than one base station.

bit error rate (BER)   The fraction of received bits on a signal path that are different from the bits sent.

block code   A *forward error correction* (FEC) system with defined blocks of k information bits mapped into n coded bits. Contrast *convolutional code.*

bursty   Having a lot of activity for short time periods. Data usage is bursty if the subscriber wants brief periods of large quantities of data, and interference can be bursty if it comes in high power spikes.

carrier   A radio wave being modulated in some way to send one or more signals.

cdma2000   The third-generation (3G) CDMA standard that is an evolution from the existing cdmaOne standard.

cdmaOne   The North American CDMA standard with over 100 million subscribers in 2002.

cell   (1) The geographic area served by a base station in a wireless telephone system. (2) The 53-byte unit of packet transport in *Asynchronous Transfer Mode* (ATM).

cellular digital packet data (CDPD)   A cdmaOne packet data service.

channel   A single signal going from one place to another. In a TDMA or CDMA system, several channels are served by a single *carrier*, whereas FDMA systems have a separate carrier for each channel. A channel may be landline or wireless and may carry voice, video, or data.

chip   A single bit of the high rate used in *direct sequence spread spectrum* (DSSS) so that there is *processing gain* for lower-rate information bits. When we use the term *bit* by itself, we are referring to the information bits rather than the chips.

chip rate or **chipping rate**   The rate of chip transmission.

closed-loop power control   The setting of radio transmit power levels by having the radio receiver measure its received power and send a stream of power-up and power-down messages. The CDMA systems uses in wireless telephony have two closed-loop power control systems, an *inner loop* and an *outer loop*.

code division multiple access (CDMA)   A multiple-access *spread-spectrum* technology where channels share the same carrier and are differentiated by pseudorandom *chip* sequences.

code-excited linear prediction (CELP)   A speech coder that uses a stored table of speech sounds called a *codebook* to recreate human speech with high quality and low bit rates. There are better-sounding speech coders that use twice as many bits, and there are lower-rate speech coders that sound a lot worse.

code rate   In a *forward error correction* (FEC) system, the ratio of information bits to coded bits.

codebook   A table of speech sounds used in a *code-excited linear prediction* (CELP) speech coder.

complex spreading   A technique used in cdma2000 to balance the *in-phase* and *quadrature* components of the modulation.

conventional reuse   Our own term for FDMA and TDMA systems and their reuse principles, in contrast to CDMA.

convolutional code   A *forward error correction* (FEC) system mapping each k information bits into n coded bits using a moving window of (uppercase) K bits. Contrast *block code*.

coverage cell   A cell in a wireless telephone system whose primary role is providing acceptable service to subscribers in the service area. Contrast *growth cell*.

delay spread   The period of time between the first and last copies of a transmitted radio signal in a multipath environment.

differential quadraphase phase shift keying (DQPSK)   A variant of *quadraphase phase shift keying* (QPSK) where the constellation of each symbol is defined by its phase difference from the previous symbol.

direct sequence spread spectrum (DSSS)   The radio technology of coding a stream of bits onto a higher-rate stream of chips using a coding sequence.

Doppler effect   The phenomenon where radio or sound frequency shifts because the transmitter and receiver are moving toward or away from each other.

downlink   Another term for *forward direction*.

E_b/N_0   The ratio between the energy per bit E_b and the effective noise level N_0, a digital radio equivalent to the analog signal-to-noise (S/N) ratio.

effective radiated power (ERP)   The power a fully omnidirectional radiator must have to equal a radio transmission power level. If a 4-W signal has an antenna gain of 10, then the ERP is 40 W.

enhanced message service (EMS)   SMS extended to allow ring tones, icons, and graphics to be sent to user terminals.

enhanced variable-rate codec (EVRC)   A more sophisticated version of *code-excited linear prediction* (CELP) that reduces the number of bits required for its linear predictor coefficients and speech synthesis so that it has higher voice quality at the same bit rate as CELP.

erlang   The telephony demand equivalent of one continuous call, k users each using the telephone $1/k$ of the time.

erlangs/MHz/cell   One measure of spectrum efficiency in wireless telephone system engineering.

error recovery   An error-correction scheme where the receiver examines the incoming data, checks them for accuracy, and requests a retransmission when an error is detected. Contrast *forward error correction* (FEC).

error resolution   The process in a *forward error correction* (FEC) scheme where the receiver determines the data sent from the data it receives, which may have errors.

forward direction   The radio direction from the base station to the user terminal, also referred to as the *downlink*.

forward error correction (FEC)   A data transmission scheme that allows the receiver not only to detect errors as in *error recovery* but also to correct them. FEC relies on the transmission of redundant data, additional significant bits.

fourth-pilot delta   The ratio, in decibels, between the highest and the fourth-highest radio path gain between a geographic location and serving antenna faces.

frame error rate (FER)   The fraction of received frames on a signal path that are different from the frames sent. This is typically a higher number than the *bit error rate* (BER) because a frame is wrong if any of its bits are incorrect.

gross chip rate   Our own measure of the CDMA resource consumed in providing a particular service at a specified quality level. We can compare different 3G CDMA services using this scale of resource usage.

growth cell   A cell in a wireless telephone system whose primary role is adding capacity in areas that already have adequate coverage. Contrast *coverage cell*.

hand down   A handoff from a higher to a lower CDMA carrier in a multicarrier CDMA system.

handoff   Service for a call changing from one cell, sector, or carrier to another while the call is in progress.

hard handoff   A handoff where one radio link is dropped before another is started. This is the normal handoff used in conventional-reuse (non-CDMA) systems.

home location register (HLR)   A computer and database that services subscribers using wireless telephone service. The HLR has some functions that are performed wherever a wireless call is placed and others that only happen when the call is in the home system.

inner-loop power control   A closed-loop system used in CDMA wireless telephony where the power-up and power-down decision is based on a measurement of the radio performance, in the form of E_b/N_0.

interference   Power received from some other radio transmission that is not the desired signal.

interworking   Communication between two different signaling systems. Used both for ANSI-41-to-GSM and ANSI-41-to-Internet specifications.

jitter   Variation in *latency*.

latency   The technical term for how long it takes data to get from one end of a data connection to the other end. In casual usage, we often use the term *delay*.

link budget   An accounting table of all the factors contributing to the radio signal quality of a radio link.

logical face   The intersection of a *sector* and a *server group*.

macroscopic diversity   Two antennas far apart used for diversity, typically more than 100 m apart. Contrast *microscopic diversity*.

microcell   A cell only a few hundred meters in radius.

microscopic diversity   Two antennas close together used for diversity, typically less than a few meters apart. Contrast *macroscopic diversity*.

mobile reported interference (MRI)   The user terminal responding to the base station and interrogating it for forward channel signal strength and *bit error rate* (BER) of the signaling stream.

mobile switching center (MSC)   (1) The central telephone switch connected to base stations and the public switched telephone network (PSTN) that handles wireless calls and handoffs. (2) The switch or switches, their supporting equipment, and the building containing them.

multimedia service (MMS)   A message system that supports a wide variety of graphic and audio formats. Unlike *short message service* (SMS) and *enhanced message service* (EMS), MMS requires the use of a data channel.

multipath   A condition where more than one radio path exists from transmitter to receiver so that several copies of the transmitted signal are received at different times.

near-far The problem of matching received power levels at a receiver where some transmitters have much higher radio path gain than others, typically because they are physically closer to the receiver.

noise Power received that is not radio transmission of a signal.

offset quadraphase phase shift keying (OQPSK) A variant of *quadraphase phase shift keying* (QPSK) where the constellation points are offset 45 degrees.

open-loop power control The setting of radio transmit power levels by measuring a signal of known transmit power in the opposite direction.

orthogonal Two vectors whose inner product is zero. When we use the term *orthogonal* for more than two vectors, we mean that every pair of them is orthogonal. Visually, we can think of orthogonal vectors as being perpendicular to each other.

orthogonal complex quadraphase phase shift keying (OCQPSK) A variant of *quadraphase phase shift keying* (QPSK) used in cdma2000 that reduces out-of-band radio emissions.

orthogonal variable spreading factor (OVSF) Codes that can share the same frequency and have different spreading factors.

outer-loop power control A closed-loop system used in CDMA wireless telephony where the setting of the target E_b/N_0 for the inner-loop power control is based on the frame error rate (FER).

over-the-air activation function (OTAF) Data exchange between a wireless telephone system and a user terminal that enables downloading information to the user terminal number assignment module to activate service.

over-the-air parameter administration (OTAPA) The process where a wireless telephone system can set user terminal parameters over the air interface. This allows customer service to solve subscriber difficulties by altering user terminal parameters remotely.

over-the-air service provisioning (OTASP) The process where a wireless telephone system can enable service for a user terminal over the air interface. This simplifies the sale of user terminals from third-party vendors.

path gain The ratio of the effective radiated power and the received power of a radio transmission. Path gain is a property of a physical radio path and does not depend on the transmitting or receiving antennas, cable, or other equipment.

phased array A cluster of radio antennas designed to use their relative phase to create a single directional transmitter or receiver.

picocell A cell confined to a very small area, typically an office complex or a single building.

pilot pollution The phenomenon in a CDMA system where multiple antenna faces have nearly equal radio signal path gains to user terminals. Pilot pollution involves the radio channels themselves and not just their pilots.

pilot signal A radio channel sent along with the information channel so the receiver can calibrate itself to the amplitude and power of the unmodulated carrier.

pipe A general term for a trunk, trunk group, or data transmission line between two switches.

power measurement report message (PMRM)   A message from a user terminal in cdmaOne informing the base station of a sequence of frames that exceeds a predetermined value for its *frame error rate* (FER).

processing gain   The effective increase in signal-to-noise (S/N) ratio or E_b/N_0 by using a high *chip* rate to send a lower bit rate in a *direct-sequence spread-spectrum* (DSSS) system.

pseudonoise (PN)   A *pseudorandom* sequence sent on a radio link.

puncturing   A sneaky way of adding a few bits on a data stream with *forward error correction*. We simply take some of the extra bits used for FEC and put other data on them. Puncturing reduces the effectiveness of FEC but allows for transmission of more bits. These extra bits can be signaling bits, power-control bits, or extra user data.

quadraphase phase shift keying (QPSK)   A form of phase shift keying (PSK) where the *constellation* has four points so that each symbol represents 2 bits.

Qualcomm code-excited linear prediction (QCELP)   The form of *code-excited linear prediction* (CELP) used by Qualcomm in its cdmaOne and cdma2000 systems.

Radio Link Protocol (RLP)   A data service in the cdmaOne standard offering rates up to 14.4 kbps.

rake filter   A continuously adapting summation of several radio paths in a multipath environment.

rate set 1 (RS1)   A low-bit-rate cdmaOne codec, 8.55 kbps.

rate set 2 (RS2)   A high-bit-rate cdmaOne codec, 13 kbps.

Rayleigh fading   Changes in received power caused by a *multipath* environment where several nearly equal radio paths interfere with one another.

repeater   A device that amplifies or remodulates a signal that otherwise would be too noisy or too weak. Repeaters are used in both radio and landline transmission.

reverse direction   The radio direction from the user terminal to the base station, also referred to as the *uplink*.

Ricean fading   Changes in received power caused by a *multipath* environment with one strong path and several weaker radio paths.

roaming   A user terminal being served in a wireless system other than its home system.

sector   The geographic area served by an *antenna face* in a wireless telephone system.

seizure message   A message from a user terminal to the wireless system base station answering a page or initiating a call.

selection diversity   A receiver with multiple antennas that monitors all the signal strengths and uses the best one. Contrast *switched diversity*.

semisoft handoff   A hard handoff for the user terminal managed as a soft handoff at the mobile switching center (MSC) so that the transition is seamless to the subscriber.

short message service (SMS)   A service initiated defined as part of GSM and now widely used on cellular systems allowing wireless subscribers to send and receive short text messages over the signaling channel.

signal A message of some kind. We typically use the term *signal* to refer to the desirable component of radio energy. See *noise* and *interference*.

signal-to-interference (S/I) The ratio, usually given in decibels, of a radio signal and all other radio sources reaching the receiver.

signal-to-noise (S/N) The ratio, usually given in decibels, of a radio signal and all the sources of noise. In casual use, we sometimes use S/N for the ratio of signal to any impairment at the radio receiver.

signaling The exchange of information between components in a telephone network that sets up, controls, and terminates telephone calls.

smart antenna Another term for *adaptive phased array*.

soft blocking The notion in CDMA planning that traffic engineering can take into account that neighbor cells and sectors have occupancies less than 100 percent.

soft handoff A handoff environment where a call is served simultaneously by two or more antenna faces.

softer handoff A soft handoff between two antenna faces on the same cell, usually managed at the base station rather than at the MSC.

space division multiple access (SDMA) Using different locations to separate signals using the same radio band at the same time. This term sounds very general but is usually used to refer to the user of *adaptive phased arrays* (smart antennas).

spread spectrum A multiple-access technology where several channels use the same frequency range in the same place at the same time.

spreading gain Another term for *processing gain*.

switched diversity A receiver with multiple antennas that uses one antenna until its signal strength falls below some threshold condition and then switches to the other. Contrast *selection diversity*.

synchronization channel (SCH) The global channels in W-CDMA that synchronize the CDMA modulation and identify each cell sector with one of 512 identifying *pseudonoise* (PN) codes.

temporary local directory number (TLDN) A temporary number assigned to a roaming user terminal and stored in the VLR so that the terminal can be located when there is an incoming call.

time division multiple access (TDMA) A multiple-access technology where channels share the same frequency carrier and are separated by placing them in different *time slots*.

time division synchronized code division multiple access (TD-SCDMA) A 3G CDMA standard combining time and code division and using smart antennas being developed for the mainland Chinese market.

tradeoff The concept of having more than one goal where improving one objective degrades another.

traffic engineering The science of calculating how much telephony resource is required to offer a statistically varying load a specific level of service.

turbo code   A *convolutional code* for *forward error correction* (FEC) with a far longer window than older systems. Turbo codes are superior in error correction at the cost of computational complexity and delay.

universal mobile telephony system (UMTS)   A European term for W-CDMA.

Universal Wireless Communications (UWC)   A consortium and its proposal (UWC-136) that meets the ITU standard for 3G cellular systems using TDMA rather than CDMA.

uplink   Another term for *reverse direction*.

user terminal   A mobile telephone or a wireless data terminal or any other equipment that receives the forward (downlink) signal and sends the reverse (uplink) signal of a wireless telephone technology. We use the general term *user terminal* because we do not know what kinds of equipment wireless subscribers will be using.

valley   In CDMA planning jargon, an area with nearly equal radio path gain to several base stations.

virtual circuit   A connection that appears to be a continuous circuit from its end points but is actually composed of discrete packets. Not only may these packets be discontinuous, they can also travel different routes to their common destination.

visitor location register (VLR)   A computer and database that services subscribers in its service area that are not in their own home systems. See *roaming*.

voice activity detection (VAD)   The determination of the presence or absence of a voice component in a signal, used to determine if transmission of voice is necessary at each moment during a call.

voice over IP (VoIP)   The use of Internet Protocol (IP) packet networks to transport voice over packet pipes or packet networks. VoIP over packet networks relies on the use of *virtual circuits*.

wideband code division multiple access (W-CDMA)   The 3G CDMA standard that is based on GSM.

wideband code division multiple access (WCDMA)   The general class of code division multiple access (CDMA) technologies using bandwidth of several megahertz (MHz) or more. The abbreviation without a hyphen (WCDMA) refers to any wideband CDMA technology.

Wireless Application Protocol (WAP)   A World Wide Web programming interface designed with small screens and minimum graphics appropriate to handheld wireless devices.

wireless data terminal   A user terminal whose primary mission is to provide data service to a wireless subscriber. We use the general term *data terminal* because new technology and new applications are appearing rapidly in the wireless world.

wireless local loop (WLL)   A fixed telephone or private branch exchange (PBX) whose access to the telephone network is wireless. We envision a normal household telephone where the wall plug goes to a user terminal box with radio transmitting and receiving equipment.

References

Antennas Wave Propagation Electromagnetics (AWE), WinProp Software Modules Web site, *http://www.awe-communications.com.*

Paul Bedell, *Wireless Crash Course,* McGraw-Hill, New York, 2001.

Bell Telephone Laboratories, *The Bell System Technical Journal,* special issue on Advanced Mobile Phone Service, Vol. 58, No. 1, American Telephone and Telegraph Company, January 1979.

Henry L. Bertoni, *Radio Propagation for Modern Wireless Systems,* Prentice Hall PTR, Upper Saddle River, NJ, 2000.

CDMA Development Group (CDG) Web site, *http://www.cdg.org.*

David T. Chen and Ivan N. Vuvovic, "CDMA 1x Radio Access Network IP Backhaul Sizing Analysis," Globecom 2002.

COperation européenne dans le domaine de la recherche Scientifique et Technique (COST), "Digital Mobile Radio Towards Future Generation Systems," COST 231 Final Report, 1994.

Digital Voice Systems, Inc., "Independent Evaluation Results" (speech coders), Web page, *http://www.dvsinc.com/eval_results.htm.*

EE Times, Commdesign Web page, *http://www.commsdesign.com.*

Vijay K. Garg, *IS-95 CDMA and cdma2000: Cellular/PCS Systems Implementation,* Prentice Hall PTR, Upper Saddle River, NJ, 2000.

Roch Guérin, Hamid Ahmadi, and Mahmoud Naghshineh, "Equivalent Capacity and Its Application to Bandwidth Allocation in High Speed Networks," *IEEE Journal on Selected Areas in Communcations,* Vol. 9, No. 7, September 1991.

Lawrence Harte, Steve Prokup, and Richard Levine, *Cellular and PCS: The Big Picture,* McGraw-Hill, New York, 1997.

Harry Holma and Antti Toskala, *WCDMA for UMTS, Radio Access for Third Generation Mobile Communications,* revised edition, Wiley, New York, 2001.

Information and Telecommunication Technology Center (ITTC) Web site, *http://www.ittc.ukans.edu.*

International Engineering Consortium, short message service (SMS) Web site, *http://www.iec.org/online/tutorials/wire_sms.*

Leonard Kleinrock, *Queuing Systems,* Vol. I: *Theory,* Wiley, New York, 1975.

Jaana Laiho, Achim Wacker, and Tomáš Novosad (eds.), *Radio Network Planning and Optimization for UMTS,* Wiley, New York, 2002.

Mobile Streams, short message service (SMS) Web site, *http://www.mobilesms.com.*

Nathan J. Muller, *Desktop Encyclopedia of Voice and Data Networking,* McGraw-Hill, New York, 2000.

Nathan J. Muller, *Bluetooth Demystified,* McGraw-Hill, New York, 2001.

Qualcomm Web site, *http://www.qualcomm.com.*

Travis Russel, *Signaling System #7,* McGraw-Hill, New York, 2000.

Andrew Seybold, "Outlook 4 Mobility," Web page, *http://outlook4mobility.com.*

Steven Shepard, *Telecommunications Convergence,* McGraw-Hill, New York, 2000.

Siemens, "White Paper: TD-SCDMA: The Solution for TDD Bands," Web page, *http://www.siemens-mobile.de/btob/external/downloads/TD-SCDMA/TD_SCDMA_White_Paper.pdf.*

Marvin K. Simon, James K. Omura, Robert A. Scholtz, and Barry K. Levitt, *Spread Spectrum Communications Handbook,* electronic edition, McGraw-Hill, New York, 2002.

Clint Smith and Curt Gervelis, *Cellular System Design and Optimization,* McGraw-Hill, New York, 1996.

Randall A. Snyder and Michael D. Gallagher, *Wireless Telecommunications Networking with ANSI-41,* 2d ed., Mc-Graw-Hill, New York, 2001.

SMS Forum Web page, *http://www.smsforum.net.*

Sprint Web page, *http://www.sprint.com.*

Jonathan Stone and Craig Partridge, "When the CRC and TCP Checksum Disagree," SIGCOMM '00, ACM, 2000.

TD-SCDMA Forum Web site, *http://www.tdscdma-forum.org*.

Verizon Communications Web site, *http://www.verizonwireless.com*.

Andrew J. Viterbi, *CDMA: Principles of Spread Spectrum Communication*, Addison, Wesley, Longman, New York, 1995.

Vodafone Group PLC Web page, *http://www.vodofone.com*.

WAP Insight, "What Is TD-SCDMA?" Web page, *http://www.wapinsight.com/what_is_tdscdma.htm*.

Westbay Engineers, Ltd., Telecom Traffic Online Web page, *http://www.erlang.com*.

Index

A

ABR (available bit rate), 228
Acceptable delay, 222
Access attempts, 129
Access links, 184
Access probes, 129
Access tandem office, 150
Access technology, 167
Access-grant channel (AGCH), 84
Accounting, 155, 156, 218, 304
Accuracy, precision vs., 550
ACELP (algebraic code excited linear prediction), 137
Acronyms (list), 587–588
ACs (authentication centers), 193
Adaptive differential pulse code modulation (ADPCM), 242–243
Adaptive matched filters, 29, 99
Adaptive multirate (AMR) speech coding, 137
Adaptive phased arrays, 52
Additive white gaussian noise (AWGN), 25–26
ADPCM (see Adaptive differential pulse code modulation)
Advanced Communications Technologies, Inc., 116
Advanced Intelligent Network (AIN), 184
Advanced Mobile Phone Service (AMPS), 71–79
  call maintenance for, 77–79
  call setup for, 76–77
  conventional reuse for, 331–333
  dialing on, 214
  economics of, 39
  as 1G technology, 118
  FM channel of, 74–76
  future of, 574
  and growth of cellular system, 72
  1AESS processor in, 151
Advertising, 527
Africa, 119
AGCH (access-grant channel), 84
AIN (Advanced Intelligent Network), 184
Air interface, 18
  planning issues for, 433–435
  and portable telephones, 36–37
Air time, 218
Air traffic controllers, 8

Aircraft radios, 5
A-law, 147, 177, 239–241
Alerting, 84
Algebraic code excited linear prediction (ACELP), 137
Algebraic fields, 93
Allocated cost, 524
Allocation:
  of dynamic channel, 141, 365
  spectrum, 115, 579–580
  of TD-SCDMA channels, 141
  of voice and data carriers, 225–226
ALOHA protocol, 138–139, 223, 471
Always-on e-mail, 572
AM (see Amplitude modulation)
AM radio stations, 4
AMA record (see Automatic message accounting record)
American National Standards Institute (ANSI), 112, 116, 117, 191
Amplifiers, 35
  base station power, 53–55
  low noise, 53
  nonlinear, 26
  power, 53–55
Amplitude, 9–10, 99
Amplitude modulation (AM), 18
  quadrature, 21
  and S/N ratios, 30
Amplitude shift keying (ASK), 23
AMPS (see Advanced Mobile Phone Service)
AMR (adaptive multirate) speech coding, 137
Analog interference, 166
Analog modulation, 18, 19, 331
Analog signal, measuring quality of, 30–31
Analog switches, 151, 153
Analog systems:
  as 1G technology, 118
  transport in, 176
Analog wireless telephony, 71–79
  and AMPS call maintenance, 77–79
  and AMPS call setup, 76–77
  and AMPS FM channel, 74–76
  and cellular principle, 72–73
  history of, 71–72
  and managed interference, 73–74

Analytical subscriber traffic models, 565–566
Anchor MSC, 170, 195
ANSI (*see* American National Standards Institute)
ANSI-41, 191–205
   automatic roaming, 198–201
   inter-MSC handoffs, 195–198
   network interaction with, 204–205
   OA&M, 203
   OTASP, 203–204
   signaling with, 179–180
   SMS, 201–203
ANSI-41 network reference model, 191, 192
ANSI-41 signaling capacity, 309–315
   calculation of, 314–315
   call processing, 311–312
   inter-MSC handoffs, 309–310
   other signaling activity of, 313–314
   roaming, 310–311
   SMS, 312–313
ANSI-41E network reference model, 193, 194
Antenna extenders, 36
Antenna faces, 67
Antenna gain, 14–17
Antennas, 3
   for base stations, 49–52
   beam, 36
   directional, 41–42
   downtilting, 421–423, 504–505
   high-gain, 421, 424
   lowering, 422–423, 505
   omnidirectional, 36
   for portable telephones, 36–37
Application-specific integrated circuits (ASICs), 37
Architecture:
   GSM, 81
   ISA, 115
   radio, 66–69
   SS7, 183–186
   wireless telephone system, 167–168
Arcs, 280, 319
ARQ (Automatic repeat request), 127, 378
ASICs (application-specific integrated circuits), 37
ASK (amplitude shift keying), 23
Assignment:
   base station, 441–442, 511–512
   carrier, 424–426
Associated signaling, 183
Asymmetric demand, 469–471
Asynchronous data, 127
Asynchronous processor chips, 38
Asynchronous Transfer Mode (ATM), 578
   classes of service in, 227–228
   voice and data sharing, 250–251
Atmospheric pressure, 5

AT&T, 78, 118, 148, 150, 151
Attenuation, 5
Auctions, 410*n2*
Australia, 119
Authentication centers (ACs), 193
Automated communications, 46–47
Automatic message accounting (AMA) record, 149, 218
Automatic repeat request (ARQ), 127, 378
Automatic roaming, 198–201
Automobiles, 47
Autonomous registration, 169
Auxiliary equipment, 515–516
Available bit rate (ABR), 228
Avaya, Inc., 148
AWGN (*see* Additive white gaussian noise)

B
Backbone backhaul networks, 449–451
Background class (W-CDMA), 136
Backhaul, 57
   design of, 317–324
   increasing capacity by adding, 517–518
   and location planning, 417–418
   measurement of, 483, 486
   planning for, 447–453
   transport of, 178–179
Backhaul design, 317–324
   fan-out, 322–324
   graph theory in, 319–321
   network flow optimization in, 321–322
   traffic engineering in, 318–319
Backhaul fade, 452*n2*
Backhaul planning, 447–453
   backbone, 449–451
   cable/point-to-point/MMDS, 453
   cost issues in, 447–448
   and data direct from base station, 451
   fan-out, 448–449
   growth anticipation in, 452–453
   and reliability engineering, 451–452
Backlobe, 16
Backlobe attenuation, 16
Backward error correction, 228
Bandwidth, 9
   future of, 578
   radio (*see* Radio bandwidth)
   and voice/data carriers, 224–226
Bandwidth territory, 9
Base station capacity, increasing, 495–500
   with carriers, 497–498
   with power, 496–497
   with repeaters, 499–500
   with sectors, 498–499
Base station controller (BSC), 57

Base station planning, 419–436
  and carrier assignment, 424–426
  and cdmaOne to cdma2000 migration, 432
  and coverage, 419–421
  and downtilting/lowering antennas, 421–423
  and FDMA/TDMA to CDMA migration, 428–432
  and peak traffic demand, 426–428
  quality/cost issues in, 433–435
  reliability in, 435–436
  and sector plans, 423–424
Base station system application part (BSSAP), 188
Base station system (BSS), 81, 84
Base stations, 49–61
  antennas for, 49–52
  architectural issues of, 66–67
  changes at, 269–270
  component reliability for, 58–61
  controller for, 57
  increasing capacity of, 495–500
  microcells/picocells/repeaters as, 56–57
  mobile, 579
  MSC assignment of, 441–442, 511–512
  planning for, 419–436
  power amplifiers for, 53–55
  power supply/environmental controls for, 56
  primary/secondary, 104
  radio receiver for, 56
  and regional growth, 525
  towers/cables for, 52–53
  transport to MSC from, 178–179, 318
  and wireless telephony, 66–67
Base transceiver station (BTS), 81
Batteries, 33–35, 37–38
BCCH (broadcast-control channel), 83
Beam antennas, 36
Bell, Alexander Graham, 10
Bell System, 71, 148, 149, 150, 433
Bell Telephone Laboratories, 13, 29n7, 71, 72, 151, 166
Bels, 10–13
BER (see Bit error rate)
Billing, 149, 155, 160, 304
Binary phase-shift keying (BPSK), 19–22
Bit error rate (BER), 31
  and cdmaOne power control, 129
  and data capacity, 373–377
  in data service demand estimation, 401
  and forward error correction, 231
  and sound quality, 166
  and W-CDMA power control, 139
Bit flow, 305–306
Bit-energy-to-noise (E_b/N_0) ratio, 88, 129, 134, 139, 373–377, 401
Bits, 19

Blank and burst messages, 75
Block code, 232
Blocked calls, 164, 286, 361–362
Blocking:
  probabilities for, 288–289
  soft, 385
Blocking rates, 285, 290, 292, 295, 307, 465
Blocking tables, 289–292
Bluetooth technology, 45
Border cells, 199–200, 311, 462
Border sectors, 309
Border zones:
  and signaling capacity planning, 456–457
Borders:
  inter-MSC, 442–443
  new MSC, 512–513
Bottlenecks, 279–280, 280, 318
Boundaries, sector, 408
BPSK (see Binary phase-shift keying)
Bridge links, 184
Broadcast licenses, 410n2
Broadcast (radio), 8
Broadcast-control channel (BCCH), 83
Broken dial tone, 211
BSC (base station controller), 57
BSS (Base station system), 81, 84
BSSAP (base station system application part), 188
BTS (base transceiver station), 81
Buffering, 153
Buildings, 14
Bursty bit errors, 166
Business:
  CDMA for, 572–573
  tools for, 274–276
Business-case models, 533–542
  cost-per-subscriber, 537–540
  costs in, 533–537
  return-on-investment, 541–542
Busy call forwarding (CFB), 312

C
Cable TV, 417–418, 453
Cables:
  base station, 53
  fiber optic, 40
Call answering, 157
Call availability, 41
Call delivery, roamer, 201, 311
Call forwarding, 157, 312
Call maintenance (AMPS), 77–79
Call management, 153–154
Call messaging, 214
Call monitoring, 76
Call notes, 214

Call processing:
  and ANSI-41 signaling, 311–312
  and signaling capacity planning, 457–458
  and switching capacity, 303–304
  as telephone switching function, 154–155
Call quality, 38–39
Call release, 198
Call setup:
  AMPS, 76–77
  GSM, 84–85
  as wireless telephony engineering issue, 169
Call states, 207–218
  defining, 207–208
  PSTN, 209–213
  wireless, 213–215
  wireless radio, 215–218
Call statistics, 160
Call transfers, 211
Call waiting, 157, 210–211, 215, 312
Called party, 163
Caller ID, 157
Calling name presentation (CNAP), 312
Calling number identification presentation (CNIP),
  312
Calling party, 163
CALLING state, 213
Canada, 119
Candlestick Park earthquake, 46
Canyons, 415
Capacity, 9, 271–393
  ANSI-41 signaling, 309–315
  base station, 495–500
  and business/engineering tools, 274–276
  CDMA data, 373–380, 540
  CDMA issues of, 381–393
  cellular network calculations for, 317–329
  and changing channel conditions, 171–172
  and constrained optimization, 280–282
  and conventional reuse, 331–353
  conventional reuse formulas for, 351–353
  Erlang-B table for, 289–290, 292
  and flow, 279–280
  increasing (see Increasing capacity)
  and multipath, 383
  of narrowband AMPS, 75
  as optimization variable, 283
  optimization variables for, 282–284
  and power control/pilots, 383–384
  and quality of service, 222–226
  and quality tradeoffs, 273–284
  quantitative optimization of, 276–278
  and radio estimation, 385–393
  and sectorized cells, 381–382
  and soft blocking, 385

Capacity (Cont.):
  and soft/softer handoff, 382–383
  and speech coding, 384–385
  switching, 303–307
  tradeoffs with, 273–284
  and tradeoffs with quality, 278–279
  and traffic engineering for voice and data, 285–302
Capital expense (CAPEX), 534
Capture ratio, 18
Carrier assignment, 424–426
Carrier asymmetry, 470
Carriers, 3, 82
  base-station capacity increases with, 497–498
  as cell count factor, 403
  higher, 425
"Carrier-to-crud ratio," 30
CBCH (see Cell broadcast-control channel)
CBR (constant bit rate), 227
CCIS (see Common channel interoffice signaling)
CCS7 (Common Channel Signaling 7), 150n7
CDG (see CDMA Development Group)
CDMA (see Code division multiple access)
CDMA Development Group (CDG), 112, 121, 122, 130,
  137, 269
cdmaOne frequency, primary, 5
cdmaOne (IS-95) standard, 23, 112, 114, 125–130,
  224–226
  cell identification of, 130
  chipping rate of, 90
  data capacity of, 378
  economics of, 39
  evolution of, 120
  forward channel coding of, 128
  migration to cdma2000 from, 432
  power control messages from, 102
  power control of, 129–130, 383
  radio bandwidth of, 128
  radio capacity estimation for, 388–389
  reverse channel coding of, 128–129
  as 2G technology, 119
  services of, 126–127
  speech coding in, 127–128, 160
cdma2000 standard, 121, 130–134
  and carrier asymmetry, 470
  cell identification of, 134
  and DO channels, 223
  economics of, 39
  forward channel coding of, 132–133
  future of, 574
  migration from cdmaOne to, 432
  multicarriers of, 123, 268
  power control of, 134, 384
  radio bandwidth of, 132
  reverse channel coding of, 133–134

cdma2000 standard (*Cont.*):
  services of, 131
  speech coding in, 160
  speech coding of, 132
  voice and data carriers in, 224–226, 267–268
cdma2000 1x, 130–131, 389–390
cdma2000 1x EV DO standard, 120, 131, 133, 223, 470, 576
cdma2000 3x services, 267–268
CDPD (cellular digital packet data), 127
Cell broadcast-control channel (CBCH), 83, 86
Cell grids:
  maintaining, 408–409, 503–504
  smaller, 343–344
Cell identification:
  cdmaOne, 130
  cdma2000, 134
  W-CDMA, 139–140
Cell phones, 33
Cell radius, 67–69, 505
Cell splitting, 343–346, 409–414, 502–503
Cells, 71
  adding to, 501–508
  architectural issues of, 66–67
  and downtilting antennas, 504–505
  estimation of required, 402–404, 501–502
  and handoff management, 506–507
  micro-, 505
  pico-, 506
  and reliability, 507–508
  and wireless telephony, 63, 66–67
Cellular backhaul network, 254–255
Cellular digital packet data (CDPD), 127
Cellular network capacity:
  and backhaul design, 317–324
  components of, 324–326
  issues with, 326–329
Cellular principle, 72–73
Cellular subscriber stations (CSSs), 192, 193
CELP (*see* Code-excited linear prediction)
Center-excited cells, 339
Central America, 119
Central office, 146
Centrex, 148
CFB (busy call forwarding), 312
CFD (default call forwarding), 312
CFNA (no answer call forwarding), 312
CFU (unconditional call forwarding), 312
Channel bandwidth, 351
Channel coding, TD-SCDMA, 141
Channels, 82, 147
  assumptions about, 363
  changing conditions of, 171–172
  digital, 82, 153

Channels (*Cont.*):
  dynamic allocation of, 141, 365
  fixed-size, 285–292
  individual radio, 69
  paging, 84
  paging and access-grant, 84
  random-access, 84
  shared packet, 470–471
  slow associated control, 84
  synchronization, 139
  traffic, 82.84
China:
  CDMA in, 576
  2G technology in, 119
  TD-SCDMA in, 43, 125
  3G technology in, 123
  voice call demand in, 399
  WLL in, 40–41
Chip transmission rate, 90
Chipping rate, 90
Chips, 90
Churn, 397, 398
Cingular, 131, 571
Circuit switching, 175, 306, 438–439
Circuits, 175
Class 5, 150
Class 4 office, 149, 150
Class 5 office, 146, 149
Class 5 switches, 65n5
Clipping, 26, 55
Cloning, 216
Closed-loop power control, 101
  cdmaOne, 129
  cdma2000, 134
  W-CDMA, 139
CNAP (calling name presentation), 312
CNIP (calling number identification presentation), 312
Coaxial cable, 40, 575–576
Cochannel interference, 73
Code division multiple access (CDMA), 3, 8
  in China, 576
  current state of, 571–578
  equations for, 355–363
  in Europe, 575–576
  and forward direction, 105
  future of, 578–583
  GSM competing with, 118
  in Japan/Korea, 576–575
  migration from TDMA/FDMA to, 428–432
  for multicellular systems, 355–371
  and multipath/rake filter, 96–100
  in North America, 571–575
  and power control, 100–102
  principles of, 87–107

Code division multiple access (CDMA) (*Cont.*):
  and processing gain, 88–90
  and pseudonoise codes, 91–96
  and reverse direction, 105–106
  and soft handoffs, 103–105
  and spread spectrum, 87–88
  and varying data rates, 102–103
Code rate, 232
Codecs, 127
Coded bits, 231
Code-excited linear prediction (CELP), 126, 243–244
Coherent modulation, 19–23
Coherent sum, 23
Cold spares, 60
Colocation:
  of antennas, 52
  of base station, 65n4
Combinatorial mathematics (combinatorics), 319
Combiners, radio, 54–55
Commercial radio stations, 8
Common channel interoffice signaling (CCIS), 150, 183
Common Channel Signaling 7 (CCS7), 150n7
Compaq, 116
Compatibility:
  international, 119–120
  of narrowband AMPS, 76
Competition, 397–398, 580
Complaints, user (*see* User complaints)
Complete graph, 319–321
Complex spreading, 134
Component reliability, base station, 58–61
Compressed voice, 305
Computer games, 136
Computer simulation, 563–565
Concave functions, 323–324
Conditional call forwarding, 312
Conferencing, 211
CONNECT event, 210
Connectionless services, 187n1
Connections, 64, 131, 152–153
Constant bit rate (CBR), 227
Constant-envelope modulations, 54
Constellations, 20, 21
Constrained optimization, 280–282
Continuous variables, 277–278
Controller, base station, 57
Conventional reuse, 331–353
  AMPS channel, 331–333
  calculation of, 353
  capacity formulas for, 351–353
  cookie-cutter hexagon models of, 350–351
  and demand growth, 342–348

Conventional reuse (*Cont.*):
  GSM channel, 333
  power balancing for, 348–349
  and power control, 340–342
  regular channel patterns of, 336–339
  repeaters in, 57
  and sectors, 339–340
  and signal-to-interference performance, 333–335
  spread spectrum vs., 88
  traffic engineering for, 349–350
Convergence, 221
Conversational class (W-CDMA), 135–136
Convex areas, 413
Convex functions, 323n7
Convolution integral, 335
Convolutional code, 232
Cookie-cutter hexagons, 350–351
Copper twisted pair wire, 40, 146, 575–576
Corner-excited cells, 339
Correlation, 23
COST 231 model, 548
Cost-per-subscriber models, 537–540
Cost(s):
  and air interface planning, 434–435
  with backhaul planning, 447–448
  in business-case models, 533–537
  of diverse routing, 452
  incremental, 524
  and location planning, 416
  and MSC planning, 437, 440, 441
  as optimization variable, 283–284
  of regional growth, 524
Country codes, 204
Coverage:
  and base station planning, 419–421
  as cell count factor, 403
  WLL, 41
Coverage cells, 53
CRC (cyclic redundancy check), 229
Critical resources, 280, 282
CRM (customer relationship management), 572
Cross links, 185
Crossbar switches, 151
Cross-interference, 16
Crosstalk, 331
CSSs (*see* Cellular subscriber stations)
Curvature, earth's, 14
Custom calling, 209, 214
Customer relationship management (CRM), 572
Customers:
  and future of CDMA, 573–574
  and quality of service experience, 221–222
Cyclic redundancy check (CRC), 229

D

D4 channelization, 147$n2$

Dahs, 3

Data, 219–270
  and backhaul planning, 451
  CDMA capacity for, 387, 388, 540
  error recovery in transmission of, 223
  hybrid voice-data networks, 247–256
  market planning for, 472
  over data links, 378
  over voice-coded links, 378
  short message service, 257–265
  speech coding, 235–245
  transmission requirements for, 222
  variation in rates of, 363–364
  wireless data services, 267–270
  and wireless data services, 268

Data calls (MSC main processing), 304

Data capacity, 373–380
  bit error rate and E_b/N_0, 373–376
  cdmaOne, 378
  E_b/N_0, 376–377
  3G, 378–380

Data links:
  data over, 378
  slow, 491–492

Data rates:
  high, 379–380
  varying, 102–103, 363–364

Data service:
  cdmaOne, 127
  demand estimation for, 400–401
  to the PPDN, 465–466

Data speeds, 378–379

Data-error complaints, 492

Data-only (DO) carriers, 221–233, 268

dB (see Decibels)

De facto standards, 112

Decibel notation, 10–13

Decibels (dB), 11–13

Decisions, measurement-based, 566–567

DECT (digital enhanced cordless telecommunications), 123

Dedicated transport, 326–328

Default call forwarding (CFD), 312

Delay:
  acceptable, 222
  measurement of, 483
  and quality of service, 226–227
  signal, 166

Delay requirements, 466

Delay spread, 28–29, 96–100

Dell Computer Company, 529

Demand:
  asymmetric, 469–471
  calculation of, 399–400
  in conventional reuse, 342–348
  data service, 400–402
  estimation of, 397–404
  handling peak, 426–428
  microcells for concentrated, 414–415, 505
  peak traffic, 426–428
  in traffic engineering, 286, 290
  voice call, 397–400, 465

Demodulation, 3

Demodulators, 36

Detail standard, 223$n2$

Diagonal links, 185

Dial tone, 211

Dialed digits, 163

Dialing:
  preorigination, 35, 64, 213–214
  roam, 156

DIALING state, 210, 213

DID (direct inward dialing), 148

Differential QPSK (DQPSK), 22

Diffie-Hoffman Key Agreement Standard, 203–204

Diffraction, 554

Digital camera, built-in, 577

Digital channels, 82, 153

Digital enhanced cordless telecommunications (DECT), 123

Digital interference, 166

Digital modulation, 18–23

Digital radio technology, 3

Digital signal, measurement of, 31

Digital subscriber line access multiplexer (DSLAM), 269, 301

Digital subscriber line (DSL), 40
  and market planning, 472
  traffic engineering issues with, 300–301

Digital switches, 151

Digital transmission technologies, 119

Digital transport protocols, 176–178

DIGITS stimulus, 213

Direct inward dialing (DID), 148

Direct sequence spread spectrum (DSSS), 90, 132$n8$

Directional antennas, 41–42

Directory number, 204

Discontinuous reception (DRx), 35, 84–85

Discontinuous transmission (DTx), 35, 83

Discrete variables, 277–278

Distance models, 546–551

Distortion, 26, 27, 166

Dits, 3

Diversity, 28

DMS-10 switch, 151
DMS-100 switch, 151
DND (do not disturb), 312
DO carriers (*see* Data-only carriers)
Do not disturb (DND), 312
Dolby systems, 74
Doppler distortion, 365
Doppler shift, 29
Dot products, 94
Downlink, 105
Downtilting antennas, 421–423, 504–505
DQPSK (differential QPSK), 22
DRx (*see* Discontinuous reception)
DS-0, 176–178
DS-1, 147, 152, 177, 178, 290, 291
DS-2, 152, 177$n5$
DS-3, 152, 178
DS-4, 178
DSL (*see* Digital subscriber line)
DSLAM (*see* Digital subscriber line access multiplexer)
DSSS (*see* Direct sequence spread spectrum)
DTMF (dual-tone multi-frequency) tones, 163
DTx (*see* Discontinuous transmission)
Dual-mode cell phones, 428
Dual-tone multi-frequency (DTMF) tones, 163
Dynamic channel allocation, 141, 365
Dynamic time alignment, 83

E
E-0, 176–178
E-1, 152, 153, 178, 290, 292
E-2, 177$n5$
E-3, 178
E-4, 178
E-911 (*see* Enhanced 911)
E lines, 150
E_b/N_0 ratio (Bit-energy-to-noise ratio), 88, 129, 134, 139, 373–377, 401
Echo, 96, 146
Echo-cancelers, 35$n2$
Economics:
    of backhaul, 324
    of base stations, 66–67
    of competing standards, 118
    of conventional reuse growth, 348
    of mobile technology, 37–39
    of number of calls, 170–171
    of 3G initiative, 121
EDGE (Enhanced Data Rate for GSM Evolution), 120
Edges, 280, 319
Effective radiated power (ERP), 10, 35
Efficiency:
    radio, 428–429
    traffic, 429

8-ary constellation, 21
EISA (extended ISA), 116
Electrical power, 4
Electromagnetic radiation, 5
Electromagnetic radio wave, 4
Electromagnetic voltage level, 9
Electronic serial number (ESN), 156
Electronic switching system (ESS), 151
Elmhurst, Illinois, 79
E-mail, 136, 572
Emergency services, 217–218, 303, 465, 574, 583
EMS (*see* Enhanced message service)
Encoded bits, 231
END keys, 35$n1$, 64$n1$
Energy, 10
Energy per unit time, 9
Engineering tools, 274–276
Engineers:
    and future of CDMA, 573
    technical issues facing, 574
Enhanced 911 (E-911), 218, 574
Enhanced Data Rate for GSM Evolution (EDGE), 120
Enhanced message service (EMS), 263
    and MSC growth, 516
    and signaling capacity planning, 459
Enhanced variable-rate codec (EVRC), 126, 127, 389–390
Environmental controls, base station, 56
Equipment:
    in business-case models, 534–535
    MSC growth and provisioning of new, 514–515
    MSC growth and upgrade of, 509–511
    MSC provisioning of, 440–441
    PSTN, 146–147
    WLL terminal, 41
Ericsson, 118
Erlang, Agner Krarup, 286
Erlang-B model, 288–292, 307
Erlangs, 286
ERP (*see* Effective radiated power)
Error correction:
    backward, 228
    in GSM voice channel, 82, 83
    for quality of service, 228–232
Error detection, 228–232
Error recovery, 222–223, 229–230
Error-correction systems, 101
Error-resolution algorithm, 231
ESME (external SME), 313
ESN (electronic serial number), 156
4ESS digital switch, 151
ESS (electronic switching system), 151
5ESS local exchange office switch, 151
1ESS switch, 151

Europe:
A-law in, 147
CDMA in, 575–576
digital transport protocols in, 176–178
ISDN Basic Rate in, 152
PSTN in, 150
2G technology in, 119
standards in, 115
3G technology in, 123
W-CDMA in, 135
European Union, 115
EV phase 1, 223
EVolution Data Only (EV-DO), 131
EVRC (see Enhanced variable-rate codec)
Extended dynamic time alignment, 83
Extended ISA (EISA), 116
Extended links, 185
Extended superframes, 177
External SME (ESME), 313

F
F1 cells, 424–426
FA (flexible alerting), 312
FAA (Federal Aviation Administration), 9
FACCH (fast associated control channel), 84
Facsimile (fax), 35n3, 127
Fading, 27–28
Fads, 574, 577
Fan-out backhaul networks, 317–318, 322–324,
    448–449
Fast associated control channel (FACCH), 84
Fax (see Facsimile)
FCC (see Federal Communications Commission)
FCS (frame checksum), 229
FDD (see Frequency division duplex)
FDMA (see Frequency division multiple access)
Feasible region, 278–279
Feature Group D protocol, 150
FEC (see Forward error correction)
Federal Aviation Administration (FAA), 9
Federal Communications Commission (FCC), 9, 53,
    71, 112, 218, 410n2, 471, 574n2
FER (see Frame error rate)
Fiber optic cables, 40
File Transfer Protocol (FTP), 223n1
Filters:
adaptive matched, 29, 99
in FDMA/TDMA receivers, 56
rake, 29, 37, 56, 98–100, 159, 368
Finding (user terminal), 65
First carriers, 425
First generation (1G) technologies, 118
Fixed costs, 533–537
Fixed-size channels (for voice calls), 285–292

Flagpoles, 417
Flags, 187
Flash, switchhook, 210
FLASH stimulus, 210
Flexible alerting (FA), 312
Flow:
of bits through switch, 305–306
network, 321, 322
Flow control, 127
Flow networks, 279–280, 280
FM (see Frequency modulation)
FM channel, 74–76
FM radio stations, 4
Formants of speech, 166
Forward channel coding:
cdmaOne, 128
cdma2000, 132–133
W-CDMA, 138
Forward direction, 100
in CDMA principle, 105
and multicellular systems, 366–368
and power control, 100–103
Forward error correction (FEC), 127, 228, 230–232
and data capacity, 375, 376
in voice transmission, 222–223
Forward link, 105
Forward pilots, 366–367
Forward power levels, 478–479
Fourier analysis, 5
Fourth world, 399
Fourth-generation (4G) technologies, 578–579
Fourth-pilot delta, 406, 408
Four-to-one cell splitting, 411–414
Frame checksum (FCS), 229
Frame error rate (FER), 31, 102, 129, 130, 134
Frames, 31, 161, 177
Fraud, 216
Fraud prevention, 313, 460
Frenkiel, Richard, 71
Frequency, 4–7, 18
Frequency division duplex (FDD), 123, 331
Frequency division multiple access (FDMA), 8
as 1G technology, 118
migration to CDMA from, 428–433
Frequency hopping, 83, 87
Frequency modulation (FM), 18, 331
and S/N ratios, 30
Frequency response, 5–6, 165–166
Frequency shift keying (FSK), 23
Frontiers (of feasible region), 279, 434
FSK (frequency shift keying), 23
FTP (File Transfer Protocol), 223n1
Fully associated links, 185
Fundamental channels, 132–134

G

1G (first generation) technology, 118
4G (fourth generation) technology, 579
2G standards, data rates in, 268
2.5G standards, 268
3G standards, 268
2G technology (*see* Second generation technology)
2.5G technology, 120
3G technology (*see* Third generation technology)
Gartner Group, 274
Gateway MSC, 195
Gateway system, 201
Gauss, Karl Friedrich, 12
Gaussian distribution, 12
Gaussian minimum phase-shift keying (GMSK), 22,
    82
Gaussian noise, 26
General Packet Radio Service (GPRS), 120
Generations (of standards), 118–123
Germany, 577
Gerrymandering, 443
Glare conditions, 167, 217
Global Positioning System (GPS), 46–47, 574
    in cell identification, 574
    complaints tracked with, 487
    and W-CDMA interactive class, 136
Global System for Mobility (GSM), 8, 81–86
    architecture of, 81
    call setup for, 84–85
    CDMA competing with, 118
    components of, 35, 36
    conventional reuse for, 333
    and fading, 28
    future of, 575
    handovers for, 85, 86
    international compatibility of, 120
    power control for, 85
    as 2G technology, 119
    short message service for, 86
    signaling with, 83–86
    and TDMA channel concept, 81–82
    TDMA channel of, 82–83
GMSK (*see* Gaussian minimum phase-shift keying)
Golay codes, 139
Government regulations, 115
GPRS (General Packet Radio Service), 120
GPS (*see* Global Positioning System)
Graph theory, 319–321
Green areas, 488
Grid maintenance:
    and adding cells, 503–504
    and base station location planning,
        408–409
Gross chip rate, 400

Growth:
    and backhaul planning, 452–453
    and conventional reuse, 342–348
    and MSC planning, 445
    reasons for change and, 492–493
    specific areas of regional, 523–530
    user complaints as basis for, 492–493
Growth cells, 53
GSM (*see* Global System for Mobility)
GSM-type user ID cards, 576
Guard bands, $29n7$, 137

H

Hadamard-Walsh orthogonal codes, 94
Hail, $5n3$
Handed down (calls), 426
Handoffs, 73
    AMPS, 78–79
    back, 197, 309
    and call release, 198
    in circuit-switching, 306
    forward, 196–197, 309
    hard (*see* Hard handoffs)
    intermode, 482–483
    inter-MSC (*see* Inter-MSC handoffs)
    management of, 506–507
    measurement of, 196, 309, 310, 480–483
    as mobile switch function, 157–158
    in MSC main processing, 305
    and MSC-to-MSC transport, 180–181, 461–462
    path minimization for, 197–198, 309–310
    and signaling capacity planning, 455–457
    soft (*see* Soft handoffs)
    softer (*see* Softer handoffs)
    as wireless telephony engineering issue, 169–170
Handovers, $65n4$, 85, 86
HANG-UP state, 210, 211
Hard dollar value, 274
Hard handoffs, 426
    in circuit-switching, 306
    as mobile switch function, 159
    in MSC main processing, 305
Hard-disk cluster size, $377n3$
Harmonics, $5n4$
Harmonized PSK, 134
Hata-Okumura propagation model, 546–547, 550, 551
Heinlein, Robert, $278n4$
Hertz (Hz), 4
Heuristics, 449
Hexagons, cookie-cutter, 350–351
Higher carriers, 425
High-gain antennas, 421, 424
High-speed data, 122
High-speed landline access, 572

Hills, 14
Hindenburg, 577
Home location register (HLR), 170
   and ANSI signaling, 180
   and ANSI-41 signaling capacity, 310–311
   in ANSI-41 wireless system, 193
   and automatic roaming, 199
   and call processing, 311–312
   capacity of, 324–325
   fault recovery for, 200–201
   in GSM call setup, 84
   and MSC growth, 515
   in MSC main processing, 303–304
   and MSC planning, 443
   and roaming registration, 156
   and signaling capacity planning, 457–458
Home MSC, 195
Home system, roaming vs., 155
Hong Kong, 399, 576
Hot spares, 60
"Hot spots," 14
HTML (HyperText Markup Language), 223*n1*
Hybrid circuits, 146
Hybrid voice-data networks, 247–256
   ATM sharing with, 250–251
   physical transport sharing with, 248–250
   quality of service for, 247–248
   voice over IP with, 251–255
HyperText Markup Language (HTML), 223*n1*
Hz (hertz), 4

I

I (*see* Interference)
I (square root of −1), 21*n11*
IBM, 38, 115–116
IBM PC, 124
IEEE (*see* Institute of Electrical and Electronics
      Engineers)
IEEE-1992, 274–275
IEEE 802.11b standard, 472–473
Image transmission, 401, 573
IMEI (international mobile equipment identity), 84
Impairments (to radio signal quality), 25–29
Impedance, 13
Improved Mobile Telephone Service (IMTS), 71, 73
Improvement, 493
IMSI (international mobile service identity), 85
IMT-2000, 123
IMT-DS (direct spread), 123
IMT-FT (frequency time), 123
IMT-MC (multicarrier), 123
IMTS (*see* Improved Mobile Telephone Service)
IMT-SC (single carrier), 123
IMT-TC (time code), 123

In-band signaling, 84
In-building models, 553–554
INCOMING-CALL stimulus, 209, 210
Increasing capacity, 475–530
   of base station, 495–500
   with cells added to CDMA system, 501–508
   with MSC growth, 509–516
   and performance measurement for growth,
      477–486
   with regional growth in specific area, 523–530
   with transport added, 517–521
   with user complaints, 487–494
Incremental cost, 524
India, 119, 399
Individual radio channels, 69
Industry standard architecture (ISA), 115
Industry standards, 112
Ineffective attempts, 164–165, 488–489
Information bits, 231
INFORMS (Institute for Operations Research and
      Management Sciences), 273
Inner loops, 102
   cdmaOne, 129
   cdma2000, 134
   W-CDMA, 139
Inner products, 94
In-phase axis, 21
Institute for Operations Research and Management
      Sciences (INFORMS), 273
Institute of Electrical and Electronics Engineers
      (IEEE), 112, 117, 273
Integrated Services Digital Network (ISDN), 40, 131,
      183
   circuit-switching capacity in, 306
   designing messaging systems in, 212–213
   traffic engineering affected by adding, 293–295
Intel, 38, 116, 580
Interactive class (W-CDMA), 136
Intercarrier handoffs, 481–482
Interexchange carriers (IXCs), 150
Interfaces, standards vs., 113
Interference (I), 26, 30, 51*n3*
   AMPS, 331–333
   assumptions about, 363
   cochannel, 73
   digital, 166
   managed, 73–74
   syllabic, 166
Interference-protection threshold, 78
Interim Standard 41 (IS-41), 191
Interim Standard 95 (IS-95), 112, 114, 125–126
Interleaving, 31, 377
Intermode handoffs, 482–483
Inter-MSC borders, 442–443

Inter-MSC handoffs:
  and ANSI signaling, 180
  and ANSI-41 signaling capacity, 309–310
  in ANSI-41 wireless system, 195–198
  measurement of, 481
  and MSC-to-MSC transport, 461–462
Inter-MSC pipes, 181
Internal noise (N), 30
International mobile equipment identity (IMEI), 84
International mobile service identity (IMSI), 85
International numbering plans, 204
International operability, 576
International standards, 112
International Telecommunications Union (ITU), 112,
    115–117, 131, 191
Internet:
  in ANSI-41 wireless system, 193
  asymmetric demand on, 469
  packet transmission on, 161
  VoIP on the, 252–253
Internet Protocol (IP), 148
Internet Protocol version 4 (IPv4), 579
Internet Protocol version 6 (IPv6), 578–579
Intersystem activity, 193
Interworking function (IWF), 193, 205
Invisible mobile, 581
IP (Internet Protocol), 148
IP services, 300–301
IP support, 269
IP-based packet communications, 160–161
IPv6 (Internet Protocol version 6), 578–579
IPv4 (Internet Protocol version 4), 579
IS-95 (see Interim Standard 95)
IS-41 (Interim Standard 41), 191
IS-126 Outdoor, 120
IS-95 standard (see CdmaOne standard)
IS-95A, 125
IS-96A, 127–128
ISA (industry standard architecture), 115
IS-95B, 126
ISDN (see Integrated Services Digital Network)
ISDN Basic Rate, 152
ISDN primary rate interface (PRI), 152
ISDN user part (ISUP), 187
Israel, 43, 119
ISUP (ISDN user part), 187
ITU (see International Telecommunications Union)
ITU-T Recommendation H.324M, 136
IWF (see Interworking function)
IXCs (interexchange carriers), 150

J
J, 21n11
Jamming, 87, 132n8

Japan:
  automated communications in, 47
  CDMA in, 576–575
  IPv6 in, 578–579
  μ-law in, 147
  2G technology in, 119
  spectrum allocation in, 115
  W-CDMA in, 135
JD (see Joint detection)
Jitter, 226–227
Joint detection (JD):
  and multicellular systems, 369
  TD-SCDMA, 142

K
Kazakhstan, 119
Keypads, 34–35, 37
Korea, 131, 576–575

L
Lamar, Heddy, 87
Landline-to-wireless signal path, 65–66
Last-number-dialed, 157
LATA (see Local access transport area)
Latency, 226–227
L-carriers, 176, 177
Lease fees, 66
Leased lines, 175–176
Leaves, 14
LECs (local exchange carriers), 150
License fees, 122
Licenses, broadcast, 410n2
Linear network flow problems, 321, 322
Linear predictive coding, 243–244
Linear programs, 281
Linear pulse code modulation, 238–239
Lines, 146, 175–176
Link budget, 17–18
Little's law, 297–298
LNA (low noise amplifier), 53
Load capacity, transport, 438
Local access transport area (LATA), 464, 466
Local exchange carriers (LECs), 150
Local exchange offices, 65, 146, 149, 183
Local loops, 40–44, 145–146
Local number portability (LNP), 157n13, 184
Local switches, 65n5
Location management, 199–200
Location planning:
  for base stations, 405–418
  and cell splitting, 409–414
  criteria for, 415–418
  and grid maintenance, 408–409
  and microcells, 414–415

Location planning (*Cont.*):
  for MSCs, 440, 513–514
  and pilot pollution, 405–408
Location-based services, 136
Locked cellular telephone, 218*n8*
Logical models, 188–190
Log-normal distribution, 12
Loop-back testing, 302, 303
Loopback trunks, 329
Loss (of information), 227, 228*n9*
Lost calls, 165, 489–490
Lousy-call complaints, 490–491
Low noise amplifier (LNA), 53
Lowering antennas, 422–423, 505

M

MAC protocol (*see* Medium access control protocol)
Macroscopic diversity, 28
MAH (mobile access hunting), 312
Main processor:
  and MSC planning, 438
  switching capacity of, 303–305
Maintenance, access to, 416
Managed interference, 73–74
MANs (metropolitan area networks), 466
Many-to-one environment, 100
MAP (mobile application part), 188
Market-entry planning, 471–473
Marketing plans, worldwide, 530
Marketing presence, regional growth and, 529–530
Markets, voice call demand in other world, 399
Maximum length linear shift register sequence, 92–93
MCs (message centers), 202
Mean time between failures (MTBF), 435
Mean time to repair (MTTR), 436
Measurement:
  of handoffs, 196, 309, 310
  of radio signal quality, 30–31
  of sound quality, 165–167
  of system performance for growth, 477–486
Measurement errors, 29–30
Measurement-based decisions, 566–567
Medium access control (MAC) protocol, 134, 470–471
Memory storage, 39, 157
Message centers (MCs), 202
Message count, handoff, 310
Message transfer part (MTP), 187
Metropolitan area networks (MANs), 466
Metropolitan areas, 574*n2*
Microcells, 56–57
  adding, 505
  and base station location planning, 414–415
  data rates in, 268
Microphones, 35

Microscopic diversity, 28
Microwave, point-to-point, 453
Microwave bands, 5–7, 9
Midamble, 138
Migration, 428–432
  to CDMA, 428–432
  to cdma2000, 432
  evolution of, 430–432
  for radio efficiency, 428–429
  for traffic efficiency, 429
Military research, 8
MIN (*see* Mobile identification number)
Miniaturization, 38
Mixed voice and data carriers, 224–225
μ-law, 147, 177, 239–241
MMDS (*see* Multichannel multipoint distribution
  system)
MMS (*see* Multimedia message service)
MMSC (multimedia message switching center), 459
Mobile access hunting (MAH), 312
Mobile application part (MAP), 188
Mobile base stations, 579
Mobile identification number (MIN), 156, 204
Mobile switch functions, 154–161
  call statistics, 160
  handoff, 157–158
  hard/semisoft handoff, 159
  paging, 154–155
  registration, 154
  roam dialing, 156
  roaming, 155–156
  roaming registration, 156
  roaming support for PCS, 157
  soft handoff, 158–159
  softer handoff, 159
  speech coding, 160–161
Mobile switching center (MSC):
  capacity of, 303–307
  changes at, 269–270
  estimation of required number of, 303–307
  growth of, 509–516
  performance measurement of, 483–484
  regional growth and capacity of, 526
  transport planning for (*see* MSC transport planning)
  in wireless-to-landline signal paths, 65
Mobile switching center (MSC) planning, 437–446
  auxiliary component, 443–445
  and base station assignments, 441–442
  and equipment provisioning, 440–441
  estimation of requirements in, 437–439
  growth anticipation in, 445
  and inter-MSC borders, 442–443
  and location, 440
Mobile telephone switching office (MTSO), 65*n3*

Mobile telephones, 33–39 (*See also* Portable telephones)
Mobile-to-mobile calls:
  capacity adjustment for, 328–329
  and MSC-to-MSC transport planning, 463
Models, 531–567
  business-case, 533–542
  future of, 575
  propagation, 543–559
  reliability, 58–60
  SS7 signaling, 188–190
  subscriber traffic, 561–567
Modem signals, 35*n3*
Modems, 56
Modulation, 3, 18–23
  analog, 18, 19
  coherent, 19–23
  constant-envelope, 54
  digital, 18–23
  frequency, 18, 30, 331
  noncoherent, 19
  quadrature amplitude, 21
Modulators, 35, 36
Morse code, 3
Motorists, 583
Motorola, 75, 118
Moving calls, 456
MR-ACELP (multirate ACELP), 137
MSC (*see* Mobile switching center)
MSC count estimation, 437–439
MSC growth, 509–516
  auxiliary equipment, 514–515
  base station assignment for new, 511–512
  and borders for new, 512–513
  equipment provisioning for new, 514–515
  equipment upgrade determination for, 509–511
  and location of new, 513–514
MSC transport planning, 461–467
  MSC-to-MSC, 461–463
  MSC-to-network, 463–467
MSC-to-MSC communication, 180–181
  increasing capacity with, 519–520
  measurement of, 485–486
MSC-to-MSC transport planning, 461–463
MSC-to-network transport planning, 463–467
MSC-to-PPDN communication:
  increasing capacity with, 520–521
  measurement of, 485
MSC-to-PSTN communication, 181
  increasing capacity with, 520
  measurement of, 485
MTBF (mean time between failures), 435
MTP (message transfer part), 186, 187
MTSO (mobile telephone switching office), 65*n3*

MTTR (mean time to repair), 436
MUD (multiuser detection), 142
Multicellular systems, CDMA equations for, 355–371
  and application to both directions, 363–366
  forward direction application of, 366–368
  reverse direction application of, 368–369
  statistical applications of, 369–371
Multichannel multipoint distribution system (MMDS):
  and backhaul network planning, 453
  future of, 579
Multimedia files, 136
Multimedia message service (MMS), 263–264, 516
Multimedia message switching center (MMSC), 459
Multimedia service (MMS):
  demand estimation for, 401
  and signaling capacity planning, 459
Multipath, 27–29
  and CDMA capacity, 383
  in CDMA principle, 96–100
  and multicellular systems, 368, 369
Multipath fading, 27
Multiple access, 7–8
Multirate ACELP (MR-ACELP), 137
Multiuser detection (MUD), 142

N
N (internal noise), 30
NAM (number-assignment module), 203
N-AMPS (*see* Narrowband AMPS)
Narrowband AMPS (N-AMPS), 75–76, 118
National standards, 112
Near-far problem, 100–101
Net present value (NPV), 448
Net revenue, 274
Network flow, 321, 322
Network operations center (NOC), 486
Network reference models, 188
Network simplex method, 322
Networks:
  capacity calculations for cellular (*see* Cellular network capacity)
  private wireless, 466–467
New York City, 72, 73, 452*n2*
900–MHz band, 9
911 calls, 303, 465, 574
1984 (George Orwell), 23
No answer call forwarding (CFNA), 312
NOC (network operations center), 486
Nodes, 280, 319
Noise, 25–26
  assumptions about, 363
  internal, 30
Nokia, 118
Noncoherent modulation, 19

Nonlinear amplifiers, 26
Nonlocal service quality, 173
Non-real-time (VBR-NRT), 227
Nonregular channel reuse, 353
Normal distribution, 12
Nortel Networks, 269
North America:
  CDMA in, 571–575
  cdmaOne in, 126
  digital transport protocols in, 176–178
  DS-1 in, 152
  international compatibility in, 119
  μ-law in, 147
  numbering plan in, 149n5
  roam dialing in, 156
  2G technology in, 119
  standards in, 115
  3G technology in, 123
Northern Telecom, 151
NPDB (number portability database), 193
NPV (net present value), 448
Number portability, 157n13, 184
Number portability database (NPDB), 193
Number-assignment module (NAM), 203
Numbering plans, international, 204

O
OA&M (*see* Operations, administration, and
    maintenance)
Occupancy:
  as cell count factor, 403
  Erlang-B table for, 290, 291, 292
OCQPSK (orthogonal complex QPSK), 134
OFDM (orthogonal frequency division multiplexing),
    578
OFF-HOOK state, 163, 208
Office code, 149
Offset QPSK (OQPSK), 22, 129
Omnidirectional antennas, 36
On cellular backhaul network, 254–255
One server, 355–358
1AESS analog switch, 306n4
1AESS processor, 151
ONE-OF-THEM-HANGS-UP state, 210
One-to-many broadcast environment, 100
1xEV-DO (*see* cdma2000 1x EV DO)
ON-HOOK state, 164, 208
Open standards, 112
Open System Interconnect (OSI) model, 113, 176, 186
Open-loop power control, 101
  cdmaOne, 129
  cdma2000, 134
  W-CDMA, 139
Operability, international, 576

Operating expense (OPEX), 534
Operations, administration, and maintenance, 203
Operations, administration, and maintenance
    (OA&M), 313
OPEX (operating expense), 534
Optimization, 273
  constrained, 280–282
  of mobile telephone quality, 74
  network flow, 321, 322
  quantitative, 276–278
  variables of, 282–284
OQPSK (*see* Offset QPSK)
Organization, PSTN, 149–150
Originating MSC, 195
Orography, 544
Orthogonal codes, 23–24, 94–96, 367
Orthogonal complex QPSK (OCQPSK), 134
Orthogonal frequency division multiplexing (OFDM),
    578
Orthogonal matrices, 95
Orthogonal sequences, 94
Orthogonal variable spreading factor (OVSF), 95, 133,
    138
Orwell, George, 23
OSI model (*see* Open System Interconnect model)
OTAF (over-the-air activation function), 180
OTAPA (over-the-air parameter administration), 203
OTASP (*see* Over-the-air service provisioning)
Outages, 435
Outer loops, 102
  cdmaOne, 129
  cdma2000, 134
  W-CDMA, 139
Out-of-band signaling, 84
Overflow, 172–173, 485
  of blocked traffic, 361–362
  traffic efficiency with, 429
Overlaid cells, 67–68, 346–348, 353
Over-the-air activation function (OTAF), 180
Over-the-air parameter administration (OTAPA),
    203
Over-the-air service provisioning (OTASP), 313
  and ANSI-41, 180, 203–204
  in ANSI-41 wireless system, 193, 195
OVSF (*see* Orthogonal variable spreading factor)

P
4π radiator, 15
Packet communications, IP-based, 160–161
Packet data:
  MSC-to-PPDN links for, 485
  traffic engineering for, 294–299
Packet delay, 226
Packet pipes, 161

Packet services:
  cdmaOne, 127
  cdma2000, 131
Packet switching:
  capacity of, 306–307
  and MSC planning, 439
Packets, 161
PAGCH (paging and access-grant channel), 84
Pagers, 8, 38
Paging, 35
  in landline-to-wireless signal paths, 65
  as mobile switch function, 154–155
  with narrowband AMPS, 76
Paging and access-grant channel (PAGCH), 84
Paging channel (PCH), 84
Pakistan, 399, 578
PANs (*see* Personal area networks)
Parallel systems, 59–60
Parity bits, 229
Password call acceptance (PCA), 312
Path, wireless signal, 63–66
Path gain, 14, 555
Path loss, 12, 14
Path minimization, 181, 197–198, 309–310
PBX (*see* Private branch exchange)
PC industry, 115–116
PCA (password call acceptance), 312
PCH (paging channel), 84
PCI (peripheral component interconnect), 116
PCM (*see* Pulse code modulation)
PCS (*see* Personal Communications Services)
Peak traffic demand, 426–428
Percussive time responses, $5n4$
Performance measurement for growth, 477–486
  backhaul, 483
  forward power level, 478–479
  handoff activity, 480–483
  MSC, 483–484
  radio signal path, 477–478
  reverse power level, 479–480
  signaling capacity, 484
  transport, 484–485
Peripheral component interconnect (PCI), 116
Permanent prevention, 494
Personal area networks (PANs), 45, 581
Personal Communications Services (PCS), 117
  roaming support for, 157
  as 2G technology, 119
Phase, 5
Phase shift, 6
Phased arrays, 51–52
Philips Electronics, 38
Physical models, 188–190
Physical units, 589–591

PICK-UP stimulus, 208, 210
Picocells, 57
  adding, 506
  data rates in, 268
Picture phones, 573
PicturePhone, 264, 401
Pilot delta maps, 566
Pilot pollution, 367
  and base station location planning, 405–408
  and carrier assignment, 425–426
  microcell patches for, 415
  and sector planning, 423
Pilot signals, 19
Pilots, 383–384
Pipes, 65
Pittsburgh, Pennsylvania, $57n8$
Plain old telephone service (POTS), $65n5$
Planning (for CDMA capacity), 395–473
  asymmetric demand, 469–471
  backhaul, 447–453
  base station, 419–436
  location of base stations, 405–418
  market-entry, 471–473
  mobile switching center, 437–446
  MSC transport, 461–467
  signaling capacity, 455–460
  wireless telephone demand estimation,
    397–404
Planning horizons, 398, 403
PMRM (power-measurement report message),
  130
PN codes (*see* Pseudonoise codes)
Points, $21n10$
Points of presence (POPs), 150, 450
Point-to-point microwave, 453
Poisson distributions, 286–288, 300
Police forces:
  and cellular phones, 583
  and real-time identification, 47
  wireless networks for, 46
Politics, standards and, 114–116
POPs (*see* Points of presence)
Portability, number, $157n13$, 184, 193
Portable telephones, 33–39
  and antennas/air interface, 36–37
  components/engineering of, 33–36
  economics of, 37–39
Porter, Philip, 71
Postage-stamp model, 536–537
POTS (plain old telephone service), $65n5$
Power, 9–10
  base-station capacity increases with, 496–497
  measurement of, 478–480
Power amplifiers, 53–55

Power balancing, 348–349
Power control:
  AMPS, 77–78
  assumptions about, 363
  CDMA, 100–102
  and CDMA capacity, 383–384
  and CDMA equations, 364
  cdmaOne, 129–130
  cdma2000, 134
  and conventional reuse, 340–342
  GSM, 85
  in multicellular systems, 364
  of radio systems, 29–30
  TD-SCDMA, 142
  W-CDMA, 139
Power sum, 23–24
Power supply, base station, 56
Power-measurement report message (PMRM),
  130
PPDN (*see* Public packet data network)
Precision, accuracy vs., 550
Preorigination dialing, 35, 64, 213–214
PRI (ISDN primary rate interface), 152
Price promotions, 527–528
Prices, 282
Primary base station, 104
Primary route, 299
Primary synchronization channel (SCH), 139
Primary threshold, 78
Private branch exchange (PBX), 147–148, 211
Private wireless networks, 466–467
Processing gain, 88–90
Project Rainbow, 580
Propagation maps, 488–489
Propagation models, 543–559
  CDMA issues/challenges for, 543–544
  distance, 546–551
  in-building, 553–554
  radio maps in, 558–559
  ray-tracing, 554–555
  and signal matrix, 557–558
  statistical view of, 555–557
  terrain-based, 551–553
  variation in radio, 544–546
Proportional-cost model, 535–536
Proprietary standards, 112
Protocols:
  SS7 signaling, 186–188
  telephone transport, 176–178
PS/2 standard, 116
PSAP (public service answering point), 465
Pseudonoise (PN) codes, 90–96
Pseudorandom sequences, 90, 92–93
PSTN (*see* Public Switched Telephone Network)

Public packet data network (PPDN), 131, 193
  packet links from MSC to, 485
  VoIP on the, 254
Public service answering point (PSAP), 465
Public Switched Telephone Network (PSTN), 145–150
  call states for, 209–213
  equipment for, 146–147
  in GSM call setup, 84
  local loop, 145–146
  local loop of, 40
  organization of, 149–150
  PBX, 147–148
  transport of MSC to, 181
  voice links from MSC to, 485
  VoIP on, 253–254
Pulse code modulation (PCM), 147, 236–243
  adaptive differential, 242–243
  linear, 238–239
  μ-law/A-law, 239–241
Pulse dialing method, 151
Puncturing, 232

Q
16QAM constellation, 21, 22
QAM (quadrature amplitude modulation), 21
QCELP (Qualcomm code excited linear prediction),
  160
QoS (*see* Quality of service)
QPSK (*see* Quadraphase phase-shift keying)
Quad links, 185
Quadraphase phase-shift keying (QPSK), 20–22, 82,
  333
Quadrature amplitude modulation (QAM), 21
Quadrature axis, 21
Qualcomm, 112, 116, 125, 580
Qualcomm code excited linear prediction (QCELP),
  160
Quality:
  and air interface planning, 433–435
  and capacity tradeoffs, 273–284
  cultural differences in, 577
  nonlocal service, 173
  as optimization variable, 282–283
  radio signal (*see* Radio signal quality)
  of service (*see* Quality of service)
Quality of service (QoS), 221–233
  and capacity, 222–226
  customer experience of, 221–222
  error detection/correction for, 228–232
  for hybrid voice-data networks, 247–248
  ineffective attempts, 164–165
  and latency/jitter/loss, 226–228
  lost calls, 165
  over the air interface, 267–268

Quality of service (QoS) (*Cont.*):
  packet data, 296
  sound quality, 165–167
  telephony engineering, 164–167
Quantitative optimization, 276–278
Queuing theory, 296–297, 307
QUIET state, 209, 210, 213

R
RACH (random-access channel), 84
Radiation hazards, 39
Radiation patterns, 16, 36, 50–51
Radio, 1–107
  analog wireless telephony, 71–79
  base station, 49–61
  basic wireless telephony, 63–69
  CDMA principle, 87–107
  radio engineering, 3–24
  signal quality, 25–31
  TDMA wireless telephony, 81–86
  user terminal, 33–48
Radio architecture, 66–69
Radio bandwidth:
  cdmaOne, 128
  cdma2000, 132
  TD-SCDMA, 140–141
  W-CDMA, 137
Radio capacity estimation, CDMA,
    385–393
  cdmaOne, 388–389
  cdma2000 1x, 389–390
  TD-SCDMA, 391–393
  W-CDMA, 390–391
Radio channels, individual, 69
Radio combiners, 54
Radio engineering, 3–24
  adding radio signals, 23–24
  amplitude/power, 9–10
  antenna gain, 14–17
  bandwidth territory, 9
  decibel notation, 10–13
  frequency, 4–7
  link budget, 17–18
  modulation, 18–23
  multiple access, 7–8
  radio, 3–4
  radio path, 13–14
Radio Link Protocol (RLP), 127, 378
Radio path, 13–14, 14
Radio propagation maps, 558–559
Radio propagation variation, 544–546
Radio receivers, base station, 56
Radio signal path measurement,
    477–478

Radio signal quality, 25–31
  impairments to, 25–29
  and measurement errors, 29–30
  measurement of, 30–31
Radio signals, adding, 23–24
Radio territory, 9
Radio wave, 3, 4
Radio frequency (RF), 53
Rain, 5
Random sequences, 92
Random-access channel (RACH), 84
Rayleigh distribution, 27, 28, 332
Rayleigh fading, 27–28, 50, 56, 97, 100
  AMPS, 332
  in GSM, 83
  and sound quality, 166
Ray-tracing models, 554–555
Received signal strength indicator (RSSI) units, 30
Receivers, 35–35
Red areas, 489
Red laser light, 5
Redundancy:
  and base station planning, 435
  telephony engineering, 167
  and traffic engineering, 301–302
Reflection, 554
Refraction, 554
Regional growth, 523–530
  determination of areas for, 527–530
  targeting, 527–530
  and worldwide marketing plans, 530
Registration:
  autonomous, 169
  as mobile switch function, 154
  roaming, 156
  user terminal, 168–169
Registration cancellation message, 311
Regular channel reuse, 336–339
Regular pulse excitation—long-term prediction (RPE-
    LTP), 82, 160, 333
Reliability:
  and adding cells, 507–508
  and backhaul planning, 451–452
  of base station components, 58–61
  in base station planning, 435–436
  in telephony engineering, 167
  and traffic engineering, 301–302
Reliability model, 58–60
Repair, 493
Repeaters, 57, 415, 499–500
Requirements specification, 275
Requirements statement, 275
Retransmission, 268
Return-on-investment (ROI) models, 274, 541–542

Reuse partitioning, 346–348
Reverse channel coding:
  cdmaOne, 128–129
  cdma2000, 133–134
  W-CDMA, 138–139
Reverse direction, 100
  in CDMA principle, 105–106
  and multicellular systems, 368–369
  and power control, 100–103
Reverse pilots, 368
Reverse power levels, 479–480
RF (radio frequency), 53
Ricean distribution, 28
Ricean multipath fading, 28
Ring, D. H., 71n1
RINGING state, 209–211, 213
RLP (*see* Radio Link Protocol)
RMS (root mean square), 55
Road rallies, 478n1
Roam dialing, 156
Roamer terminations:
  capacity for, 328
  and MSC transport planning, 462–464
Roamers, 458
Roaming, 65
  and ANSI-41 signaling, 310–311
  automatic, 198–201
  and call delivery, 201
  HLR/VLR, 199
  and HLR/VLR fault recovery, 200–201
  and location management, 199–200
  as mobile switch function, 155–156
  registration of, 156
  service qualification, 199
  and support for PCS, 157
  and user terminal state management, 200
  as wireless telephony engineering issue, 170
Robbed-bit signaling, 152, 177
ROI models (*see* Return-on-investment models)
Root mean square (RMS), 55
Root-cause analysis, 494
Rotary dial, 151, 163n2
Routing, network, 299–300, 324
RPE-LTP (regular pulse excitation—long-term
  prediction), 82
RS1 standard, 127, 128, 130, 389
RS2 standard, 127–130, 389
RSSI (received signal strength indicator) units, 30
Rural areas, 574n2
Russia, 119

S
SACCH (slow associated control channel), 84
Sales forces, 572

SAT (*see* Supervisory audio tone)
Satellite communications, 46, 47
Satellite transmission, 5, 46
SCA (selective call acceptance), 312
SCCP (signaling connection control part), 187
SCH (synchronization channel), 139
SCP (*see* Service control point)
SDCCH (*see* Stand-alone dedicated control channel)
SDMA (space division multiple access), 134
Seamless handovers, 86
Second generation (2G) technology, 119–120, 378
Second world, 399
Secondary base stations, 104
Secondary cells, 345
Secondary route, 299
Secondary threshold, 78
Sector boundaries, 408, 423
Sectors, 16
  architectural issues of, 67
  base-station capacity increases with, 498–499
  and CDMA capacity, 381–382
  as cell count factor, 403
  and conventional reuse, 339–340, 352
  planning for, 423–424
  and wireless telephony, 67
Security systems:
  future of, 581–582
  IBM PC, 124
Seizure messages, 76–77
Selection diversity, 28
Selective call acceptance (SCA), 312
Semisoft handoff, 159
SEND keys, 35, 64
September 11, 2001 terrorist attacks, 46, 47,
  582–583
Series systems, 58–59
Server groups, 67–69
Server(s):
  multiple, 358–360
  one, 355–358
Service control point (SCP), 184, 185
Service qualification, 199, 311
Service switching point (SSP), 183–184
Service-level agreements (SLAs), 452, 467
Services:
  cdmaOne, 126–127
  cdma2000, 131
  regional growth and mix of, 528–529
  TD-SCDMA, 140
  W-CDMA, 135–136
Serving MSC, 170, 195
Seven-layer OSI model, 186
Seybold, Andrew, 580
Shared packet channels, 138–139, 470–471

Sharing, tower, 416–417
Short message entities (SMEs):
    and ANSI-41 signaling capacity, 313
    in ANSI-41 wireless system, 193
    short message service (SMS), 201–202
Short message service (SMS), 257–265
    advanced uses of, 262
    ANSI-41, 201–203
    and ANSI signaling, 180
    and ANSI-41 signaling capacity, 312–313
    benefits of, 258–259
    cdmaOne, 126, 378
    EMS/MMS, 263–264
    GSM, 86
    history of, 257–258
    in Japan, 576
    MCs, 202
    with narrowband AMPS, 76
    processing, 202–203
    and signaling capacity planning, 458–459
    SMEs, 201–202
    and standards, 118
    and switching capacity, 304
    technical specifications for, 259–262
    and W-CDMA background class, 136
    wireless data terminals using, 45
Shrubbery, 14
S/I ratio (see Signal-to-interference ratio)
SID (silence descriptor), 137
SID (system identification), 155
Sidetones, 35
Signal delay, 166
Signal distribution, 334–335
Signal matrix, 557–558
Signal quality, radio (see Radio signal quality)
Signal transfer point (STP), 184–185
    and MSC growth, 516
    and MSC planning, 444
Signal variability, 14
Signaling, 66
    with ANSI-41, 179–180
    associated, 183
    in-band, 84
    out-of-band, 84
    with SS7, 183–190
Signaling capacity, 484
Signaling capacity planning, 455–460
    call processing, 457–458
    and fraud prevention, 460
    handoffs, 455–457
    roamers, 458
    short message service, 458–459
Signaling connection control part (SCCP), 187
Signaling links, 183

Signaling System No. 7 (SS7), 150, 183–190, 191
    and GSM SMS, 86
    models of, 188–190
    network architecture for, 183–186
    protocol for, 186–188
    reliability in, 302
    for wireless system transport, 179–180
Signals, 3
Signal-to-interference (S/I) ratio, 10, 12
    AMPS, 332–333
    and conventional reuse, 333–335
Signal-to-noise (S/N) ratio, 10, 12, 13
    AMPS, 332
    as quality measurement, 30–31
Silence descriptor (SID), 137
Simplex method, 281
6 dB of S/N, 351–352
Sixth-pilot delta, 406–407
Size (of mobile telephones), 38
Slack resources, 280, 282
SLAs (see Service-level agreements)
Sleep mode, 74, 84, 200
Slotted ALOHA protocol, 138–139
Slow associated control channel (SACCH), 84
Slow-data-link complaints, 491–492
Smart antennas, 51–52, 424
    benefits of, 269–270
    and CDMA equations, 364–365
    future of, 574–575
    in multicellular systems, 364–365
    TD-SCDMA space division with, 141
Smart cards, 47, 120
SMEs (see Short message entities)
SMS (see Short message service)
SMS message center (SMSC):
    and ANSI-41 signaling capacity, 313
    in ANSI-41 wireless system, 193
    capacity of, 325–326
    and MSC growth, 516
    in MSC main processing, 304
    and MSC planning, 444
    and signaling capacity planning, 459
S/N ratio (see Signal-to-noise ratio)
Snell's law, 554
Soft blocking, 385, 427
Soft dollar value, 274
Soft handoffs:
    and CDMA capacity, 382–383
    in CDMA principle, 103–105
    in circuit-switching, 306
    measurement of, 480–481
    as mobile switch function, 158–159
    in MSC main processing, 305
    and multicellular systems, 367, 368

Softer handoffs, 104–105
  as mobile switch function, 159
  and multicellular systems, 367, 369
SONET (Synchronous Optical Network), 519
Sound, 4
Sound pressure level (SPL), 12
Sound quality, 165–167, 490–491
Sound shapes of speech, 166
South America, 119
South Korea, 119
Southeast Asia, 119
Space division multiple access (SDMA), 134
Speakers, 35, 36
Specifications, standards vs., 113–114
Spectrum, 9
Spectrum allocation:
  future of, 579–580
  and politics, 115
Speech coding, 235–245
  and CDMA capacity, 384–385
  cdmaOne, 127–128
  cdma2000, 132
  CELP, 126
  in digital telephony, 74
  in GSM voice channel, 82
  linear predictive, 243–244
  as mobile switch function, 160–161
  and MSC planning, 439
  PCM, 236–243
  performance of, 244–245
  switching capacity of, 307
  W-CDMA, 137
Speeds, data, 378–379
SPL (sound pressure level), 12
Spread spectrum, 8, 87–88
Spreading, 28–29
Spreading gain, 88
Sprint, 580
Sprint PCS, 121, 126, 131, 155, 156, 204, 269, 571
SS7 (see Signaling System No. 7)
SSP (see Service switching point)
Stable call, 164
Staggered carriers, 426
Stand-alone dedicated control channel (SDCCH), 84, 86
Standards (for cellular systems), 109–142
  benefits of, 114
  comparison of, 126
  general, 111–124
  generations of, 118–123
  and interfaces, 113
  and politics, 114–116
  role of, 109–118
  specifications vs., 113–114

Standards (for cellular systems) (Cont.):
  and standards organizations, 116–117
  technology affected by, 117–118
  types of, 112
  worldwide CDMA (see Worldwide CDMA standards)
Star graph, 319, 320
Stars, 21n10
State diagrams, call, 207–209, 211–213, 215–216, 287, 288, 293–295
State police, 583
States, call (see Call states)
Statistical multiplexing, 175
Statistics, propagation model, 555–557
Storage:
  memory, 39, 157
  voice message, 445
Store-and-forward technology, 226, 296
STP (see Signal transfer point)
Streaming, 227
Streaming class (W-CDMA), 136
Subscriber traffic modeling, 561–567
  analytical, 565–566
  CDMA-specific factors in, 562–563
  and measurement-based decisions, 566–567
  and pilot delta maps, 566
  simulation of, 563–565
  and traffic tally calculations, 561–562
Subscribers:
  competition for, 397–398
  SMS benefits for, 258
Subscriber-screened calls, 312
Subscriber-to-base-station ratios, 39n9
Superframes, 177
Supervision, 75, 76
Supervisory audio tone (SAT), 75, 76, 77, 331
Supplemental channels, 132–134
Switched diversity, 28
Switches, digital, 151
Switchhook flash, 210
Switching, 151–161
  mobile functions of, 154–161
  technology of telephone, 151
  telephone functions of, 152–154
Switching capacity, 303–307
  and bit flow, 305–306
  circuit, 306
  main processor, 303–305
  packet, 306–307
  speech coding, 307
Syllabic interference, 166
Symbols, 19
Synchronization channel (SCH), 139
Synchronous Optical Network (SONET), 519
System identification (SID), 155

T

TALKING state, 210, 213
Tandem MSC, 195
Tandem route, 299, 300
Tandem switches, 151
Target MSC, 170, 195
Taxicabs, 8
TCAP (transaction capabilities application part), 187
TCH (*see* Traffic channel)
TCO (total cost of ownership) model, 274
TCP/IP routers, 269
TDD (*see* Time division duplex)
TDMA (*see* Time division multiple access)
TDMA-EDGE, 120
TD-SCDMA standard (*see* Time division synchronized CDMA standard)
Technology:
  and standards, 117–118
  telephone switch, 151
Telegraph, 3
Telephone call sequence, 163–164
Telephone switches:
  functions of, 152–154
  technology of, 151
Telephone transport, 175–182
  protocols for, 176–178
  wireless system, 178–181
Telephony, 143–218
  ANSI-41, 191–205
  call states, 207–218
  engineering of, 163–173
  PSTN, 145–150
  SS7 signaling, 183–190
  switching for, 151–161
  transport of, 175–182
Telephony engineering, 163–173
  quality of service, 164–167
  reliability/redundancy, 167
  telephone call sequence, 163–164
  and wireless architecture, 167–168
  wireless issues with, 168–173
Television stations, 5
Temporary local directory number (TLDN), 201, 464
Terminal equipment, WLL, 41–42
Terminal equipment unit, 42–43
Terminals without human interface, 46–47
Terminated calls, roamer, 311, 328
Terrain obstruction, 13
Terrain-based models, 551–553
Tertiary cells, 345
Tesselations, 336
"There ain't no such thing as a free lunch" (TANSTAAFL), 278*n4*
Thermal noise, 25

THEY-HANG-UP state, 210
Third generation (3G) technology, 112, 120–123
  data capacity of, 378–380
  in Europe, 575
  future of, 578–579
  spectrum allocation in, 115
  voice/data carriers in, 224–226
Third world, 399
"38.4 dB per decade," 13
G/UMTS standards initiative, 45
3.5G technologies, 578
Three-to-one cell splitting, 411–413
Three-way calling, 157, 211–212, 214, 215, 218, 312
Thresholds, 78
TIA/EIA/IS-756, 193
Time, 572
  in business-case models, 533–534
  and packet switching capacity, 307
  and queuing theory, 296–297
Time division duplex (TDD), 123
  asymmetric demand on, 471
  and CDMA equations, 365–366
Time division multiple access (TDMA), 8, 81–86
  channel concept of, 81–82
  and 4G technologies, 578
  future of, 575–576
  future of CDMA vs., 574
  GSM channel of, 82–83
  migration to CDMA from, 428–433
  as 2G technology, 119
  See also Global System for Mobility
Time division synchronized CDMA (TD-SCDMA) standard, 123, 140–142
  and asymmetric demand, 470, 471
  CDMA radio capacity estimation for, 391–393
  channel coding of, 141
  in China, 43, 125, 576
  and DO channels, 223
  dynamic channel allocation of, 141
  joint detection/power control of, 142
  radio bandwidth of, 140–141
  services of, 140
  space division with smart antennas of, 141
  voice channel of, 140
Time response, 5–6
Time slots, 147
Time-frequency relationship, 6
Time-of-day usage, 398–399, 427, 463, 466
TLDN (*see* Temporary local directory number)
Total cost of ownership (TCO) model, 274
Touch Tone service, 163*n2*, 214
Towers:
  base station, 52–53
  sharing of, 416–417

Tradeoffs, 278–279
Traffic channel (TCH), 82, 84
Traffic engineering, 148, 285–302
  of backhaul, 324
  in backhaul design, 318–319
  for conventional reuse, 349–350
  and IP services, 300–301
  ISDN service effects on, 293–295
  and network routing, 299–300
  for packet data, 294–299
  and reliability/redundancy, 301–302
  voice calls assuming fixed-size channels, 285–292
Traffic intensity, 297
Traffic tally calculations, 561–562
Transaction capabilities application part (TCAP), 188
Transceivers, 35–36
Transistors, 151
Translations, 149
Transmission, 3
Transmission Control Protocol (TCP), 228n9, 229
Transmission license, 9
Transmit (Tx), 35n4
Transmitter amplifiers, 34
Transmitters, 35
Transport, 517–521
  backhaul added to, 517–518
  existing routes of, 417–418
  measurement of, 484–486
  and MSC planning, 438
  MSC-to-MSC capacity added to, 519–520
  MSC-to-PPDN capacity added to, 520–521
  MSC-to-PSTN capacity added to, 520
  and regional growth, 525–526
  voice and data sharing physical, 248–250
Tree graph, 321
Trees, 14
Triple amplitude, 99
Tromboning, 168, 181
Trunk groups, 146
Trunks, 57n9, 146–147, 151, 152, 164
Tunnels, 415
Two-to-one compandor system, 74
Two-way communication, 8
Tx (see Transmit)
Type 1, intra-BTS handovers, 85
Type 2, intra BSS, inter-BTS handovers, 85
Type 3, intra-MSC, inter-BSS handovers, 86
Type 4, inter-MSC, inter-BSS handovers, 86

U
Ubiquitous computing, 581–582
UBR (unspecified bit rate), 227
UDP (User Datagram Protocol), 187n1

UHF (ultrahigh frequency) band, 6
"UHF wasteland," 9
Ultrahigh frequency (UHF) band, 6
UMTS (Universal Mobile Telephony System), 135
Unconditional call forwarding (CFU), 312
Unification (of standards), 122–123
United States:
  approach to quality in, 577
  international compatibility in, 119
Units, physical, 589–591
Universal Mobile Telephony System (UMTS), 135
UNIX, 112
Unspecified bit rate (UBR), 227
Uplink, 105–106
U.S. National Geological Survey (USGS), 551
U.S. Postal Service, 536n2
User complaints, 487–494
  data-error, 492
  ineffective-attempt, 488–489
  lost-call, 489–490
  lousy-call, 490–491
  and reasons for change/growth, 492–493
  slow-data-link, 491–492
  symptoms/causes of, 487–488
User Datagram Protocol (UDP), 187n1
User ID cards, 576
User terminal authentication, 313
User terminals, 33–48
  future of, 581
  mobile/portable telephones, 33–39
  movement of, 365
  registration of, 168–169
  state management of, 200
  wireless data terminals, 44–46
  wireless local loop, 40–44
  without human interface, 46–47
Users, assumptions about, 363
USGS (U.S. National Geological Survey), 551
Utility companies, 47
Utilization, 297, 403
UWC-136 TDMA, 123

V
VAD (see Voice-activity detection)
Valley, 406–407
Variable bit rate (VBR), 227
Variable costs, 533–537
VBR (variable bit rate), 227
VBR-NRT (non-real-time), 227
Vehicular telephone technology, 72
Vendors, 258–259
Verizon, 121, 571
Verizon Wireless, 126, 131

Vertical radiation patterns, 51
Vertical services, 209
Vertices, 280, 319
Very high frequency (VHF) band, 5
Video phones, 573
Videoconferencing, 131
Videotelephony, 401
    demand estimation for, 401–402
    W-CDMA, 136
Visitor location register (VLR), 170
    and ANSI signaling, 180
    and ANSI-41 signaling capacity, 310–311
    in ANSI-41 wireless system, 193
    and automatic roaming, 199
    capacity of, 324–325
    fault recovery for, 200–201
    in GSM call setup, 84
    and MSC growth, 515–516
    and MSC planning, 443–444
    and roaming registration, 156
    and signaling capacity planning, 458
    and SMSC planning, 444
VMACs (voice mobile attenuation codes), 77
Vocoders, 244
Voice, compressed, 305
Voice calls:
    blocking of, 288–292
    CDMA capacity for, 387, 388, 540
    and data carriers, 224–226
    demand estimation for, 397–400
    fixed-size channels for, 285–292
    in MSC main processing, 303–304
    MSC-to-PSTN links for, 485
    to the PSTN, 464–465
    quality of service for, 267–268
    TD-SCDMA, 140
    in traffic engineering, 285–293
    transmission requirements for, 222
    W-CDMA, 135–136
    and wireless data services, 267–268
Voice mobile attenuation codes (VMACs), 77
Voice over IP (VoIP), 251–255
    on cellular backhaul network, 254–255
    on Internet, 252–253
    on PPDN, 254
    on PSTN, 253–254
Voice radios, 56
Voice services:
    cdmaOne, 126
    future of, 580
Voice signal (on FM channel), 74
Voice-activated dialing, 35, 38
Voice-activity detection (VAD), 83, 102–103, 137

Voice-coded links, data over, 378
Voice-mail systems, 214
    capacity of, 326
    and MSC growth, 516
    and MSC planning, 445
    with narrowband AMPS, 76
VoIP (see Voice over IP)
Voltage, 5, 13

W
Waiting probabilities, 297–299
Walfisch-Ikegami propagation model, 548–551
Walkie-talkies, 8
Walsh codes, 128, 129, 133
Water, 14
Wavelength, 5
W-CDMA frequency division duplex, 123
W-CDMA standard (see Wideband CDMA standard)
W-CDMA time division duplex, 123
Western Electric, 151
White noise, 26
Wideband CDMA (W-CDMA) standard, 121–122, 125, 135–140
    and carrier asymmetry, 470
    CDMA radio capacity estimation for, 390–392
    cell identification of, 139–140
    and DO channels, 223
    and fading, 28
    forward channel coding of, 138
    future of, 575
    power control of, 139, 384
    radio bandwidth of, 137
    reverse channel coding of, 138–139
    services of, 135–136
    speech coding in, 160
    speech coding of, 137
    two receivers used in, 56
Wi-Fi (see Wireless fidelity)
Wireless call states, 213–215
Wireless data services, 267–270
    and base station/MSC changes, 269–270
    quality of service for, 267–268
Wireless data terminals, 44–46
Wireless fidelity (Wi-Fi), 473
    competition using, 580
    future of, 579
    spectrum allocation for, 579
Wireless local loop (WLL), 40–44
    capacity for, 360–361
    demand for, 399
    future of, 580–581
Wireless radio states, 215–218
Wireless service providers, 573, 574

Wireless system(s), 8
    architecture of, 167–168
    definition of, 193
    transport in, 178–181
Wireless telephone demand estimation, 397–404
    cell count, 402–404
    data service, 400–401
    MMS/videotelephony, 401–402
    voice call, 397–400
Wireless telephony, 63–69
    analog, 71–79
    CDMA principle of, 87–107
    and radio architecture, 66–69
    signal path for, 63–66
    TDMA, 81–86
Wireless telephony engineering issues, 168–173
    call setup, 169
    changing channel conditions, 171–172
    handoff, 169–170
    nonlocal service quality, 173
    number of calls, 170–171
    overflow characteristics, 172–173
    roaming, 170
    tromboning, 168
    user terminal registration, 168–169
Wireless-to-landline signal path, 63–65

Wireless-to-wireless signal path, 66
WLL (*see* Wireless local loop)
Woods, 14
World Cup Soccer Finals, 576
World markets:
    and marketing plans, 530
    2G technology in, 119
    voice call demand estimates in, 399
World Trade Center, 44, 46, 47, 582–583
World Wide Web, 136
Worldwide CDMA standards, 125–142
    cdmaOne, 125–130
    cdma2000, 130–134
    comparison of, 126
    TD-SCDMA, 140–142
    W-CDMA, 125–140
Wrist cell phones, 38

X
X.25 packet protocol, 191
Xerox, 529*n4*
1Xtreme, 133

Y
Yellow areas, 489

About the Authors

Adam Rosenberg is an industrial mathematician who has worked in wireless and landline telephony, airline planning, railroad line optimization, used car value prediction, printed circuit board design and manufacture, image processing of flood plain maps, and, currently, hotel revenue management. He worked with the cellular founders at Bell Telephone Laboratories on Advanced Mobile Phone Service (AMPS) and, more recently, on Broadband Code Division Multiple Access (B-CDMA) at Inter-Digital Communications Corporation. Dr. Rosenberg earned his A.B. in Mathematics from Princeton University, cum laude, and his M.S. and Ph.D. in Operations Research from Stanford University. He can be reached at adam@the-adam.com.

Sid Kemp is an expert in architecture, standards, quality management, and organizational structure for information technology and telecommunications. As president of Quality Technology & Instruction (QTI), he provides clients with consulting and training that increase quality of service especially by improving teamwork between technical engineers and business experts and executives. He appreciates comments on his writing and can be reached at sid@qualitytechnology.com. For updates and additional information about this book, see his corporate Web site, www.qualitytechnology.com/qti/cdma.